SAN FRANCISCO BAY: THE URBANIZED ESTUARY
Investigations into the Natural History of San Francisco Bay and Delta
With Reference to the Influence of Man

SAN FRANCISCO BAY AND ENVIRONS
Courtesy of the National Aeronautics and Space Administration

FIFTY-EIGHTH ANNUAL MEETING
of the
PACIFIC DIVISION/AMERICAN ASSOCIATION FOR THE ADVANCEMENT OF SCIENCE
held at
SAN FRANCISCO STATE UNIVERSITY, SAN FRANCISCO, CALIFORNIA
June 12-16, 1977

SAN FRANCISCO BAY: THE URBANIZED ESTUARY
Investigations into the Natural History of San Francisco Bay and Delta
With Reference to the Influence of Man

Editor
T. John Conomos
U. S. Geological Survey

Executive Editor
Alan E. Leviton
*Pacific Division, AAAS and
California Academy of Sciences*

Assistant Editor
Margaret Berson
California Academy of Sciences

San Francisco, California
1979

ISBN 0-934394-01-6

This volume has been typeset in *Press Roman* type on an IBM Mag Card Composer in the Department of Herpetology, California Academy of Sciences.

Copyright © 1979 by the Pacific Division of the
American Association for the Advancement of Science
c/o California Academy of Sciences
Golden Gate Park
San Francisco, California 94118

Manufactured in the United States of America by the Allen Press, Lawrence, Kansas 66044

TABLE OF CONTENTS

	Page
FOREWORD *J. J. Goering*	5
INTRODUCTION *T. J. Conomos*	7

HISTORICAL OVERVIEW

San Francisco Bay: The Unsuspected Estuary 9
J. W. Hedgpeth

PHYSICAL PROCESSES

Ancient Processes at the Site of Southern San Francisco Bay: Movement of the
Crust and Changes in Sea Level 31
B. F. Atwater

Properties and Circulation of San Francisco Bay Waters 47
T. J. Conomos

Sedimentation in the San Francisco Bay System 85
R. B. Krone

The Movement and Equilibrium of Bedforms in Central San Francisco Bay 97
D. M. Rubin and D. S. McCulloch

WATER PROPERTIES AND QUALITY

Processes Affecting Seasonal Distributions of Water Properties in the San Francisco
Bay Estuarine System 115
T. J. Conomos, R. E. Smith, D. H. Peterson, S. W. Hager, L. E. Schemel

Factors Influencing the Entrapment of Suspended Material in the San
Francisco Bay-Delta Estuary 143
J. F. Arthur and M. D. Ball

Sources and Sinks of Biologically Reactive Oxygen, Carbon, Nitrogen, and
Silica in Northern San Francisco Bay 175
D. H. Peterson

Distribution and Stable-Isotope Composition of Carbon in San Francisco Bay 195
E. C. Spiker and L. E. Schemel

The Use of Radon-222 to Estimate Benthic Exchange and Atmospheric
Exchange Rates in San Francisco Bay 213
D. E. Hammond and C. Fuller

	Page
Fluctuations of Copper, Zinc, and Silver in Tellenid Clams as Related to Freshwater Discharge—South San Francisco Bay *S. N. Luoma and D. J. Cain*	231

THE ECOSYSTEM

Phytoplankton Ecology of the San Francisco Bay System: The Status of our Current Understanding *J. E. Cloern*	247
Planktonic Chlorophyll Dynamics in the Northern San Francisco Bay and Delta *M. D. Ball and J. F. Arthur*	265
The Benthic Algal Flora of Central San Francisco Bay *P. C. Silva*	287
History, Landforms, and Vegetation of the Estuary's Tidal Marshes *B. F. Atwater, S. G. Conard, J. N. Dowden, C. W. Hedel,* *R. L. MacDonald, W. Savage*	347
Some Legal Problems of Tidal Marshes *J. Briscoe*	387
The Role of Mysid Shrimp in the Sacramento-San Joaquin Estuary and Factors Affecting their Abundance and Distribution *J. J. Orsi and A. C. Knutson, Jr.*	401
Natural and Anthropogenic Influences on Benthic Community Structure in San Francisco Bay *F. H. Nichols*	409
Introduced Invertebrates of San Francisco Bay *J. T. Carlton*	427

FISHERIES RESOURCES

The Fisheries of San Francisco Bay: Past, Present, and Future *S. E. Smith and S. Kato*	445
Environmental Factors Affecting Striped Bass (*Morone saxatilis*) in the Sacramento-San Joaquin Estuary *D. E. Stevens*	469
San Francisco Bay: Critical to the Dungeness Crab? *R. N. Tasto*	479

SUMMARY

T. J. Conomos	491

FOREWORD

Uniquely productive, the San Francisco Bay bountifully provides living resources, recreation and waste assimilation to the many people that inhabit its shores. Its stock of fin- and shell-fish are important sources of food and recreation, and its scenic beauty provides pleasure to the masses that are fortunate enough to be able to partake of it.

Although its shores have been populated since before the turn of the century, San Francisco Bay has received surprisingly little scientific study until the environmental awareness movement of the late 1960's and 1970's. Additionally, few, if any, scientific meetings and symposia devoted to the Bay's ecosystem have been held. It, therefore, seemed very appropriate to hold a symposium on the San Francisco Bay system in conjunction with the annual meeting of the Pacific Section of the American Society of Limnology and Oceanography held June 12-16, 1977 at San Francisco State University, San Francisco, California. The symposium was held as a plenary session on 13 June 1977 under the sponsorship of the American Society of Limnology and Oceanography and in affiliation with the Pacific Division, American Association for the Advancement of Science.

This volume includes most of the papers presented at the symposium and some additional chapters written since then. It is hoped it will serve as a useful reference on how the San Francisco Bay system functions biologically, chemically, geologically, and physically, both for scientists and for the persons responsible for managing the Bay's resources.

I should like to express my appreciation to T. John Conomos for organization of the symposium and for efficiently conducting the difficult task of extracting the written material from authors, and editing it. Without his perseverance, this volume would never have come about. Further, I wish to take note of the decision of the Executive Committee of the Pacific Division, AAAS for undertaking the publication of this volume, which initiates the Division's new "Symposium Series." The members of the Pacific Division, AAAS and of its affiliated societies are the longterm beneficiaries of this action.

John J. Goering, *President*
Pacific Section of the American Society
of Limnology and Oceanography
June, 1977

INTRODUCTION

T. JOHN CONOMOS
U. S. Geological Survey, 345 Middlefield Road, Menlo Park, CA 94025

San Francisco Bay and Delta, one of the world's largest estuarine systems, profoundly influences and enhances the economic, climatic and aesthetic quality of the surrounding urban-suburban region to the great benefit of its 5 million inhabitants. With its strategic location on the coast of central California and its huge natural harbor, San Francisco Bay serves as a major center for commerce and industry and as a gateway to the Far East. Its Mediterranean climate and beautiful setting attract people from around the world both to visit and to establish residence. The estuarine system itself is a natural habitat for fish and wildlife and as such provides abundant quantities of edible fish and shellfish and supports a wide variety of water-oriented recreation.

With the progress of its energetic urban-suburban society, however, have come major changes to the estuarine system. The Bay and Delta have been heavily modified by man since the arrival of the Argonauts in the mid-19th century. As the margins have been filled and diked, the overall size of the Bay and Delta has greatly shrunk. The result has been the loss of wildlife habitats and a reduction of tide-related flushing, which in turn has led to progressive deterioration of the quality of Bay waters. Water-quality degradation from wastes discharged by a rapidly growing population has undoubtedly altered the indigenous ecosystem. Unfortunately, growing demands to reduce or eliminate waste discharges have been accompanied by the reduction, through massive diversions of river inflow, of the ability of the system to flush itself naturally.

In response to environmental concerns during the last few decades, legislative committees have agreed that this estuarine system should be protected against further indiscriminate and unrestrained exploitation. These committees and subsequent Federal and State legislation have mandated that sound plans for long-term intelligent and rational management of this valuable resource be formulated and implemented. There is, unfortunately, little scientific data on which to base these plans. Our knowledge of the complex physical, chemical, biological, and sedimentological estuarine processes is relatively primitive. This is surprising, considering the importance and irreplaceable nature of the system, the magnitude and cost of the public works already built or in the planning stages, and the demands and standards imposed by environmental and regulatory agencies.

Our purpose in this volume is to summarize in individual chapters our knowledge of the natural processes that contribute to the maintenance of the estuary as we see it. These discussions include, of necessity, some emphasis on the influence of man. Half of the chapters were presented during the course of a symposium, sponsored by the Pacific Section of the American Society of Limnology and Oceanography, held at San Francisco State University, 13 June 1977. Because of the enthusiasm generated by this symposium, papers covering additional topics were solicited for inclusion in this enlarged volume.

We attempted to be as comprehensive as possible, bringing together reports dealing with the many interrelated aspects of estuarine research ongoing in San Francisco Bay and Delta. The chapters vary in their content: some are summaries of established published and unpublished work, others are syntheses of our knowledge of a given topic, and still others are research papers reporting results of promising ongoing research. It is our hope that this volume will serve as a timely and useful reference for those planning and conducting estuarine research and as a status report for legislators, planners and coastal-zone managers.

This contribution is the work of many dedicated persons. I thank the contributors, who enthusiastically and freely gave of their time and energies in preparing their chapters, and particularly Brian F. Atwater, James E. Cloern, Douglas E. Hammond, Samuel N. Luoma, David S. McCulloch, Frederic H. Nichols, and David H. Peterson who, additionally, served as peer reviewers of many of the chapters. Sally M. Wienke prepared many of the figures and Kaye M. Walz provided secretarial and editorial assistance. I also wish to thank Janice R. Conomos, David S. McCulloch, and Frederic H. Nichols for their persistent encouragement throughout the preparation of this book, and to the Pacific Division, American Association for the Advancement of Science, the U. S. Geological Survey, and the California Academy of Sciences for financial support and/or other courtesies without which this volume could not have been produced.

SAN FRANCISCO BAY—THE UNSUSPECTED ESTUARY
A History of Researches

JOEL W. HEDGPETH
5660 Montecito Avenue, Santa Rosa, California 95404

This brief historical account of research in San Francisco Bay and the Delta of California indicates that motivation for the study of the system has been primarily related to economic aspects of the alteration of the environment by acts of man, e.g. the study of hydraulic mining debris by G. K. Gilbert, or the stresses associated with drought periods, exemplified by the marine borer studies of the 1920's. Although there was an active interest in making a baseline study of the entire system around 1911-12 by members of the departments of zoology and geology of the University of California, the effort was soon abandoned and the Bay and its delta lapsed into academic oblivion. By the 1930's there was such general unawareness of the environment or its significance that major decisions for diversion of water from the system were made as if the Bay simply did not exist, and serious consideration was given to schemes, especially the Reber Plan, that would have destroyed San Francisco Bay completely.

Of all the great estuarine systems of the world, San Francisco Bay and the delta of the Sacramento and San Joaquin rivers (Fig. 1) have been among the last to be critically studied by scientists. Indeed, for many years this region had not even been thought of as an estuary, and from the viewpoint of politicians and water engineers it was not a great natural California phenomenon to be proud of like Yosemite or Mount Lassen, but a hindrance to progress. The Bay, of course, was a natural resource whose shores had been built up as one of the world's great harbors, and was incidentally a beautiful part of the world. But that was taken as a sort of fringe benefit that made progress pleasing to the eye. No one knew very much about the Bay, about what kinds of plants grew in its waters and what sorts of animals lived in the muds and sand of the bottom or what fishes swam in its water. And no one had tried to put this information together, especially to think of the Bay and its delta as interrelated. Yet there had been for more than 3,000 years a colonization, if not civilization, of the Bay by the native "Indians" adjusted to their environment and flourishing upon its resources, even to the extent of trading their surplus of food with those living in the hinterland of the Delta (Nelson 1909; Cook 1964).

For nearly 200 years, Spanish navigators sailed the Manila galleons from Acapulco, northward past the unseen Golden Gate of San Francisco Bay, to Cape Mendocino whence they turned westward across the Pacific. It was not until 1769 that European explorers and "men of God" discovered the Bay, not from the sea by entering the Golden Gate, but from a hill to the south. Six years later, in 1775, the Spanish at last found their way by ship into San Francisco Bay. Among the early voyagers who apparently missed San Francisco Bay was Francis Drake, who passed by in 1579 and laid over for repairs on the shore of what is now Marin County. Scientific explorers did no better. Captain James Cook sailed, according to plan, northeast from Hawaii in 1778 and made his landfall of western North America at Cape Foulweather on the Oregon coast. Expedition after expedition, many with naturalists aboard, bypassed San Francisco Bay, although most of them stopped at Monterey. Jean François de Galaup de la Perouse visited Monterey in 1786, and Alejandro Malaspina stopped there for two weeks in 1791 on his way southward from

SAN FRANCISCO BAY

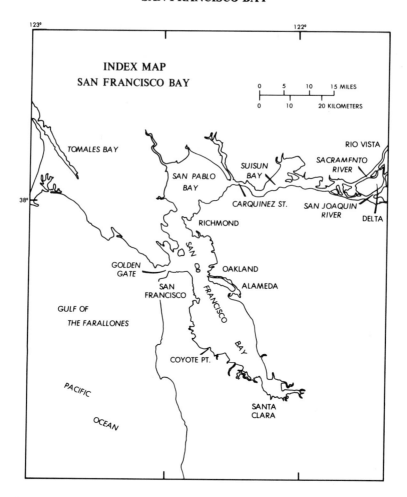

Fig. 1. San Francisco Bay and Delta.

what is now British Columbia. Juan Francisco de la Bodega y Quadra might have entered San Francisco Bay in 1775 had he not lost his shore boat in a mishap with the sneaker wave at the mouth of Tomales Bay, and in his later years of sailing to and from Nootka Sound he apparently made no effort to sail into San Francisco Bay. George Vancouver visited the California coast, including Monterey, in 1793 and 1794 but he did not attempt San Francisco Bay.

It was not until the Russian expedition of the *Rurik* under Captain Otto von Kotzebue sailed into San Francisco Bay in 1816 for a month's stay that natural history observations of any significance were made in the Bay area. The naturalists were the poet-botanist Adelbert von Chamisso and the ship's doctor and zoologist Dr. Ivan Eschscholtz, accompanied by the artist Louis Choris. Little was done in the Bay itself as collecting and observation were restricted for the most part to the vicinity of the Presidio. Eight years later (1824) von Kotzebue again visited San Francisco Bay, with Eschscholtz as naturalist. On this visit they made a trip by small boat to Santa Clara, and in November they made a small boat voyage up the Bay as far as the Sacramento River near Rio Vista, making significant observations of the natural history and agricultural potential of the region (Essig 1933).

At last, in 1826, San Francisco Bay was visited by a competent cartographer when Captain

HEDGPETH: THE UNSUSPECTED ESTUARY

Frederick William Beechey entered the Bay on November 6 with H.M.S. *Blossom* to explore and survey the Bay, secure supplies and allow the crew to rest. The *Blossom* stayed until December 28, and returned again in November of 1827 after a second attempt to find the Northwest Passage. The mission was primarily that of surveying the Bay, and the chart that resulted from their efforts was excellent and reliable as far as Carquinez Strait. It has often been used since to settle land claims involving shorelines and the status of islands (Lincoln 1969).

History began to accelerate after the Beechey survey; first the Mexican War and the acquisition of California by the United States, then the Gold Rush, and, by 1851, 25 years after Captain Beechey's first view of San Francisco Bay, California was a state. San Francisco became a city overnight; the once pleasant shores and coves of the northern end of the peninsula of San Francisco were filled with debris, old ships and earth from the hillsides to provide docks and land for warehouses and the financial district. The Indians were extirpated and the rich bird and animal life of the delta region was levied upon by commercial hunters. A man thought to be the last descendant of the aboriginal settlers of the bay region lived at Coyote Point until about 1942 (Stanger 1963). From accounts in the pioneer press of the years following the Gold Rush, especially Hutchings' California Magazine (Olmsted 1962) one can get some idea of what the Bay and Delta must have been like, but there was no naturalist who made more than casual mention of natural phenomena. There is a vivid account of the nature of the great flood of 1862, when most of the Sacramento Valley was a vast lake, in Brewer's "Up and Down California" (Farquhar 1966). There are also some selections about the wealth of fish and game resources in Neasham (1973). For an account of the tule elk, the most abundant and characteristic large animal of the delta region, see McCullough (1971).

The great flood of 1862 is vividly described in Brewer's "Up and Down California" (Farquhar 1966). This pluvial demonstration has never been adequately considered by ecologists. The rains were so heavy from November 1861 through January 1862 that most of the city of Sacramento was under water, and the legislature was forced to hold its sessions in San Francisco. Much of the Central Valley was under water; in some places the tops of the telegraph poles were submerged. Water flowed for at least 10 days through the Golden Gate in a steady torrent, blocking tidal reversal. During that time the "null zone" may have been somewhere out beyond the Golden Gate (Peterson et al. 1975). Such a flood today would cause billions of dollars damage and utterly wipe out many valley towns. San Francisco Bay was a fresh-water lake. Ecological questions were not asked in those days, so no one wondered what effect such an episode had on the life of the Bay and the Delta. It probably took several years, possibly a decade, for the native estuarine species to re-establish themselves. By that time it would have been too late, for the completion of the continental railroad in 1869 made introduction of eastern species of fishes and invertebrates possible. Almost immediately eastern oysters were brought to San Francisco Bay, along with sundry invertebrates as inadvertent stowaways, and an intensive effort to establish east coast and midwestern fishes was begun (Smith 1896). At that time so little was known about the natural history of the state's most abundant commercial fish, the salmon, that no one knew where they spawned, and Livingston Stone was told they probably spawned near Rio Vista (Hedgpeth 1941). In 1872 shad were planted in the Sacramento River, and the striped bass was introduced in 1879 by Livingston Stone (Hedgpeth 1941). Many of the fish introduced in those days have become major additions to the fauna of California waters, most notably the striped bass, which is the principal recreational fish in the delta region and upper bay. One can only conjecture about the events of the decades 1860-1880; whether, for example, the striped bass would have succeeded so easily at any other time (Fig. 2).

Most of the marine and estuarine invertebrates that have become established in the Bay were not deliberately introduced; those that were, e.g. oysters and lobsters, did not become naturalized

(Hedgpeth 1968). All of these introductions and immigrations not only altered the nature of the ecological communities in the waters of the Bay and the streams; they have also posed unsuspected complications for investigators who have required identifications of species in recent years (Carlton 1975).

The very nature of San Francisco Bay itself was altered by the physical process of massive sedimentation induced by the development of hydraulic mining in the gold districts. This began in 1853, soon after the Gold Rush of 1849 (May 1970). Hydraulic mining was ended by judicial decision in 1884 not because it was endangering San Francisco Bay by rapid siltation (shoaling the upper arms of the Bay by a few meters) but because it was destroying agricultural land in the

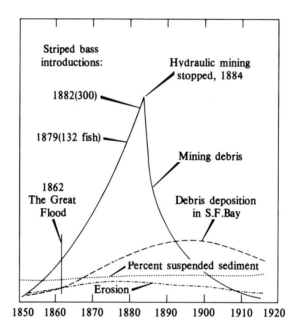

Fig. 2. Certain events associated with hydraulic mining debris deposition in San Francisco Bay. Compiled in part from Gilbert 1917.

Sacramento Valley. The effect on San Francisco Bay was not to be evaluated until long after hydraulic mining had ceased.

In all these years between the onslaughts of mining, the development of agriculture with extensive irrigation diversions and the urbanization of the San Francisco Bay (by 1910 much of the Bay was noticeably polluted and oyster farming was on the decline), there were only casual observations of its natural history. One man who might have been a great naturalist of the Bay was diverted to the higher calling of philosophy, although he had entered the University of California in 1871 with an interest in natural science. This was Josiah Royce, who found the doctrinaire Darwinian approach of Joseph LeConte, the chief natural scientist of the university in those days, uncongenial, although he admired LeConte as a teacher. For several years Royce taught English at Berkeley, but he was called to Harvard as a philosopher in 1882. We have his perceptive essay on the effect of climate on life and civilization in California (Royce 1898), and a poetic meditation on San Francisco Bay, written in 1879, but published many years later in a posthumous gathering of essays (Royce 1920):

"Meditation Before the Gate"

I am a Californian, and day after day, by the order of the World Spirit (whose commands we all do ever obey, whether we will it or no), I am accustomed to be found at my tasks in a certain place that looks down upon the Bay of San Francisco and over the same out into the water of the Western Ocean. The place is not without beauty, and the prospect is far-reaching. Here as I do my work I often find time for contemplation...

That one realizes the greatness of the world better when he rises a little above the level of the lowlands, and looks upon the large landscape beneath, this we all know; and all of us, too, must have wondered that a few feet of elevation should tend so greatly to change our feeling toward the universe. Moreover the place of which I speak is such as to make one regret when he considers its loveliness that there are not far better eyes beholding it than his own. For could a truly noble soul be nourished by the continual sight of the nature that is here, such a soul would be not a little enviable. Yet for most of us Nature is but a poor teacher.

Still even to me, she teaches something. The high dark hills on the western shore of the Bay, the water at their feet, the Golden Gate that breaks through them and opens up to one the view of the sea beyond, the smoke-obscured city at the south of the Gate, and the barren ranges yet farther to the left, these are the permanent background whereon many passing shapes of light and shadow, of cloud and storm, of mist and of sunset glow are projected as I watch all from my station on the hillside. The seasons go by quietly, and without many great changes. The darkest days of what we here call winter seem always to leave not wholly without brightness one part of the sky, that just above the Gate. When the rain storms are broken by the fresh breezes from the far-off northern Sierras, one sees the departing clouds gather in threatening masses about the hilltops, while the Bay spreads out at one's feet, calm and restful after its little hour of tempest. When the time of great rains gives place to the showers of early spring one scarcely knows which to delight in the more, whether in the fair green fields, that slope down gently to the water, or in the sky of the west, continually filled with fantastic shapes of light and cloud—nor does even our long dry summer, with its parched meadows and its daily sea winds leave this spot without beauty. The ocean and the Bay are yet there; the high hills beyond change not at all for any season; but are ever rugged and cold and stern; and the long lines of fog, borne in through the Gate or through the depressions of the range, stretch out over many miles of country like columns of an invading host, now shining in innocent whiteness as if their mission were but one of love, now becoming dark and dreadful, as when they smother the sun at evening. So, while the year goes by, one is never without the companionship of Nature. And there are heroic deeds done in cloud-land, if one will but look forth and see them.

But I have here... to speak not so much of Nature as of Life. And I shall undertake to deal with a few problems such as are often thought to be metaphysical (whereby one means that they are worthless), and are also often quite rightly called philosophical (whereby one means that it were the part of wisdom to solve them if we could). With these problems I shall seek to busy myself earnestly, because that is each one's duty; independently, because I am a Californian, as little bound to follow mere tradition as I am liable to find an audience by preaching in this wilderness; reverently, because I am thinking and writing face to face with a mighty and lovely Nature, by the side of whose greatness I am but as a worm.

The naturalist who finally did come to San Francisco Bay was a very different sort of person, a classical geologist trained in the field under John Wesley Powell, with an infinite capacity for meticulous details. This was Grove Karl Gilbert (Fig. 3). He was called to undertake the study of the effect of hydraulic mining in California when he was the senior geologist of the U.S. Geological Survey, long past the time when he needed such an assignment in his career. The project was generated in response to a memorial or resolution in 1904 by the California Miners' Association requesting an investigation into the possibility of resuming hydraulic mining. G. K. Gilbert came to

SAN FRANCISCO BAY

California in 1905 to undertake the required study and was based at the Department of Geology of the University of California; he worked much of the time in the field until 1908.

Gilbert quickly realized that an adequate study of the mining debris problem required understanding the processes of sedimentation and tidal action in San Francisco Bay. The result was that his study, finally published in 1917 as Professional Paper 105 of the U. S. Geological Survey, was also a study of San Francisco Bay; it constitutes the nearest thing we have to a base line analysis of the physical characteristics of San Francisco Bay although it bears the title "Hydraulic-Mining Débris in the Sierra Nevada":

> Through the interlocking of subjects my attention has been drawn to matters apparently remote from problems of mining débris. Mining débris merged, both bodily and in its effects, with débris sent to the streams by agriculture and other industries; the aggravation of valley floods due to the clogging of channels by débris was inseparable from the aggravation due to the exclusion of floods from lands reclaimed for agriculture; the weakening of tidal currents at the Golden Gate by the deposition of débris in the bays is inseparable from the weakening by the reclamation of tide lands; and the attention given to these collateral subjects has not only delayed the completion of the report but has added materially to its volume.

Fig. 3. Grove Karl Gilbert. Photograph courtesy Harry S. Ladd.

The Bay cast its spell over Gilbert in an even more romantic way than it had over Josiah Royce. After the publication of Professional Paper 105, and although he was 74 years old, once again he prepared to travel west, to take up a new life in San Francisco, to marry Alice Eastwood and to raise a grandson. But his health had not been good for many years, and he died just as he packed to leave (Pyne 1975).

Gilbert's presence at Berkeley evidently aroused interest in the study of San Francisco Bay among the geologists, especially the paleontologist John C. Merriam and the geologist George D. Louderback, who in turn stimulated others. There was also another contributing factor in arousing interest in San Francisco Bay at the time, the obvious decline of the oyster industry

Fig. 4. Charles Atwood Kofoid. Photograph by Alden E. Noble.

(Barrett 1963). In any event, the stimulus for a comprehensive study of San Francisco Bay seems to have originated with the geologists (E. Packard pers. comm.) and finally reached the organizational stage early in 1911 although at the time Professor Charles A. Kofoid (Fig. 4) was considered the prime mover, at least for the biological survey:

> In February, 1911, The Biological Division of the University of California prepared a series of recommendations looking to a biological survey of San Francisco Bay. Correspondence was entered into with Stanford University and with the State Fish and Game Commission of California, both of which organizations pledged their support to the undertaking. A carefully prepared plan was at length drawn up by those chiefly interested in the project, and this plan was submitted to the Bureau of Fisheries and to the California delegation in Congress.
>
> Attention was therein called to the value of the fisheries of San Francisco Bay, and to the scanty knowledge of the biological and physical conditions upon which their existence depended. From the more purely scientific standpoint, the importance was urged of obtaining data upon which to base a handbook or series of monographic

papers dealing with the local marine fauna and flora. Such a work would be of great value to naturalists, as well as to those having economic problems more clearly in view.

The presence of the United States Fisheries steamer "Albatross" in the neighborhood of San Francisco during a considerable part of each year suggested the most practicable means by which such a survey could be conducted, and it was therefore proposed that the federal Bureau of Fisheries should undertake the execution of this project, in co-operation with the institutions named.

That Bureau assented cordially to the proposal, which had already been discussed informally at a considerably earlier date. In October, 1911, formal approval was given to the plan, under conditions prescribed by the Bureau of Fisheries. The execution of the project was later vested in a board, consisting of the commanding officer and the naturalist of the "Albatross", and a third member to be designated by the committee representing the local institutions above mentioned. Professor C. A. Kofoid, of the University of California, was chosen as the representative of the latter body, while the other members were Commander G. H. Burrage, U.S.N., succeeded by Lieutenant-Commander H. B. Soule, U.S.N., together with the senior author of the present report.

A definite programme of work was formulated, and some important additions to the equipment of the "Albatross" were decided upon. Field operations were commenced on January 30, 1912.

This survey has been concerned almost wholly with San Francisco Bay, including San Pablo Bay, though a considerable number of stations were dredged outside of the Golden Gate, even to a point beyond the Farallon Islands. . .

In the deeper waters the "Albatross" herself was employed in these operations, in the shallower waters a launch was used. In either case, however, the position of the vessel was determined at various points in the course of a haul by means of a sextant or an azimuth compass. With the launches it was, of course, impossible to employ any of the heavier types of apparatus, so that the exploration of the extensive areas of shoal water, so characteristic of San Francisco Bay, has necessarily been much less thorough than that of the navigable waters.

Even less attention has been devoted to the littoral (intertidal) zone, though collecting parties visited the following points and obtained considerable material: Bonita Point (piles of pier and on beach), Red Rock, Richmond, Key Route pier (off Oakland).

This limitation of the scope of our collecting operations has resulted from the inadequacy of the force available for such work, as well as from the necessity of restricting the amount of material accumulated for subsequent examination. It was early decided that we must resist the temptation, to which so many collectors yield, of continuing indefinitely the gathering of specimens, without regard for the likelihood of compiling any scientific results of value.

It is fully realized, however, that a complete biological survey of these waters requires the exploration of certain fields as yet scarcely touched, and it is hoped that some of the more important of these gaps may be filled in before the preparation of the final report. (Sumner et al. 1914).

The *Albatross* (Fig. 5), which was made available for these studies, was the first vessel built specifically for research for any nation (if we except the interesting example of the two small exploring vessels built by Spain for the Malaspina expedition in the last decade of the 18th Century) and was designed primarily for oceanic research (Hedgpeth and Schmitt 1945). She drew too much water for inshore purposes, as indicated by the above quotation.

Among the members of the field teams to study San Francisco Bay were Waldo L. Schmitt (1887-1977) and Earl L. Packard; the latter, who had come to the geology department to study paleontology, was assigned the molluscs of San Francisco Bay. And he remembers that it was Merriam who called attention to the innovative work of C. G. J. Petersen in Denmark. This attempt to make a quantitative study of the benthos of San Francisco Bay, inspired by Petersen's work, was the first such application of the idea outside Denmark. Unfortunately, the gear selected,

Fig. 5. The United States Fisheries Commission Steamer *Albatross* (from Hedgpeth 1945).

a commercial orange peel grab, was inappropriate, as it is even less reliable quantitatively than the rectangular mouthed Petersen Grab (see Thorson 1957). However, it was selected primarily because it was available commercially and in those days of ecological innocence, was judged completely satisfactory:

> So far as we know, ours is the first application to biological exploration of this type of apparatus. After considerable experience, we can unreservedly recommend its use for such purposes, at least in relatively shallow waters. Its chief advantage lies in the taking of comparatively large masses of mud from a single spot, and particularly in the penetrating power of the apparatus which renders possible the capture of deeply burrowing annelids, lamellibranchs, etc. (Sumner et al. 1914).

The harm done by this selection and resulting enthusiastic recommendation has been incalculable. A careful reading of Petersen's classic papers should have alerted even inexperienced investigators. A further disadvantage of the investigations was the restricted location of the sampling stations. Because of the 12-ft draft of the *Albatross,* these were, for the most part, in the deeper channels of the Bay, and even the use of a shallow draft launch left extensive areas of the Bay unsampled (Fig. 6).

Professor Kofoid had visions of a comprehensive series of monographs on the various organisms of San Francisco Bay, together with the studies of physical-chemical conditions. Several of these were accomplished, notably Waldo Schmitt's Marine Decapod Crusacea (1921), the report on physical conditions (Sumner et al. 1914), and Packard's studies of mollusca (1918a,b). Because of his interest in river plankton, Kofoid encouraged W. E. Allen, then a high school teacher in Stockton, to undertake a study of the plankton of the San Joaquin River (Allen 1920).

The grand scheme of studies and monographs, however, was not carried through. The chief

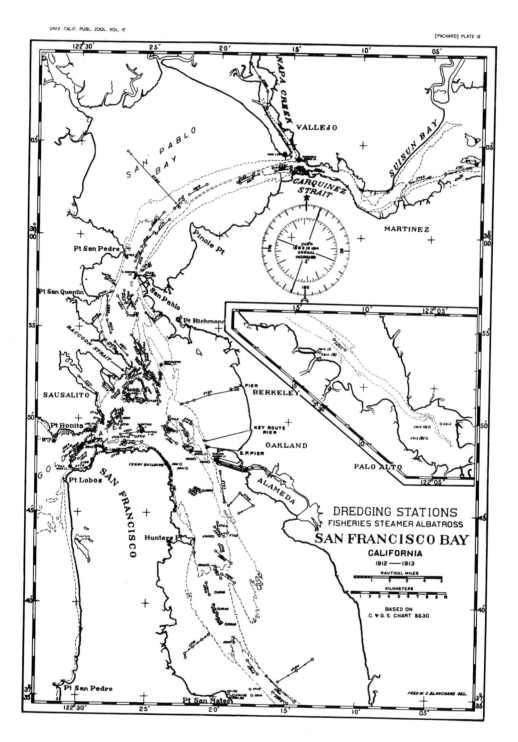

Fig. 6. Station map of the *Albatross* investigations in San Francisco Bay (from Packard 1918a).

The chief naturalist of the effort, Francis B. Sumner, had been delegated to the task because he had just finished the biological study of Woods Hole (Sumner et al. 1913); he had no enthusiasm for the work, and regarded it "from the standpoint of scientific achievement. . . was [sic] distinctly low-grade ore. Not that the work was carelessly done. The methods employed were, I think, reasonably accurate, and our labors were performed on the whole conscientiously. But the results hardly ever rise above the purely descriptive plane. There is little in them on which to base a scientific generalization of more than very limited scope." (Sumner 1945). Sumner maneuvered, successfully, to move to Scripps Institution for Biological Research at La Jolla, under W. E. Ritter. The active participation of the *Albatross* and her staff ended in 1914 when the ship was returned to sea duty for studies in Alaska, the Pribilofs and the Gulf of California.

Apparently there was no momentum to carry on the original plan for comprehensive studies, despite the flurry of activity associated with the marine borer outbreak of 1919-20 which resulted in the appointment of a committee to study the marine borer problem in San Francisco Bay. C. L. Hill and Charles A. Kofoid were the chairmen of this committee and the editors of the final report (1927). Several papers were also published in University of California Publications in Zoology, including analysis of hydrographic data in the Bay (Miller et al. 1928). The borer outbreak was associated with drought conditions in California, which also stimulated agricultural interests and politicians in Sacramento to consider measures to prevent future droughts. These concerns gained considerable momentum by 1930 without any clear idea on the part of the agencies responsible for water and fish conservation that San Francisco Bay, and its delta, was a large estuary.

After the salinity incursion of the late 1920's several comprehensive and extensive engineering studies were conducted by the Division of Water Resources, and a joint state-federal committee was set up to look into problems of water conservation, flood control or navigation. Biologists were notably lacking from these deliberations, many of which were about the need to build some kind of barrier to prevent the incursion of salt water into the agricultural lands of the Delta. The fish could shift for themselves, according to a brief statement in Department of Water Resources Bulletin 22:

> Fishing industries above the barrier, if constructed, should not suffer for the reason that, even though the fish ladder which is an integral part of the structure, should fail to function, the fish would not be prevented from entering the fresh water reservoir because they would have free access to it through the ship locks which, under normal conditions, would be operated many times throughout each day and night. (Young 1929:33).

Two years later, in Bulletin 28 on the Economic Aspects of the Salt Water Barrier, the Fish and Game people were allowed the last five pages in a 445-page volume to state the case for fish and to warn of the potential impact of such a structure. Not only would it alter salinity conditions from a more or less gradual change or gradient to those of drastic and sudden change to which most anadromous species could not adapt; it would also imperil the feeding grounds of the Delta and upper bays:

> The food supply for the young salmon and both the young and adult striped bass and shad might be seriously affected by the elimination of the tidal flow of brackish water over the shallow mud flats and in the sloughs of the upper bays. Adult salmon do not feed after entering the bay so a barrier would not affect them as far as food is concerned. This matter of food supply should be the subject of study, as it is possible the rich feeding grounds in San Pablo and Suisun bays and tributary sloughs might be eliminated if the present brackish water areas over the shallow mud flats and in the sloughs were changed to fresh water or salt water above and below a barrier, respectively. Such a change in conditions would have a profound effect on the minute

marine life which furnishes the basic food supply for these migrating fish. (Scofield 1931).

Significantly, however, no studies were instituted and by the late 1930's the decisions were made to build Shasta Dam and, to enable the resumption of limited hydraulic mining, construct debris dams as the rivers were assumed to be barren of fish, at least of desirable salmonids. Field studies of the streams were begun as the keyways of the dams were being blasted out. There turned out to be more salmon than most people thought, and hatcheries had to be built to replace the loss of spawning grounds upstream of the damsites.

Other plans were developed; about 1930 the idea of a Delta cross channel, a bypass around the Delta for irrigation water to be diverted to the San Joaquin Valley, emerged. This was the ancestor of the Peripheral Canal. Yet, it was to be more than 30 years before investigations of the Delta environment were to be funded, and we began to obtain the ecological information that should have been on hand before the engineering decisions were made (Kelley 1966; Turner and Kelley 1966).

In 1945 Governor Earl Warren called a conference on water developments in California. Most of the people who attended were businessmen, directors of irrigation districts, engineers, labor union representatives and the like. Among those agencies invited to participate were the Fish and Game Commission, and under "miscellaneous" the University of California and Stanford University. Apparently no one from the universities except perhaps agricultural engineers, came to the meeting (Warren 1945). There was no formal statement from any scientist directly about the potential effects on the aquatic environment of all the massive dams, diversions and in particular the Delta cross channel. The position in behalf of the fish was stated by the chairman of the San Francisco Tyee Club on pp. 479-480 of the 510-page printed report. Part of this statement consists of the direct quotations from biologists of the Fish and Game Department and the U. S. Fish and Wildlife Service although neither the individuals nor the agency submitted anything for the record. Another statement, specifically concerning the proposed Iron Canyon Dam was presented for the record by the Salmon Conservation League. The water juggernaut continued to accelerate, its road paved by environmental ignorance.

Among those respectfully heard at the 1945 water conference was John Reber (Fig. 7), a producer of amateur theatricals, who from early youth had developed an elaborate plan to destroy San Francisco Bay. His plan would simply dam off the north and south arms of the Bay and turn the impoundments ultimately into fresh water that could be sent all over California, even through the mountains to Los Angeles. The broad dams would accommodate railroads and highways from the east bay shore to San Francisco and Marin County. Indeed, as he stated at the meeting, his ultimate plan was to "streamline" California.

John Reber was a familiar figure throughout California:

> . . . Maps underneath his arm, he hiked from office to office, enlisting businessmen, publishers, political officials, military leaders. The plan was his Bible and he was its prophet. In an emotion-charged voice he presented his idea to hundreds of clubs, schools, churches, and civic groups. Chambers of commerce, women's clubs, granges, and farm bureaus passed resolutions in favor of the plan, and it became the subject of countless laudatory editorials and articles. (Gilliam 1957:220).

All of this ultimately had one concrete result, the San Francisco Bay Hydraulic Model, constructed in Sausalito by the Corps of Engineers, in part to test the Reber Plan but still actively used as a research tool. Although Reber was its most avid sidewalk superintendent during construction, the bay model became the instrument, after its completion in 1960, that finally set the Reber Plan to rest.

Fig. 7. John Reber, preaching his gospel. Courtesy San Francisco Chronicle.

It is difficult, at this late date, to understand why or how this lack of interest in the Bay and of understanding of its nature came about. Apparently Kofoid was unable to generate the interest among his colleagues at Stanford that he had hoped, while those interested in the marine environment in his own department shifted their work to Scripps Institution. It is probably due, to no small degree, to the famous "gentlemen's agreement" of 1912 which resulted in emphasis of the study of fresh water fish at Stanford and upland birds and mammals at Berkeley. According to George S. Myers' recollection (pers. comm.), this agreement was made between Joseph Grinnell (Fig. 8), who was not even noticeably interested in shore birds and perhaps reluctantly included the water ouzel in his purview since it did live in the mountains and was, after all, an aberrant wren, and James Otterbein Snyder of Stanford whose principal concern was trout. A third party of the agreement, John van Denburgh of the California Academy of Sciences, settled for the pickled vertebrates, traditionally the reptiles and amphibians.

While one would not expect to find definite documentary evidence of such a *modus vivendi*, a letter from Joseph Grinnell to J. O. Snyder, in the Museum of Vertebrate Zoology files under the date of December 20, 1912, does suggest a certain degree of conspiratorial understanding:

> I am sending you by Wells Fargo Express a can of fishes. These are to be added to your collections at Stanford.
> This transfer must be considered as unofficial. Our policy does not countenance exchanges; but I am deeply appreciative of your, also unofficial, act in giving us the condor bones!

SAN FRANCISCO BAY

> The fishes sent are more or less well preserved, some doubtless of no value whatever--to be thrown away. The specimens hail from various out of the way places so may include some good things.
>
> Should you wish to know more in regard to the exact locality, such information can be furnished, in some cases, by reference to our field note books. I will gladly look up any points you wish to know more about. (J. Grinnell to John O. Snyder, December 20, 1912).

This amiable division of material did not mean, however, that Grinnell opposed any possibly competing activity at Stanford. Quite the contrary; 12 years later he wrote a strong letter to Ray

Fig. 8. Joseph Grinnell. Photograph by Alden E. Noble.

Lyman Wilbur, then President of Stanford University, endorsing the establishment of a museum at Stanford, naturally along the systematic-faunistic lines of the Museum of Vertebrate Zoology:

> Is there not a really profound need that permanent provision be made, of an interdepartmental curator (and of funds at his disposal), who has the "museum conscience"--whose *primary* duty it would be to preserve the wealth of natural science materials which Stanford possesses, and who would promote their continued use by research students at home and elsewhere?
>
> I will go one step farther and urge what I believe would be a justifiable move on the part of your administration. Particularly with the higher vertebrate animals (fresh water fishes to mammals); our native fauna is *going*. Drainage, irrigation, the pumping of water from beneath seepage areas, deforestation, afforestation, limit grazing by cattle and especially by sheep,--all those processes accompanying close settlement of our territory--are resulting in profound faunal changes. Very many species are doomed to extinction (dozens have already disappeared); right now is the time that record should be kept of what is going on. In the interests of future intensive study of the problems involved, there ought to be preserved now, in more than one place, as complete representations as possible of the original, endemic animal life.

> You may say that the Museum of Vertebrate Zoology here at Berkeley is performing this function for California. I would say to this that we here, although I believe we are doing about the best we can with our relatively limited means, can only do a *part* of what demands immediate attention. Then there is another, and very important, factor–that of insuring future scientific needs by having *several* centers of accumulation. This means insurance against catastrophe. Treasure should not be housed all in one place.
>
> My second concern, then, after that for the safe preservation of what Stanford University already owns in the way of natural history materials, is for renewed activity in the accumulation of appropriate materials, in the interests of historical faunistics. (J. Grinnell to Ray L. Wilbur, March 7, 1924).

The present status of Stanford University's biological effort would sadden Grinnell. All the collections have been consigned to other institutions and the Nobel prize syndrome flourishes on the Farm. This reductionist trend was set in motion by C. V. Taylor, a product of the Berkeley department of zoology, who was chairman of Stanford's biology department during the 1930's.

Much of the correspondence in the Museum of Vertebrate Zoology files indicates that Snyder and his ichthyological candidates at Stanford were relied upon for identifications of fish, and matters continued that way until retirements and departures at Stanford left no systematist on the staff. Grinnell then had a meeting with President Wilbur to insist that someone in California had to teach ichthyology, and that Stanford should keep up the tradition established by David Starr Jordan or the University of California would have to take up the torch. He insisted upon the appointment of George S. Myers or someone "equally good." Wilbur, who had had some experience as a vertebrate zoologist, agreed.

Later, after the death of Grinnell in 1939, Tracy Storer began his ultimately successful campaign to have an ichthyologist on the staff of the University of California, Berkeley. It had become an intolerable situation when graduates of the state university were not qualified to take civil service examinations for some positions in the state Department of Fish and Game because appropriate courses were not offered at Berkeley. Unfortunately the person selected, Paul R. Needham, a specialist in trout and mountain streams, could not provide the broad background needed in classical ichthyology as well as applied fisheries. After Needham's death the effort lapsed and the main branch of the state university system still is unable to offer a program for students who want to learn about fish, and tangentially, about aquatic biology.

For many years the only course in the zoology department that might have an application to the analysis of aquatic organisms in their environment was the one unit service course in identification of organisms in domestic water for engineering students. This was usually conducted by some deserving graduate student.

Some time in 1950 Alex Calhoun of the California Department of Fish and Game approached the Department of Zoology at Berkeley about undertaking a research project in San Francisco Bay that would provide funding for some students, but he met with active lack of interest in the chairman's office. When I happened by in the hall just after this episode, he was bewildered by the interview and asked me if I knew of anyone who might be interested in working with Fish and Game on the project. I told him that Frank Filice, like myself a somewhat belated student in the department, had just asked me if I knew of such a possibility for supporting his students at the University of San Francisco. I introduced them, and thus was begun the second era of benthic studies in San Francisco Bay.

Shortly after these studies by Filice in 1952-52 (1954a,b, 1958, 1959), the California Department of Public Health instigated studies in the same part of the Bay, along the Richmond shoreline. The benthic survey, conducted in cooperation with the Department of Fish and Game, was carried out by Meredith L. Jones of the Department of Zoology in 1953 and in 1955-56.

Jones became concerned with the accuracy and dependability of sampling gear and with the need for sampling the same populations throughout the year to take account of seasonal changes (Jones 1961).

About this same time the Leslie Salt Company financed a fellowship at Stanford University for the study of the biological aspects of its salt evaporation ponds in the southern part of the Bay, studies carried out by Lars Carpelan (1955, 1957). As a result, Carpelan developed a life-long interest in the biological aspects of high salinity environments, although the biology department at Stanford has remained predominantly non-environmental in its approach to the phenomena of nature.

It is unfortunate that there was such a lag between Jones' critical studies of sampling problems and their publication, for in 1958 the Sanitary Engineering Laboratory began its work under contract with the California Water Quality Control Board. The investigators relied for the most part on "Standard Methods for the Examination of Water and Wastewater," a well known *vade mecum* not distinguished for awareness of problems of ecological sampling, nor for that matter, of the need for critical taxonomic (and curatorial) control of the organisms collected, in the stern tradition of Joseph Grinnell.

The studies of the Sanitary Engineering Research Laboratory (SERL; see Pearson et al. 1970) have become, unfortunately, a classic example of the results of haste combined with lack of ecological sophistication, engendered by limited time and the requirement to produce paper. They were also unnecessarily expensive. Nichols (1973) has examined the results of these investigations in some detail and it seems unnecessary to repeat the analysis. The SERL studies merged into subsequent studies by Kaiser Engineers to develop a comprehensive waste water plan for the Bay. The adoption of the Environmental Protection Act in 1970, followed by California's similar legislation, has stimulated all sorts of investigations, many of them to provide information for environmental impact reports. Some of the more important benthic studies have been summarized by Nichols (Table 1). Unfortunately environmental impact assessment studies are not necessarily addressed to the most interesting and significant questions, or at least the contracts and budgets do not permit seeking out the questions. They are not research but compiling and cataloguing, of the sort that F. B. Sumner so quickly tired of after having done such a job once. Nevertheless, studies motivated by the need to assess the effects of waste water and chemical discharges in San Francisco Bay will continue to be funded and will constitute the principal research activity for at least a decade. The difficulties of studying the effects of domestic waste water discharge, in particular, are emphasized by an anonymous drawing received in a plain brown envelope from Sacramento (Fig. 9)! The situation is not enhanced by a misplaced faith in the predictive value of such all-encompassing models as that proposed by Haven (1975). Such models, when prematurely applied to the "real world" are simply muddles (Hedgpeth 1977).

The sanitary engineers did attempt one application of ecological theory, the diversity index as a measure of the "health" of the environment (Pearson et al. 1967). What seemed to be a useful, comparatively simple number turned out to be useless because there is no demonstrable relation between "health" and diversity, and the index per se was invalid because of the variable number of samples and uncertain identifications of taxa. Yet the index almost became a requirement of waste-water discharge.

While efforts to study all aspects of San Francisco Bay have increased these last 20 years, it is startling to realize that the term estuary was not applied to San Francisco Bay and Delta until the House Subcommittee of Government Operations held its hearings on the "Nation's Estuaries" (Hedgpeth 1969). In Gilliam's graceful profile of the Bay (Gilliam 1957), the term estuary is applied only to the tidal creek between Oakland and Alameda, a local, traditional usage. Perhaps this has been part of the problem; we define terms according to our immediate experience and the

TABLE 1. SUMMARY OF METHODS USED IN SURVEYS OF THE BENTHIC ORGANISMS OF SAN FRANCISCO BAY.[a]

—Summary of methods used in surveys of the benthic organisms of San Francisco Bay, arranged alphabetically by author

Reference	Number of stations; number of samples at each station	Sampler (area)	Screen mesh size (mm)	Count data	Identification level	Biomass estimate
Allen (1971)	23;3–7	Ekman (~0.02 m²)	1.5	All species	All species	Biovolume of species.
Aplin (1967)	6;1	100 in.³ orange-peel (~0.07 m²)	Not given	All but small organisms.	Dominant species.	None.
Brown and Caldwell Engineers (1954).	27;1	Ekman (~0.02 m²)	do	None	Faunal types (crude).	Do.
Brown and Caldwell Engineers (1971).	10;2, samples combined.	Petersen (0.5 m²) or orangepeel (?).	do	All species	All species	Do.
Burton (1972)	45;5	Ekman (~0.02 m²)	1.0	do	do	Do.
California Department of Public Health (1954).	65;1	do	.42	All large; subsamples of small.	do	Biovolume of species.
California Water Resources Control Board (1972a).	6 areas, 8–14 stations per area; 4.	6.4–1 Ponar (~0.05 m²).	.6	All species	do	Dry weight of species.
Dederian (1966)	147;1	Petersen (~0.1 m²).	1.6	do	do	Biovolume of total sample.
Filice (1954a, b, 1958, 1959).	460;3, samples combined (shallow). 1 (deep)	Ekman (~0.02 m²) Petersen (?)	~1.0 ~1.0	do do	do do	Do. Do.
Jones (1961)	4;30	22-mm corer (~3 cm²).	.417	All species	All species	Biovolume of species.
	4;1	Ekman (~0.02 m²).	.417	do	do	Do.
Kaiser Engineers (1968b).	20;2–4, samples combined and inconsistently subsampled.	100 in.³ orange-peel (~0.07 m²).	.5	All species in subsamples.	All species in subsamples	Biovolume of phyla.
Packard (1918b)	27;1 or 2	2.5 ft³ orange-peel (~0.7 m²).	~1.0	All large; subsamples of small.	All species of mollusca.	None.
Painter (1966)	27;2 or 3	Petersen (~0.09 m²).	.6 or .3	All species	All species	Biovolume of selected species.
Pintler (1958)	24;1	Ekman (~0.02 m²).	.6	Major faunal "components."	Major faunal "components."	None.
Storrs, Pearson, and Selleck (1966a).	[1958–59] 10;2.	100 in.³ orange-peel (~0.07 m²).	1.3	All species sample data combined.	All species	Biovolume of species.
	[1960–61] 28;2.	do	~1.0	do	do	Do.
	[1961–62] 28;2.	do	.5	do	do	Do.
	[1962–63] 29;2, samples halved and combined.	220 in.³ orange-peel(?); 300 in.³ Petersen (0.065 m²) for sand stations.	.5	All species	do	Do.
	[1963–64] 18;2, samples halved and combined.	do	do	do	do	Do.
U.S. Fish and Wildlife Service (1970).	17;(?)	Not specified	.5	All species except polychaetes and small gastropods.	All species except polychaetes and small gastropods.	None.
Vassallo (1969a, 1971).	63;1	Box corer (94 cm²).	.297 or .500	All large; subsamples of small.	All species	None.
Yancey and Wilde (1970).	50;1	Pipe dredge	.5	All hard parts.	All hard parts.	Dry weight.

[a] Reproduced from Nichols 1973, p. 11.

"Oakland Estuary" is of course no estuary at all, not even an estero. (See also Hedgpeth 1973, 1975). When Conomos et al. (1970) first demonstrated the upstream movement of seabed drifters into the Bay, the State Director of Water Resources, William Gianelli, said at the time that there must be some mistake, as water does not move uphill. But of course it does in an estuary. Congressman Hosmer of the Joint Atomic Energy Commission commented that the findings had not been examined by a congressional committee. This suggests an interesting but impractical criterion for verification of scientific data.

Although they may not have had the word for it, the location of the null zone, or region of greatest productivity in the estuarine system of Bay and Delta in the region between the confluence of the rivers and Carquinez Strait would not have surprised such investigators as N. B.

Scofield. However, finding a word for it does increase understanding of the processes as indicated in the work of Peterson et al. (1975). The concurrent investigations of the same phenomenon by the U.S. Bureau of Reclamation are summarized elsewhere in this volume (Arthur and Ball 1979).

When the San Francisco Bay Conservation and Development Commission (BCDC) was established in 1964, its first task was to study the Bay and to summarize knowledge of its physical and biological characteristics in order to develop management plans. The reports generated by this effort still remain for the most part in the grey literature of mimeographed or xerographed documents, possibly available at libraries. Most of them, especially the Fish and Game contribution on

Fig. 9. The defecators (received in plain brown envelope from Sacramento).

fish and shellfish resources are based on extant published material. Some parts have been made more available, especially the general ecological sections, in the handbook compiled by Dreisbach (1969).

It is disconcerting to realize from examining this material that the basic ecological information was not on hand at the time to justify some of the management decisions being made. Some of the critical knowledge of current systems in the Bay as related to salinity exchange and dispersal of pollutants has only recently been gained (McCulloch et al. 1970; Conomos and Peterson 1974; Imberger et al. 1977). In H. T. Harvey's report for the BCDC, "Some Ecological Aspects of San Francisco Bay," as quoted by Dreisbach, there is an implicit reliance on the venerable ecological concept that primary productivity must be greater than secondary productivity, which should be suspected as an assumption in estuaries because they are known to be nutrient traps. This is obvious from the importance attributed to cordgrass and the recommendation that marshlands should be increased in the Bay (with the welcome caveat that only native species should be used in such restoration). At the same time, however, Harvey cautions that "mud flats should be retained in their entirety at least until further studies point to their relative significance." The somewhat exaggerated faith in the significance of *Spartina* is based on the principal literature available at that time, generated from the studies of Georgia salt marshes, which are very unlike the shoals and flats of San Francisco Bay. It is possible that in such a system as San Francisco Bay, mud flats are more significant than cordgrass marshes; at least they appear to be major factors in the secondary productivity of the Bay as suggested by work of Nichols (1977). In a complex bay like San Francisco, the primary production may be inadequate for all that thrives in the Bay itself, and hence it is obvious that what is going on is related to exchange, accumulation, and retention of materials. The same is evidently true for the Limfjord, site of the classic investigations of C. G. J. Petersen, and it is these aspects of the estuarine ecosystem that are now the predominant concern of investigators, as exemplified by the papers in this volume. The unfortunate aspect of this delayed understanding is that most of the major management decisions that will affect the operation of this complex natural system have already been made; indeed the seeds for such decisions as the mass diversion of fresh water and discharge of waste water within the Bay, the Peripheral Canal, and the deeper dredging of ship channels were planted by the ignorance of nearly 50 years ago.

ACKNOWLEDGEMENTS

I am indebted to E. L. Packard and George S. Myers for personal reminiscences, and to O. P. Pearson of the Museum of Vertebrate Zoology at the University of California for access to the museum's files. I also wish to thank Harry S. Ladd of the U.S. Geological Survey for the photograph of G. K. Gilbert and the San Francisco Chronicle for the photograph of John Reber.

LITERATURE CITED

Allen, W. E. 1920. A quantitative and statistical study of the plankton of the San Joaquin River and its tributaries in and near Stockton, California, in 1913. Univ. Calif. Publ. Zool. 22:1-292.

Arthur, J. F., and M. D. Ball. 1979. Summary of factors influencing the entrapment of suspended material in the San Francisco Bay-Delta estuary. Pages 143-174 *in* T. J. Conomos, ed. San Francisco Bay: The Urbanized Estuary. Pacific Division, Amer. Assoc. Advance. Sci., San Francisco, Calif.

Barrett, E. M. 1963. The California oyster industry. Calif. Fish Game Fish Bull. 123. 103 pp.

Carlton, J. T. 1975. Introduced intertidal invertebrates. Pages 17-25 *in* R. I. Smith and J. T. Carlton, eds. Light's manual, intertidal invertebrates of the central California coast. University of California Press, Berkeley, Calif.

Carpelan, L. H. 1955. Tolerance of the San Francisco topsmelt, *Atherinops affinis affinis*, to conditions in salt-producing ponds bordering San Francisco Bay. Calif. Fish Game 41(4):279-284.

Conomos, T. J. 1975. Movement of spilled oil as predicted by estuarine nontidal drift. Limnol. Oceanogr. 20(2):159-173.

Conomos, T. J., and D. H. Peterson. 1974. Biological and chemical aspects of the San Francisco Bay turbidity maximum. Mem. Inst. Geol. Bassin Aquitaine 1974 (7):45-52.

Conomos, T. J., D. H. Peterson, P. R. Carlson, and D. S. McCulloch. 1970. Movement of seabed drifters in the San Francisco Bay estuary and the adjacent Pacific Ocean. U.S. Geol. Surv. Circ. 637B. 8 pp.

Cook, S. F. 1964. The aboriginal population of upper California. Pages 397-403 *in* Actas y Memorias XXXV Congreso Internacional de Americanistas, Mexico, 1962.

Dreisbach, R. H. 1969. Handbook of the San Francisco region. Environment Studies, Palo Alto, Calif. 564 pp.

Essig, E. O. 1933. The Russian settlement at Ross. Calif. Histor. Soc. Quart. 12(3):3-21.

Farquhar, F. P. 1966. Up and down California in 1860-1867. The journal of William H. Brewer, etc. University of California Press, Berkeley, Calif. 583 pp.

Filice, F. P. 1954a. An ecological survey of the Castro Creek area in San Pablo Bay. Wasmann J. Biol. 12(1):1-24.

Filice, F. P. 1954b. A study of some factors affecting the bottom fauna of a portion of the San Francisco Bay estuary. Wasmann J. Biol. 12(3):257-292.

Filice, F. P. 1958. Invertebrates from the estuarine portion of San Francisco Bay and some factors influencing their distribution. Wasmann J. Biol. 16(2):159-211.

Filice, F. P. 1959. The effect of wastes on the distribution of bottom invertebrates in the San Francisco Bay estuary. Wasmann J. Biol. 17(1):1-17.

Gilbert, G. K. 1917. Hydraulic-mining débris in the Sierra Nevada. U.S. Geol. Surv. Prof. Paper 105. 154 pp.

Gilliam, H. 1957. San Francisco Bay. Doubleday and Company, Garden City, N.Y. 366 pp.

Haven, K. F. 1975. A methodology for impact assessment in the estuarine/marine environment. University of California Lawrence Livermore Lab. 51949. 43 pp.

Hedgpeth, J. W. 1941. Livingston Stone and fish culture in California. Calif. Fish Game 27(3): 126-148.

Hedgpeth, J. W. 1945. The United States Fish Commission Steamer Albatross. With an appendix by W. L. Schmitt. Amer. Neptune 5(1):5-26.

Hedgpeth, J. W. 1968. Newcomers to the Pacific Coast: The estuarine itinerants. Pages 376-380 *in* E. Ricketts, J. Calvin, and J. W. Hedgpeth, eds. Between Pacific Tides. Stanford University Press, Stanford, Calif.

Hedgpeth, J. W. 1969. The San Francisco Bay-Delta estuary, an ecological system. Pages 367-386 *in* The Nation's Estuaries, San Francisco Bay and Delta, California. Part 2. Hearing, House Subcommittee of the Committee on Government Operations, U.S. 91st Congress, First Session.

Hedgpeth, J. W. 1973. Protection of environmental quality in estuaries. Pages 233-249 *in* C. R. Goldman, James McEvoy III, and P. J. Richerson, eds. Environmental Quality and Water Development. W. H. Freeman, San Francisco, Calif.

Hedgpeth, J. W. 1975. San Francisco Bay. Geoscience and Man 12:23-30.

Hedgpeth, J. W. 1977. Models and muddles. Some philosophical observations. Helgoländer wiss. Meeresunters 30:92-104.

Hill, C. L., and C. A. Kofoid. 1927. Marine borers and their relation to marine construction on the Pacific Coast. Final Report, San Francisco Bay Marine Piling Committee. San Francisco, Calif.

Imberger, J., W. B. Kirkland, and H. B. Fischer. 1977. The effect of delta outflow on the density stratification in San Francisco Bay. ABAG (Assoc. San Francisco Bay Area Govern.) Rep. HBF-77/02. 109 pp.

Jones, M. L. 1961. A quantitative evaluation of the benthic fauna off Point Richmond, California. Univ. Calif. Publ. Zool. 67(3):219-320.

Kelley, D. W. 1966. Ecological studies of the Sacramento-San Joaquin estuary. Part I. Zooplankton, zoobenthos, and fishes of San Pablo and Suisun bays, zooplankton and zoobenthos of the Delta. Calif. Fish Bull. 133. 133 pp.

Lincoln, A. 1969. The Beechey expedition visits San Francisco. Pacific Discovery 22(1):1-8.

McCullough, D. R. 1971. The tule elk, its history, behavior, and ecology. Univ. Calif. Publ. Zool. 88:1-191.

McCulloch, D. S., D. H. Peterson, P. R. Carlson, and T. J. Conomos. 1970. A preliminary study of the effects of water circulation in the San Francisco Bay estuary. U.S. Geol. Surv. Circ. 637A. 27 pp.

May, P. R. 1970. Origins of hydraulic mining in California. The Holmes Book Company, Oakland, Calif. 88 pp.

Miller, R. C., W. D. Ramage, and E. L. Lazier. 1928. A study of physical and chemical conditions in San Francisco Bay especially in relation to the tides. Univ. Calif. Publ. Zool. 31(11):201-267.

Neasham, V. A. 1973. Wild legacy. California hunting and fishing tales. Howell-North Books, Berkeley, Calif. 178 pp.

Nelson, N. C. 1909. Shellmounds of the San Francisco Bay region. Univ. Calif. Publ. Amer. Arch. Ethn. 7(4):309-348.

Nichols, F. H. 1973. A review of benthic faunal surveys in San Francisco Bay. U.S. Geol. Surv. Circ. 677. 20 pp.

Nichols, F. H. 1977. Infaunal biomass and production on a mudflat, San Francisco Bay, California. Pages 339-357 *in* B. C. Coull, ed. Ecology of Marine Benthos. University of South Carolina Press, Columbia, S.C.

Olmsted, R. R. 1962. Scenes of wonder and curiosity from Hutchings' California Magazine 1856-1861. Howell-North, Berkeley, Calif. 413 pp.

Packard. E. L. 1918a. Molluscan fauna from San Francisco Bay. Univ. Calif. Publ. Zool. 14: 199-452.

Packard, E. L. 1918b. A quantitative analysis of the molluscan fauna of San Francisco Bay. Univ. Calif. Publ. Zool. 18:299-336.

Pearson, E. A., P. N. Storrs, and R. E. Selleck. 1967. Some physical parameters and their significance in marine waste disposal. Pages 297-315 *in* T. A. Olson and F. J. Burgess, ed. Pollution and Marine Ecology. Interscience-Wiley & Sons, New York.

Pearson, E. A., P. N. Storrs, and R. E. Selleck. 1970. A comprehensive study of San Francisco Bay, final report; VIII. Summary, conclusions and recommendations. University of California Sanitary Engineering Res. Lab. Rep. 67(5):1-85.

Peterson, D. H., T. J. Conomos, W. W. Broenkow, and P. C. Doherty. 1975. Location of the non-tidal current null zone in northern San Francisco Bay. Estuarine Coastal Mar. Sci. 3:1-11.

Pyne, S. 1975. The mind of Grove Karl Gilbert. Pages 277-298 *in* Theories of Landform Development. Publ. Geomorphology No. 6. State University of New York, Binghampton, N.Y.

Royce, J. 1898. The Pacific Coast: A psychological study of the relations of climate and civilization. Prepared for the National Geographic Society.

Royce, J. 1920. Fugitive essays. Harvard University Press, Cambridge, Mass. 432 pp.

Schmitt, W. L. 1921. The marine decapod crustacea of California. Univ. Calif. Publ. Zool. 23. 470 pp.

Scofield, N. B. 1931. The fishing industry. Appendix F, Economic aspects of a salt water barrier. Calif. Div. Water Resources Bull. 28. pp. 435-445.

Smith, H. M. 1896. A review of the history and results of the attempts to acclimatize fish and other animals in the Pacific states. U.S. Fish Comm. Bull. 15 (1895):379-472, pls. 73-83.

Stanger, F. M. 1963. South from San Francisco. San Mateo County, California. Its history and heritage. San Mateo County Historical Association. 214 pp.

Sumner, F. B. 1945. The life history of an American naturalist. Jacques Cattell Press, Lancaster, Pa. 298 pp.

Sumner, F. B., G. D. Louderback, W. L. Schmitt, and G. E. Johnston. 1914. A report on the physical conditions in San Francisco Bay, based upon the operations of the United States Fisheries Steamer Albatross, 1912-1913. Univ. Calif. Publ. Zool. 14:1-198.

Sumner, F. B., R. C. Osburn, L. J. Cole, and B. M. Davis. 1913. A biological survey of the waters of Woods Hole and vicinity. Bull. U.S. Bur. Fish. 31(1&2).

Thorson, G. 1957. Sampling the benthos. Pages 61-73 *in* Mem. Geol. Soc. Amer. 67, vol. 1.

Turner, J. L., and D. W. Kelley. 1966. Ecological studies of the Sacramento-San Joaquin Delta. Part II. Fishes of the delta. Calif. Fish Game Fish Bull. 136:1-168.

Warren, E. 1945. Proceedings of California Water Conference. State Capitol, Sacramento. 510 pp.

Young, W. R. 1929. Report on salt water barrier below confluence of Sacramento and San Joaquin Rivers, California. Calif. Div. Water Resources Bull. 22, vol. 1. 667 pp.

ANCIENT PROCESSES AT THE SITE OF SOUTHERN SAN FRANCISCO BAY: MOVEMENT OF THE CRUST AND CHANGES IN SEA LEVEL

BRIAN F. ATWATER
U. S. Geological Survey, 345 Middlefield Road, Menlo Park, CA 94025

Several kilometers of sea water probably covered the site of San Francisco Bay 150 million years ago. Motion of lithospheric plates fostered sedimentary and structural accretion that subsequently transformed this oceanic abyss into a landmass. The current episode of right-handed movement along the San Andreas fault reflects sideways motion between lithospheric plates that began locally during the past 10-15 million years. Concurrent vertical movement in the vicinity of the Bay, though at least 20 times slower than displacement along the San Andreas fault, probably created the bedrock trough that now contains much of the Bay and closed a strait that once linked the site of the Bay with the Pacific Ocean. Sediment beneath the floor of the Bay suggests that at least four ephemeral estuaries have occupied the site of the Bay during the past 700,000 years. These estuaries presumably reflect global fluctuations in sea level caused by exchange of water between oceans and continental glaciers. The present estuary originated when the Pacific Ocean entered the Golden Gate about 10,000 years ago. Most of the growth of this estuary occurred during the next 5,000 years. Sites of human habitation contemporaneous with this episode of rapid submergence have not yet been discovered in central California, perhaps because they now lie beneath the mud and water of San Francisco Bay.

Though recently transfigured by bridges, levees, and fill, the site of San Francisco Bay[1] has undergone far greater changes during the geologic past. Events recorded by local rocks begin 100-200 million years ago when the oldest of these rocks accumulated beneath several kilometers of sea water. Since that time, the site of the Bay has hosted deep and shallow seas, stream valleys, and hills, as well as estuarine embayments such as we have today (Louderback 1951; Taliaferro 1951; Howard 1951).

Many processes contributed to ancient geographic changes at the site of the Bay. This chapter relates the evolution of southern San Francisco Bay to two major agents of change: (1) movement of the Earth's crust during the past 150 million years, which largely built the bedrock foundation that transformed the site of the Bay from ocean abyss to continental hills and valleys; and (2) worldwide sea-level fluctuations during the past few million years, which have caused episodic submergence and emergence of low-lying valleys and thereby created such ephemeral embayments as the present San Francisco Bay estuary.

CRUSTAL MOVEMENT

Many geologists now attribute movement of the Earth's crust near San Francisco Bay to the

---

[1] The "San Francisco Bay estuary" refers herein collectively to San Francisco, San Pablo, and Suisun bays; Carquinez Strait; tidal marshes surrounding these bodies of water; and the Sacramento-San Joaquin Delta (see Atwater et al. 1979; Fig. 2). San Francisco Bay, abbreviated "the Bay," borders San Pablo Bay about 7 km southwest of Pinole Point. The part of San Francisco Bay located south of the latitude of the Golden Gate is informally designated southern San Francisco Bay.

SAN FRANCISCO BAY

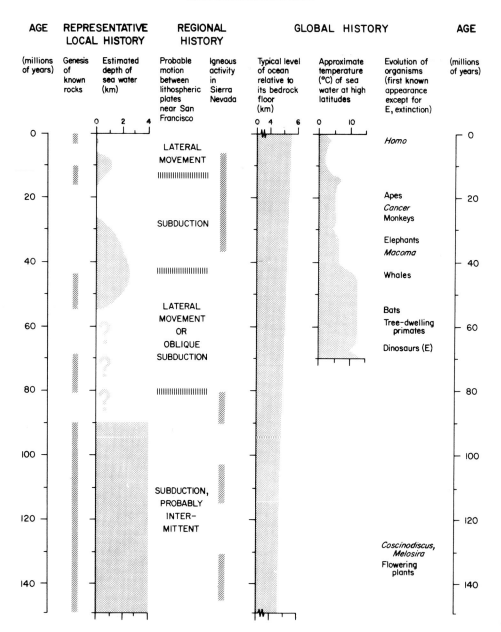

Fig. 1. Elements of local, regional, and global history for the past 150 million years. Local history refers to events and conditions recorded by rocks between southern San Francisco Bay and the San Andreas fault (Graham and Church 1963; Dibblee 1966; Page and Tabor 1967; Clark 1968; Beaulieu 1970: 41, 82, 97). Time periods indicated by bars give maximum probable ranges in age for these rocks. Most of the rocks originated on or beneath the ocean floor. Depths of overlaying water, inferred from such features as fossils and layering, imply multiple episodes of submergence and emergence. No attempt is made to estimate local depths from the 60- to 80-million-year-old rocks because, like some other rocks in western North America (Swe and Dickinson 1970; Blake and Jones 1974; Jones et al. 1977), they may have migrated from a distant place of deposition.

motion of slablike pieces of crust and upper mantle called lithospheric plates (T. Atwater 1970; Blake and Jones 1974). A boundary between plates has probably spanned or flanked the site of San Francisco Bay during most of the past 100-200 million years. For much of this time a western plate (or plates), composed partly of oceanic basalt and a veneer of oceanic sediment, may have slid beneath a continental plate along a roughly north-south-trending zone of convergence. This process, labeled "subduction" (Fig. 1), would have telescoped and imbricated large quantities of crust and upper mantle at or near the site of the Bay. Resulting accumulations of rock would have added substantially to the thickness of the local crust and helped to transform the site of the Bay from an abyssal ocean to shallower seas and dry land.

Today's right-handed displacement along the San Andreas fault and its branches reflects sideways motion between plates rather than subduction. In the vicinity of San Francisco, the current episode of sideways motion (lateral movement, Fig. 1) probably began within the past 15 million years.

Not all recent crustal movement, however, mimics sideways motion between plates. During the past few million years, for example, the crust has risen along the Pacific coast and subsided at the site of southern San Francisco Bay (Lawson 1914; Radbruch 1957; Christensen 1965; Schlocker 1974: 72-74; Bradley and Griggs 1976; Atwater et al. 1977; Helley et al. in press). It is not evident how much, if any, of this vertical motion occurred along known lateral faults. The reasons for uplift and subsidence in the vicinity of the Bay likewise remain mysterious, although one strong possibility is that lateral motion causes compression or extension of the crust where faults diverge from the trend of horizontal plate motion (Crowell 1974; T. Atwater 1970).

It is tempting to depreciate the effects of vertical movement near San Francisco Bay over the past 1 million years, because horizontal movement has been so much faster. While lateral offset along the San Andreas fault has averaged about 10-30 m per millennium (see caption to Fig. 2), subsidence at the site of southern San Francisco Bay and uplift along the Pacific coast to the west have averaged no more than 0.5 m per millennium (Table 1). Nevertheless, vertical movement has contributed to geographic changes. Crustal subsidence, for example, largely created the bedrock trough that contains much of the Bay (Lawson 1914; Louderback 1951; Atwater et al. 1977). A likely consequence of the uplift was the closure of former connection between the Pacific Ocean and the site of the Bay. This connection, located 15-20 km south of the Golden Gate, is evidenced by marine sediments younger than about 0.5 million years that are situated as much as 50 m above modern sea level (the Colma Formation of Bonilla 1971). If, as seems probable from oxygen-isotope records (Fig. 3), sea levels of the past 0.5 million years have not reached such high elevations, then the former strait marked by these sediments has been uplifted. These examples imply that

Depths for long periods lacking a local rock record are estimated from depositional environments of immediately older and younger rocks. Zero depth indicates emergence from the sea. Important processes of regional significance include motion of lithospheric plates (T. Atwater 1970; Travers 1972; Blake and Jones 1974) and igneous activity in the Sierra Nevada (Evernden and Kistler 1970: 17; Lipman et al. 1972). Global changes include a gradual rise in sea level (Vasil'kosskiy 1973), a sporadic decline in temperature of sea water at high latitudes (Savin et al. 1975), and appearance or extinction of organisms (McAlester 1968:87, 112, 123, 130). Italicized names denote some of the genera whose living representatives are mentioned elsewhere in this volume: *Coscinodiscus* and *Melosira*, common among estuarine diatoms; *Macoma*, the genus of clams now including the mudflat-dwelling *M. balthica; Cancer*, generic name of the dungeness crab; and *Homo*, an evolutionary late-comer now represented by approximately 5 million humans near the San Francisco Bay estuary. Times of first appearance for these genera follow Wornardt (1972), Cox et al. (1969), Brooks et al. (1969:509), and Leakey (1976).

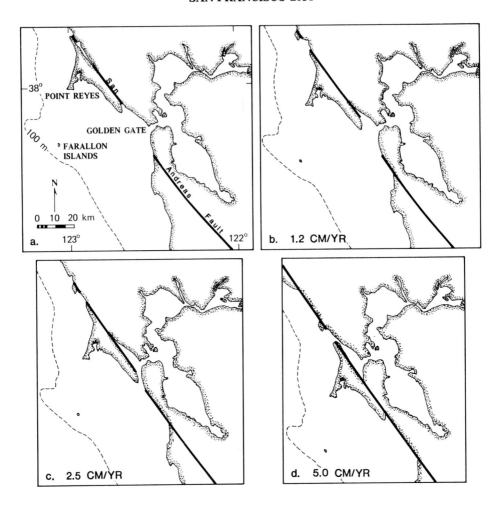

Fig. 2. Hypothetical geography of the San Francisco area 1 million years ago, assuming various rates of right-lateral movement along the San Andreas fault. Geographic reconstructions (B-D) were prepared by cutting a modern map (A) and undoing the lateral displacement that would result from 1 million years of movement at indicated rates. Except where modified for cartographic simplicity, diagrams B-D make no other changes in shorelines and do not compensate for deposition, erosion, sea-level changes, vertical crustal movement, or horizontal offset along other faults. Rates of displacement assumed in diagrams B and C span most of the range of plausible values for movement along the San Andreas fault over the past several million years. Offset during the past 30 million years, as determined by matching distinctive rocks on opposite sides of the fault, averages about 1-3 cm·yr^{-1} (compilations by Grantz and Dickinson 1968; and Dickinson et al. 1972). These rates overlap with estimates of historic offset, which average about 2.5-3.5 cm·yr^{-1} (Nason and Tocher 1970; Savage and Burford 1973). Appreciably faster movement (D) requires that the San Andreas fault accommodate most of the relative motion between bounding plates, which averages about 6 cm·yr^{-1} (Vine 1966; Graham and Dickinson 1978). Such a monopoly seems unlikely because the boundary between plates is considerably more diffuse than a single fault (T. Atwater 1970). Diagram D neglects relocation of fault movement on the San Francisco peninsula; about 10-25 km of indicated offset would probably have occurred along an ancestral trace of the San Andreas fault located as much as 7 km to the west, the present trace having displaced 100-million-year-old rocks by only 25-40 km (Bailey et al. 1964:160; Dibblee 1966).

the hypothetical maps (Fig. 2) may badly misrepresent the ancient landscape by ignoring the vertical component of crustal movement.

CHANGES IN SEA LEVEL

People have long surmised that relative motion between land and sea helped to create San Francisco Bay. Perhaps the oldest surviving statement of this hypothesis is the aboriginal tradition recorded in the diary of Mariano Payeras (1769-1823), a California missionary:

> "The day, May 28 [1818]. I left the presidio of N. P. [San Francisco] accompanied by P. Luis Gil, and Captain Don Luis Arguello in a launch at about 11 o'clock in the morning and from the wharf we came down as far as the middle of the port[.] [W]hat is now the port of San Francisco was formerly according to the tradition of the old ones an oak grove, and without water other than of a river that crossed at its foot, and in evidence of this tradition, they say you still find in the port and marsh, trunks and roots of oak trees" (Payeras 1818).[2]

Pioneer English-speaking geologists also surmised submergence of a bygone valley. Andrew C. Lawson (1894) interpreted the numerous islands, peninsulas, and small embayments near San Francisco as former hills, ridges, and stream valleys drowned by the sea. Inferring that San Francisco Bay did not exist before this submergence, Lawson proposed that the ancestral drainage of the San Joaquin and Sacramento Rivers must have flowed through the Golden Gate to a coastline situated some distance to the west. Grove Karl Gilbert (1917:16-24) deduced submergence not only from drowned topography but also from eroded shorelines and submerged aboriginal middens.

During the past 30 years, geologists have learned about additional motion between sea and land by searching beneath the floor of the Bay. Most of the evidence has come from core samples, plugs of sediment that were collected by engineers to assist in the design of footings for bridges and buildings (Trask and Rolston 1951; Treasher 1963; Goldman 1969). Constituents and properties of these samples, particularly their fossils, grain size, color, and density, commonly indicate whether the sediment accumulated in an estuary or in a stream valley (Atwater et al. 1977). Interpreted in this manner, core samples suggest that estuaries and stream valleys—at least four of each—have alternately occupied the site of the Bay during the past 1 million years (Figs. 3,4; Wagner 1978:137-138). Thus, no fewer than three cycles of submergence and emergence preceded the episode of inundation that created the present estuary.

Origins of Relative Motion between Sea and Land

Global fluctuations in sea level caused by the exchange of water between oceans and continental glaciers are principally responsible for episodic submergence and emergence of the site of the Bay over the past million years. During glacial ages, sheets of ice as thick as several kilometers covered large areas of land at northerly latitudes, particularly in northern Europe, Canada, and northernmost parts of the continental United States (Flint 1971:73-80). In addition, smaller glaciers occupied alpine areas further south, including parts of the Sierra Nevada such as Yosemite Valley (Wahrhaftig and Birman 1965). When this land ice formed, it withdrew large quantities of

---

[2] Bancroft (1884:247) equates aboriginal peoples with "the old ones", Spaniards with the antecedent of "they", and the whole of San Francisco Bay with the "port of San Francisco". Payeras implies that the trunks and roots came from both evergreen and deciduous oaks ("encinos" and "robles").

water from the oceans, and when it melted, the water returned. Resulting fluctuations in sea level measured as much as 100-150 m, spanned many thousands of years, proceeded as rapidly as 10-20 m per millennium, and probably overshadowed other kinds of motion between sea and land near San Francisco (Table 1).

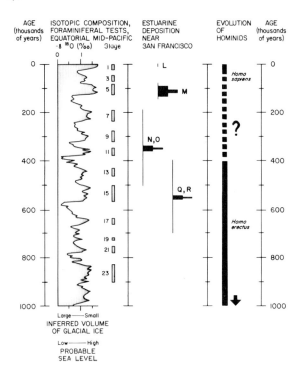

Fig. 3. Probable ages of known estuarine sediment between San Francisco and Oakland. Letters identify units of sediment labeled in Fig. 4. Thin vertical lines show range of likely ages as determined from radiocarbon dates (unit L; Atwater et al. 1977), as estimated from the amino-acid stereochemistry of fossil clams and oysters (unit N; Brian Atwater and John F. Wehmiller, unpublished data), as limited by the magnetic polarity and radiometric ages of volcanic ash (unit Q; C. Naser pers. comm.); G. Dalrymple and M. Lanphere pers. comm.; Sarna-Wojcicki 1976; J. Hillhouse pers. comm., and as required by superposition (unit M underlies sediment older than 40,000 years [Atwater et al. 1977], units N and O underlie unit M, as do units Q and R [Ross 1977]). Bars attached to vertical lines represent time required to build thickest remaining part of unit(s) at average sedimentation rates of 1 m (thick bars) and 3 m (thin bars) per millennium. Measured thicknesses have been multiplied by 1.4 to correct for post-depositional compaction. Assumed rates of deposition are typical average rates for unit L, computed from radiocarbon dates and sediment thicknesses reported by Atwater et al. (1977). The graph at left (Shackleton and Opdyke 1976) shows the relative proportion of heavy (^{18}O) and light (^{16}O) oxygen in the fossil shells (tests) of one-celled marine animals (foraminifera). Variable rates of accumulation have allowed biological mixing and burrowing of bottom sediment to squash some peaks and accentuate others (Schackleton and Opdyke 1976) so that changes in oxygen-isotope composition merely suggest the approximate frequency and relative magnitude of glaciation and sea-level change. Except where dated by reversals in direction of the Earth's magnetic field, isotopic changes are correlated with time by assuming a constant sedimentation rate of 1 cm per millenium (Shackleton and Opdyke 1976). Ranges in age for hominids follow summaries by Leakey (1976) and Tattersall and Eldredge (1977).

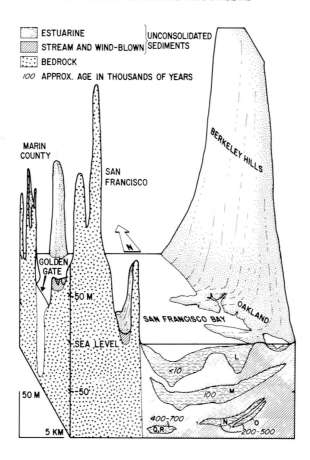

Fig. 4. Generalized cross section of some rocks and sediment near San Francisco. Front panel modifies and combines two slightly different cross sections: one by Ross (1977), and the other by Atwater et al. (1977, sections A-A'). Both sections rely on samples and descriptions from boreholes that explored the foundations of proposed bridges, moles, and tunnels. Estuarine sediment is keyed by letter to Fig. 3. Ranges in age for estuarine sediment below unit M reflect uncertainties in measurement rather than long episodes of deposition. Shorelines of the bay denote reach of highest tides circa 1850 (Nichols and Wright 1971).

If ice ages account for the fluctuating sea levels, then what accounts for the ice ages? Currently foremost among a multitude of theories are motion of lithospheric plates and changes in the Earth's rotation and orbit.

Motion of lithospheric plates promoted the ice ages by rafting continents to high latitudes (Ewing and Donn 1956; Donn and Shaw 1977). About 50 million years ago, oceans rather than continents covered the geodetic poles of the Earth. It seems unlikely that large ice sheets routinely covered these polar seas because the surface water was warm—perhaps 10-12°C (Savin et al. 1975; Fig. 1)—owing to communication with low-latitude oceans. Later, however, drifting continents gradually displaced these mild seas and disrupted interchange of polar and equatorial water. Resulting insolation and refrigeration of polar regions apparently prepared these areas for continental glaciation.

Cycles in the orientation of the Earth's axis and the shape of its orbit probably triggered the

TABLE 1. SOME KINDS OF RELATIVE CHANGE IN LEVEL BETWEEN LAND AND SEA. MAGNITUDES AND RATES GIVE ESTIMATES FOR THE VICINITY OF SAN FRANCISCO BAY DURING THE PAST ONE MILLION YEARS.

Principal Causes	Maximum Probable Magnitude (m)	Typical Duration	Typical Average Rate (m·1000 yr^{-1})	Oscillatory (O) or Uni-directional (D)	References
Rise or fall of the sea[1]					
Waves, swells	5	seconds			
Astronomical tides	3	hours			
Meteorological tides atmospheric pressure, wind, river discharge	5	hours, days, years	NA[2]	O	Lisitzin (1974)
Astronomical cycles of tides	1	weeks, years			
Exchange of H_2O between oceans and glaciers	150	thousands of years	1-20	O	Flint (1971: 315-342)
Reduction in area of oceans because of continental accretion; increase in volume of ocean water because of liberation of water from the earth's interior	1	billions of years	0.001	D	Vasil'kovskiy (1974); partial record shown in Fig. 1, this chapter
Subsidence or uplift of coastal land[1]					
Adjustment of the earth's crust to addition or removal of sea water and glacial ice	10	thousands of years	0.1-1.0	O	Bloom (1971); Clark et al. (1978)
Vertical crustal movement probably related to motion of lithospheric plates	300	thousands and millions of years	0.5-0.5	D,O	Bradley and Griggs (1976); Atwater et al. (1977)

[1] Motion referenced to a stable plane or point such as the center of the earth

[2] NA - Not applicable

principal episodes of glaciation and deglaciation that have occurred since polar lands approached their present positions. According to a theory named after one of its advocates, Mulutin Milankovich, these astronomical cycles have periodically prevented the summer sun from melting all of a winter's snow at high northerly latitudes. Considerable support for the Milankovich theory comes from changes in the ratio of heavy (^{18}O) to light (^{16}O) oxygen in fossil cells of one-celled marine animals (Hays et al. 1976). It seems likely that, during the past one million years, these changes have mainly reflected the oxygen-isotope composition of sea water (Shackleton and Opdyke 1976: 459). The ratio of ^{18}O to ^{16}O in sea water depends on evaporation and fresh-water input because water containing ^{18}O evaporates less readily than water containing ^{16}O. Thus, a high concentration of ^{18}O in fossil shells ($-\delta^{18}O$ between 0 and 1, Fig. 3) implies the large glaciers and low sea levels of the ice ages, and lower concentrations of ^{18}O ($-\delta^{18}O$ greater than 1) imply small glaciers and high sea levels, such as we have today.

Legacies of Ancient Sea Levels

Ephemeral estuaries probably occupied low-lying areas at or near the site of the Bay each time the sea approached its present level during the past 0.5-1.0 million years (that is, during most odd-numbered stages in Fig. 3). Sediment from at least four of these estuaries has been found beneath the site of the Bay (Ross 1977; Fig. 4), and evidence of other estuaries probably remains to be discovered or confirmed. The youngest known estuarine sediment corresponds to the current high stand of the sea (Figs. 3, 4: unit L, stage 1) and its immediate predecessor (unit M, stage 5). Older sediment (units N, O, Q, P) cannot yet be assigned to a single isotopic stage because of uncertainty about its age (Fig. 3, thin vertical lines).[3]

Crustal subsidence appears to have moved some estuarine deposits below the reach of glacial-age streams, thereby limiting the erosion of units M, N, O, Q, and R during low stands of the sea. The continuity, thickness, and depth of unit M, for instance, seem best explained by a downward crustal movement of 20 m (Atwater et al. 1977). Older estuarine sediment also appears to have subsided, though not fast enough to escape considerable erosion. Largely as the result of such erosion, known sediment beneath the floor of the Bay records neither the full number nor the full duration of Pleistocene high stands of the sea (Fig. 3).

Growth of the Most Recent Estuary

The episode of submergence that created San Francisco Bay began about 15,000-18,000 years ago, when glaciers of the last ice age started their retreat (Prest 1969). At the onset of glacial retreat, the Pacific Ocean lapped against a shoreline located near the Farallon Islands (Fig. 6). In order to meet this shoreline, the combined Sacramento and San Joaquin Rivers must have flowed through the Golden Gate and traversed an exposed continental shelf. Some of the riverborne sand that reached flood plains and beaches on the shelf was probably swept by westerly winds into the ancient dunes that covered much of the site of San Francisco and extended across the site of the Bay to Oakland (Atwater et al. 1977). Southeast of these dunes was a broad stream valley in which roamed now-extinct species of camel, horse, bison, and ground sloth (Helley et al. in press).

Most of the submergence that transformed this landscape occurred earlier than 5,000 years ago (Fig. 6). Initial migration of shorelines brought the rising sea into the Golden Gate about 10,000 years ago. During the next few thousand years, the newborn estuary spread as rapidly as 30 m·yr^{-1} across low-lying areas in response to a rise in relative sea level that averaged nearly 2

---

[3] J. F. Wehmiller, J. W. Hillhouse, Andrei Sarna-Wojcicki, and I are refining the chronology of sediments below unit M. Ages shown in Figs. 3 and 4 will probably be revised by us and by others.

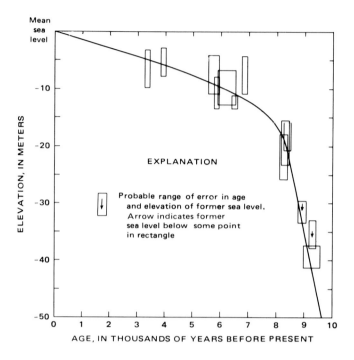

Fig. 5. Changes in sea level relative to land during the past 10,000 years at the site of southern San Francisco Bay (after Atwater et al. 1977). Boxes represent uncertainty in reckoning former sea levels from radiocarbon ages and elevations of plant fossils from unit L (Fig. 4). Most of the dated fossils originated as vascular plants in ancient tidal marshes. By analogy with their modern counterparts (Atwater et al. 1979, Figs. 7, 10), these plants probably grew very close to contemporaneous sea levels. See Kvenvolden (1962) and Storey et al. (1966) for additional radiocarbon dates pertaining to changes in sea level at the site of San Francisco Bay.

cm·yr^{-1} (Fig. 5; Atwater et al. 1977). Thereafter, relative sea level changed more slowly because, by 5,000 years ago, glaciers had reached approximately their present size (Bloom 1971). Submergence since that date has averaged only 0.1-0.2 cm·yr^{-1} and probably includes a large component of crustal subsidence (Atwater et al. 1977).

The difference in rates of submergence before and after 5,000 years ago may influence the apparent antiquity of human habitation in the vicinity of San Francisco Bay. Settlement of the Americas began at least 20,000 years ago, and by 10,000 years ago people inhabited much of the continental United States (Haynes 1969). However, no known archeological site in central California appears much older than 5,000 years (Gerow and Force 1968:174). One way to approach this problem is to assume that traces of the earliest central Californians have been covered by the rising sea. Given the rapidity of changes in sea levels and shorelines 5,000-10,000 years ago, sites of habitation located at that time along the shores of estuaries must now lie beneath mud and tidal water. Sites younger than 5,000 years, alternatively, postdate rapid submergence and would therefore more likely escape total inundation (K. Lajoie pers. comm.).

How old, then, is the aboriginal tradition recorded by Mariano Payeras? If originated by people who actually saw the site of the Bay before widespread submergence, this tradition must be nearly 10,000 years old. Such antiquity, though improbable, cannot be ruled out in light of Lajoie's hypothesis. Alternatively, the tradition originated as entertainment or science among people of the past 5,000 years. Too late to observe a stream at the site of the Bay, some of these people may nevertheless have deduced its former presence; perhaps, like Gilbert and Lawson, they read ancient history from soggy middens and drowned topography.

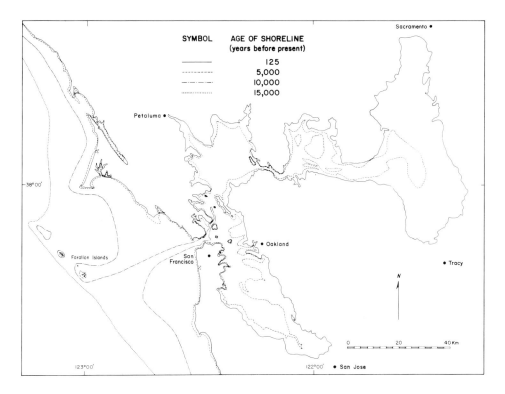

Fig. 6. Approximate high-tide shorelines near San Francisco during the past 15,000 years. The 125-year-old shoreline, based on compilations by Gilbert (1917:76) and Nichols and Wright (1971), denotes the landward edge of tidal marshes before human encroachment or, where no marsh was present, the high-water line circa 1850. Locations of older shorelines are estimated by projecting sea levels of the past 15,000 years onto the land surface inundated by the growing estuary during this time. We assume the following sea levels, expressed relative to present mean sea level (Fig. 5; Flint 1971:321): 5,000 years ago, -8 m; 10,000 years ago, -55 m; and 15,000 years ago, -100 m. Topography of the ancient land surface east of the Golden Gate follows reconstructions by Goldman (1969, pl. 3), the U. S. Army Corps of Engineers (1963, pls. 6-7), Carlson et al. (1970), and B. Atwater, S. D. McDonald, and D. R. Nichols (unpublished data). Because of variations in abundance and quality of boreholes and acoustic profiles, these topographic reconstructions are most accurate for the southern arm of the estuary and least accurate for open-water areas of the northern part of the estuary. Topography of the ancient land surface west of the Golden Gate is inferred mostly from modern water depths as shown on NOS Nautical Charts 5402 and 5502. Local adjustments uncontrolled by boreholes or subbottom profiles attempt to correct for differences between modern bathymetry and ancient topography. Location of the 10,000-year-old shoreline between San Francisco and the Farallon Islands depends greatly on such adjustments and is therefore extremely speculative.

SAN FRANCISCO BAY

ACKNOWLEDGMENTS

Illustrators for this chapter include Sara Boore (Fig. 4), Yosh Inouye (Fig. 2), Bruce Rogers (Fig. 6), and Steven Talco (Figs. 1, 3). Frances DeMarco translated Payeras' diary. D. F. Belknap, D. G. Herd, M. R. Hill, and J. C. Yount reviewed the text. Persons who provided information not acknowledged elsewhere include W. O. Addicott, R. E. Arnal, J. A. Barron, James Buckley, Gregory Frantz, E. J. Helley, K. R. Lajoie, R. D. Nason, Wayne Savage, and the hundreds of drillers, engineers, and geologists who generated borehole records.

LITERATURE CITED

Atwater, B. F., C. W. Hedel, and E. J. Helley. 1977. Late Quaternary depositional history, Holocene sea-level changes, and vertical crustal movement, southern San Francisco Bay, California. U. S. Geol. Surv. Prof. Paper 1014. 15 pp.

Atwater, B. F. et al. 1979. History, landforms and vegetation of the estuary's tidal marshes. Pages 347-386 in T. J. Conomos, ed. San Francisco Bay: The Urbanized Estuary. Pacific Division, Amer. Assoc. Advance. Sci., San Francisco, Calif.

Atwater, T. 1970. Implications of plate tectonics for the Cenozoic tectonic evolution of western North America. Geol. Soc. Amer. Bull. 81:3513-3536.

Bailey, E. H., W. P. Irwin, and D. L. Jones. 1964. Franciscan and related rocks and their significance in the geology of western California. Calif. Div. Mines and Geol. Bull. 183. 173 pp.

Bancroft, H. H. 1884. History of California, 1542-1800, vol. 1; volume 18 of The Works of Hubert Howe Bancroft. A. L. Bancroft & Co., San Francisco, Calif. 744 pp.

Beaulieu, J. D. 1970. Cenozoic stratigraphy of the Santa Cruz Mountains and inferred displacement along the San Andreas fault. Ph.D. Thesis. Stanford University, Stanford, Calif. 101 pp.

Blake, M. C., and D. L. Jones. 1974. Origin of Franciscan melanges in northern California. Pages 345-357 in R. H. Dott, Jr. and R. H. Shaver, eds. Modern and Ancient Geosynclinal Sedimentation. Soc. Econ. Paleont. and Mineral., Special Publication 19.

Bloom, A. L. 1971. Glacial-eustatic and isostatic controls of sea level since the last glaciation. Pages 355-379 in K. K. Turekian, ed. The Late Cenozoic Glacial Ages. Yale University Press, New Haven, Conn.

Bonilla, M. G. 1971. Preliminary geologic map of the San Francisco South Quadrangle and part of the Hunters Point Quadrangle, California. U. S. Geol. Surv. Misc. Field Invest. Map MF-311, scale 1:24,000.

Bradley, W. C., and G. B. Griggs. 1976. Form, genesis, and deformation of central California wave-cut platforms. Geol. Soc. Amer. Bull. 87(3):443-449.

Brooks, H. K. et al. 1969. Arthropoda 4. Pages 400-641 in R. C. Moore, ed. Treatise on Invertebrate Paleontology, part R. Geol. Soc. Amer. and University of Kansas, Lawrence, Kan.

Carlson, P. R., T. R. Alpha, and D. S. McCulloch. 1970. The floor of San Francisco Bay. Calif. Geol. 23:97-107.

Christensen, M. N. 1965. Late Cenozoic deformation in the Central Coast Ranges of California. Geol. Soc. Amer. Bull. 76:1105-1124.

Clark, J. A., W. E. Farrell, and W. R. Peltier. 1978. Global changes in postglacial sea level: a numerical calculation. Quat. Res. 9(3):265-287.

Clark, J. C. 1968. Correlation of the Santa Cruz Mountains Tertiary—implications for San Andreas history. Pages 166-180 in W. R. Dickinson and Arthur Grantz, eds. Proceedings of Conference on Geologic Problems of San Andreas Fault System. Stanford University Pubs. Geol. Sci., vol. 11.

Cox, L. R. et al. 1969. Mollusca 6, Bivalvia. Pages 1-951 in R. C. Moore, ed. Treatise on Invertebrate Paleontology, part N. Geol. Soc. Amer. and University of Kansas, Lawrence, Kan.

Crowell, J. C. 1974. Origin of late Cenozoic basins in southern California. Pages 190-204 *in* W. R. Dickinson, ed. Tectonics and Sedimentation. Soc. Econ. Paleont. Mineral. Special Publication 22.

Dibblee, T. W., Jr. 1966. Geologic map of the Palo Alto 15-minute quadrangle, California. Calif. Div. Mines and Geol. Map Sheet 8, scale 1:62,500.

Dickinson, W. R., D. S. Cowan, and R. A. Schwickert. 1972. Test of the new global tectonics—discussion. Amer. Assoc. Petrol. Geol. Bull. 56:375-384.

Donn, W. L., and D. M. Shaw. 1977. Model of climate evolution based on continental drift and polar wandering. Geol. Soc. Amer. Bull. 88:390-396.

Evernden, J. F., and R. W. Kistler. 1970. Chronology of emplacement of Mesozoic batholith complexes in California and Western Nevada. U. S. Geol. Surv. Prof. Paper 623. 42 pp.

Ewing, M., and W. L. Donn. 1956. A theory of ice ages. Science 123:1061-1066.

Flint, R. F. 1971. Glacial and Quaternary geology. John Wiley, New York. 892 pp.

Gerow, B. A., and R. W. Force. 1968. An analysis of the University Village complex; with a reappraisal of central California archeology. Stanford University Press, Stanford, Calif. 109 pp.

Gilbert, G. K. 1917. Hydraulic-mining debris in the Sierra Nevada. U. S. Geol. Surv. Prof. Paper 105. 154 pp.

Goldman, H. B. 1969. Geology of San Francisco Bay. Pages 9-29 *in* H. B. Goldman, ed. Geologic and Engineering Aspects of San Francisco Bay Fill. Calif. Div. Mines and Geol. Special Rep. 97.

Graham, J. J., and C. C. Church. 1963. Campagnian foraminifera from the Stanford University campus, California. Stanford University Pubs. Geol. Sci. 8: 107 pp.

Graham, S. A., and W. R. Dickinson. 1978. Evidence for 115 kilometers of right slip on the San Gregorio-Hosqri fault trend. Science 199(13):179-181.

Grantz, A., and W. R. Dickinson. 1968. Indicated cumulative offsets along the San Andreas fault in the California Coast Ranges. Pages 117-119 *in* W. R. Dickinson and Arthur Grantz, eds. Proceedings of Conference on Geologic Problems of San Andreas Fault System. Stanford University Pubs. Geol. Sci., vol. 11.

Haynes, C. V. 1969. The earliest Americans. Science 166:709-715.

Hays, J. D., J. Imbrie, and N. J. Shackleton. 1976. Variations in the earth's orbit—pacemaker of the ice ages. Science 194:1121-1132.

Helley, E. J., K. R. Lajoie, W. E. Spangle, and M. L. Blair. In press. Flatland deposits of the San Francisco Bay region, California—their geology and engineering properties, and their importance to comprehensive planning. U. S. Geol. Surv. Prof. Paper 943. 290 pp.

Howard, A. D. 1951. Development of the landscape of the San Francisco Bay counties. Pages 95-106 *in* O. P. Jenkins, ed. Geologic Guidebook of the San Francisco Bay Counties. Calif. Div. Mines Bull. 154.

Jones, D. L., N. J. Silberling, and J. Hillhouse. 1977. Wrangellia—a displaced terrane in northwestern North America. Canadian J. Earth Sci. 14(11):2565-2577.

Kvenvolden, K. A. 1962. Normal paraffin hydrocarbons in sediments from San Francisco Bay, California. Amer. Assoc. Petrol. Geol. Bull. 46(9):1643-1652.

Lawson, A. C. 1894. The geomorphogeny of the coast of Northern California. University of California Dep. Geol. Bull. 1:241-271.

Lawson, A. C. 1914. Description of the San Francisco district (Tamalpais, San Francisco, Concord, San Mateo, and Hayward quadrangles). U. S. Geol. Surv. Geol. Atlas, Folio 193. 25 pp.

Leakey, R. B. 1976. Hominids in Africa. Amer. Sci. 64(2):174-178.

Lipman, P. W., H. J. Prosktka, and R. L. Christiansen. 1972. Cenozoic volcanism and plate tectonic evolution of the western United States, Part I, early and middle Cenozoic. Phil. Trans. R. Soc. London, Series A. 271:217-248.

Lisitzin, E. 1974. Sea-level changes. Elsevier Oceanography Series, 8. 286 pp.

Louderback, G. D. 1951. Geologic history of San Francisco Bay. Pages 75-94 *in* O. P. Jenkins, ed. Geologic Guidebook to the San Francisco Bay Counties. Calif. Div. Mines Bull. 154.

McAlester, A. L. 1968. The history of life. Foundation in earth science series. Prentice-Hall,

Englewood Cliffs, N. J. 151 pp.
Nason, R. D., and D. Tocher. 1970. Measurement of movement on the San Andreas fault. Pages 246-254 in L. Mansinha et al., eds. Earthquake Displacement Fields and the Rotation of the Earth. D. Reidel Publishing Co., Dordrcht, Holland.
Nichols, D. R., and N. A. Wright. 1971. Preliminary map of historic margins of marshland, San Francisco Bay, California. U. S. Geol. Surv. Open-File Map, scale 1:125,000.
Page, B. M., and L. L. Tabor. 1967. Chaotic structure and decollment in Cenozoic rocks near Stanford University, California. Geol. Soc. Amer. Bull. 78:1-12.
Payeras, M. 1818. Noticia de un viage a San Rafael (MS). Page 153 in Espediciones y Caminatas, 1806-1821, in Archivo de la Santa Barbara, tom. IV.
Prest, V. K. 1969. Retreat of Wisconsin and recent ice in North America. Geol. Surv. Canada Map 1257A, Scale 1:5,000,000.
Radbruch, D. H. 1957. Areal and engineering geology of the Oakland West Quadrangle, California. U. S. Geol. Surv. Misc. Geol. Invest. Map. I-239, scale 1:24,000.
Ross, B. E. 1977. The Pleistocene history of San Francisco Bay along the Southern Crossing. M.S. Thesis. San Jose State University, San Jose, Calif. 121 pp.
Sarna-Wojcicki, A. M. Correlation of the late Cenozoic tuffs in central California by means of trace and minor element chemistry. U. S. Geol. Surv. Prof. Paper 972. 30 pp.
Savage, J. C., and R. O. Burford. 1973. Geodetic determination of relative plate motion in central California. J. Geophys. Res. 78:832-845.
Savin, S. M., R. G. Douglas, and F. G. Stehli. 1975. Tertiary marine paleotemperatures. Geol. Soc. Amer. Bull. 86:1499-1510.
Schlocker, J. 1974. Geology of the San Francisco North Quadrangle, California. U. S. Geol. Surv. Prof. Paper 782. 109 pp.
Shackleton, N. J., and N. D. Opdyke. 1976. Oxygen-isotope and paleomagnetic stratigraphy of Pacific core V28-239 late Pliocene to latest Pleistocene. Pages 449-464 in R. M. Cline and J. D. Hays, eds. Investigation of Late Quaternary Paleooceanography and Paleoclimatology. Geol. Soc. Amer. Mem. 145.
Storey, J. A., V. E. Wessels, and J. A. Wolfe. 1966. Radiocarbon dating of recent sediments in San Francisco Bay. Calif. Div. Mines and Geol. Min. Inf. Service 19(3):47-50.
Swe, W., and W. R. Dickinson. 1970. Sedimentation and thrusting of late Mesozoic rocks in the Coast ranges near Clear Lake, California. Geol. Soc. Amer. Bull. 81:164-188.
Taliaferro, N. L. 1951. Geology of the San Francisco Bay counties. Pages 117-150 in O. P. Jenkins, ed. Geologic Guidebook of the San Francisco Bay Counties. Calif. Div. Mines Bull. 154.
Tattersall, I., and N. Eldredge. 1977. Fact, theory, and fantasy in human paleontology. Amer. Sci. 65:204-211.
Trask, P. D., and J. W. Rolston. 1951. Engineering geology of San Francisco Bay, California. Geol. Soc. Amer. Bull. 62:1079-1110.
Travers, W. B. 1972. A trench off central California in Late Eocene-Early Oligocene time. Pages 173-182 in R. Shagam et al., eds. Studies in Earth and Space Sciences, Harry Hess volume. Geol. Soc. Amer. Mem. 132.
Treasher, R. C. 1963. Geology of the sedimentary deposits in San Francisco Bay, California. Calif. Div. Mines and Geol. Special Rep 82, pp. 11-24.
U. S. Army Corps of Engineers. 1963. Technical report on San Francisco Bay barriers, Appendix E. U. S. Army Engineer District, San Francisco. 136 pp.
Vasil'kovskiy, N. P. 1973. Sea level changes in the geological past. P. P. Shirsov Inst. Oceanology, U.S.S.R. Acad. Sci. pp. 847-859.
Vine, F. J. 1966. Spreading in the ocean floor—new evidence. Science 154:1405-1415.
Wagner, D. B. 1978. Environmental history of central San Francisco Bay with emphasis on foraminiferal paleontology and clay mineralogy. Ph.D. Thesis. University of California, Berkeley, Calif. 274 pp.
Wahrhaftig, C., and J. H. Birman. 1965. The Quaternary of the Pacific Mountain system in Cali-

fornia. Pages 299-340 *in* H. E. Wright, Jr. and D. G. Frey, eds. The Quaternary of the United States. Princeton University Press, Princeton, N. J.

Wornardt, W. W., Jr. 1972. Stratigraphic distribution of diatom genera in marine sediments of the western United States. Palaeogeog., Palaeoclim., Palaeoecol. 12:49-74.

PROPERTIES AND CIRCULATION
OF SAN FRANCISCO BAY WATERS

T. JOHN CONOMOS
U. S. Geological Survey, 345 Middlefield Road, Menlo Park, CA 94025

Differences in river and waste-water inflow and wind stress create contrasting environments and dissimilar distributions of properties between the northern and southern reaches of the San Francisco Bay system. A conceptual framework describing the physical processes which control these distributions, although still incomplete, is outlined.

The northern reach receives 90% of the mean annual river inflow and 24% of the waste-water inflow. It changes from a partially mixed estuary, with a vertical salinity gradient of 10 °/oo during high river inflow, to a well-mixed estuary with a vertical salinity gradient of 3 °/oo during low summer inflow. The southern reach also has seasonally varying water properties. There the variations are determined by water exchange from the northern reach and the ocean and by direct waste inflow (76% of total Bay waste inputs). Salinity stratification is present during winter, whereas during summer the water is nearly isohaline because of wind and tidal mixing.

Our knowledge of transport mechanisms is fragmentary. The northern reach has a permanent estuarine circulation cell that is largely maintained by the salinity-controlled density differences between river and ocean waters. Although wind variations alter this circulation, it is largely modulated by the timing and magnitude of the highly seasonal river inflow. This nontidal circulation is nearly equivalent to tidal diffusion in controlling the water-replacement rates in the channels, which vary from weeks (winter) to months (summer). The southern reach, in contrast, has seasonally reversing but sluggish near-bottom and surface nontidal currents that are generated by prevailing summer and episodic winter-storm winds and by winter flows of Delta-derived low-salinity water from the northern reach. Although the diffusion of substances by the strong tidal currents is notable, the relative importance of diffusion by strong tidal currents and the episodic advective processes in controlling water replacement mechanisms and rates has not yet been fully determined.

Studies of transport processes in San Francisco Bay, an estuary surrounded by a heavily urbanized area, could now most profitably focus on water-replacement mechanisms and rates because inflowing water dilutes unfavorable anthropogenic substances and flushes them from the system. Of greatest importance are studies defining the effects of river inflow in modulating water-residence time, not only because inflow is perhaps the dominant agent in this modulation but also because man is able to control the inflow through massive river diversions.

San Francisco Bay waters are mixtures of ocean, river, and waste waters. The compositions and relative fractions of these mixtures in the Bay change rapidly in space and time with changes in the amounts and character of the source waters and with differences in the depth and degree of circulation and mixing. The constantly changing process of mixing is, in addition, modified by evaporation and precipitation.

SAN FRANCISCO BAY

Knowledge of the distributions of water properties and of circulation and mixing of these waters is essential to the understanding of environmental aspects of estuarine systems, particularly pollution problems. Our meager understanding of the hydrodynamics of the Delta-Bay system (Fig. 1) has not come primarily from purely scientific investigations, but as an indirect result of

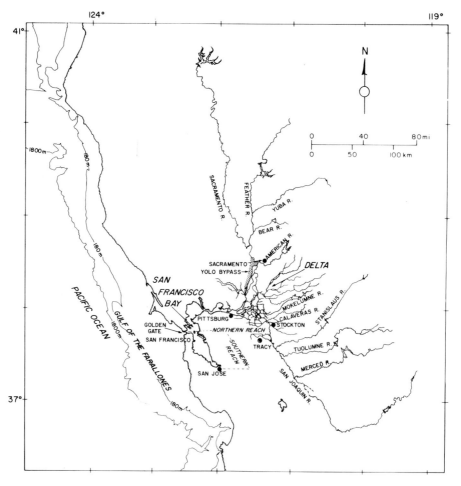

Fig. 1. The San Francisco Bay-Delta system, its drainage basin, and the adjacent ocean.

man's seeking to solve specific engineering problems such as arresting landward salt-water intrusion through the construction of barriers, defining levels to which fresh water discharges can be economically reduced without harmful effects, and defining dispersal characteristics of receiving waters for the inflow of sewage from point sources (Table 1).

The purpose of this chapter is to serve as a background for the following chapters by synthesizing our knowledge of the climate, water properties (such as temperature and salinity) and hydrodynamics of the Bay system and adjacent ocean. Specific emphasis is placed on the effects of riverflow, tides, and winds on circulation, mixing, and water-replacement time. Sedimentological, chemical, and biological consequences of these physical processes are discussed elsewhere (Arthur and Ball 1979; Cloern 1979; Conomos et al. 1979; Nichols 1979; Peterson 1979; Peterson et al. 1975b; Rubin and McCulloch 1979).

TABLE 1. PREVIOUS STUDIES OF WATER PROPERTIES AND OF HYDRODYNAMICS OF THE SAN FRANCISCO BAY

Topic	Reference
Historical account of exploration	Hedgpeth 1979
Bay-wide surveys of water properties	Sumner et al. 1914; Miller et al. 1928; Storrs et al. 1963; Selleck et al. 1966; Conomos et al. 1979
Salt-intrusion and salt barrier studies	Young 1929; Edmonston and Matthew 1931; Arthur and Ball 1979
Current-metering studies	Disney and Overshiner 1925; Marmer 1926; Grimm 1931; Corps of Engineers 1963; Peterson et al. 1975a; Cheng 1978
Bottom and surface drift studies	Conomos et al. 1970, 1971; Conomos 1975; Conomos and Peterson 1977
Dispersion studies (in field and/or physical models)	Bailey et al. 1966; Lager and Tchobanoglous 1968; Nelson and Lerseth 1972; Kirkland and Fischer 1976; Imberger et al. 1977; O'Connell and Walter 1963
Numerical modelling studies	Nelson and Lerseth 1972; O'Connor and Lung 1977; Walters and Cheng 1978; Festa and Hansen 1976, 1978; Peterson et al. 1978; King and Norton 1978

PHYSICAL SETTING

The San Francisco Bay estuarine system is a complex of interconnected embayments, sloughs, marshes, channels, and rivers (Fig. 2). In the context of this chapter, the Bay system is comprised of the Delta, receiving the waters of the Sacramento and San Joaquin river systems, and the Bay proper, into which the Delta waters flow. Geographically and hydrodynamically the Bay can be thought of as two reaches, the northern, which passes south and westward from the Delta through Suisun and San Pablo bays, and the southern (also called South Bay), which extends southeastward towards San Jose. They join in the Central Bay near the Golden Gate, the connection with the ocean.

The water properties of these interconnected features, as well as the nature and mode of circulation and mixing, change continuously in response to the topographical features of the system and to the effects of wind, river inflow, salt and heat input, and tides.

Bathymetry

The Bay is relatively shallow, having an average depth of 6 m at mean lower low water (MLLW) (Table 2; Fig. 3). Broad expanses of the bay floor are incised by narrow channels that are typically 10 to 20 m deep. The deepest sections of channel, such as the Golden Gate (110 m) and Carquinez Strait (27 m), are topographic constrictions whose depths tend to be maintained by tidal currents.

SAN FRANCISCO BAY

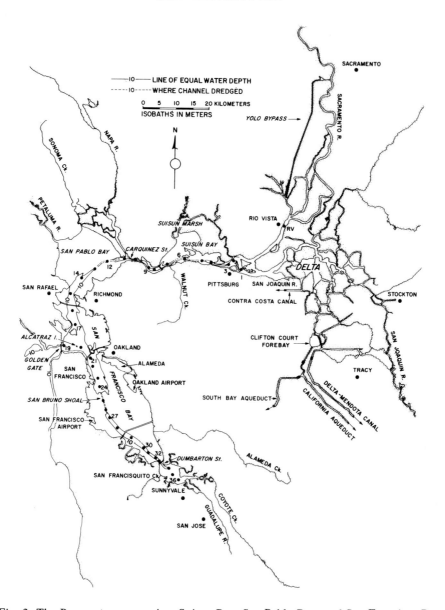

Fig. 2. The Bay system comprises Suisun Bay, San Pablo Bay, and San Francisco Bay, but is herein termed San Francisco Bay. The northern reach is Suisun Bay, San Pablo Bay and the northern part of San Francisco Bay (to the Golden Gate). The southern reach is San Francisco Bay south of the Golden Gate. Station numbers are established hydrographic stations occupied by the Geological Survey at near-monthly intervals from 1969 to present (Conomos et al. 1978).

In its natural state, 100 years ago, the Bay had a surface area of approximately 1.97×10^9 m^2. Shoaling caused by the influx of hydraulic-mining debris (Gilbert 1917; Atwater et al. 1979; Krone 1979) and the diking and filling of marshes (Nichols and Wright 1971) have decreased its surface area by 37% to its present 1.24×10^9 m^2. Recent concern for the shrinking Bay has stimulated control over additional proposed filling.

TABLE 2. GEOSTATISTICS OF SAN FRANCISCO BAY[a]

Statistic	Value
Area (MLLW)[b]	1.04×10^9 m^2
Including mudflats	1.24×10^9 m^2
Volume	6.66×10^9 m^3
Tidal prism[c]	1.59×10^9 m^3
Average depth[d]	6.1 m
From hypsometric curve[e]	2 m
River discharge (annual)	20.9×10^9 m^3
Delta outflow[f]	19.0×10^9 m^3
All other streams	1.9×10^9 m^3

[a] Taken in part from Conomos and Peterson (1977)
[b] Planimetered from Fig. 3; at MLLW
[c] From Edmonston and Matthew (1931)
[d] Volume divided by area; at MLLW
[e] Obtained graphically from hypsometric curve and includes mudflats (Fig. 3)
[f] From Federal Water Pollution Control Administration (1967)

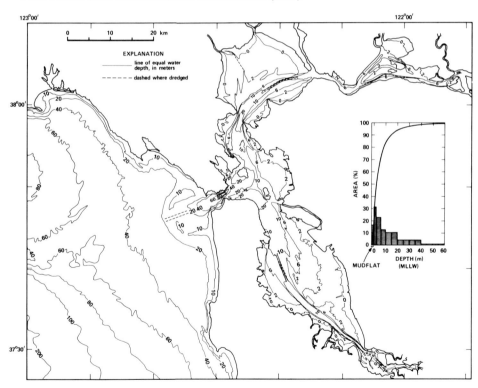

Fig. 3. Bathymetric chart of San Francisco Bay compiled from National Ocean Survey charts 18651, 18649, 18654, 18656, and 18645. Datum is mean lower low water. Hypsometric curve (inset) constructed from bathymetric contours, and includes mudflats. (After Conomos and Peterson 1977).

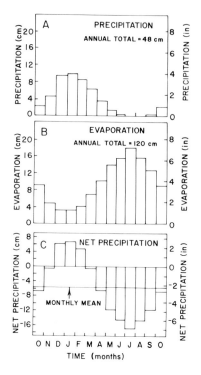

Fig. 4. Monthly distribution of precipitation (A), evaporation (B), and net precipitation (C). Data averaged from various stations in the Bay region during various time spans (modified from Selleck et al. 1966).

Climate

The Bay waters and the biota they support are greatly affected by the sun that promotes photosynthesis and warms the large shallow areas (Fig. 3) and by the wind that blows over their surface. Thus, to understand some of the basic biological and physical mechanisms that are present one needs some idea of the climate, particularly how the wind changes in strength, direction and duration.

Summer in central California is dry because the migrating Pacific high pressure area deflects storms to the north. In winter this high decreases in intensity and moves southward so that it no longer blocks the intrusion of moisture-laden lows (Fig. 4A). Occasionally, these winter lows arrive as a series of storms that move in from the southeast and produce gale winds, heavy rains, and large changes in barometric pressure. These successive lows produce 2- to 5-day periods of stormy weather that alternate with 7- to 14-day periods of pleasant weather. Occasionally, however, heavy rains persist over 7- to 10-day periods.

In the San Francisco Bay and Delta region the climate is more variable because of the effects of local topography and the continuous interaction of maritime and continental air masses (Elford 1970). The climate of San Francisco is dominated by the ocean, and so it enjoys warm winters, cool summers, and small seasonal temperature changes (Fig. 5; Gilliam 1962). In contrast, inland areas have a continental climate with warmer summers and colder winters. The climate over the Bay is transitional between these extremes, but it is also affected by the influence of local topography on air-circulation patterns.

The regional airflow is from the west or northwest during summer (Fig. 6). During winter,

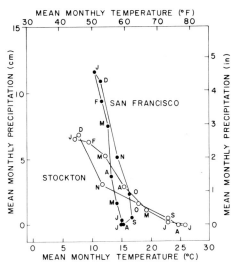

Fig. 5. Temperature-precipitation relationships at San Francisco and Stockton. Monthly means, 1931 through 1960. Data from National Weather Service.

however, when storm centers pass to the south, the winds often blow from the east or southeast. The local terrain funnels and deflects these winds from the prevailing direction (see, for example, deviation of Pittsburg [PITT] data from averages for other stations). Wind speeds increase in the funnels but decrease markedly downwind (compare the wind velocities at the San Francisco Airport [SFO], which lies at the end of the Colma Gap wind funnel, with those at Oakland [OAK] or Sunnyvale [SNVL] [Fig. 6]).

The prevailing southwest summer wind is reinforced by a pressure-gradient-induced movement of air caused by the solar heating of the air masses in the interior (e.g. Stockton, Fig. 5). This heating effect is greatest during the day and causes a marked diurnal as well as a seasonal pattern in the wind strength (Fig. 7).

These strong winds are important within the Bay, for they exert stress on the water surface which transports water and creates waves. The waves, in turn, resuspend sediment (Krone 1979), and mix and oxygenate the water (Peterson 1979). Prevailing summer winds generate waves, with largest of the waves having periods of 2 to 3 seconds (Putnam 1947) and wave heights exceeding one meter (Fig. 8). During winter storms, however, in areas with long fetch, 5-sec waves can be generated. Seaward of the Golden Gate, swells with periods 8 to 12 sec are common during summer, whereas during winter, 18-sec waves are common (National Marine Consultants 1960).

Evaporation from the surface of the water is very effective in raising the salinity in the shallow areas of the Bay. It is of such magnitude that salt production is economically feasible in the southern reach. The approximate annual evaporation rate is 120 cm·yr^{-1}, with the greatest monthly rates (16 to 18 cm) occurring during summer (Fig. 4B). This summer maximum is promoted by the high insolation (Conomos et al. 1979, Fig. 3), air temperature (Fig. 3), and wind speeds (Fig. 5) found during these months. The high evaporation and low precipitation create a large net water loss (negative net precipitation rate; Fig. 4C) and increase the salinity of the surface waters.

River Inflow

San Francisco Bay receives runoff from a 163,000-km^2 drainage basin (Fig. 1) which covers 40% of the land area of California. Ninety percent of the water enters the Bay through the Delta because most of the Bay is separated from the basin by a mountain range (Porterfield et al. 1961).

SAN FRANCISCO BAY

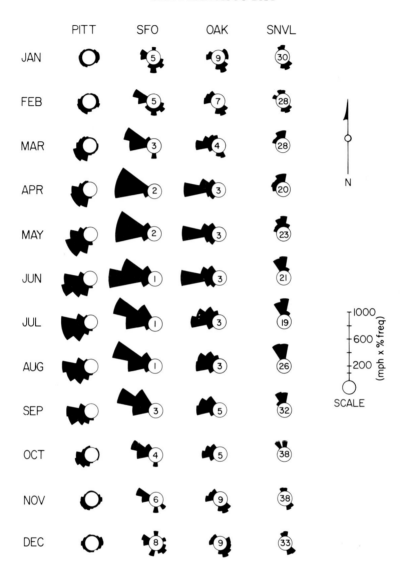

Fig. 6. The cross product of wind direction and frequency at Pittsburg (PITT), San Francisco International Airport (SFO), Oakland International Airport (OAK), and Sunnyvale (SNVL), 1969 through 1975. Vectors point to the direction from which the wind was blowing. SFO, OAK and SNVL data, from the National Weather Service, indicate the percent time of calm (in circles). Pittsburg data, courtesy of Dow Chemical Company, do not include calculations of calm periods.

This basin is drained in the north by the Sacramento River and its tributary streams (Feather, Yuba, Bear, and American), in the south by the San Joaquin River system (includes Merced, Tuolumne, and Stanislaus), and in the east by the east-side streams (Mokelumne and Calaveras river groups; Fig. 1). These river systems contribute 80, 15 and 5%, respectively, of the inflow to the Delta. The inflow is highly seasonal (Figs. 9, 10A) and is composed of rain runoff during winter and snowmelt runoff during early summer. The major rivers are dammed for flood control, water storage, and hydroelectric power. The water that spills over the dams or that is released then

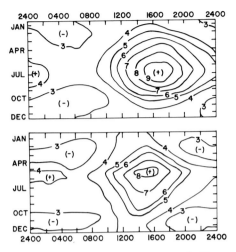

Fig. 7. Average hourly wind speeds expressed in m·sec^{-1}. Upper panel: San Francisco, 1891 through 1910. Lower panel: Oakland, 1930 through 1936. Figures modified from unpublished municipal studies. Data from National Weather Service.

flows into the Delta, where it is then controlled by internal use and massive diversions for irrigation, industrial and domestic uses (Fig. 11).

Direct measurements of outflow from the Delta into the northern reach of the Bay cannot be made because of difficulties introduced by tidally reversing flows and the complex geometry of the channels (Smith 1969). In lieu of actual measurements, the U. S. Bureau of Reclamation calculates a Delta Outflow Index (DOI) by subtracting water exported, used within the Delta, and lost by evaporation from measured flow to the Delta (Table 3). The resulting outflow value (DOI) can be in significant error during both high and low river inflows (Fig. 12) because of the exclusion of inputs from Yolo Bypass and east-side streams. During summer, when the inflow and removal terms are nearly of the same magnitude and the excluded east-side stream discharge becomes relatively

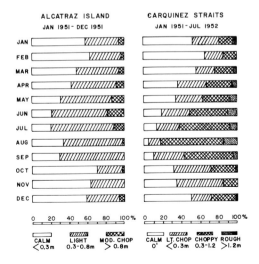

Fig. 8. Seasonal variations in heights on wind waves, Alcatraz Island and Carquinez Straits. Figures modified from unpublished U. S. Coast Guard data.

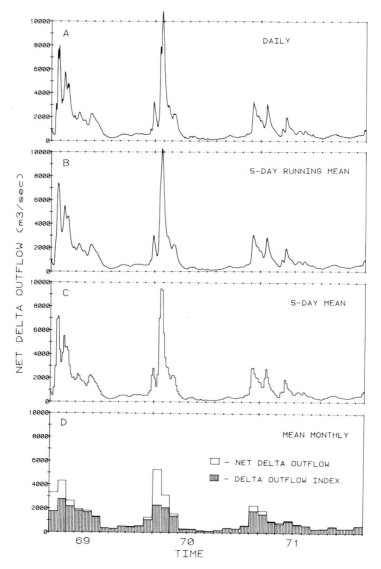

Fig. 9. Net Delta Outflow (NDO) as expressed statistically in daily (A), 5-day running mean (B), 5-day mean (C), and monthly mean (D) outflows. Panel D also compares the mean monthly NDO index with the Delta Outflow Index (DOI): large differences during peak flows are caused by the Yolo-Bypass contributions that are not included in DOI. Data sources are listed in Table 3.

important, the error may be as great as 50% (Fig. 13; see also Arthur and Ball 1979 and Ball and Arthur 1979). In winter, during sustained high flows (>1500 $m^3 \cdot sec^{-1}$), the water flowing into the Delta through the Yolo Bypass, a flood-control canal (Figs. 2, 11), is nearly equal to flow in the river channel (Figs. 9D, 12A). A more comprehensive measure of Delta outflow, termed Net Delta Outflow (NDO), is presented herein (Table 3). It consists of the data used in the DOI and includes discharge of Yolo Bypass and east-side streams.

As noted above, non-Delta inflow to the Bay is about 10% of the annual river inflow (Fig. 9B). The Napa and Petaluma rivers (Fig. 2) provide local drainage to the northern reach

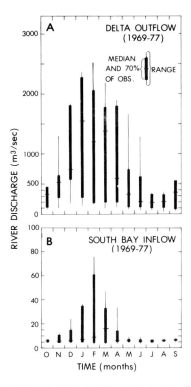

Fig. 10. Monthly means of river inflow: Delta Outflow Index (DOI), 1969-1977 (A), calculated by the U. S. Bureau of Reclamation. Southern-reach streams (Fig. 2), 1969-1977, including San Jose sewage inflows, which average 5 m^3·sec^{-1} (B). Stream discharge from U. S. Geological Survey; sewage data courtesy of San Jose-Santa Clara Water Pollution Control Facility.

(Porterfield et al. 1961). Because their discharge is small compared to Delta outflow, and because their effects on the Bay waters are commonly masked by those of the Delta, they are often ignored in calculations of total inflow to the Bay. In the southern reach, all tributary streams are of

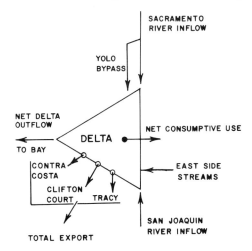

Fig. 11. Schematic diagram of Delta water balance modified after Orlob 1977.

SAN FRANCISCO BAY

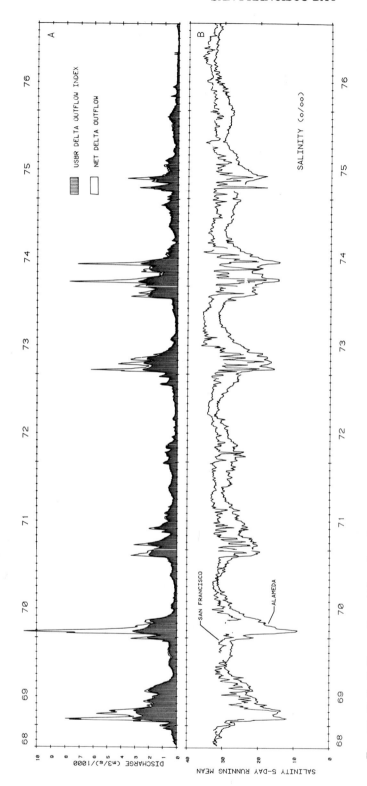

Fig. 12. Daily Delta outflow (A) and surface salinity at Alameda and San Francisco (Fort Point) (B) for 1968 through 1976. USBR Delta Outflow Index (DOI) and Net Delta Outflow (NDO) data sources listed in Table 3. Salinity data were collected by National Ocean Survey and are not corrected for tidal variations.

TABLE 3. FLOW DATA USED IN COMPUTING DELTA OUTFLOW INDICES[a]

Data source	Flow contribution
Delta Outflow Index—DOI (U. S. Bureau of Reclamation)	
River discharge	
Sacramento River (at Sacramento)	+
San Joaquin River (near Vernalis)	+
Exports	
Tracy pumping plant	−
Contra Costa Canal	−
Clifton Forebay	−
Net local use	±
Net Delta Outflow—NDO (This study)	
River Discharge	
Sacramento River (at Sacramento)	+
Sacramento Weir	+
San Joaquin River (near Vernalis)	+
Yolo Bypass	+
Putah Creek	+
East-side streams	+
Cosumnes River	
Dry Creek	
Mokelumne River	
Calaveras River	
Exports	
Tracy pumping plant	−
Contra Costa Canal	−
Clifton Forebay	−
Putah South Canal	−
Net local use	±

[a] River discharge data from U. S. Geological Survey (1977a, b); U. S. Bureau of Reclamation furnishes unpublished export and local-use data.

local drainage and are intermittent. During summer, sewage water inflows to the southern reach exceed the natural stream inflows (see Conomos et al. 1979, Fig. 10). Indeed, while the southern reach receives 10% of the mean annual river runoff it also receives 76% of the total waste-water inflow to the Bay.

Tides

The tides in the Bay are mixed and semidiurnal: two cycles (two low and two high tides) occur each tidal day (24.84 hrs), and the highs in each cycle are usually quite different in height and the lows also differ in height (Fig. 14B; Disney and Overshiner 1925). Tide-height differences vary considerably within a lunar month, from nearly equal tides (equatorial tides created when there is no lunar declination with the earth's equator) to a maximum difference (tropic tides because of maximum lunar declination) of over 1.5 m within a lunar day. In addition to this daily inequality in successive high tides or low tides, there is a fortnightly change in absolute tidal range.

SAN FRANCISCO BAY

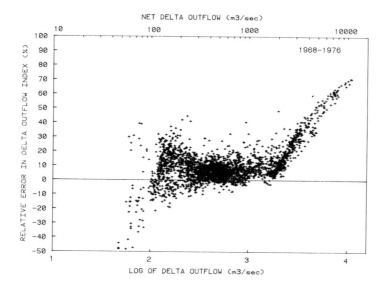

Fig. 13. Percent relative error in Delta Outflow Index (DOI) as a function of Net Delta Outflow (NDO). Daily values are from sources listed in Table 3 and are plotted in Fig. 12A. The large positive error at high NDO values is caused primarily by the exclusion of Yolo Bypass flow from the DOI computations (Table 3). The positive error at low NDO is caused partly by the exclusion of east-side streams whose contributions can total 10 to 25% of the total inflow to the Delta during summer. The negative error at low NDO may be attributed to the preliminary nature of some of the data: the USBR may subsequently revise the most recent data by as much as 25 $m^3 \cdot sec^{-1}$ before final publication.

Spring tides occur near the times of full and new moons and have the highest tidal range of the lunar month. During the first and third quarter, the tidal range is least; these are called neap tides.

Because of the complex geographic and bathymetric configuration of the Bay, there is a spatial variation in mean tide elevation and tidal range (Fig. 14A). The mean tide level is 0.2 m higher in the northern reach, but the tidal range is greater in the southern reach (2.6 m at the southward boundary as compared to 1.7 m at the Golden Gate and 1.3 m at Pittsburg). This Bay-wide tidal range, large relative to the average water depths, creates a tidal prism that is 24% of the Bay volume (Table 2). The configuration also creates different types of tidal waves in the different reaches. A standing wave oscillates in the southern reach, whereas a progressive wave propagates through the northern sector. Because a standing wave propagates faster than a progressive wave, high tides appear earlier at a given distance from the Golden Gate in the southern reach than the northern (see also Dyer 1973, Fig. 2.5).

The Coastal Ocean

Together with river inflow, exchange across the ocean boundary profoundly affects the properties and circulation of Bay waters. Like the rivers, the coastal ocean experiences seasonal changes. The characteristics of the adjacent ocean are largely determined by the strength and location of the Pacific high-pressure cell. The prevailing northwest summer winds drive the meandering, diffuse California Current as it flows southeastward at the surface (Reid et al. 1958). At depth (> 200 m) a countercurrent flows northward. These summer winds transport surface water offshore, inducing deeper waters to upwell to the surface. During winter, when the prevailing wind direction shifts to the south and southeast, the northward flowing Davidson Current moves adjacent and opposite to the California Current, which now is located farther offshore. The surface currents and upwelling introduce and mix several water masses with differing characteristics, and thus determine the character or ambience of the coastal water that is carried into the Bay by tidal and estuarine circulation.

Temperature and salinity are nearly constant throughout the year below 400 m, whereas at 300 m, development of the thermocline (a relatively strong vertical [negative] temperature gradient) during summer causes seasonal temperature variations (Fig. 15). The base of the permanent halocline (a well-defined vertical [positive] salinity gradient), at 200 m, marks the lower boundary of the integrated effects of river runoff from western North America (Roden 1967). Well-defined seasonal variations, modulated by solar heating, upwelling, and dilution by coastal-river inflow and by precipitation, are readily apparent at 50 m.

During summer the salinity is high (33.2 °/oo) because of the low river discharge and the lack of precipitation. Water temperatures are high (13-14°C) and density stratification caused by summer solar heating persists through fall (October-December). During winter (January-March) temperatures are lowered (12.5°C) and the thermocline is destroyed by wind mixing. A well-mixed surface layer is created and a halocline is formed by a slight depression of surface salinity by precipitation and high coastal-river runoff (particularly Delta outflow). Continual runoff during spring

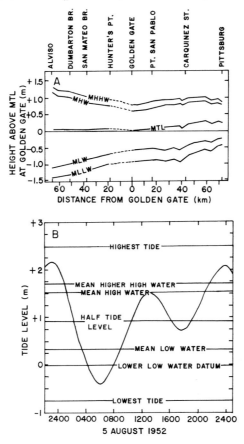

Fig. 14. Tidal characteristics of San Francisco Bay. Mean tidal ranges (A), and Golden Gate (Presidio) tidal curve showing mixed semi-diurnal tide, 5 August 1952 (B). Both panels modified after U. S. Army Corps of Engineers (1963).

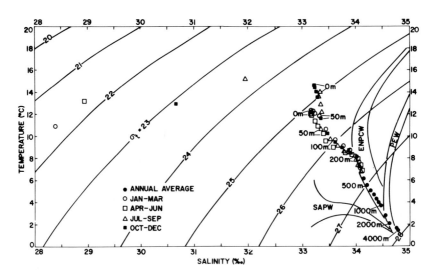

Fig. 15. Seasonal temperature-salinity (T-S) relationships for the ocean adjacent to San Francisco Bay (1920 through 1970) and at the Golden Gate (1949 through 1970). This plot, a T-S diagram, specifies the water density at atmospheric pressure (σ_t) and is routinely used to distinguish water masses and to indicate the relative water-column stability. Ocean data (salinities >33 $^\circ/_{\circ\circ}$) are from Churgin and Halminski (1974) and include the area bounded by 36-38°N, 121-126° W. The four Golden Gate data points (salinities ≤32 $^\circ/_{\circ\circ}$) are from National Ocean Survey unpublished data. Positions of water types SAPW (Subarctic Pacific Water), ENPCW (Eastern North Pacific Central Water) and PEW (Pacific Equatorial Water) are taken from Sverdrup et al. 1942.

(April-June) maintains and deepens the halocline, but solar warming of the surface layers and maintenance of lower temperatures in the deeper (40-50 m) waters by coastal upwelling begin to create a thermocline. Continuing solar heating further develops the thermocline during summer (July-September). These temperatures are relatively lower than those during fall because of the continued upwelling. The halocline dissipates because of the diminution of river inflow and precipitation.

Because of the shallow (15-m) sill formed by the bar at the Bay entrance (Fig. 3), the uppermost waters of this 50-m surface layer are most readily available for exchange with Bay waters. Water movements in the Gulf of the Farallones are complicated and undefined; they are affected by tidal currents, local upwelling, and the irregular bottom and coastline. The temperature and salinity of these waters during the annual cycle are intermediate between the averaged values at the Golden Gate and in the ocean (Fig. 15).

WATER PROPERTIES AND STRUCTURE

For simplification, the seasonal changes of the salinity and temperature fields in the Bay are discussed separately, and the typical summer and winter distributions of each property are presented in a generalized longitudinal section. Because of the highly variable character of Delta outflow, which primarily affects salinity, atypical conditions in "dry" and "wet" years are also discussed. Although water temperature has considerable seasonal variation, it is of little importance in water circulation because it has only a small effect on water density.

Salinity Field

Both ocean-Bay exchange through the Golden Gate and river inflow determine seasonal

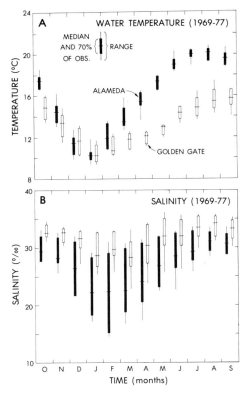

Fig. 16. Mean monthly near-surface water temperatures (A) and salinities (B) at Alameda and the Golden Gate, 1969 through 1977. These unpublished data, from National Ocean Survey, are not corrected for diurnal tidal variations.

changes in the salinity distribution. River inflow has a greater Bay-wide seasonal effect, however, because the river inflow varies widely (Fig. 12A) and the ocean salinities vary by only 3 °/oo (Fig. 15). The salinity of the near-surface water in the central part of the Bay varies approximately inversely with river inflow and particularly that from the Delta (McCulloch et al. 1970; Fig. 12B). Salinity data (Figs. 12B, 16B) at Alameda and the Golden Gate show the relation between station location and the relative influences of river inflow and ocean-Bay exchange. The consistently higher salinities at the Golden Gate are controlled more by exchange with high salinity ocean water than by dilution by Delta outflow. More surprising, however, this ocean influence is relatively local and does not have as much effect on Alameda. Alameda, although much farther from the Delta than from the ocean, is strongly influenced by Delta outflow (Fig. 17).

The dominance of Delta outflow over streams in the southern reach in determining the salinity field throughout the Bay is apparent in longitudinal sections (Fig. 18). The northern reach has a longitudinal salinity gradient ranging from less than 1 °/oo at the Delta to 32 °/oo at the Golden Gate. Its vertical salinity gradient typically has differences of 5 °/oo during winter (Fig. 18A) and 3 °/oo during summer (Fig. 18D-F). During a wet winter, however, vertical differences of more than 10 °/oo have been recorded (Fig. 18B). Seawater intrudes landward of the confluence of the Sacramento and San Joaquin rivers during dry summer periods, whereas during the peak winter inflows, water of these salinities is forced seaward over a 50-km distance to eastern San Pablo Bay (Peterson et al. 1975a). Waters of intermediate salinity (15 °/oo) are found from Carquinez Strait (summer) to the Golden Gate (winter).

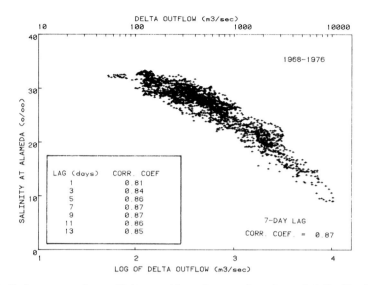

Fig. 17. Daily near-surface salinity at Alameda as a function of daily Net Delta Outflow (NDO) from 1968 through 1976 (see also Fig. 12). The daily salinity values were progressively lagged (from 1 to 14 days) behind the corresponding daily NDO and regressed to determine the best correlation coefficient (see inset). A 7-day lag time yields the best correlation and is considered to be the average travel time needed for a low-salinity water parcel to travel from the Delta southward to Alameda regardless of outflow level.

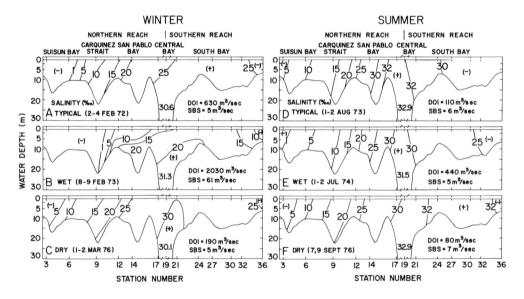

Fig. 18. Vertical distribution of salinity during winter and summer periods. Value above station 19 is averaged Golden Gate salinity and is not included in contoured data. Typical-, wet- and dry-period terminology determined by the mean monthly Delta Outflow Index (Fig. 9D) and correspond to water-temperature distributions (Fig. 15). Data (uncorrected for tidal variations) were obtained at hydrographic stations plotted on Fig. 2 and with methods described by Smith et al. 1979.

In contrast, the southern reach has near-oceanic salinities during much of the year; but as in the northern reach, the salinity and other properties are controlled largely by water exchanges from the Pacific Ocean and by Delta outflow. Intrusion of Delta-derived low-salinity water is particularly evident during winter periods of wet years (Fig. 18B), when relatively pronounced stratification occurs. The mechanisms of this Delta-derived low-salinity intrusion are discussed in detail below. During summer the water is nearly isohaline with depth because of tidally and wind-induced vertical mixing. During dry summers the elevation of salinity caused by evaporation is measurable, and during extended droughts the salinities can exceed those of the adjacent ocean. The intrusion of water of lower salinity at the southern boundary of the southern reach is caused by sewage discharge from San Jose as well as by local stream discharge (see also Conomos et al. 1979). The water volume south of station 32 is small because of the narrowing and shoaling of the southern reach (Fig. 3).

Temperature Field

Temperature does not exhibit as definitive a distribution as does salinity. Although water temperature is directly coupled to the atmosphere, the heat-exchange rate is different for the heat reservoirs of the Delta, the Bay, and the ocean. River temperatures vary most widely, in response to the rate of flow and temperature of tributary waters. Ocean temperature varies the least, being moderated by the buffering effect of the large ocean volume and of vertical mixing; Bay temperatures are controlled by local weather conditions and by local discharge of waste heat, as well as by the temperatures of the river and ocean.

Water temperatures at the surface near Alameda (Fig. 16A), indicative of the central part of the Bay, follow a smooth curve varying from 20°C in summer to 10°C during winter. Water temperatures at Golden Gate, in addition to reflecting the moderating influence of the ocean (annual range 10-15°C), show the effects of coastal upwelling. The spring (May-June) upwelling of cold water can be seen in the (negative) 2°C deviation from a smooth curve (compare annual temperature curve with Lynn 1967, Fig. 14, section I, stations 67.50, 67.55).

During summer, water temperature generally decreases from the Delta to the Golden Gate as a result of the inflow of warmer river water (22°C) and the presence of colder (15-17°C) upwelled ocean water (Fig. 19D, E, F). The southern reach, although maintaining oceanic salinities, is warmed 4-5°C by solar heating of the shallow water. Warming is enhanced by the long residence time of water in the southern reach. This heating is especially evident during dry summers when a warm-water lens is formed and maintained at the water surface despite vertical mixing (Fig. 19D, F).

In contrast to summer distribution of temperatures, only small temperature gradients persist during winter. Water temperatures in the northern reach generally decrease toward the Delta as a result of the flow of warmer ocean water through the Golden Gate and of the cooling of river water (Figs. 19A-C). The southern reach usually has temperatures intermediate between those of the Delta and the ocean, implying that even during winter heat is gained from the atmosphere. During wet winters the intrusion of colder Delta-derived water into the southern reach may lower ambient water temperatures slightly (Fig. 19B).

DISPERSION CHARACTERISTICS

Advective (circulation) and diffusive (mixing) processes in estuaries are often indistinguishable

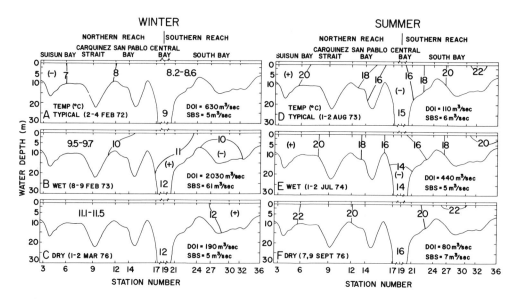

Fig. 19. Vertical distribution of water temperature during winter and summer periods. Value above station 19 is averaged Golden Gate water temperature and is not included in contoured data. These sections are not meant to portray the temperatures indicative of typical-, wet-, and dry-period conditions, but rather to correspond to salinity distributions in Fig. 14. Data (uncorrected for tidal variations) are obtained at hydrographic stations (Fig. 2) and with methods described by Smith et al. 1979.

or inseparable, and together they are termed dispersion.[1] Dispersion is controlled by a number of factors—physical dimensions, river inflow, tidal conditions, and often by winds. As was discussed above, these parameters are quite different in the northern and southern reaches, as are the seasonal changes of most of them. Discussion of these differences follows.

Circulation

Some inference of circulation has been drawn from field observations which include (1) salinity measurements and (2) direct current measurements by either current meters or drifters. Complementary information has been gained from physical- (hydraulic-) model experiments and mathematical-model simulations.

Tidal currents. Although the marked seasonal differences in wind and river inflow alter water-mass movement, the basic flow patterns are tidally induced and remain relatively unchanged throughout the year.

Within the Bay the tides create reversing currents that are strongest in the channels (typical maxima of 60-90 cm·sec^{-1}) and weaker in the shoals (35 cm·sec^{-1}) (Disney and Overshiner 1925). At the Golden Gate, maximum ebb-current speeds of 280 cm·sec^{-1} are typical (see also Rubin and McCulloch 1979). Within the Bay, other constrictions such as the Carquinez and Dumbarton straits (Fig. 2) exhibit current maxima. The tidal excursion, the distance which a parcel of water

---

[1] For the purpose of this chapter, the following definitions are used (Officer 1976): *Diffusion* is the turbulent mixing of a water property (such as salinity) without any net mass transfer of the water itself. *Advection* is the circulation (or flow) of the water that causes a net transfer of the water mass.
"*Dispersion*" is the undifferentiated result of both (1) diffusion effects (mostly from tidal mixing) and (2) current-velocity shear or circulation effects.

might travel through a tidal cycle, varies within the Bay but is typically about 10 km (Fig. 20C, D).

The differences in tide phasing, coupled with the large tidal excursion, cause transport of water between the northern and southern reaches. Because of the phase differences between the high-tide elevations and tidal-current velocities in the different parts of the Bay, the water mass of the southern reach begins to ebb while the water mass of the northern reach is flooding. The converse is also true later in the tidal cycle when the low-salinity water of the northern reach during ebbing is carried into the southern reach while it is still flooding (Carlson and McCulloch 1974). This interchange occurs near the time of low slack at the Golden Gate (3 hours after maximum ebb).

Seaward of the Golden Gate, where the tidal currents are not restricted by the physical dimensions of the Bay, they have a clock-wise rotation (Fig. 21). The diurnal inequality of the tides (Fig. 14B) creates great differences in current velocity. The typical current velocity (seaward)

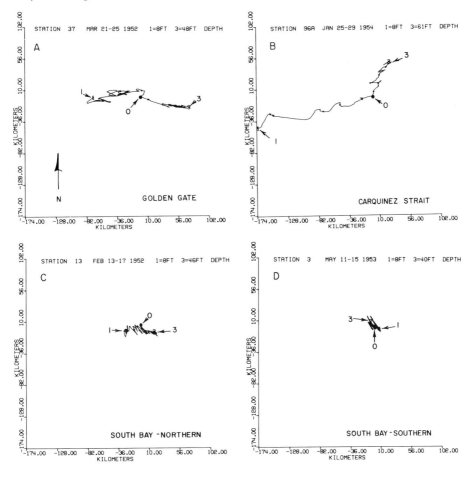

Fig. 20. Progressive vector diagrams of surface and near-bottom currents measured with current meters at Golden Gate (A); Carquinez Strait (B); northern part of the southern reach (C); and the southern part of the southern reach (D). O is the origin of the 5-day records and asterisks indicate midnight of each succeeding day; 1 is the surface meter (at 2.4 m depth) and 3 is the meter located within 1 m of the bottom. Plots are made from unpublished data of the National Ocean Survey (NOS); station numbers refer to NOS locations.

during an equatorial tide is about 15 cm·sec^{-1}, whereas during a tropic tide (two lunar weeks later), the velocities increase twofold.

Nontidal currents. Nontidal currents, generated by winds and river flow, perturb the velocity of tidal currents. Although nontidal-current velocities are one-tenth those of tidal currents, these residual currents are important in transporting dissolved and particulate substances into and from the Bay. The current patterns are better defined in the channels than in the broad shallow areas of the Bay (Fig. 3). Vector addition of current-meter measurements over 5-day periods show great differences in the flow characteristics in the channels of different areas of the Bay (Fig. 20). The southern reach has oscillatory flow with little net geographic displacement (Figs. 20C, D). The northern reach (Fig. 20B) and the Golden Gate (Fig. 20A) also have oscillatory flow, but with large net displacements which are typically 10 to 20 km·day^{-1} (about 10 to 20 cm·sec^{-1}).

There are no similar field data available for shallow water. The large-scale residual-transverse circulations computed by Nelson and Lerseth (1972) with a two-dimensional numerical program show a counterclockwise gyre (northward along the eastern shore and southward in the channel) in the southern reach, and clockwise gyres in the San Pablo and Suisun bays. Their results are not conclusive, however, as the variation between the computed velocity data and the few existing current-meter measurements (Nelson and Lerseth 1972, Figs. 6, 7) create a strong bias in their calculated net flows.

Most residual circulation observed in the deeper channels shows strong vertical variations. In the northern reach the nontidal currents are principally caused by river inflow, and are thus density induced. Typical current measurements show ebb dominance of the surface water and flood dominance of the near-bottom water (Fig. 22A; Simmons 1955). When these data are averaged over a tidal cycle, the landward-flowing density current can be readily demonstrated (Fig. 22C; Peterson et al. 1975). This phenomenon is termed estuarine or gravitational circulation (Pritchard 1956). Within San Francisco Bay it has been defined in the channels, over weekly and bimonthly time scales, using current meters (Fig. 19) and drifters (Fig. 23). Nontidal current speeds, estimated by drifter movements, average 4 cm·sec^{-1} for the landward-flowing density current and 5 cm·sec^{-1} for the seaward-flowing surface current (Conomos and Peterson 1977). Speeds determined by

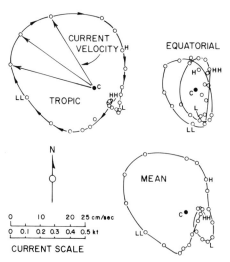

Fig. 21. Near-surface semidiurnal rotary tidal current velocities measured 16 km west of the Golden Gate showing high (H), higher high (HH), low (l) and lower low (LL) tide levels for tropic, equatorial, and mean tidal conditions modified from Marmer (1926).

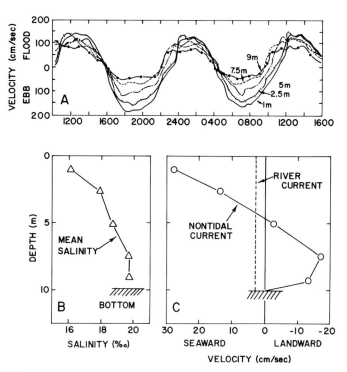

Fig. 22. Tidal-current characteristics near station 9 (Fig. 2) of the northern reach. Current measurements with depth, 21-22 September 1965 (A), modified after U. S. Army Corps of Engineers 1963. Mean salinity (B) and nontidal and river current (C) after Peterson et al. 1975a. The mean monthly Delta Outflow Index was 350 m^3·sec^{-1}:

current meters (Figs. 20A, B; 22A) are approximately double these values.

In contrast, the southern reach receives too little river inflow from the local streams for creation of such an estuarine circulation cell. Available current-meter data (Figs. 20C, D), and dye-dispersion studies in the Bay (Nelson and Lerseth 1972) and in the physical model (U. S. Army Corps of Engineers 1963; Lager and Tchobanoglous 1967; Kirkland and Fischer 1976) over a dozen or more tidal cycles, show little or no net tidal movement. Further, floats placed in the physical model by Lager and Tchobanoglous (1967) and Kirkland and Fischer (1976) similarly show little net motion. However, longer term (bimonthly) field observations using drifters have suggested seasonally reversing nontidal movement of surface and near-bottom water (Fig. 23) with averaged speeds of about 1 to 2 cm·sec^{-1}. These apparent movements have been attributed to wind stress on the water surface (Conomos et al. 1971) and are seen most clearly during summer when prevailing winds are stronger and there is little density-induced motion. During summer, when the prevailing northwest winds are strongest (Fig. 6) and the water column is nearly isohaline (Figs. 18A-C), the transport of surface waters is generally in the direction of the wind (Fig. 22A, inset). This southeastward displacement of surface water apparently generates compensating currents of near-bottom water which flow in an opposing (northwestward) direction (Fig. 23B, inset).

The drifter and salinity data collected during winter disagree and suggest different flow patterns. The drifters indicate that two-layer drift sets opposite to that of summer (Fig. 23), implying surface-water displacement northwestward and near-bottom compensating flows in response to episodic southwest gale winds. McCulloch et al. (1970), using salinity data and considering only longitudinal density-induced effects, suggested that a reverse estuarine-circulation cell could be generated during the peak Delta outflow: low-salinity water moves southward into the southern reach and displaces a more saline, denser water northward. This simple two-layer model is greatly complicated if the salinity data (in the channel) during the wettest months of two successive years are considered representative of the entire lateral cross-section. The salinity distribution during these two months (Figs. 24B, C) is similar to that of the three-layer Baltimore Harbor case (Fig.

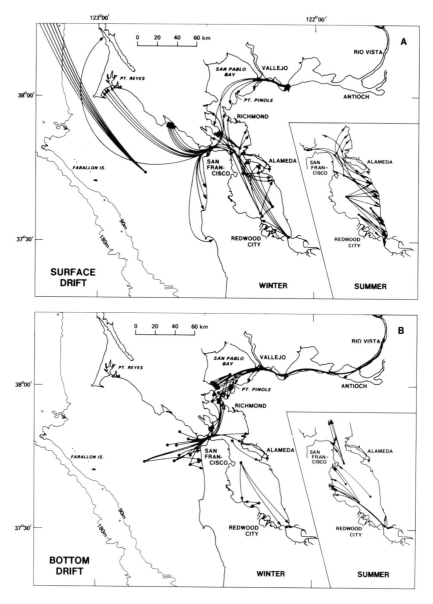

Fig. 23. Release and recovery points for surface (A) and seabed (B) drifters in the Bay and adjacent ocean, after Conomos and Peterson 1977. Drifter movements are shown as arrows drawn from release points to recovery locations and portray simplified paths of movement occurring within 2 months of release. Winter release: Winter 1970. Summer release (southern reach only shown as inset): September 1971. Data are typical of 18 releases over a 3-year period (1970-1973).

24A) in which the surface and near-bottom layers are thought to advect landward with a return seaward flow at mid-depth (Cameron and Pritchard 1963).

These examples are given to emphasize the presence of sluggish, transient (and undefined) three-dimensional circulation in the southern reach that is apparently both density- and wind-induced. Features of Delta-induced flushing of the southern reach are discussed below (WATER RENEWAL PROCESSES AND RATES).

Mixing

Mixing (turbulent diffusion) is created primarily by tidal currents, wind, and river inflow. The effect of the current-induced mixing is most dramatic at constrictions such as the Golden Gate where the high current speeds cause visible eddies and boils. Wind generates waves (compare Figs. 7 and 8) that break down river-induced stratification and tend to make the water column isohaline. These wind effects are most evident in the southern reach during summer. The river-induced mixing effects are complex interactions between turbulent mixing and current shear and together with tidal mixing, control dispersion, particularly in the longitudinal direction.

Because of the paucity of long-term salinity and current meter data, the longitudinal dispersion characteristics of the Bay system can be only grossly estimated. Glenne and Selleck (1969), with hydrographic data from the SERL study, determined the general distribution of the longitudinal dispersion coefficient, K_x, using cross-sectionally averaged mass and salt continuity equations.[2] They found values of about 0.5×10^6 cm$^2 \cdot$sec^{-1} in the southern reach, which is well mixed and has a relatively small nontidal flow, and values of 2.0×10^6 cm$^2 \cdot$sec^{-1} in the northern reach,

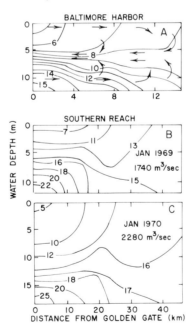

Fig. 24. Longitudinal salinity section of Baltimore Harbor (A), after Cameron and Pritchard 1963, compared with those of the southern reach (B, C). Vectors in (A) are inferred nontidal circulation patterns.

[2] Although the equations were cross-sectionally averaged, the data were collected only in the channels.

SAN FRANCISCO BAY

which has larger nontidal flow (Table 4). The longitudinal dispersion coefficients decreased towards the Golden Gate in both reaches. These and other dispersion coefficients calculated from modelling studies (Table 4) are of the same order of magnitude as those reported in other temperate-zone estuaries (Officer 1977).

TABLE 4. DIFFUSION/DISPERSION COEFFICIENTS MEASURED
AND/OR USED IN SAN FRANCISCO BAY STUDIES

Place	Coefficients (10^6 cm^2·sec^{-1})			Method	Reference
	K_x	K_y	K_z		
Sacramento, San Joaquin	0.09-0.9			instantaneous dye experiments	Bailey et al. 1966
Suisun Bay	0.6-14			instantaneous dye experiments	Bailey et al. 1966
Northern reach	0.1-10			fraction of freshwater	Glenne and Selleck 1969
Southern reach	0.1-1			fraction of freshwater	Glenne and Selleck 1969
Southern reach		0.01[a]		dye experiments	Ward 1974
Southern reach				numerical calculation	Lager and Tchobanoglous 1968
Northern reach	4		0.000004	2-D (vertical) numerical model	Peterson et al. 1978

[a] Reported as $E_z/d\bar{U}$ where E_z is the transverse dispersion coefficient and $\bar{U}$ is the shear velocity.

Circulation-Mixing Relationships

To understand the dispersive characteristics of estuaries, it is necessary to quantify and separate diffusive from advective processes. Hansen and Rattray (1966) introduced an empirical way to classify estuaries that is based on an analytical approach for quantifying the processes causing longitudinal, gravitationally driven circulation.

In their method two dimensionless parameters are used: a stratification parameter, $\delta S/S_0$ (ratio of surface-to-bottom difference in salinity $[\delta S]$ to the mean cross-sectional salinity, $[S_0]$), and a circulation parameter, U_s/U_f (the ratio of the net surface current $[U_s]$ to the mean cross-sectional velocity $[U_f]$; Fig. 25). This circulation parameter expresses the ratio between a measure of the mean fresh-water flow plus the flow of water mixed into it by entrainment or eddy diffusion, to the river flow. The parameter ν represents the diffusive fraction (nongravitational fraction) of the total upstream salt flux due to river flow and $1-\nu$ is that fraction accounted for explicitly by the gravitational circulation. That is, when $\nu = 1$, gravitational circulation ceases and landward salt transport is entirely by diffusion (mixing); as ν approaches zero, diffusion becomes less important and the upstream salt flux is almost entirely accomplished by advection. Available field and

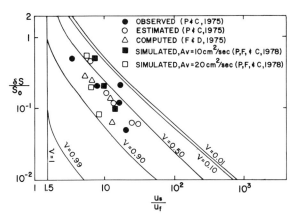

Fig. 25. Fraction of horizontal salt balance by diffusion (ν) as a function of salinity stratification ($\delta S/S_0$) and convective circulation (U_s/U_f) in a rectangular channel (Hansen and Rattray 1966). Data sources include physical- and numerical-model studies (Fischer and Dudley 1975), field observations (Peterson and Conomos 1975) and numerical model data (Peterson et al. 1978); calculated data of Peterson and Conomos (1975) follow method of Bowden and Gilligan (1971).

modelling data from the northern reach are in general agreement, and suggest that about 60 to 70% of the upstream salt flux is due to diffusion and 30 to 40% to advection (Fig. 25).

The values of ν so defined identify quantitatively four types of estuaries; these are subdivided into strongly (b-type) and weakly (a-type) stratified estuaries. In this scheme, the northern reach behaves as a partially mixed estuary; that is, it is strongly stratified during the high Delta outflows of winter (mean monthly NDO = 1000 m³·sec⁻¹) and weakly stratified during lower (100 m³·sec⁻¹) summer flows (Fig. 26).

This treatment of longitudinal dispersion is ideal for a narrow channel of uniform depth. In wider systems like San Francisco Bay, however, lateral (transverse) dispersion may be important. Dyer (1977) has shown that in partially mixed estuaries the net transverse and vertical components of gravitational circulations are similar in magnitude, and that the lateral effects are important in maintaining the dynamic balance and salt transport. Bowden (1977) states that mixing across an estuary may be characterized by mixing coefficients (K_y) which are often 10^2-10^3 times greater than the vertical coefficient (K_z) (compare, for example, the data in Table 4) because of the presence of transverse circulations. Thus, because the northern reach has variations in both width and depth and lateral variations (gradients) of salinity, diffusion and current velocity, lateral dispersion may be significant (Fischer 1972; Okubo 1973). Unfortunately, adequate field data for calculating lateral dispersion are lacking.

Lateral mixing may also interact with transverse variations in the longitudinal velocity to add to the longitudinal dispersion (Fischer 1972). In addition to this lateral shear effect, the irregular topographic features of the northern reach may favor complicated "trapping" and "tidal pumping" mechanisms which Fischer and Dudley (1975) believe may contribute substantially to longitudinal dispersion and may be relatively more important than gravitational circulation in contributing to salt intrusion. Trapping occurs along a series of shoreline embayments in which portions of a moving water mass are temporarily trapped and gradually released to the tidal channel after the main body of the water mass has passed. This release thus promotes longitudinal dispersion. In tidal pumping, the tidal wave interacts with an irregular channel bottom, creating a nontidal flow (net circulation) whose ebb and flood flows favor different channels. Fischer and Dudley's argument is based on analytical data (using the two-dimensional numerical model of

SAN FRANCISCO BAY

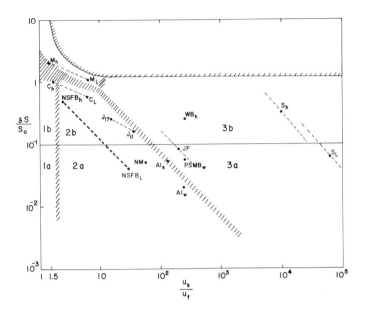

Fig. 26. Classification diagram (Hansen and Rattray 1966) of estuarine types incorporating data from the northern reach and other estuarine systems. Parameters used are defined in Fig. 25. Shaded areas are boundaries. In type 1, the net flow is seaward at all depths and upstream salt transport is by diffusion. In type 2, the flow reverses at depth and both advection and diffusion contribute to upstream salt flux. In type 3, the salt transfer is primarily advective. Type 4 has intense stratification and is a salt wedge type. These types are further sub-classified into strongly (b-type) and weakly (a-type) stratified. Station codes as follows. From Fig. 25: NFSB, northern San Francisco Bay (northern reach). From Hansen and Rattray (1966): M, Mississippi River mouth; C, Columbia River Estuary; J, James River Estuary; JF, Strait of Juan de Fuca; NM, Narrows of the Mersey Estuary; S, Silver Bay. From Barnes and Ebbesmeyer (1978): Al, Admiralty Inlet; PSMB, Puget Sound Main Basin; WB, Whidbey Basin. Subscripts: h and l refer to high and low river discharge; w and s to winter and summer; numbers indicate distance (in miles) from mouth of the James River estuary.

Nelson and Lerseth [1972]) and physical-model experimental results. They reasoned that because their data yielded values for ν of about 0.7 (Fig. 25), vertical gravitational circulation was unlikely to be responsible for the observed length of salinity intrusion. Rather, the combined effects of trapping, tidal pumping, and wind were entirely capable of producing the observed results.

Virtually nothing is known of the separation of dispersion characteristics of the southern-reach water mass. Description of low salinity water intrusions are discussed in detail below.

WATER RENEWAL PROCESSES AND RATES

It is apparent that increased Delta outflow causes both a seaward movement of the salinity gradient and a more rapid circulation and intensive mixing of the water mass. This increased Delta outflow is accompanied by a more rapid exchange of river water with the ocean, the volume of river water accumulation in the Bay increasing relatively less that of the Delta outflow. A knowledge of flushing time (i.e., the time required to replace the existing fresh water in the estuary at a rate equal to the river inflow) is important in estuarine studies, as it is a fundamental parameter used for predicting dilution and removal of pollutants. There have been no rigorous measures or calculations of flushing time made in the Bay. Instead, there have been several different estimates of "water-renewal times": residence time, hydraulic displacement time, and water-parcel displacement time. These methods, applied only to the northern reach, are defined below because each is calculated differently.

In the most general cases considering the entire Bay, however, the residence time of the entire water mass could be approximated by comparing the Bay volume with the annual river inflow volume (Table 2). Since the river-inflow volume is typically about three times the Bay volume, residence time in a salt-free Bay would be about four months, if calculations were based on a simple volumetric replacement. A slightly more refined estimate could be made by correcting for the volume of the Bay occupied by seawater. Assuming a mean volume-weighted salinity for the Bay of 20 °/oo and an ocean value of 30 °/oo, the residence time would be two-thirds of 4 months, or about 2.7 months. These calculations, however, assume the water in all portions of the Bay is mixed and exchanged uniformly in space and time. Salinity reduction controlled by this inflow, however, was seen to vary greatly with season (rate of river inflow) and with location within the Bay (the northern reach experiences most of the dilution).

The Northern Reach

Delta outflow has a direct effect on water residence time, particularly during winter. Glenne (1966) and Selleck et al. (1966) have calculated both water-residence and hydraulic-displacement times from field data collected during 1961-62. Water-residence time is based on the arithmetic mean of dissolved-constituent mass (silica or salinity) in a given segment of estuary divided by the net transport of constituent mass through that segment. Their mathematical expression contains diffusion terms as well as advection terms because of the silica and salinity concentration gradients present. Hydraulic-displacement time is based on the arithmetic mean of water volume through that segment. This expression contains only an advection term. The mean residence time of a water parcel is 60 days during a dry summer (DOI = 90 $m^3 \cdot sec^{-1}$) whereas it is 14 days during a relatively dry winter (DOI = 420 $m^3 \cdot sec^{-1}$; Fig. 27). The hydraulic-displacement time for this winter is nearly identical to the residence time, and implies that river-induced advection is the dominant process controlling residence time; during summer, however, diffusion becomes relatively more important, and the hydraulic-displacement time is much greater than the replacement time.

Travel-time calculations based on SERL data and the link-node numerical model of Nelson and Lerseth (1972) indicate that a water parcel flows from the Delta to the Golden Gate in a month during summer whereas it takes about 2 weeks during winter (J. B. Gilbert and Associates 1977; Fig. 28).

Peterson et al. (1978), using a two-dimensional (longitudinal and vertical) steady-state numerical model, calculated advective-water replacement times at DOI of 400, 200, and 100 $m^3 \cdot sec^{-1}$ to be less than 25, 45, and 75 days, respectively (see Table 4 and Fig. 25 for additional information on these results). These independent estimates are in fair agreement and suggest that water residence times during winter are typically 2 to 3 weeks, whereas during summer, they can be on the order of 2 months.

The Southern Reach

Prior to the late 1960's, the southern reach was believed to be flushed by tidal action and by the dilution effects of the local streams. A study of existing salinity data led Peterson and Carlson (1968) to suggest that Delta outflow controlled the salinity of the southern reach and to infer that this outflow was a dominant factor influencing seasonal circulation there. McCulloch et al. (1970) presented data from the Geological Survey's first (1969-70) year of near-monthly salinity measurements that showed well-defined salinity stratification (McCulloch et al. 1970, Fig.

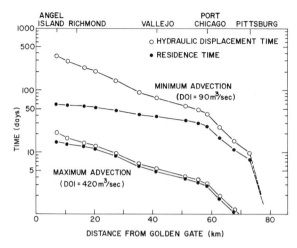

Fig. 27. Residence and hydraulic-displacement time in the northern reach during high (19 June through 31 July 1962) and low (12 December 1961 through 14 February 1962) Delta outflow levels. The outflow values are the mean monthly Delta Outflow Index during those months. The residence time is the arithmetic mean time the mass of dissolved silica contained in a given segment of the northern reach divided by the mass flux of silica through that segment. The hydraulic-displacement time is the arithmetic mean time a parcel of water is retained in a given segment of the reach divided by the flux of the water parcel through that segment. Modified after Glenne (1966).

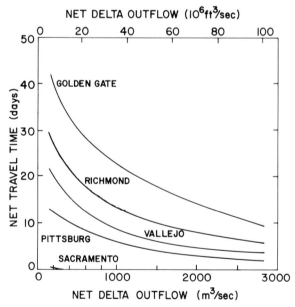

Fig. 28. Net travel time of a water parcel from Sacramento to various locations in the northern reach as a function of "Daily Delta Outflow", modified after J. B. Gilbert and Associates (1977). Daily Delta Outflow was calculated by adjusting the Delta Outflow Index with monthly Delta outflow data published by the California Department of Water Resources.

4). It was a wet year (Fig. 12) and the low-salinity source clearly appeared to be the Delta outflow[3]. They showed that the winter flushing of southern-reach waters could not be explained by tidal flushing, but rather by dilution resulting from the high Delta outflow (see, for example, Fig. 12B). Their salinity data further indicated that density-induced circulation was present and together with the tidal current patterns, was the mechanism transporting this low-salinity water into the southern reach. Although their report met with skepticism by some who maintained that seasonal flushing of the southern reach could only be caused by abnormally high Delta outflows, McCulloch (1972) demonstrated a similar winter flushing of the southern reach accompanying the normal and less than normal winter Delta outflow[4] of the succeeding two years (1970-72; Fig. 12).

More recently, Imberger et al. (1977) conducted a series of experiments with the Corps of Engineers physical model in which they compared averaged daily Delta outflow[5] with salinity data collected during 1972-73 by the Geological Survey (see Smith et al. 1979). Their results verified the concepts and conclusions presented by McCulloch et al (1970). To summarize Imberger et al. (1977):

(1) Typical winter Delta outflows affect the salinity of the southern reach shortly (~5 days) after the start of the winter flow (see also Fig. 17). The amount of change in the salinity is dependent on the magnitude of the flow (Table 5). Flows as low as 400 m^3·sec^{-1} depress the salinities in the northern portion by 1 to 2 °/oo. Flows of 1100 m^3·sec^{-1} significantly affect salinity structure throughout the southern reach. The rate at which the low-salinity water intrudes the southern reach depends upon the mechanics of the density stratified flow.

(2) Successive high outflows have differing effects on the salinity field. A second flood is less effective than the first if the second intrudes before the system has had a chance to recover.

(3) Changes in the southern-reach salinity field depend more on the magnitude of the peak Delta outflow and on the history of previous flood events than on the total outflow volume during a given period.

(4) The recovery to near steady-state summer conditions of high salinities (Fig. 18DEF) is accomplished by tidal and wind mixing. The rate is 2 °/oo per month until an ambient salinity of 29 °/oo is reached. Thereafter, the final recovery to 31 °/oo takes as long as four months.

Field data (Imberger et al. 1977; Table 5) show that the flushing time of the southern-reach water mass varies greatly throughout the year. During winter, when the first large flood flow of the Delta intrudes, the initial flushing of a substantial fraction of the water mass may take one to two weeks (Figs. 12, 17). As the winter progresses and high inflow is maintained, diffusive processes become greater relative to river-induced advection. Dilution effects of Delta-derived low-salinity waters control the flushing of the southern reach, and residence times are probably on the order of a few months. After Delta outflow decreases to low summer levels, exchange of water is controlled by tidal currents and further promoted by wind- and tidally-induced mixing. Typical residence times approach five months.

Exchange with Ocean

Exchange through the Golden Gate is of prime importance in understanding the ultimate flushing rates of the Bay system. Hydrographic data indicate that although the tidal currents mix

---

[3] McCulloch et al. (1970) used discharge of the major tributaries of the Sacramento and San Joaquin river drainage basins for their qualitative discussions.

[4] McCulloch (1972) used the mean monthly DOI (Fig. 9D).

[5] Imberger et al. (1977) used 5-day averages of an outflow index similar to Fig. 9C.

TABLE 5. THE EFFECTS OF VARIOUS NET DELTA OUTFLOWS (NDO)
ON THE SALINITY FIELD ON THE SOUTHERN REACH

Delta Outflow[a] ($m^3 \cdot sec^{-1}$)	Frequency[b] (% of year)	Salinity conditions in the southern reach[a]
≤140[c]	8	oceanic salinities present (31-32 ‰)
140-280	18	measureable change of 1-2 ‰ in northern part; weak vertical differences of 1-2 ‰
280-390	15	central and southern parts (south of San Bruno Shoal) affected only if outflow maintained for a long period
390-840	32	surface salinities throughout reach noticably depressed
840-1120	5	salinity near San Bruno Shoal reduced to about 26 ‰
1120-2800	16	salinity structure throughout the southern reach is profoundly affected
2800-3360	2	stratifies entire reach with surface salinities about 15 ‰ and bottom about 25 ‰
3360-9350	4	lowered salinity in the central part by >4 ‰ for 8 days
≥9350	0.2	lowered salinity in the central part to below 10 ‰

[a] Taken in part from Imberger et al. 1977.

[b] Taken from 5-day running mean of Net Delta Outflow from 1969-1977 (Fig. 12A).

[c] 140 $m^3 \cdot sec^{-1}$ = 5000 $ft^3 \cdot sec^{-1}$ = 10,000 acre-ft·day^{-1}

large volumes of water daily over the shallow bar (sill depth 15 m) and through the mile-wide constriction (110-m depth), much of the stratification is maintained (see for example Peterson 1979, Fig. 8). Current-meter (Fig. 20) and drifter (Fig. 23; Conomos and Peterson 1977) data indicate that the Bay system maintains a pronounced estuarine circulation cell during typical river-discharge conditions. The upper portion of the water column has a net transport seaward, while the lower layer has transport landward.

Parker et al. (1972) attempted to measure the tidal-exchange ratio (the portion of new ocean water entering the Bay system) using salinity and temperature as tracers. Recognizing the importance of nontidal effects during high river inflow periods, they conducted their 25-hr experiments on days of equilibrium tides (see, for example, Fig. 21) during two successive summer months. From their findings, they considered a "reasonable and conservative" value for tidal ratio during summer (with little stratification) to be 24%, and were of the opinion that during winter conditions of maximum stratification, it may be higher than 80%. These values correspond to 0.34×10^9 and 1.12×10^9 m^3, respectively, of new water entering the Bay system during each tidal cycle, and are about 21 and 70%, respectively, of the tidal prism volume (Table 1). These results suggest a large degree of exchange with the ocean and rapid flushing of at least the central part of the Bay.

FUTURE RESEARCH NEEDS

In San Francisco Bay, a very complex estuarine system, our knowledge of hydrodynamic processes is primitive, and our efforts thus far must be considered preliminary. A basic goal in the studies of estuaries is to devise practical models (conceptual and numerical) of circulation and mixing processes that can be used for predictive purposes (see, for example, National Academy of Science 1977; Kinsman et al. 1977; Kjerfve 1978). Major applications of such models in the Bay would be the prediction of conditions resulting from the works of man. These include decreased Delta outflow following massive river diversions (Gill et al. 1971), effects of proposed deepening (25%) of navigation channels of the northern reach, and effects of increases in agricultural, industrial, and domestic waste inputs (Federal Water Pollution Control Administration 1967; Hines 1973).

Because the three-dimensionality of the Bay is reflected in its physical processes, the models that may best approximate real conditions are three-dimensional in space as well as time-dependent. These models must be based on, and consistent with, field observations of the following types:

(1) Variability of Bay-wide circulation in time and space. Continuous long-term (months) current-meter records are necessary to document daily changes in river inflows, daily and fortnightly tidal current variations, and episodic (3-4 day) wind events. In addition, near-synoptic measurements of the three-dimensional salinity field are needed, with these observations being conducted on daily time scales immediately before, during and after episodic weather and Delta-outflow events.

(2) Measurement of the boundary conditions. These measurements include long-term, but short-time scale monitoring of river inflow, exchange through the Golden Gate (rise and fall of water-surface elevation as well as spatial distribution of salinity and velocity with time), and wind stress on the water surface.

ACKNOWLEDGMENTS

I thank D. S. McCulloch, D. H. Peterson, and R. A. Walters for valuable discussions, and W. W. Broenkow, R. T. Cheng, D. S. McCulloch, F. H. Nichols, D. H. Peterson, R. E. Smith, J. K. Thompson, F. W. Trainer, R. A. Walters, and S. M. Wienke for their helpful reviews of this manuscript.

R. E. Smith compiled, computed, and plotted the data shown in Figures 9, 12, 13, and 17. S. M. Wienke prepared most of the illustrations and K. M. Walz provided secretarial and editorial assistance.

LITERATURE CITED

Arthur, J. F., and M. D. Ball. 1979. Factors influencing the entrapment of suspended material in the San Francisco Bay-Delta estuary. Pages 143-174 *in* T. J. Conomos, ed. San Francisco Bay: The Urbanized Estuary. Pacific Division, Amer. Assoc. Advance. Sci., San Francisco, Calif.

Atwater, B. F., et al. 1979. History, landforms, and vegetation of the estuary's tidal marshes.

SAN FRANCISCO BAY

Pages 347-386 *in* T. J. Conomos, ed. San Francisco Bay: The Urbanized Estuary. Pacific Division, Amer. Assoc. Advance. Sci., San Francisco, Calif.

Bailey, T. E., C. A. McCullough, and C. G. Gunnerson. 1966. Mixing and dispersion studies in San Francisco Bay. Proc. Amer. Soc. Civil Eng. J. Sanitary Eng. Div. 92:23-45.

Ball, M. D., and J. F. Arthur. 1979. Planktonic chlorophyll dynamics in the northern San Francisco Bay and Delta. Pages 265-285 *in* T. J. Conomos, ed. San Francisco Bay: The Urbanized Estuary. Pacific Division, Amer. Assoc. Advance. Sci., San Francisco, Calif.

Barnes, C. A. and C. C. Ebbesmeyer. 1978. Some aspects of Puget Sound's circulation and water properties. Pages 209-228 *in* B. Kjerfve, ed. Estuarine Transport Processes. The Belle W. Baruch Library in Marine Science, No. 7. University of South Carolina Press, Columbia, S. C. 331 pp.

Bowden, K. F. 1967. Circulation and diffusion. Pages 15-36 *in* G. H. Lauff, ed. Estuaries. Amer. Assoc. Advance. Sci. Pub. 83.

Bowden, K. F. 1977. Turbulent processes in estuaries. Pages 46-56 *in* Estuaries, Geophysics and the Environment. National Research Council, National Academy of Science. 127 pp.

Bowden, K. F., and R. M. Gilligan. 1971. Characteristic features of estuarine circulation as represented in the Mercey estuary. Limnol. Oceanogr. 16:490-502.

Cameron, W. M., and D. W. Pritchard. 1963. Estuaries. Pages 306-324 *in* M. N. Hill, ed. The Sea. Vol. 2. Interscience, New York. 554 pp.

Carlson, P. R., and D. S. McCulloch. 1974. Aerial observations of suspended sediment plumes in San Francisco Bay and the adjacent Pacific Ocean. J. Res. U. S. Geol. Surv. 2(5):519-526.

Cheng, R. T. 1978. Comparing a few recording current meters in San Francisco Bay. Working Conference on Current Measurement, 11-13 January 1978, University of Delaware, Newark, Del. (Abstr.)

Churgin, J., and S. J. Halminski. 1974. Key to oceanographic research documentation, no. 2. Temperature, Salinity, Oxygen and Phosphate in Waters off the United States. Vol. 3. Eastern North Pacific. National Oceanic and Atmospheric Administration, U. S. Department of Commerce, Washington, D. C. 259 pp.

Cloern, J. E. 1979. Phytoplankton ecology of the San Francisco Bay system: The status of our present understanding. Pages 247-264 *in* T. J. Conomos, ed. San Francisco Bay: The Urbanized Estuary. Pacific Division, Amer. Assoc. Advance. Sci., San Francisco, Calif.

Conomos, T. J. 1975. Movement of spilled oil as predicted by estuarine nontidal drift. Limnol. Oceanogr. 20(2):159-173.

Conomos, T. J., D. H. Peterson, P. R. Carlson, and D. S. McCulloch. 1970. Movement of seabed drifters in the San Francisco Bay estuary and the adjacent Pacific Ocean: A preliminary report. U. S. Geol. Surv. Circ. 637B. 8 pp.

Conomos, T. J., D. S. McCulloch, D. H. Peterson, and P. R. Carlson. 1971. Drift of surface and near-bottom waters of the San Francisco Bay system: March 1970 through April 1971. U. S. Geol. Surv. Open-File Map.

Conomos, T. J., F. H. Nichols, R. T. Cheng, and D. H. Peterson. 1978. Field and modeling studies of San Francisco Bay. Pages 1917-1927 *in* Vol. 3, Coastal Zone '78. Symp. on Technical, Environmental, Socioeconomic and Regulatory Aspects of Coastal Zone Management, March 14-16, San Francisco, Calif.

Conomos, T. J., and D. H. Peterson. 1977. Suspended particle transport and circulation in San Francisco Bay: An overview. Pages 82-97 *in* M. Wiley, ed. Estuarine Processes. Vol. 2. Academic Press, New York.

Conomos, T. J., R. E. Smith, D. H. Peterson, S. W. Hager, and L. E. Schemel. 1979. Processes affecting seasonal distributions of water properties in the San Francisco Bay estuarine system. Pages 115-141 *in* T. J. Conomos, ed. San Francisco Bay: The Urbanized Estuary. Pacific Division, Amer. Assoc. Advance. Sci., San Francisco, Calif.

Disney, L. P., and W. H. Overshiner. 1925. Tides and currents in San Francisco Bay. U. S. Dep. Com., U. S. Coast Geod. Surv., Spec. Pub. 115. 125 pp.

Dyer, K. R. 1973. Estuaries: A physical introduction. Interscience, New York. 140 pp.

Dyer, K. R. 1977. Lateral circulation effects in estuaries. Pages 22-29 *in* Estuaries, Geophysics and the Environment. National Research Council, National Academy of Science.

Edmonston, A. D., and R. Matthew. 1931. Variation and control of salinity in Sacramento-San Joaquin Delta and upper San Francisco Bay. California Dep. Public Works Bull. 27. 440 pp.

Elford, C. R. 1970. The climate of California. Pages 538-546 *in* Climates of the States, vol. II. Western States including Alaska and Hawaii. National Oceanic and Atmospheric Administration, U. S. Department of Commerce.

Federal Water Pollution Control Administration. 1967. Effects of the San Joaquin Master Drain on water quality of the San Francisco Bay and Delta. Central Pacific Basins Comprehensive Water Pollution Control Project Report. 101 pp.

Festa, J. F., and D. V. Hansen. 1976. A two-dimensional numerical model of estuarine circulation: the effects of altering depth and river discharge. Estuarine Coastal Mar. Sci. 4:309-323.

Festa, J. F., and D. V. Hansen. 1978. Turbidity maxima in partially mixed estuaries: a two-dimensional numerical model. Estuarine Coastal Mar. Sci. 7:347-359.

Fischer, H. B. 1972. Mass transport mechanisms in partially stratified estuaries. J. Fluid Dynamics 53:671-687.

Fischer, H. B. 1976. Mixing and dispersion in estuaries. Ann. Rev. Fluid Mechanics 8:107-133.

Fischer, H. B., and E. Dudley. 1975. Salinity intrusion mechanism and San Francisco Bay, California. Proc. 16th Congr. Int. Assoc. Hydraul. Res. 1:124-133.

Gilbert, J. B., and Associates, Inc. 1977. Effects of Delta outflow on the San Francisco Bay system. Prepared for Association of Bay Area Governments (ABAG). 139 pp.

Gilbert, G. K. 1917. Hydraulic-mining debris in the Sierra Nevada. U. S. Geol. Surv. Prof. Paper 105. 154 pp.

Gill, G. S., E. C. Gray, and D. Seckler. 1971. The California water plan and its critics: A brief review. Pages 3-27 *in* D. Seckler, ed. California Water; A Study in Resource Management. University of California, Berkeley, Calif.

Gilliam, H. 1962. Weather of the San Francisco Bay region. University of California Press, Berkeley, Calif. 72 pp.

Glenne, B. 1966. Diffusive processes in estuaries. Sanitary Engineering Research Laboratory Pub. No. 66-6. University of California, Berkeley, Calif. 78 pp.

Glenne, B., and R. E. Selleck. 1969. Longitudinal estuarine diffusion in San Francisco Bay, California. Water Research 3:1-20.

Grimm, C. K. 1931. Study of tidal currents and silt movements in the San Francisco Bay area with particular reference to the effect of a salt water barrier upon them. Unpublished report, U. S. Army Corps of Engineers, Appendix D to House Document 191, 71st Congress, 3rd Session.

Hansen, D. V. 1965. Currents and mixing in the Columbia River Estuary. Pages 943-955 *in* Ocean Science and Ocean Engineering Transactions. Joint Conference Marine Technology Society. American Society of Limnology and Oceanography.

Hansen, D. V., and M. Rattray, Jr. 1966. New dimension in estuary classification. Limnol. Oceanogr. 11:319-326.

Hansen, D. V., and M. Rattray, Jr. 1972. Estuarine circulation induced by diffusion. J. Mar. Sci. 30(2):281-294.

Hedgpeth, J. W. 1979. San Francisco Bay—the unsuspected estuary. A history of researches. Pages 9-29 *in* T. J. Conomos, ed. San Francisco Bay: The Urbanized Estuary. Pacific Division, Amer. Assoc. Advance. Sci., San Francisco, Calif.

Hines, W. G. 1973. A review of wastewater problems and wastewater management planning in the San Francisco Bay region, California. U. S. Geol. Surv. Open-File Report. 46 pp.

Imberger, J., W. B. Kirkland, Jr., and H. B. Fischer. 1977. The effect of delta outflow on density stratification in San Francisco Bay: Report to ABAG (Assoc. of Bay Area Governments). Rep. HBF-77/02. Berkeley, Calif. 109 pp.

King, I. P., and W. R. Norton. 1978. Recent application of RMA's finite element models for two-dimensional hydrodynamics and water quality. Pages 2.81-2.99 *in* C. A. Brebbia et al., eds. Finite Elements in Water Resources. Pentech Press, London.

Kinsman, B., et al. 1977. Transport processes in estuaries: recommendations for research. Marine Sciences Research Center, State University of New York, Stony Brook, Spec. Rep. 6. 21 pp.

Kirkland, W. B. Jr., and H. B. Fischer. 1976. Hydraulic model studies (San Francisco Bay-Delta model) for East Bay Dischargers Authority. March 31, 1976. Water Front Associates, Alameda, Calif. 44 pp.

Kjerfve, B., ed. 1978. Estuarine transport processes. The Belle W. Baruch Library in Marine Science, No. 7. University of South Carolina Press, Columbia, S. C. 331 pp.

Krone, R. B. 1979. Sedimentation in the San Francisco Bay system. Pages 85-96 *in* T. J. Conomos, ed. San Francisco Bay: The Urbanized Estuary. Pacific Division, Amer. Assoc. Advance. Sci., San Francisco, Calif.

Lager, J. A., and G. Tchobanoglous. 1968. Effluent disposal in South San Francisco Bay. Proc. Amer. Soc. Civil Eng. J. Sanitary Eng. Div. 94:213-236.

Lynn, R. J. 1967. Seasonal variation of temperature and salinity at 10 meters in the California Current. California Cooperative Oceanic Fisheries Investigations Reports 11:157-174.

Marmer, H. A. 1926. Coastal currents along the Pacific Coast of the United States. U. S. Coast Geod. Surv. Spec. Pub. No. 121. 80 pp.

McCulloch, D. S. 1972. Seasonal flushing of South San Francisco Bay: 1969-1972. Pages 39-46 *in* V. A. Frizzell, ed. Progress Report on U. S. Geological Survey Quaternary Studies in the San Francisco Bay Area. Guidebook for Friends of the Pleistocene.

McCulloch, D. S., D. H. Peterson, P. R. Carlson, and T. J. Conomos. 1970. Some effects of fresh-water inflow on the flushing of south San Francisco Bay: A preliminary report. U. S. Geol. Surv. Circ. 637A. 27 pp.

Miller, R. C., W. D. Ramage, and E. L. Lazier. 1928. A study of physical and chemical conditions in San Francisco Bay specially in relation to the tides. Univ. Calif. Publ. Zool. 31:201-267.

National Academy of Sciences. 1977. Estuaries, geophysics, and the environment. Studies in geophysics. 127 pp.

National Marine Consultants. 1960. Wave statistics for seven deep water stations along the California Coast. Prepared for U. S. Army Corps Eng. District, Los Angeles. 20 pp.

Nelson, A. W., and R. J. Lerseth. 1972. A study of dispersion capability of San Francisco Bay-Delta waters. California Department of Water Resources, Sacramento, Calif. 89 pp.

Nichols, F. H. 1979. Natural and anthropogenic influences on benthic community structure in San Francisco Bay. Pages 409-426 *in* T. J. Conomos, ed. San Francisco Bay: The Urbanized Estuary. Pacific Division, Amer. Assoc. Advance. Sci., San Francisco, Calif.

Nichols, D. R., and N. A. Wright. 1971. Preliminary map of historic margins of marshland, San Francisco Bay, California. Open-File Rep. U. S. Geol. Surv. 10 pp.

O'Connell, R. L., and C. M. Walter. 1963. Hydraulic model tests of estuarial waste dispersion. Proc. Amer. Soc. Civil Eng. J. Sanitary Eng. Div. 89:51-65.

O'Connor, D. J. and W-S Lung. 1977. Preliminary report on two-dimensional hydrodynamic, salinity transport and suspended solids distribution in western Delta. Hydroscience, Inc., Westwood, N. J. 66 pp.

Officer, C. B. 1976. Physical oceanography in estuaries (and associated coastal waters). Interscience, New York. 465 pp.

Okubo, A. 1973. Effect of shoreline irregularities on streamwise dispersion in estuaries and other embayments. Netherlands J. Sea Res. 6:213-224.

Orlob, G. T. 1977. Impact of upstream storage and diversions on salinity balance in estuaries. Pages 3-17 *in* M. Wiley, ed. Estuarine Processes. Vol. 2. Academic Press, New York.

Parker, D. S., D. P. Morris, and A. W. Nelson. 1972. Tidal exchange at Golden Gate. Proc. Amer. Soc. Civil Eng. J. Sanitary Eng. Div. 98:305-323.

Peterson, D. H. 1979. Sources and sinks of biologically reactive oxygen, carbon, nitrogen, and silica in northern San Francisco Bay. Pages 175-193 *in* T. J. Conomos, ed. San Francisco Bay:

The Urbanized Estuary. Pacific Division, Amer. Assoc. Advance. Sci., San Francisco, Calif.

Peterson, D. H., and P. R. Carlson. 1968. Influence of runoff on seasonal changes in salinity in San Francisco Bay, California. Amer. Geophys. Union Trans. 49:704. (Abstr.)

Peterson, D. H., and T. J. Conomos. 1975. Implications of seasonal chemical and physical factors on the production of phytoplankton in northern San Francisco Bay. Pages 147-165 in R. L. Brown, ed. Proc. Workshop on Algae Nutrient Relationships in the San Francisco Bay and Delta (8-10 November 1973, Clear Lake, California): The San Francisco Bay and Estuarine Assoc.

Peterson, D. H., T. J. Conomos, W. W. Broenkow, and P. C. Doherty. 1975a. Location of the non-tidal current null zone in northern San Francisco Bay. Estuarine Coastal Mar. Sci. 3:1-11.

Peterson, D. H., T. J. Conomos, W. W. Broenkow, and E. P. Scrivani. 1975b. Processes controlling the dissolved silica distribution in San Francisco Bay. Pages 153-187 in L. E. Cronin, ed. Estuarine Research. Vol. 1. Chemistry and Biology. Academic Press, New York.

Peterson, D. H., J. F. Festa, and T. J. Conomos. 1978. Numerical simulation of dissolved silica in the San Francisco Bay. Estuarine Coastal Mar. Sci. 7:99-116.

Porterfield, G., N. L. Hawley, and C. A. Dunnam. 1961. Fluvial sediments transported by streams tributary to the San Francisco Bay area. U. S. Geol. Surv. Open-File Rep. 70 pp.

Pritchard, D. W. 1956. The dynamic structure of a coastal plain estuary. J. Mar. Res. 15(1):33-42.

Putnam, J. A. 1947. Estimating storm-wave conditions in San Francisco Bay. Trans. Amer. Geophys. Union 28(2):271-278.

Reid, J. L. Jr., G. I. Roden, and J. G. Wyllie. 1958. Studies of the California Current system. Pages 29-57 in Calif. Coop. Ocean Fish. Invest. Prog. Rep., 1 July 1956 to 1 January 1958.

Roden, G. I. 1967. On river discharge into the northeastern Pacific Ocean and the Bering Sea. J. Geophys. Res. 72(22):5613-5629.

Rubin, D. M., and D. S. McCulloch. 1979. The movement and equilibrium of bedforms in central San Francisco Bay. Pages 97-113 in T. J. Conomos, ed., San Francisco Bay: The Urbanized Estuary. Pacific Division, Amer. Assoc. Advance. Sci., San Francisco, Calif.

Selleck, R. E., B. Glenne, and E. A. Pearson. 1966. A comprehensive study of San Francisco Bay. Final report. Vol. VII. A model of mixing and dispersion in San Francisco Bay. SERL Rep. No. 67-1. 111 pp.

Selleck, R. E., E. A. Pearson, B. Glenne, and P. N. Storrs. 1966. A comprehensive study of San Francisco Bay. Final report. Vol. IV. Physical and hydrological characteristics of San Francisco Bay. SERL Rep. No. 65-10. 99 pp.

Simmons, H. B. 1955. Some effects of upland discharge on estuarine hydraulics. Amer. Soc. Civil Eng. Proc. 81 (Separate 792). 20 pp.

Smith, R. E., R. E. Herndon, and D. D. Harmon. 1979. Physical and chemical properties of San Francisco Bay waters, 1969-1976. U. S. Geol. Surv. Open-File Rep. 79-511. 630 pp.

Smith, W. 1969. Feasibility study of the use of the acoustic velocity meter for measurement of net outflow from the Sacramento-San Joaquin Delta in California. U. S. Geol. Surv. Water-Supply Paper 1877. 54 pp.

Storrs, P. N., R. E. Selleck, and E. A. Pearson. 1963. A comprehensive study of San Francisco Bay, 1961-62. Sanitary Engineering Research Laboratory, University of California, Berkeley, Calif. 221 pp.

Sumner, F. G., G. D. Louderback, W. L. Schmitt, and E. C. Johnston. 1914. A report upon the physical conditions in San Francisco Bay, based upon the operations of the United States Fisheries steamer *Albatross* during the years 1912 and 1913. Univ. Calif. Publ. Zool. 14(1):1-198.

Sverdrup, H. U., M. W. Johnson, and R. H. Fleming. 1942. The oceans. Prentice-Hall, New York. 1087 pp.

U. S. Army Corps of Engineers. 1963. Comprehensive survey of San Francisco Bay and tributaries, California. Appendix H. Hydraulic model studies, vol. 1. U. S. Army Eng. District, San Francisco, California. 339 pp.

U. S. Geological Survey. 1977a. Water Resources Data for California. Water year 1976. Vol. 3. Southern Central Valley Basins and The Great Basin from Walker River to Truckee River. U. S.

Geol. Surv. Water-Data Rep. CA-76-3. 397 pp.
U. S. Geological Survey. 1977b. Water Resources Data for California. Water year 1976. Vol. 4. Northern Central Valley Basins and the Great Basin from Honey Lake Basin to Oregon State Line. U. S. Geol. Surv. Water-Data Rep. CA-76-4. 389 pp.
Ward, P. R. B. 1974. Transverse dispersion in oscillatory channel flow. Proc. Amer. Soc. Civil Eng. J. Hydraulic Eng. Div. 100:755-772.
Walters, R. A., and R. T. Cheng. 1978. A two-dimensional hydrodynamic model of a tidal estuary. Pages 2.3-2.21 *in* C. A. Brebbia et al., eds. Finite Elements in Water Resources. Pentech Press, London.
Young, W. R. 1929. Report on salt water barrier below confluence of Sacramento and San Joaquin rivers, California. California Dep. Public Works Bull. 22, no. 1. 667 pp.

SEDIMENTATION IN THE SAN FRANCISCO BAY SYSTEM

RAY B. KRONE
Department of Civil Engineering, University of California, Davis, Davis CA 95616

Sediment inflows to the San Francisco Bay system have been significantly affected by man since the 1860's. Mining and agriculture caused large increases in sediment inflows during the late 1800's, and rapidly increasing fresh water diversions for irrigation are now causing depleted sediment inflows. In addition, maintenance dredging within the system alters sediment transport.

Sediments entering the system with land drainage consist largely of silts and clay minerals. These materials enter with high winter river flows and settle initially in the upper bays. Daily spring and summer onshore winds generate waves that suspend fine materials and hold them in suspension while tidal- and wind-driven currents circulate the suspended material throughout the system and to the ocean.

The effectiveness of waves in suspending deposited material increases rapidly with decreasing water depth. As the upper bays fill with sediment to depths where wave action resuspends the annual load at the same rate as the supply, the water depths tend to remain constant there, and further accumulation of sediment in the system occurs seaward. Evaluation of historical bathymetric surveys, including the effects of rising sea level, shows progressive sedimentation in the system that is now approaching Central Bay. Future fresh water diversions will materially slow this trend and will cause reduced turbidity from sediment particles.

The processes of aggregation, deposition, suspension, erosion, and circulation of sediment materials in the San Francisco Bay system have been described by Einstein and Krone (1961), Peterson et al. (1975), Krone (1976), Conomos and Peterson (1977), and the U. S. Army Corps of Engineers (1977). These descriptions have time scales on the order of a year or less, and while they are largely qualitative, the descriptions illuminate the sediment movements in the system. Longer term descriptions of sedimentation were made by Gilbert (1917), who described the excessive accumulations in the Bay system that resulted from hydraulic mining activities during the period 1850 to 1884, and by Smith (1965) who extended Gilbert's calculations of changes in Bay water volumes using more recent bathymetric survey data.

Both the short and long term sediment studies and studies of hydrodynamics by McCulloch et al. (1970) and Imberger et al. (1977) show that the Bay system is dominated by effects of changing sediment and water discharges through the Sacramento-San Joaquin River system (Fig. 1) from Great Central Valley drainage. There has been for decades a continuing trend of increasing consumptive use of water in the Central Valley and of export of water — and sediments — that would otherwise flow through the San Francisco Bay system. The diversion of fresh water has accelerated during recent years, and during the past two years of drought the fresh water flow to the Bay system has been the lowest in history. In view of the decreases in fresh water flows that the system is experiencing and the possibility of further decreases, it is appropriate that the existing information on sedimentation be re-examined to identify areas that may concern those responsible for management of sediments and Bay system water quality. This chapter presents an overall description of sediment movements and identifies areas of concern.

Copyright © 1979, Pacific Division, AAAS.

SAN FRANCISCO BAY

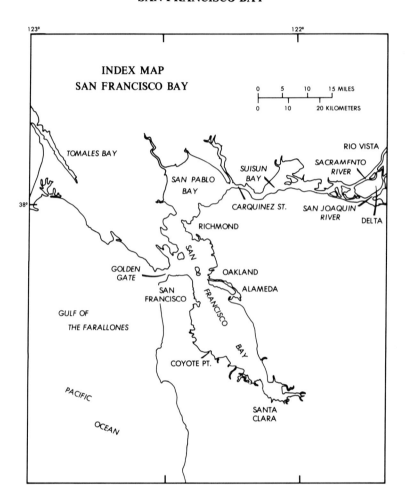

Fig. 1. San Francisco Bay system and environs.

SEDIMENT SUPPLY

Roughly 80 to 90% of the sediment entering the system is the product of soil erosion in the 163,000-km² inland drainage basin. The remainder comes from erosion of lands adjacent to the Bay system. The rate of sand transport by river flows of the Central Valley is diminished in the lower reaches of the rivers, so that the material entering San Pablo Bay is the remnant of the eroded soil and consists of clay and silt minerals carried in suspension as wash load, with only a small amount of fine sand. Most of the sediment enters with the higher winter and spring flows that result from rainfall and snowmelt.

Very large amounts of clay and silt were carried by Central Valley streams during the hydraulic-mining era in the Sierra Nevada, and this material remained in suspension in stream waters until the water velocity slowed in the broad shallow expanses of Suisun and San Pablo bays. Deposition in these upper bays was enhanced further by the increased salinities of these bays, which made suspended particles cohesive, and by the waves and gentle turbulence that caused suspended particles to collide repeatedly and form aggregates. Such aggregates have very greatly enhanced settling velocities. These "schlickens" created huge deposits in the upper bays and all but obliterated Vallejo Bay (now Carquinez Strait) at Martinez. Gilbert (1917) estimated the clay and silt deposit in the upper bays from mining debris by calculating the change in water volume. He believed that mining debris was still entering the Bay system at the time of his study, and he calculated that during the period 1849 to 1914 a total of 1.146×10^9 yd^3 was deposited. Undoubtedly additional amounts were lost to the ocean.

Hydraulic mining was stopped in 1884. Fresh water diversions for irrigation gradually increased until the early 1940's when the Central Valley Project and the federal dams in the San Joaquin Valley streams were built (see, for example, Gill et al. 1971). Very rapid decline in fresh water and sediment outflows occurred thereafter. A program for measuring suspended sediment outflows was initiated by the U.S. Geological Survey in 1957 (Porterfield et al. 1961), and estimates of sediment production are limited to calculations using subsequent suspended solids data and historic water outflows.

Estimates of sediment inflow to the Bay system were made by establishing a relation between annual water flow and annual sediment production during later years (Krone 1966). Annual production is useful because the long dry summers return the drainage system to virtually the same condition by 1 October, the start of the "water year," and each year's runoff can be considered independent of preceding years (Fig. 2). The data (Fig. 2) include those both for the San Joaquin

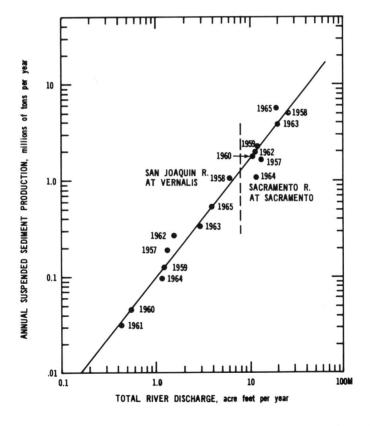

Fig. 2. Relation of annual suspended sediment production to river discharge. Reproduced from Krone (1966).

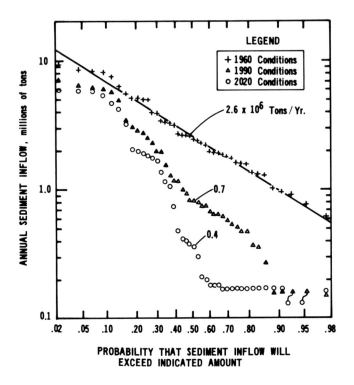

Fig. 3. Annual suspended sediment inflow from the Delta to the Bay system. Reproduced from Krone (1966).

and Sacramento rivers; the plots show that the relation indicated by the line represents both rivers reasonably well.

This relation was applied to the historic fresh water flows, modified by the U.S. Bureau of Reclamation (USBR) to the fresh water flows to the Bay system that would have occurred if the facilities and demands of 1960 existed throughout the period of record. The resulting sediment production for the wide range of flows that occurred between 1921 and 1971 is plotted on logarithmic-probability ordinates (Fig. 3). This plot, shown by the crosses, shows a 20-fold range, with a median annual production of 2.6 million t. The USBR projected water development and water demands for the years 1990 and 2020 and appropriate water management operations are applied to the same historic data, leading to the other two plots. These plots show that if such plans are realized the Bay system will experience "droughts" of sediment inflow a larger and larger fraction of the time.

The estimated annual *average* sediment inflows are presented in Table 1. If the bed load is taken to be 0.065 of the total and the dry unit weight of the sediment deposit is 33 lbs·ft^{-3} (Schultz 1965), the average annual volume of sediment "deposit" under 1960 conditions would be 10.5 million yd^3 which is close to Schultz' (1965) estimated of 11.1 million yd^3.

Decisions on future water diversions are lacking, and the projections shown for 1990 and 2020 are subject to decisions between competing political pressures for fresh water diversion and Bay system water quality. Figure 3 does show, however, the drastic reductions in sediment inflow to the Bay system that would result from future diversions planned by USBR in 1966.

TABLE 1. ESTIMATED ANNUAL AVERAGE SUSPENDED SEDIMENT PRODUCTIONS ($10^6 \, t \cdot yr^{-1}$)[a]

Sediment Source	Stream Conditions		
	1960	1990	2020
Sediment supplied to the Delta	3.75	3.42	3.34
Sediment from the Delta to San Francisco Bay system	3.35	1.79	1.22
Sediment from local streams to:[b]			
Suisun Bay	0.23	0.23	0.23
San Pablo Bay	0.29	0.29	0.29
San Francisco Bay	0.51	0.51	0.51
Total sediment to Bay system	4.38	2.82	2.25

[a] $1.0 t$ (common short ton) = $0.907 t$ (metric tonnes)

[b] From Smith (1965, Table 5). 1957 to 1959 values, measured and estimated, are data of Porterfield et al. (1961).

SEDIMENT CIRCULATIONS

The fine cohesive particles that comprise most of the material are transported in suspension. Their transport throughout the system is determined by the water movements, and by the local hydraulic conditions that facilitate deposition, erosion, and aggregation. The water movements in the Bay system are exceedingly complex and are very strongly affected by fresh water flows, the distribution of salinity, and wind stresses. A description of the general character of water movements will serve to explain sediment movements, however.

The Bay system consists of a number of broad shallow bays interconnected by narrow openings (Fig. 1). The western part of North San Francisco Bay and the narrow opening to the Pacific Ocean (Golden Gate) are quite deep, however, and water depths are maintained by the strong tidal currents. The large surface areas of the bays, combined with the restricted connections, cause progressive delays in the tides with distance from the ocean and relatively deeper channels at the narrow openings. The system is resonant to the tides, with the result that the mean range of the tide at the southern tip of South Bay is 2.2 m, 1.0 m greater than the mean ocean tide range of 1.2 m at the Golden Gate (U. S. Army Corps of Engineers 1961). This resonance causes north-south tidal currents in the central portion of the Bay system that is out of phase with flows through the Golden Gate, with the result that there is circulation between San Pablo Bay and South San Francisco Bay.

Fresh water outflows from Central Valley drainage superimpose another circulation system on the oscillatory tidal flows. More dense ocean waters tend to move upstream under the seaward flowing fresh waters (see also Conomos 1979). The oscillatory flows that result from tidal motion, combined with irregularities of the bed, cause vertical mixing with the result that there is an oscillatory but net landward movement of saline water near the bed, and this water dilutes fresh seaward flowing water above. The location and length of this mixing zone depend strongly on the fresh water flow and the flow history (McCulloch et al. 1970; Peterson et al. 1975; Imberger et al. 1977; Conomos 1979). During extremely large discharges the mixing zone extends out into the

Pacific Ocean. More commonly during winter flows the mixing zone extends from the Golden Gate to Carquinez Strait. As winter fresh water flows decrease the mixing zone moves landward, and during typical summer flows during the period 1943 to 1970, the mixing zone extended from mid San Pablo Bay to Antioch.

South San Francisco Bay is also strongly affected by high fresh water flows, with fresh water "lenses" developing during high flows (McCulloch et al. 1970; Imberger et al. 1977; Conomos 1979).

Winds affect water circulations, particularly in the broad bays where the fetch is appreciable. The winds of greatest importance appear to be the daily onshore breezes that blow from the ocean to the hot Central Valley during spring and summer. These winds also generate waves in the shallow bays every day during these months (see also Conomos 1979).

There is strong evidence that large amounts of sediment are deposited in Suisun and San Pablo bays during winter runoff (U. S. Army Corps of Engineers 1977). Waves that appear daily on the bays suspend this material and hold it in suspension while slow tidal currents transport the material to channels (Einstein and Krone 1961). During flood tides this material moves upstream through Carquinez Strait, and because the particles aggregate rapidly at the high suspended sediment concentrations that prevail, the aggregates tend to settle and there is a higher concentration near the bed (Arthur and Ball 1979). These particles move upstream with the net upstream flow near the bed, mixing vertically upward with the more saline waters. Aggregates whose settling velocities approximate or exceed the upward velocity of the more saline waters accumulate in the mixing zone and cause the well-known "turbidity maximum" there (Conomos and Peterson 1977).

Little deposition occurs in this mixing zone now because the large tidal prisms of Suisun, Grizzly, and San Pablo bays, combined with the narrow channels, cause high velocity currents that keep the channels scoured to their self-maintained depths. An attempt to cut an 11-m deep approach basin for a wharf at Benicia, however, resulted in the formation of a 5-m deep deposit in three months. Large amounts of sediment are in motion there.

Material suspended by waves in San Pablo Bay continually feeds this net upstream flow. The vertical density gradient in the mixing zone causes the velocity profile there to have exceptional velocity gradients. These gradients promote the collision of suspended particles and thereby promotes their aggregation (Krone 1972). Particles and aggregates from San Pablo Bay mix upward with riverborne dispersed particles and "scavenge" them. Algae are also scavenged this way. Aggregates carried seaward in the upper portions of the flow settle as the tidal current slows in San Pablo Bay, to either be carried back upstream for another cycle or to circulate further in San Pablo Bay.

Suspension of deposited material by waves is a process that has several important aspects. For a given wave, the maximum bed shear stress is very sensitive to water depth and is proportional to the square of the maximum near-bed orbital velocity, u^2_{max},

$$u^2_{max} = [\pi H/(T \cdot \sinh \cdot 2\pi h/L)]^2, \text{ approximately}$$

where H is the wave height, T is the wave period, L is the wave length, and h is the water depth (Komar and Miller 1973). Since $1/(\sinh \cdot 2\pi h/L)$ falls off very rapidly with depth, and its square falls even faster, it is evident that the applied stress is sensitive to depth. The suspending force is periodic, and upward diffusion is weak. The result is that fine particles are winnowed from coarser particles, so that over a period of time when suspended material is transported away by tidal currents, the remaining material is coarser than the original deposit. The applied shear stress must exceed the shear strength of the deposit before there is any suspension (Alishahi and Krone 1964).

Winnowing of sediments in San Pablo Bay is shown by the data of Storrs et al. (1963) (Fig. 4). The surface of the bed was coarser during the summer and fall than it was after the high fresh water flows. An armor having a high content of fine sand and silt is found in wind-swept shallow parts of San Pablo Bay. Below this armor, which can support a person, reside large depths of the mud from hydraulic mining.

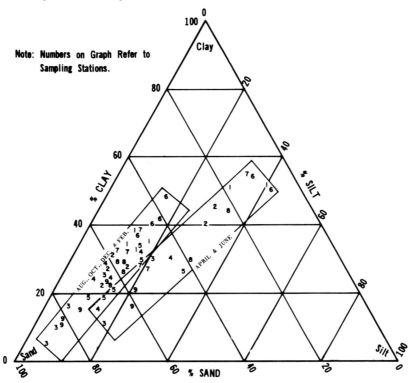

Fig. 4. Surface sediment particle-size distribution, San Pablo Bay, August 1961-June 1962. Data from Storrs et al. 1963. Numbers refer to Sanitary Engineering Research Laboratory (SERL) sampling stations.

Material suspended in San Pablo Bay and carried southward with tidal currents contains less fine sand and is easily carried with tidal currents as they circulate throughout the system. Suspended material settles wherever the water is quiet, such as in shallow areas at night when the wind dies, or in navigation facilities. Where subsequent wave action or tidal currents are insufficient to resuspend deposited material, it accumulates. Material from San Pablo Bay may deposit and be resuspended many times as it circulates and finds its way to a resting place or is carried to sea, progressively becoming finer-grained and more easily transported. Marsh areas now diked off once accumulated this fine material and probably reduced the loss to the ocean.

The scenario is repeated in miniature for each of the streams tributary to the bays, and areas of sandy material can be found near their mouths and near eroding banks. Bay waters are muddy during periods of high runoff and progressively become clearer during each year as the quantity remaining in suspension diminishes.

SAN FRANCISCO BAY

LONGER-TERM SEDIMENT DEPOSITION

The surface of the oceans is rising at an approximate rate of 0.2 m·century^{-1} (Fig. 5). If there were no deposition of sediment, the bays would continually deepen. Alternately, if deposition is so rapid that the water depths become shallow, wave action erodes the new deposit down

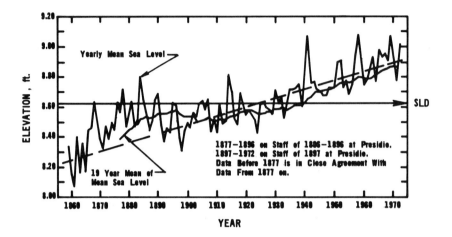

Fig. 5. Yearly mean sea level changes at San Francisco, California, 1860-1970. Data from National Ocean Survey.

to a depth where wave erosion compensates deposition, assuming no armoring by coarse material. When the supply of sediment inflow is adequate, therefore, water depths would tend to remain constant, and the rate of deposition would follow sea level rise. When the sediment supply is inadequate, the water depths would increase.

Smith (1965) reported calculations made by the U. S. Army Corps of Engineers that used averages of water depths over 1/8-min quadrangles. Averages for successive bathymetric surveys were compared to obtain changes in water depths with no allowance made for sea level rise. Tables 2 and 3 were constructed from Smith's data using linear interpolation where necessary to determine the changes over comparable time periods. The tables are arranged with areas in their geographical sequence from the Delta southwestward through the system. A pattern of deposition becomes apparent from the data in Table 2 when the accumulation or loss during successive periods is compared for successive bays. Suisun and Grizzly bays filled during the first periods and lost relatively small amounts during the second and third periods. San Pablo Bay accumulated a large amount during the first period when hydraulic mining provided a supply and successively smaller amounts during the next two periods. North San Francisco Bay showed negligible accumulation during the first period and markedly increasing amounts during the later periods. South San Francisco Bay showed deepening water during all three periods.

The rise in sea level was evidently sufficiently rapid prior to 1870 so that the sediment accumulation rate in most of the system was not sufficient to compensate the increase in water volume due to sea level rise. The very large amount of finer-grained material produced by hydraulic mining and by poor agricultural practices caused very rapid deposition in Grizzly, Suisun, and San Pablo bays. The erosion in Suisun and Grizzly bays shown to occur during the subsequent two periods indicates that the rate of supply during the first period exceeded the transport capacity of waves and currents. In fact, erosion during the second and third periods indicates that some of the

TABLE 2. APPARENT SEDIMENT ACCUMULATION (10^6 yds^3)[a,b]

Area	1870-1896 (27 years)	1897-1922 (26 years)	1923-1950 (28 years)
Suisun & Grizzly Bays & Carquinez Strait	64.3	-17.2	-4.7
San Pablo Bay	181.3	60.2	17.4
North San Francisco Bay	0.66	67.4	106.4
South San Francisco Bay	-36.1	-51.1	-55.0

[a] Data from Smith (1965)
[b] 1.0yd^3 = 0.76m^3

material deposited during the first period was transported toward San Pablo Bay in addition to the river-borne material that entered during these later periods. As this material eroded under wave action, the finer-grained fraction was washed out, and the bed became progressively more resistant to erosion. The water depths in Grizzly and Suisun bays probably now are approaching values that can be expected to remain constant unless the supply of sediment is stopped.

San Pablo Bay continued to accumulate sediment during the second and third periods, but at decreasing rates.

North San Francisco Bay did not receive much material during the period of hydraulic mining discharge, which suggests that the upper bays trapped most of the sediment that remained in the system. The losses of water volumes during the second and third periods show that the capacities of the upper bays to store material were decreasing, and the material worked its way through the upper bays until it found a permanent resting place in North San Francisco Bay. This interpretation is strengthened by the presentation of average annual apparent deposition rates presented in Table 3.

The average annual loss in water volume (the apparent sediment accumulation rate) for the total of Suisun, Grizzly, San Pablo, and North San Francisco bays is 4.2 million yd^3·yr^{-1} for both second and third periods.

The figures for South San Francisco Bay in Tables 2 and 3 are strikingly different: the water volume increased during all three periods. The average annual rate of sea level rise was used to calculate the annual change in water volume. As shown in Table 3, the observed increase in water volume exceeded that due to sea level rise during all three periods. Considering the uncertainties in the determination of sea level in the earliest bathymetric surveys it can be concluded that there were comparable rates of slow erosion in South San Francisco Bay during all three periods. Hydraulic mining debris did not cause accumulations in the open water areas of South San Francisco Bay, nor did material from Central Valley appear to be accumulating there even in 1950.

The data in Tables 2 and 3 are temporal and spatial averages. While they are valuable for the analysis given above, they are not suitable for descriptions of local areas. Sediment can move around within each of the areas used in Tables 2 and 3 without affecting the average. Smith's report shows the local changes in 1/8 min quadrangles between bathymetric surveys. Further, the data do not preclude the movement of fine material into and back out of South San Francisco Bay each year. All present evidence indicates that annual supplies of Central Valley sediments do in fact enter South San Francisco Bay under present conditions. Only the finest material, clay and silt, remains in the suspended load, however, and deposits are stable only in areas protected from

SAN FRANCISCO BAY

TABLE 3. AVERAGE ANNUAL SEDIMENT ACCUMULATION RATES (10^6 yds$^3 \cdot$yr^{-1})[a]

Area	1897-1922 (26 years)	1923-1950 (28 years)
Suisun & Grizzly Bays & Carquinez Straits	-0.66	-0.17
San Pablo Bay	2.31	0.62
North San Francisco Bay	2.59	3.80
Loss of Water Volume	4.24	4.25
Volume of Sealevel Rise	1.29	1.29
TOTAL	5.5	5.5
South San Francisco Bay		
Loss of Water Volume	-1.96	-1.96
Volume of Sealevel Rise	1.07	1.07
TOTAL	-0.91	-0.91

[a] Calculated from data in Table 2.

waves and currents. I believe that a much smaller portion of the fine-grained sediments were transported into South San Francisco Bay during the hydraulic-mining era because the trap efficiency of the upper bays was much greater then.

If the 1960 condition data (Fig. 3) can be compared with the data for 1923 to 1950 in Table 3, it is possible to calculate a sediment balance that shows the average budget (Fig. 6). The 5.5 million yd^3 deposited in the upper bays is assumed to have come entirely from inflow. The increase in water depths in South San Francisco Bay, which would result largely from erosion of very fine-grained material that would be carried to the ocean or to marsh areas, is assumed not to have contributed to deposition in the northern bays.

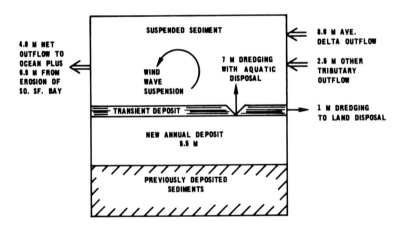

Fig. 6. Average annual San Francisco Bay sediment deposition budget. Values in millions of cubic yards of deposit.

FUTURE CHANGES

Projected fresh water diversions indicate that future supplies of suspended sediment to the Bay system will be less than historic supplies and will vary over a much wider range. At present, the Bay system steadily removes fine suspended material during each year following the winter and spring runoffs. If less sediment is supplied as the result of increased water diversion, Bay system waters will clear. Several successive years of very low flows will surely result in greatly reduced suspended sediment concentrations in Bay waters and reduced turbidity. The ample nutrient levels in Bay waters, particularly with low fresh water flows, will promote growth of algae to objectionable levels.

The clay minerals sorb toxic materials from waste discharges and thereby remove such materials from the water column. The sorbed materials are removed from the water column when the sediments are removed from suspension and thereby provide an assimilative capacity for such undesirable substances. If significant reductions of sediment inflows are to be made, either waste discharges into the Bay system will have to be greatly modified or water quality will deteriorate.

We have not acquired the necessary field data nor made quantitative calculations that show the effects of changing fresh water and sediment flows to the Bay system on the quality of Bay waters. Decisions on water diversions are being made without such information. Detailed descriptions of the water and sediment transport are needed in order to predict the effects of various fresh water and sediment outflows on the Bay system.

New bathymetric surveys will be made in a few years. These data, combined with continuing measurements of suspended sediments, should enable more precise descriptions of trends in sediment accumulation in the Bay system.

LITERATURE CITED

Alishahi, M. R., and R. B. Krone. 1964. Suspension of cohesive sediment by wind-generated waves. University of California, Berkeley, Calif. IER Tech. Rep. HEL-2-9. 23 pp.

Arthur, J. F., and M. D. Ball. 1979. Factors influencing the entrapment of suspended material in the San Francisco Bay-Delta estuary. Pages 143-174 in T. J. Conomos, ed. San Francisco Bay: The Urbanized Estuary. Pacific Division, Amer. Assoc. Advance. Sci., San Francisco, Calif.

Conomos, T. J. 1979. Properties and circulation of San Francisco Bay waters. Pages 47-84 in T. J. Conomos, ed. San Francisco Bay: The Urbanized Estuary. Pacific Division, Amer. Assoc. Advance. Sci., San Francisco, Calif.

Conomos, T. J., and D. H. Peterson. 1977. Suspended-particle transport and circulation in San Francisco Bay: An overview. Pages 82-97 in L. E. Cronin, ed. Estuarine Processes. Vol. 2. Academic Press, New York.

Einstein, H. A., and R. B. Krone. 1961. Estuarial sediment transport patterns. J. Amer. Soc. Civil Eng. H12:51-59.

Gilbert, G. K. 1917. Hydraulic mining debris in the Sierra Nevada. U. S. Geol. Surv. Prof. Paper 105. 148 pp.

Gill, G. S., E. C. Gray, and D. Seckler. 1971. The California water plan and its critics: A brief review. Pages 3-27 in D. Seckler, ed. California Water; A Study in Resource Management. University of California Press, Berkeley, Calif.

Imberger, J., W. B. Kirkland Jr., and H. B. Fischer. 1977. The effect of delta outflow on the density stratification in San Francisco Bay. ABAG (Assoc. San Francisco Bay Area Govern.) Rep. HBF-77/02. 109 pp.

Komar, P. D., and M. C. Miller. 1973. The threshold in sediment movement under oscillatory water waves. J. Sed. Petrol. 43(4):1101-1110.

SAN FRANCISCO BAY

Krone, R. B. 1966. Predicted suspended sediment inflows to the San Francisco Bay system. Report to Central Pacific River Basins Comprehensive Water Pollution Control Proj., Federal Water Pollution Control Proj., Federal Water Pollution Control Admin. (Southwest Region). 33 pp.

Krone, R. B. 1972. A field study of flocculation as a factor in estuarial shoaling processes. U. S. Army Eng. Comm. on Tidal Hydraulics. Tech. Bull. 19. Waterways Expt. Sta., Vicksburg, Miss. 61 pp.

Krone, R. B. 1976. Ultimate fate of suspended sediment in estuaries. Pages 180-201 in Proc. Specialty Conf. on Dredging and its Environmental Effects. Amer. Soc. Civil Eng., Mobile, Ala.

McCulloch, D. S., D. H. Peterson, P. R. Carlson, and T. J. Conomos. 1970. A preliminary study of the effects of water circulation in the San Francisco Bay estuary. U. S. Geol. Surv. Circ. 637(A, B). 34 pp.

Peterson, D. H., T. J. Conomos, W. W. Broenkow, and P. C. Doherty. 1975. Location of the nontidal current null zone in northern San Francisco Bay. Estuarine Coastal Marine Sci. 3(1):1-11.

Porterfield, G., N. L. Hawley, and C. A. Dunnam. 1961. Fluvial sediments transported by streams tributary to the San Francisco Bay area. Open-file rep. U. S. Geol. Surv. 70 pp.

Schultz, E. A. 1965. San Francisco Bay dredge spoil disposal. Prepared for presentation to the Comm. on Tidal Hydraulics, 53rd meeting, San Francisco, May 1965. 48 pp.

Smith, B. J. 1965. Sedimentation in the San Francisco Bay system. Pages 675-708 in Proc. Federal Interagency Sedimentation Conf. 1963. U. S. Dep. Agr. Misc. Pub. 970.

Storrs, P. N., R. E. Selleck, and E. A. Pearson. 1963. A comprehensive study of San Francisco Bay 1961-62. Sanitary Eng. Res. Lab. and School Publ. Health, University of California, Berkeley, Calif. 220 pp.

U. S. Army Corps of Engineers. 1963. San Francisco Bay and tributaries, California. Appendix H. Hydraulic Model Studies. Tech. Rep. San Francisco Bay Barriers, vol. 1. U. S. Army Eng. District, San Francisco, Calif. 280 pp.

U. S. Army Corps of Engineers. 1977. Dredge disposal study, San Francisco Bay and estuary. Appendix E. Material Release. U. S. Army Eng. District, San Francisco, Calif. 623 pp.

THE MOVEMENT AND EQUILIBRIUM OF
BEDFORMS IN CENTRAL SAN FRANCISCO BAY

DAVID M. RUBIN AND DAVID S. McCULLOCH
U. S. Geological Survey, 345 Middlefield Road, Menlo Park, CA 94025

The sand-covered floor of Central San Francisco Bay is molded by tidal currents into a series of bedforms, each of which is stable through a discrete range of tidal velocity, grain size, and water depth. Many of the bedforms move during average tide cycles, and do not require storms, floods or abnormal flow conditions to be active. The net direction of bottom sediment transport has been deduced from bedform asymmetry. The geometry of Central Bay exerts considerable control on the sediment transport pattern. Tidal flows accelerate as they pass through the narrow Golden Gate and produce ebb and flood jets that transport sediment away from the Gate. Lower velocity flows that occur between the shoreline and the jets are ebb dominant within the Bay, and flood dominant outside the Gate, and these flows transport sediment toward the Gate.

In Central Bay, where many of the bedforms are active during average tide cycles, sediment turnover, which is important in organic and inorganic exchange between the sediment and the water column, results largely from bedform migration. This rigorous hydraulic regime also acts to reduce biological turnover by benthic organisms by producing an environment more suited to animals that extract nutrients from the water column and surface and suspended sediment, rather than from buried sediment.

Sediment on the floor of San Francisco Bay is an active element in the organic and inorganic processes that go on in the estuary. The sediment provides a repository through which there is a flux of plant and animal nutrients, trace metals (McCulloch et al. 1971; Peterson et al. 1972; Moyer and Budinger 1974; Girvin et al. 1978; Luoma and Cain 1979), and man-made synthetic organic compounds. It also provides a domicile for a varied benthic community (Nichols 1979).

The sediment is dynamic; it responds to physical stirring by organisms, to estuarine circulation, to oscillating tidal flows, and to wind-generated waves. In some areas the tidal flow produces hydraulic sorting of the bed load which largely determines the local grain-size of the sediment, and also the pattern, rate, and total flux of the sediment through that reach of the estuary. This hydraulic sorting, which separates mudflats from sandy bottoms, exerts marked control on the distribution of benthic species. Hydraulic conditions also affect the distribution of benthic species by controlling the duration of bottom stability (Nichols 1979).

PURPOSE

In Central Bay, the area described in this chapter, where tidal currents are strong and the Bay reaches its greatest depth, the sediment is generally sandy. The sandy sediment responds to the local hydraulic regime by forming several distinct types of bedforms, each of which is stable, or in equilibrium, for some discrete range of water depth, flow velocity and grain size. Thus, knowledge of the distribution of these bedforms not only indicates the local hydraulic environment that prevailed when the bedforms were produced, but with some reservations, can be used to estimate sediment transport rates.

SAN FRANCISCO BAY

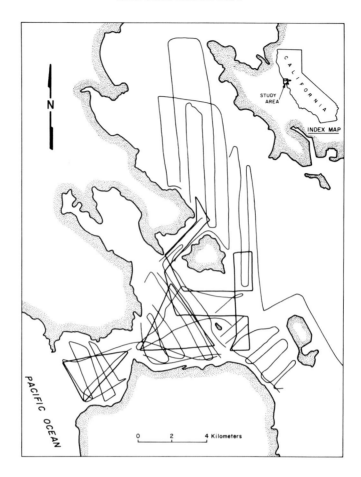

Fig. 1. Track-line locations. Light lines were run once. Heavy lines were run throughout tidal cycles.

This chapter discusses the kinds and distribution of bedforms in Central Bay, the directions of bedload transport, some rates of sand-wave migration, how often the sand waves move, and the hydraulic factors that control bedform distribution. A more detailed quantitative discussion is given elsewhere (Rubin and McCulloch in press).

PROCEDURE

Bedforms were mapped with a side-scanning sonar system that bounces a high frequency (100 kHz) sonic signal off the seafloor and produces a continuously recorded oblique view of the sea floor on both sides of the survey vessel. The vessel location (Fig. 1) was continuously recorded with an electronic dual transponder range-range system that has a precision of about 5m. Bedload transport directions were inferred from the orientation of the crests and the asymmetry of sand waves. Sand-wave movement was studied by resurveying selected sonar lines through several tide cycles, and by making longer term observations from a fixed point with a bottom-mounted rotating side-scan sonar system (Rubin et al. 1977). Bedforms in any given area were related to sediment grain size, water depth, and depth-averaged current velocity in order to determine equilibrium bedform conditions. Depth-averaged velocities were calculated from unpublished National Ocean Survey current velocity data by plotting semilog velocity profiles and averaging the velocities read from the plot at each tenth of the flow depth from 1/10 to 9/10.

SETTING

Central Bay is one of four bays in the San Francisco Bay complex. Within this complex the deepest water (Fig. 2), the coarsest sediment (Fig. 3), and the highest velocities (Fig. 4) occur where flow is constricted by bedrock at the Golden Gate, the entrance to Central Bay. The eastern and northern margins of Central Bay are lined with broad muddy flats in relatively shallow slow-moving water. Between these broad, shallow, muddy flats, and the steep, rocky Golden Gate, the Bay is floored with sand and has tidal currents that peak at about 70-200 cm·s^{-1} during average tides.

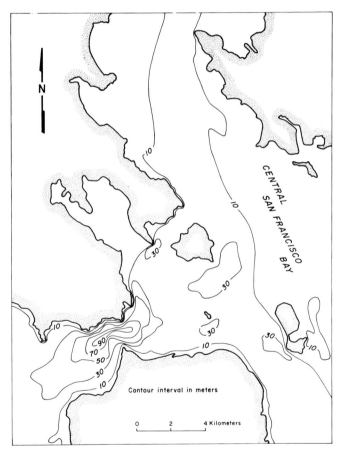

Fig. 2. Bathymetry of Central Bay.

BOTTOM TYPES

As sandy sediment is exposed to increasingly strong flows it responds by forming a progression of bedforms, from low regime flat beds or ripples, to sand waves, and finally, when bed

SAN FRANCISCO BAY

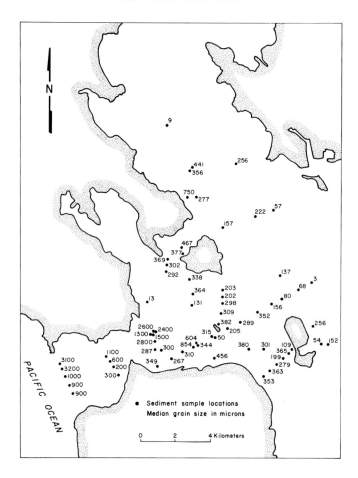

Fig. 3. Median grain sizes in Central Bay (P. R. Carlson, unpublished U. S. Geological Survey data).

irregularities are unstable, to upper regime flat beds.[1] The progression is well demonstrated in laboratory flumes and is observed in natural flows (see later discussion). Limited observations by divers and underwater television indicate that these same bedforms occur on the Central-Bay floor. Unfortunately, for mapping purposes we were limited by the resolution of the sonar system that cannot differentiate lower from upper regime flat beds, and has insufficient definition to recognize ripples. Thus, only sand waves and flat beds are shown in Fig. 5.

Sand Waves

Sand waves, first reported in Central Bay by Gibson (1951), cover approximately half of the area surveyed in Central Bay (Fig. 5). In plan view as seen by the side-scan sonar, the sand waves are straight-crested (Figs. 6, 7) sinuous, catenary, or barchan-like (Fig. 8). In cross section, they are triangular (Fig. 6) or convex upstream (Fig. 8). Their heights range from less than 20 cm to more than 8 m, and height-to-wave length ratios are typically 1/10 to 1/40. Although the sand waves

---

[1] We recognize that some workers subdivide the larger bedforms (e.g., megaripples, dunes, etc.) but for simplicity these additional forms are included with sand waves in this discussion.

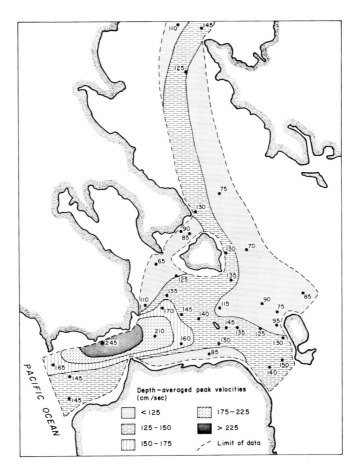

Fig. 4. Peak depth-averaged tidal currents in the Central Bay during an average tide. Current velocities were measured by the National Ocean Survey (unpublished data).

vary in size, in any given area they are approximately the same size and can be divided and mapped by size (Fig. 4).

Flat Beds

Beds that appear flat are the second most abundant bottom type observed in Central Bay. The theoretical resolution of the side-scan system is approximately 10 cm, and the smallest sand waves visible on the side-scan records are 10-20 cm high. Beds with bedforms less than 10 cm high, small sand waves and all current ripples, would therefore be expected to appear flat. Despite this lack of resolution, the extensive flat bed areas can be grossly separated into two types; one contains boulders and bare bedrock and lies in the high-velocity tidal currents in and adjacent to the Golden Gate, the other is free of bedrock and boulders, and lies in an area of weaker tidal currents along the eastern fringe of Central Bay. Upper regime flat beds occur in the former, and current ripples predominate in the latter.

Bedrock and Boulders

As noted above, bedrock and boulders occur mostly in the Golden Gate area where fast currents keep the bed swept clear of sand. In some areas, boulders are numerous enough to form a boulder pavement. In other areas, boulders and bedrock locally protrude through a sand veneer. Turbulence developed at these protuberances downmixes fast-moving water and, in some places, the resulting increase in flow velocity at the bottom is sufficient to scour away the thin sand veneer downstream from obstructions and to produce rock or boulder-floored sharp pointed depressions (Fig. 9) called comet marks (Werner and Newton 1975).

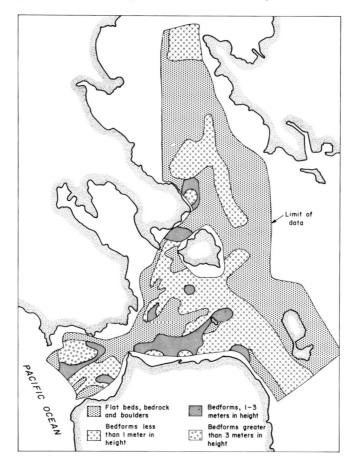

Fig. 5. Distribution of bedforms mapped by side-scan sonar. Flat-appearing beds in the Golden Gate area are upper flow regime flat beds, and in this same area boulders and bare bedrock occur on the Bay floor. Other flat-appearing beds are inferred from Fig. 12 to be rippled.

Other Bottom Features

At several locations in Central Bay the bottom is dominated by forms produced by processes other than flowing water. On Southampton Shoal in the northeast corner of Central Bay there is a field of holes up to 4 m in diameter and 0.5 m in depth that resemble holes formed by feeding bat rays in nearby Bodega Bay (Nichols 1979, Fig. 7). In this same area, adjacent to an oil refinery pier off Richmond, the bottom has been grooved by numerous ships as they plow their keels through the sediment. Our side-scan profiles show that similar keel plow marks are common off the mouth, and throughout the Oakland Estuary, and in this same area a long chain of large cycloidal bottom gouges appear to have been made by the propeller of a large ship (Hartman 1976, Figs. 1.2 and 1.3).

Bedload Transport Directions

Directions of bedload transport can be estimated by assuming that bedload sediment is transported normal to a sand-wave crest. Although the current reverses direction during each tide cycle, and transport may occur in both directions, the net transport occurs down the steeper slope of the sand waves. Net bedload transport directions inferred from sand-wave geometry are shown in Fig. 10. Because the sediment transport rate increases as a high power of the velocity of a flow, sediment transport is strongly biased in the direction of the peak-velocity tidal current. Consequently, sediment transport directions inferred from sand-wave orientation are generally within 15° of the directions measured by current meters of the strongest near-bottom currents (Fig. 10).

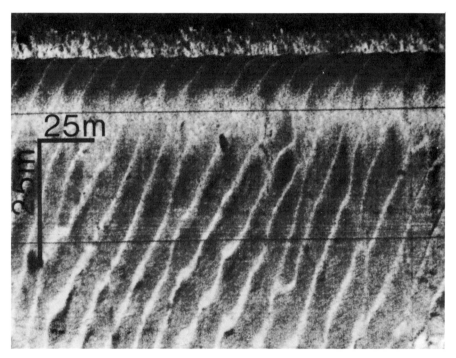

Fig. 6. Side-scan record of straight-crested sand waves. Sand wave heights are 0.5 m; wavelength is 5-10 m; depth is 20 m. Transport is from left to right.

The bedload transport pattern in the Golden Gate area is dominated by high velocity flows generated by jet currents that are formed by both the ebb and flood tides as they flow through the Golden Gate. During flood flow (Fig. 11A), ocean water enters the Gate and accelerates because the channel decreases in cross-sectional area. This jet current enters the Bay with depth-averaged velocities of more than 200 cm·s^{-1} and is maintained for some distance by its momentum. Flood velocities north and south of the jet current are lower, or about 50-100 cm·s^{-1}. During ebb flow (Fig. 11B), Bay water converges radially toward the Golden Gate with peak velocities of 100-150 cm·s^{-1}. Consequently, east of the Gate, where the jet current flows, flood velocities exceed ebb velocities, but in the adjacent areas north and south of the jet, ebb velocities exceed flood velocities (Fig. 11C).

East of the Golden Gate the jet current maintains a channel that is deeper than the adjacent ebb-dominated shoals. The resulting bedload transport pattern forms a cell similar to that observed in other tidal inlets where transport in jet-dominated channels is away from inlet openings (Dean and Walton 1975; Wright and Sonu 1975) and transport in adjacent shoals is toward inlet openings (Fig. 11C).

During ebb flow (Fig. 11B) the velocity is also increased by the constriction of the Golden Gate, and a high velocity flow is formed along the north side of the Gate west of the constriction. North and south of the jet, surface current data (U. S. Coast and Geodetic Survey 1964) indicate counter-rotating eddies that are driven by the jet flow. Bedforms beneath the northern eddy were not observed, but bedforms beneath the southern counterclockwise eddy indicate that the counter-clockwise circulation dominates. Because of the formation of the jet flow and the fact that the most rapid changes in sea level occur in the downward direction, ebb flow dominates, and the directions of sand transport closely resemble the directions of ebb circulation. Thus, within the Golden Gate, as within the adjacent Bay, sand is circulated in a cell, with jet transport away from, and adjacent transport toward the inlet opening.

Between the inward and outward flowing jets lies the deepest channel of San Francisco Bay, at a depth of about 110 m. Side-scan sonar profiles and high resolution sub-bottom acoustic profiles (Carlson and McCulloch 1970) indicate that the bedrock channel floor is swept partially free of sediment. Flat beds accompanied by boulders also reflect the high velocities of the jets. Although the bedforms define the transport cells as both sides of the Golden Gate, they do not

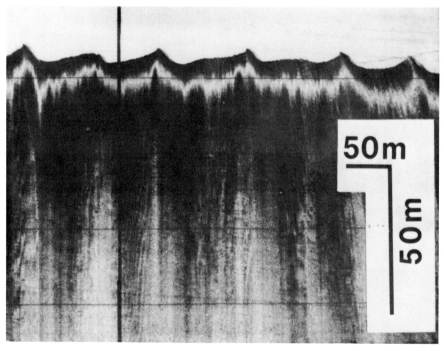

Fig. 7. Side-scan record of sand waves with reversed crests. Dominant transport is left to right. Record shows the reversal of the sand-wave crests caused by right-to-left flow. Maximum sand-wave height is 5 m, and water depth is 40 m.

define the net direction of bedload transport through the Gate.

Because sand is moved mainly by peak-flow currents, its net transport direction may vary from the net transport direction of near bottom suspended sediment that depends solely on net water circulation. The San Francisco Bay estuary enjoys estuarine circulation with a seaward flow of brackish surface water over a compensating landward flow of more saline bottom water (Conomos 1979). As demonstrated by the movement of seabed drifters (Conomos 1979, Fig. 23) the bottom flow transports the near-bottom suspended sediment landward (Conomos and Peterson 1977). Thus, in ebb-dominated areas of Central Bay the tractive bedload and the near bottom suspended sediment have opposite net transport directions.

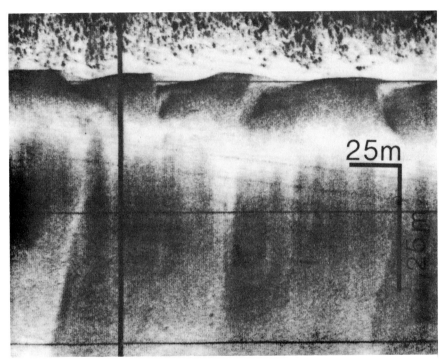

Fig. 8. Side-scan record of barchan-like sand waves. Maximum bedform height is 2.5 m; water depth is 30 m, and transport is from left to right.

SAND-WAVE MOVEMENT RATES

Sand-wave migration rates are very difficult to measure from a vessel, because even with very precise navigation the error in determining the vessel's position may exceed the distance that sand waves move in many months. However, one can determine whether or not sand waves move without measuring their migration by observing short-term changes in sand-wave shape. During average and spring tides, the crests of many sand waves in Central Bay reverse orientation daily in response to the oscillating tidal currents (Fig. 7), indicating that these sand waves are active under average tidal conditions. Sand-wave movement, or lack of movement, during neap tides has not been studied.

As noted above, the measurement of sand-wave migration rates is difficult. Tidal sand waves typically move only a few centimeters or tens of centimeters per day (Jones et al. 1965; Boothroyd and Hubbard 1975; Bokuniewicz et al. 1977). With commonly available navigation systems one would need a long elapsed interval of months or years between repeated observations, and

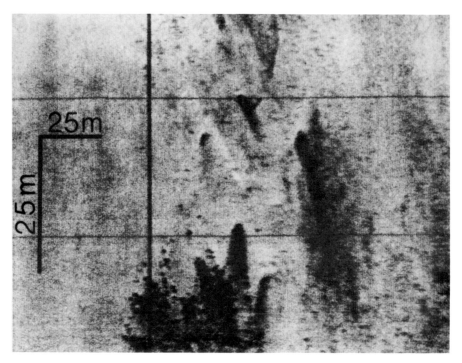

Fig. 9. Side-scan record of comet marks; dark areas are boulders, light areas are sand. Turbulence downstream from obstructions has scoured the sand veneer, producing comet-shaped boulder exposures. Transport is from top to bottom.

even then the navigation problem may make a significant uncertainty in measured migration rates. In order to overcome the navigational uncertainty, sand-wave migration in a portion of Central Bay was measured with a rotating side-scan sonar system (Rubin et al. 1977) that was placed on the Bay floor about 2.5 km east of the Golden Gate and about 400 m offshore of San Francisco. The sonar system was wired to shore where a recorder provided images of a 400-m-diameter circular area of the Bay floor. The side-scan system was in water about 20 m deep and sand waves in the field of view were approximately 0.6 m in height and 18 m in wavelength. Sediment at the site had a median grain size of approximately 0.3 mm. During a two-month observation period from this fixed position sand-wave crests moved 0.6 to 2.4 m, with average rates of 1 to 4 cm·d^{-1}. The site of this study was chosen for its proximity to shore, and current velocities at the site are lower than at many other sand-wave fields. Consequently, sediment transport rates and bedform migration rates at the side-scan site may be relatively low.

CONTROL OF BED CONFIGURATION AND DISTRIBUTION

Sedimentologists and engineers working with flumes have found that specific kinds of bedforms reproduced with a given flow (velocity, depth, and viscosity) and sediment (grain size, density, and sorting). This has led to many studies designed to determine how flow and sediment parameters are related to bed configuration (Allen 1963, 1968; Raudkivi 1963, 1966; Yalin 1964; Harms and Fahnestock 1965; Simons et al. 1965; Guy et al. 1966; Hill 1966; Znamenskaya 1966; Harms 1969; Hill et al. 1969; Kennedy 1969; and Southard 1971, 1975). These experimental

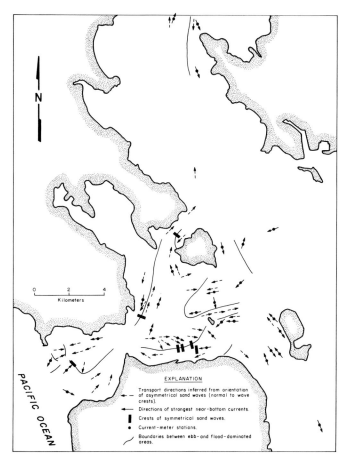

Fig. 10. Map showing the orientation of sand waves and the orientation of the strongest near-bottom currents. Symmetrical sand waves occur along the boundaries between ebb- and flood-dominant areas.

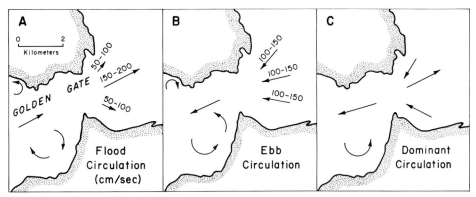

Fig. 11. A. Generalized diagram showing water-circulation pattern in the Golden Gate area during maximum flood. Water enters the Bay in a jet current. B. During ebb circulation west of the Golden Gate water exits the Bay as a jet with adjacent counter-rotating eddies. C. Flow directions of the strongest currents beneath the jets are flood oriented in the Bay and ebb oriented west of the Golden Gate. Strongest currents adjacent to the jets are ebb oriented in the Bay and flood oriented west of the Golden Gate.

studies were generally limited to flow depths of less than 0.5 m, and extrapolation of bed phase boundaries to greater depths was largely conjectural. In 1975 Boothroyd and Hubbard extended these studies to flows of several meters depth in describing bed phases in two shallow New England tidal estuaries. Rubin and McCulloch (1976) described bed-phase boundaries in San Francisco Bay where flow depths reach 80 m, and with these deeper flow data they proposed an extension of previously established shallow flow bed phase boundaries to deeper flows (Fig. 12). The phase

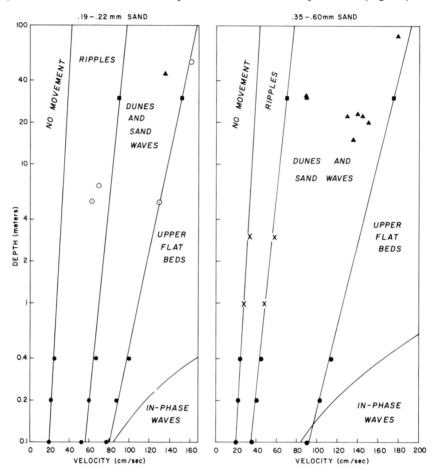

Fig. 12. Plots of bed configuration as a function of depth and velocity for two sediment-size ranges. Triangles and circles represent San Francisco Bay sand waves and flat-appearing beds respectively; squares are points from bed-phase boundaries on grain size vs. velocity plots for additional bay data; x's are points on bed-phase boundaries determined by Boothroyd and Hubbard (1975); dots are points on bed-phase boundaries measured or interpolated from Southard (1975, Figs. 2-2, 2-3, and 2-5); circles with dots are points from boundaries separating ripple, dune, and flat-bed phases determined by Dalrymple, et al. (1978). Flume velocities are depth-averaged for steady flow; bay velocities are depth-averaged during peak flow of average tides. High-velocity, shallow flows that produce in-phase waves (also called antidunes) have not been seen in San Francisco Bay.

boundaries they proposed are shown in Fig. 13 where phase boundaries and sand-wave height are related to sediment size, depth, and velocity. More recent work by Dalrymple et al. (1978) describes bed phases in intermediate depths up to about 14 m in the Bay of Fundy. Allowing for differences in bedform nomenclature, and the fact that smaller straight-crested bedforms they call

Type I megaripples would appear as flat beds to the side-scan sonar, their data fit the trends of the phase boundaries based on flume and Bay data.

The construction of Fig. 13 is not meant to imply more than correlation between the related parameters. It should be clearly understood that although it is valid to draw such a figure, and that such a figure might be used for predictive purposes, there is very little that is understood about the details of the specific mechanisms that generate bedforms.

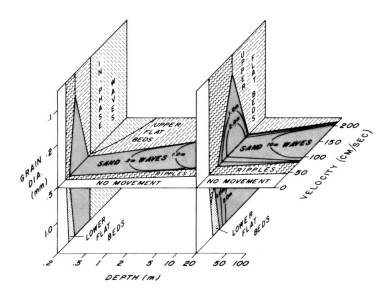

Fig. 13. Plot of bed-phase and sand-wave height as a function of velocity, depth, and sediment size, generalized from Bay and flume data cited in text.

SEDIMENT TURNOVER

For some biological and chemical processes that involve exchange between sediment on the floor of the Bay and the water column, sediment turnover rates are more important than the rate of net sediment transport over the Bay floor. Physical overturning of sediment accompanies bedform migration, but is also caused by benthic organisms. In physically stable environments such as deep subtidal areas, equilibrium faunas can be effective in turning over the sediment (Nichols 1974). But in Central Bay, where the benthic population is dominated by siphon feeders that extract nutrients from surface or suspended sediment and from the water column, rather than from buried sediment (Nichols 1979), benthic stirring is of less importance than bedform migration. Central Bay sediment is highly dynamic. Where tidal-current velocities are high, sand waves as much as 1 m in height have been observed to reverse their asymmetry in a single tide cycle. This requires turning over the bulk of the sediment within the sand wave. Thus, locally, turnover may reach a depth of 1 m·d^{-1}. In general, however, physical turnover is lower, the principal process being the migration of current ripples that turn over only the upper 2-5 cm of sediment each day.

Although difficult to quantify, a total Central Bay turnover rate estimated from these observations would be consistent with a rate calculated from the amount of radon contributed to the water column by Central Bay sediment (Hammond and Fuller 1979). By assuming a rate of radon production for sandy sediment (1000 atoms·m^{-3}·s^{-1}) they estimate that approximately 40 cm of

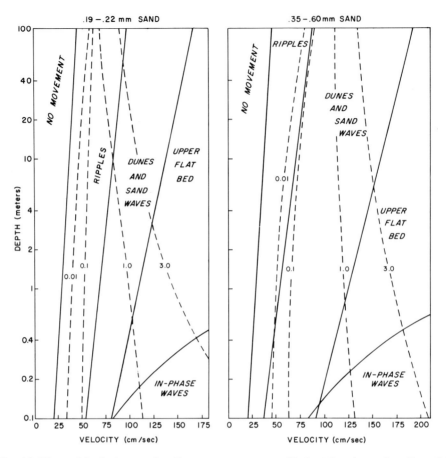

Fig. 14. Plots of bed phase and sediment transport rate (dashed lines) as a function of depth and velocity for two sediment sizes. Transport rates in kg·m^{-1}·s^{-1} are from Colby (1964).

turnover must occur every few days.

As noted earlier, sediment dynamics are important in determining the composition of the benthic fauna. In the Bay mudflats, where sediment is generally fine-grained, episodes of physical turnover or disruption are related to storms or seasonal wind patterns. Although less dynamic on a daily basis than Central Bay, sediment turnover in South Bay may be caused largely by physical stirring (storm resuspension) rather than biological stirring. The South Bay benthic community is composed of opportunistic pioneering species adapted to fine-grained sediment that can re-establish themselves rapidly after disruption (Nichols 1979). Rhoads et al. (1978) have shown that pioneering species like those in South Bay have little effect below the surface of the bottom, and make only minimal contribution to sediment turnover.

SEDIMENT TRANSPORT RATES

Because both the rate of sediment transport and bed phase are functions of velocity, depth, and sediment size, they can be directly compared. In Fig. 14, empirically determined and extrapolated sediment transport rates from Colby (1964) are superimposed on bed-phase plots. These plots give an estimate of the rate of sediment transport for flows in *equilibrium* with specified beds. Equilibrium is stressed because bedforms can persist in flow velocities higher and lower than those which produced them, and the transport rates apply only to the time interval during which the form was in equilibrium with the flow. In tidally oscillating flows, where most transport occurs during peak flow, this interval may be short, and total transport over long time periods is difficult to establish.

FUTURE RESEARCH

The present study was limited to sedimentary processes active in sandy areas of Central Bay. Little is known about those processes active in other sandy areas in San Francisco Bay, and even less is known about sedimentary processes active in the muddy areas that predominate in the Bay. In general, the sedimentary processes that have been identified are known only qualitatively, and the rates at which they operate are only beginning to be quantified.

ACKNOWLEDGMENTS

H. Edward Clifton, Ralph E. Hunter, and R. Lawrence Phillips made suggestions that aided this study. Arnold H. Bouma and David A. Cacchione read the manuscript. Paul R. Carlson supplied grain-size data. Part of this work was supported by a National Research Council Postdoctoral Research Associateship for David M. Rubin.

LITERATURE CITED

Allen, J. R. L. 1963. Asymmetrical ripple marks and the origin of water-laid cosets of cross-strata. Liverpool Manchester Geol. J. 3:187-236.

Allen, J. R. L. 1968. Current ripples. North-Holland Publ. Co., Amsterdam. 433 pp.

Bokuniewicz, H. J., R. B. Gordon, and K. A. Kastens. 1977. Form and migration of sand waves in a large estuary, Long Island Sound. Mar. Geol. 24:185-199.

Boothroyd, J. C., and D. K. Hubbard. 1975. Genesis of bedforms in mesotidal estuaries. Pages 217-234 *in* L. E. Cronin, ed. Estuarine Research. Vol. 2. Academic Press, New York.

Carlson, P. R., and D. S. McCulloch. 1970. The floor of Central San Francisco Bay. California Division of Mines and Geology, Mineral Information Service 23(5):97-107.

Colby, B. R. 1964. Discharge of sand and mean-velocity relationships in sand-bed streams. U. S. Geol. Surv. Prof. Paper 462-A, pp. A1-A47.

Conomos, T. J. 1979. Properties and circulation of San Francisco Bay waters. Pages 47-84 *in* T. J. Conomos, ed. San Francisco Bay: The Urbanized Estuary. Pacific Division, Amer. Assoc. Advance. Sci., San Francisco, Calif.

Conomos, T. J., D. S. McCulloch, D. H. Peterson, and P. R. Carlson. 1971. Drift of surface and near-bottom waters of the San Francisco Bay system. March 1970 through April 1971. U. S. Geol. Surv. Open-File Map.

Conomos, T. J., and D. H. Peterson. 1977. Suspended-particle transport and circulation in San Francisco Bay: An overview. Pages 82-97 *in* L. E. Cronin, ed. Estuarine Processes. Vol. 2. Academic Press, New York.

Dalrymple, R. W., R. J. Knight, and J. J. Lambiase. 1978. Bedforms and their hydraulic stability relationships in a tidal environment, Bay of Fundy, Canada. Nature 275:100-104.

Dean, R. G., and T. L. Walton, Jr. 1975. Sediment transport processes in the vicinity of inlets with special reference to sand trapping. Pages 129-149 *in* L. E. Cronin, ed. Estuarine Research. Vol. 2. Academic Press, New York.

Gibson, W. M. 1951. Sand waves in San Francisco Bay. J. Coast Geod. Surv. 4:54-58.

Girvin, D. C., A. T. Hodgson, M. E. Tatro, and R. N. Anaclerio. 1978. Spatial and seasonal variations of silver, cadmium, copper, nickel, lead and zinc in South San Francisco Bay water during two consecutive drought years. U. S. Dep. Energy, Contract W-7405-ENG-48. Lawrence Berkeley Lab. UCID-8008. 117 pp.

Guy, H. P., D. B. Simons, and E. V. Richardson. 1966. Summary of alluvial channel data from flume experiments, 1956-1961. U. S. Geol. Surv. Prof Paper 462-I. 96 pp.

Hammond, D. E., and C. Fuller. 1979. The use of Radon-222 to estimate benthic exchange and atmospheric exchange rates in San Francisco Bay. Pages 213-230 *in* T. J. Conomos, ed. San Francisco Bay: The Urbanized Estuary. Pacific Division, Amer. Assoc. Advance. Sci., San Francisco, Calif.

Harms, J. C. 1969. Hydraulic significance of some sand ripples. Geol. Soc. Amer. Bull. 80:363-396.

Harms, J. C., and R. K. Fahnestock. 1965. Stratification, bed forms, and flow phenomena (with an example from the Rio Grande). Pages 84-115 *in* G. V. Middleton, ed. Primary Sedimentary Structures and Their Hydrodynamic Interpretation. Soc. Econ. Paleontol. Mineral., Spec. Pub. No. 12.

Hartman, G. L. 1976. Evaluation of estuarine channel conditions in Coos Bay, Oregon, using side-scan sonar. U. S. Army Corps of Engineers and Oregon State University, Corvallis, Ore. Figs. 1.2 and 1.3, pp. 5 and 7.

Hill, H. M. 1966. Bed forms due to a fluid stream. Proc. Amer. Soc. Civil Eng. J. Hydraul. Div. 92:127-143.

Hill, H. M., V. S. Srinivasan, and T. E. Unny, Jr. 1969. Instability of flat bed in alluvial channels. Proc. Amer. Soc. Civil Eng. J. Hydraul. Div. 95:1545-1558.

Jones, N. S., J. M. Kain, and A. H. Stride. 1965. The movement of sand on Warts Bank, Isle of Man. Mar. Geol. 3:329-336.

Kennedy, J. F. 1969. The formation of sediment ripples, dunes, and anti-dunes. Ann. Rev. Fluid Mech. 1:147-168.

Luoma, S. N., and D. J. Cain. 1979. Fluctuations of copper, zinc, and silver in tellenid clams as as related to fresh-water discharge—South San Francisco Bay. Pages 231-246 *in* T. J. Conomos, ed. San Francisco Bay: The Urbanized Estuary. Pacific Division, Amer. Assoc. Advance. Sci., San Francisco, Calif.

McCulloch, D. S., T. J. Conomos, D. H. Peterson, and K. W. Leong. 1971. Distribution of mercury in surface sediments in San Francisco Bay estuary. U. S. Geol. Surv. Open-File Rep.

Moyer, B. R., and T. F. Budinger. 1974. Cadmium levels in the shoreline sediments of San Francisco Bay. U. S. Atomic Energy Contract W-7405-ENG-48. Lawrence Berkeley Lab. 2642. 37 pp.

Nichols, F. H. 1974. Sediment turnover by a deposit-feeding polychaete. Limnol. and Oceanogr. 19(6):945-950.

Nichols, F. H. 1979. Natural and anthropogenic influences on benthic community structure in San Francisco Bay. Pages 409-426 *in* T. J. Conomos, ed. San Francisco Bay: The Urbanized Estuary. Pacific Division, Amer. Assoc. Advance. Sci., San Francisco, Calif.

Peterson, D. H., D. S. McCulloch, T. J. Conomos, and P. R. Carlson. 1972. Distribution of lead and copper in surface sediments in San Francisco Bay estuary, California. U. S. Geol. Surv. Misc. Field Studies Map MF-323.

Raudkivi, A. J. 1963. Study of sediment ripple formation. Proc. Amer. Soc. Civil Eng. J. Hydraul. Div. 89:15-33.

Raudkivi, A. J. 1966. Bedforms in alluvial channels. J. Fluid Mech. 26:507-514.

Rhoads, D. C., P. L. McCall, and J. Y. Yingst. 1978. Disturbance and production on the estuarine seafloor. Amer. Sci. 66:577-586.

Rubin, D. M., and D. S. McCulloch. 1976. Bedform dynamics in San Francisco Bay, California. Geol. Soc. Amer. Abstr. and Prog. 8:1079. (Abstr.)

Rubin, D. M., and D. S. McCulloch. In press. Single and superimposed bedforms: A synthesis of

San Francisco Bay and flume observations *in* A. H. Bouma, D. S. Gorsline, G. P. Allen, and C. Monty, ed. Shallow Marine Processes and Products. Special issue Sed. Geol.

Rubin, D. M., D. S. McCulloch, and H. R. Hill. 1977. Bedform observations with a bottom-mounted rotating side-scan sonar in San Francisco Bay, California. Trans. Amer. Geophys. Union 58:1162. (Abstr.)

Simons, D. B., E. V. Richardson, and C. F. Nordin, Jr. 1965. Sedimentary structures formed by flow in alluvial channels. Pages 34-52 *in* G. V. Middleton, ed. Primary Sedimentary Structures and Their Hydrodrynamic Interpretation. Soc. Econ. Paleontol. Mineral., Spec. Pub. No. 12.

Southard, J. B. 1971. Representation of bed configurations in depth-velocity-size diagrams. J. Sed. Petrol. 41:903-915.

Southard, J. B. 1975. Bed configurations. Pages 5-43 *in* Depositional Environments as Interpreted from Primary Sedimentary Structures and Stratification Sequences. Soc. Econ. Paleontol. Mineral. Short Course No. 2, Dallas, 1975. Ch. 2.

U. S. Coast and Geodetic Survey. 1964. Tidal current charts, San Francisco Bay. U. S. Dep. Com. Washington, D. C. 12 pp.

Werner, F. and R. S. Newton. 1975. The pattern of large-scale bed forms in the Langeland Belt (Baltic Sea), Marine Geol. 19:29-59.

Wright, L. D. and C. H. Sonu. 1975. Processes of sediment transport and tidal delta development in a stratified tidal inlet. Pages 63-76 *in* L. E. Cronin, ed. Estuarine Research. Vol. 2. Academic Press, New York.

Yalin, M. S. 1964. Geometrical properties of sand waves. Proc. Amer. Soc. Civil Eng. J. Hydraulic Div. 90:105-119.

Znamenskaya, N. S. 1966. Experimentation on erodible models. Soviet Hydrology: Selected Papers 6:477-486.

PROCESSES AFFECTING SEASONAL DISTRIBUTIONS OF WATER PROPERTIES IN THE SAN FRANCISCO BAY ESTUARINE SYSTEM

T. JOHN CONOMOS, RICHARD E. SMITH, DAVID H. PETERSON,
STEPHEN W. HAGER, LAURENCE E. SCHEMEL
U.S. Geological Survey, 345 Middlefield Road, Menlo Park, CA 94025

    The timing and general location of major processes modifying the distributions of conservative (temperature, salinity, alkalinity) as well as biologically reactive (oxygen, carbon, nutrients, pH) water properties can be inferred from the seasonal features of the Delta-outflow hydrograph. River-modulated physical effects on these distributions in near-surface midchannel water are characteristically defined by season (high versus low river inflow) and geographic region (northern versus southern reach of the estuarine system).

    Delta outflow directly controls and often dominates the spatial and temporal distributions of most properties and biological processes in the northern reach. The outflow contributes suspended particles, dissolved oxygen, and silicate, and generates an estuarine circulation cell and a turbidity maximum. The circulation pattern and associated features largely dictate spatial distributions. Seasonal changes, however, are caused by relative changes in outflow (which determine water-residence time and thus flushing rates) and light-limited biological activity (photosynthesis, nutrient uptake, and oxygen production): during winter, mixing and advection control biological activity, whereas during summer, both biological activity and physical processes are important.

    The relation between Delta outflow and biological processes in the southern reach, however, is less direct: biological activity has a relatively greater effect on the spatial and temporal distributions of these properties. Distributions of properties are dominated by the perennial inflow of detritus and nutrient-rich waste water at the southern boundary. These inputs are augmented during winter by discharges from local intermittent streams that may contribute large amounts of nitrogenous compounds. The substrate is the major source of particles and dissolved silicate. Greatest biological activity apparently takes place during spring rather than summer as in the northern reach. This increased activity in the southern reach is caused in part by Delta-outflow induced stratification that tends to maintain algal cells in the photic zone.

Dissolved and particulate properties constituting the bulk composition of water in an estuarine system are fundamental to the biological processes occurring within such a system. This set of properties determines which species are found, and influences seasonal and annual patterns in their abundance and distribution. The presence of man along the San Francisco Bay shorelines has caused inevitable changes in the concentrations and relative ratios of these properties. Some of these changes, in turn, have adverse impact on the biota and have led to the establishment of water quality criteria (Regional Water Quality Control Board 1975).

Our purpose is to summarize the distributions of several basic conservative[1] (temperature,

---

[1] A conservative property is one whose chemical and biological activity is so low relative to the rate of physical processes (mixing and advection) that its distribution is essentially controlled by these physical processes.

salinity, alkalinity) and nonconservative[2] (hydrogen-ion activity—pH, turbidity, chlorophyll *a*, suspended particulate matter, plant nutrient, and dissolved-oxygen concentrations) properties in near-surface, mid-channel water throughout the entire Bay system by season (Fig. 1; Table 1) and to identify sources and sinks of some of these properties. We place particular emphasis on describing the dominant processes controlling their seasonal variations. To identify the effects of river-modulated physical processes (mixing and advection) on the supply, dilution and removal (flushing) rates of most of these properties, we divide the year into a winter (high river inflow) and a summer (low river inflow) period. This division further permits us to make preliminary inferences as to the timing and general locations of major biological processes modifying these distributions.

We will show here that Delta outflow directly controls and often dominates the spatial and temporal distributions of most properties and biological processes in the northern reach. In the southern reach, however, the dependence of biological processes on Delta outflow is less direct, and biological processes have a relatively greater effect on the spatial and temporal distributions of these properties.

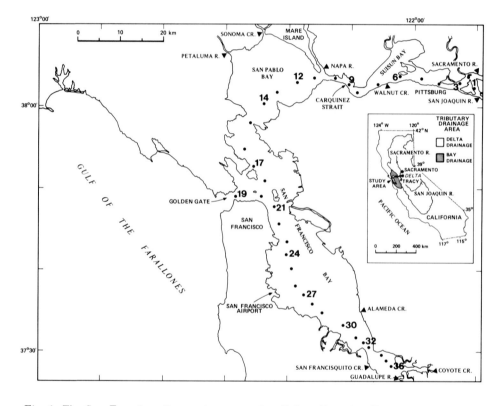

Fig. 1. The San Francisco Bay system comprises Suisun Bay, San Pablo Bay, and San Francisco Bay, but is termed San Francisco Bay herein. The northern reach is Suisun Bay, San Pablo Bay and the northern part of San Francisco Bay (to the Golden Gate). The southern reach is San Francisco Bay south of the Golden Gate. Station numbers are established hydrographic stations occupied near-monthly from 1969 to the present. The drainage basins of the Sacramento-San Joaquin River system and of the peripheral streams are in inset.

---

[2] A nonconservative property is one whose distribution is determined by the effects of short-term chemical and biological activity as well as by the effect of physical processes (mixing and advection).

TABLE 1. HYDROGRAPHIC STATION LOCATIONS AND NUMBERS OF OBSERVATIONS[a]

Station Number	Location[b] N. Lat.	W. Long	Observations Summer	Winter
1	38° 2.4'	121° 50.4'	18	7
3	38 3.0	121 52.7	57	38
6	38 3.9	122 2.1	59	47
9	38 3.0	122 10.4	61	52
12	38 3.1	122 18.7	59	48
14	38 0.5	122 24.1	56	47
17	37 52.9	122 25.6	55	53
19	37 49.1	122 28.3	31	40
21	37 48.0	122 22.2	55	48
24	37 42.0	122 20.3	50	44
27	37 37.1	122 17.5	43	40
30	37 33.3	122 11.5	50	42
32	37 31.1	122 8.1	41	32
36	37 28.3	122 3.8	41	34
		TOTAL	676	572

[a] Observations reported are water samples collected at ≤2-m water depths during winters (December through April) and summers (July through October), 1969-1977.

[b] Locations shown on Fig. 1.

PREVIOUS WORK

Early Bay-wide surveys of salinity and temperature were conducted by Sumner et al. (1914) and Miller et al. (1928). These surveys were followed 30 years later by a comprehensive 5-year water-quality study by the Sanitary Engineering Research Laboratory (SERL) (Harris et al. 1961; McCarty et al. 1962; Pearson et al. 1967, 1969; Selleck et al. 1966a, b; Storrs et al. 1966, 1968, 1969).

Our work began in 1969 and initially considered the relation of Bay-wide flushing with Delta outflow (McCulloch et al. 1970; McCulloch 1972; Carlson and McCulloch 1974). Later publications have emphasized the distributions and interrelations of plant nutrients and suspended particles in the northern reach (Conomos and Peterson 1974, 1975, 1977; Peterson et al. 1975a, b; Peterson 1979). Spiker and Schemel (1979) have summarized seasonal distributions of organic and inorganic carbon, and Hammond and Fuller (1979) have reported initial findings of substrate-water exchanges of several parameters. Temporal and spatial distributions of suspended particles, salinity, and plant nutrients in the Delta and the landward part of the northern reach are summarized by Arthur and Ball (1979) and Ball and Arthur (1979). Cloern (1979) has reviewed our present understanding of phytoplankton ecology in the Bay and has presented interesting new data.

THE DATA

Data were collected from 1969 through 1977 during near-monthly surveys in order to relate changes in water properties in the channels to seasonal variations of river discharge and insolation (Conomos et al. 1978). Water pumped to the vessel was continuously analyzed for temperature,

TABLE 2. SUMMARY OF ANALYTICAL METHODS

Variable	Primary Reference	Precision[a]	Principal Analysts
position		±0.2 km	1
depth	Schemel and Dedini (1979a)	±0.2 m	2, 3
temperature	Schemel and Dedini (1979a)	±0.1°C	2, 3
salinity	W. Peterson (consultant)	0.05 °/∞	2, 3
light transmission	Hydroproducts Corp.[b]	5 units (day to day) 10 units (cruise to cruise)	2, 3
dissolved oxygen	Carpenter (1965)	±4 μg-atoms·liter^{-1}	4, 5, 6
reactive silicate	Technicon (1976)[b]	3% at 20 to 300 μg-atoms·liter^{-1}	7, 8, 9
orthophosphate	Atlas et al. (1971)	3% at 2 to 50 μg-atoms·liter^{-1}	7, 8, 9
nitrate+nitrite	Technicon (1973)[b]	3% at 1 to 80 μg-atoms·liter^{-1}	7, 8, 9
nitrate	Kahn and Brezenski (1967)	5% at 20 to 40 μg-atoms·liter^{-1}	7, 8, 9
nitrite	Technicon (1973)[b]	3% at 1 to 40 μg-atoms·liter^{-1}	7, 8, 9
ammonia	Solorzano (1969)	3% at 5 to 100 μg-atoms·liter^{-1}	7, 8, 9
chlorophyll *a* fluorescence	Lorenzen (1966)	±5%	2, 3, 10, 11
turbidity	Turner Designs[b]	±5%	2, 3
particulate organic matter (POC)	Schemel and Dedini (1979b)	±5% at 5 to 250 μg-atoms·liter^{-1}	2, 9
alkalinity	Culberson et al. (1969)	±0.02 meq·liter^{-1}	2, 5, 6
pH		±0.025 units	2, 3, 5
suspended particulate matter (SPM)	Strickland and Parsons (1972)	20%	5, 9, 12

[a] Where possible, estimates of precision are based on splits of discrete samples.
[b] The mention of brand names is for identification purposes and does not constitute endorsement by the U.S. Geological Survey.
[c] Principal analysts: 1, F. Lewis; 2, L. Schemel; 3, L. Dedini; 4, A. Hutchinson; 5, R. Smith; 6, G. Massoth; 7, S. Hager; 8, D. Harmon; 9, S. Wienke; 10, B. Cole; 11, A. Alpine; 12, R. Herndon.

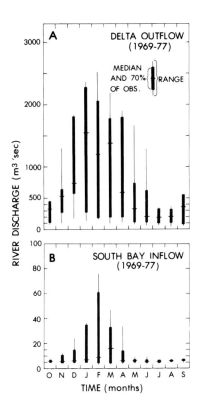

Fig. 2. Monthly means of river inflow: Delta Outflow Index (DOI), 1969 through 1977 (A), calculated by the U.S. Bureau of Reclamation; southern-reach streams (Fig. 1), 1969 through 1977, including San Jose sewage inflows, which average 5 $m^3 \cdot sec^{-1}$ (B). Stream discharge from U.S. Geological Survey; sewage data courtesy of San Jose-Santa Clara Water Pollution Control Facility. After Conomos (1979).

salinity, light transmission, plant nutrients (silicate, phosphate, nitrate+nitrite, ammonia), pH, and chlorophyll *a* fluorescence (Schemel and Dedini 1979a; Smith et al. 1979; Table 2). Discrete samples were taken for measurements of dissolved oxygen, alkalinity, particulate organic carbon (POC), and suspended particulate matter (SPM).

The near-surface (≤ 2 m) data gathered at 14 of the 36 mid-channel stations (Fig. 1) are discussed herein. These data are separated by seasons defined on the basis of river inflow (Fig. 2), with "winter" consisting of the data collected from December through April and "summer" July through October (Table 1). The data are grouped by station but they are not tidally averaged. The stations chosen for discussion are approximately 10 to 12 km apart, a distance comparable to the typical tidal excursion (Conomos 1979). The median value and the range of the central 70% of the observations made from 1969 through 1977 are used here to represent a "typical" value and the variation about this value.

ENVIRONMENTAL SETTING

The Bay is influenced by strong winds and tidal currents and highly seasonal river inflow (Conomos 1979). Winds generate large waves which resuspend sediments, vertically mix the water column, and exert stress on the water surface creating important nontidal currents. The Sacramento

SAN FRANCISCO BAY

and San Joaquin rivers introduce large volumes of dissolved and particulate substances, generate an estuarine (gravitational) circulation cell in the northern reach, and create density-induced advection in the southern reach. Tidal currents, together with wind and river inflow, cause mixing.

The rates of movement and mixing of dissolved and suspended substances are seasonally modulated, with river inflow being dominant during winter and wind effects dominant during summer. Tides, although changing character fortnightly, are relatively constant throughout the year. Incident solar radiation (insolation) varies seasonally (Fig. 3), modulating growth rates of plants. This production, in turn, results in alteration of the concentrations and distributions of dissolved and particulate substances. Insolation also heats the water masses and promotes evaporation during summer. The water masses in the northern and southern reaches each have a unique character because they respond differently to seasonal changes in the rates of physical and biological processes.

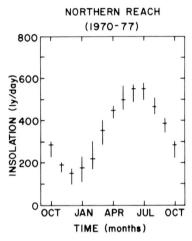

Fig. 3. Monthly distributions of insolation in the northern reach. Data averaged from Richmond (1970 through 1973) and San Rafael (1975 through 1977), courtesy of the Bay Area Air Pollution Control Board.

A Typical Winter

High river inflow (1500 $m^3 \cdot sec^{-1}$; 80% of annual total) carries large quantities of riverborne suspended sediment (3.2 x 10^6 metric tonnes; 80% of annual total) to the Bay during winter (Conomos and Peterson 1977). The largest fraction of this influx of water and sediment passes through the Delta into the northern reach. Although some sediment escapes to the ocean, most deposits in the northern reach (Fig. 4). The large volumes of Delta outflow control the salinity distribution in the Bay (Fig. 5B) and slightly lower the surface salinities of the coastal waters (Fig. 6). The longitudinal salinity gradient in the northern reach is compressed slightly and pushed seaward (Conomos 1979, Fig. 18). Because the vertical salinity gradient is strong (often 10 °/oo from surface to bottom), this reach becomes a partially mixed estuary. Some of the Delta-derived low-salinity water intrudes into the southern reach, stratifying the water with salinity gradients of 5 °/oo from surface to bottom, the high-salinity water present there from the preceding summer. The cold (10°C) water of the Delta outflow depresses slightly the ambient Bay-water temperatures (Fig. 5A). The exchange of slightly warmer ocean water (Fig. 6) through the Golden Gate is not sufficient to raise the Bay-water temperatures noticeably.

During winter the Bay is flushed relatively rapidly by Delta outflow: water parcels remain

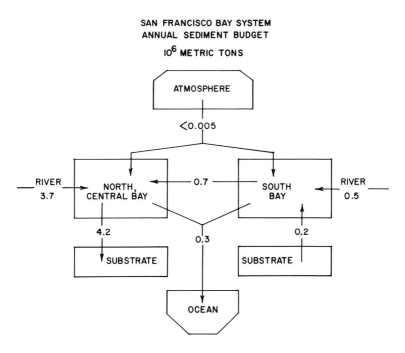

Fig. 4. Annual sediment budget of the San Francisco Bay system. Calculations made from sediment data of Porterfield et al. (1961) and Smith (1965), and from unpublished air-quality data of the Environmental Protection Agency.

in the northern reach for only a few weeks and in the southern reach for a few months. Insolation is low, and together with the relatively low residence times of the cold water masses, the algal biomass is relatively low, and some biologically reactive substances behave conservatively (Peterson et al. 1975b). Although the prevailing winds are weak, winter storms are often accompanied by episodic southeast gales which mix and move water masses (Conomos 1979).

Offshore, the seasonal Davidson Current (Conomos 1979) flows slowly northward. The cold (12°C) surface layers are wind-mixed to 50 meters (Fig. 6) and the nutrient levels are relatively high.

A Typical Summer

Summer is marked by a much decreased and warmer river inflow (Figs. 2, 5A) and diminished sediment influx. The salinity gradients weaken, and the northern reach becomes a well-mixed estuary (Conomos 1979, Fig. 18). The Delta outflow is not of sufficient magnitude to notably depress the salinities of the southern reach (Fig. 5B). Here the water mass approaches a near-steady-state condition, the salt from the ocean having diffused landward primarily by tidal mixing. The prevailing winds shift to the northwest and become substantially strengthened, vertically mix the water column and break down the salinity stratification. The winds also resuspend large quantities of bottom sediment in the shallower areas of the Bay. The sediment in suspension is transported northward by nontidal currents and redeposited (Conomos and Peterson 1977).

With decreased river inflow, residence times of the water masses are increased to a few months in the northern reach and several months in the southern. Away from the cooling effects of the relatively colder ocean water (15°C), the summer water temperatures are 10° warmer than in winter (Fig. 5A). In the northern reach, the warming effect of the Delta outflow dominates the effect of solar heating. In the southern reach, however, solar heating dominates, and water temperatures are often 5°C warmer than Golden Gate waters. The surface warming, in concert with increased wind speed and long water-residence times, facilitates evaporation of surface waters and sometimes causes thermal stratification (Conomos 1979). The long residence times also permit sewage-derived substances to accumulate.

The high summer insolation (Fig. 3) results in increased light availability, which is the primary factor promoting plant growth throughout the Bay (Cloern 1979; Peterson 1979). The increased light, warmer temperatures, longer water-residence times, and reduced river supply of some dissolved substances cause the distributions of biologically reactive substances to be noticeably affected by the production of plants and cause an increase in the mineralization of organic matter by bacterial decomposition.

Seaward of the Golden Gate, the Davidson Current is replaced in summer by the southward-flowing California Current. The surface waters are warmer, saltier and stripped of nutrients by biological production (Fig. 6). The northwest winds induce shallow (≤ 200 m) coastal upwelling, bringing cool nutrient-rich water closer to the surface.

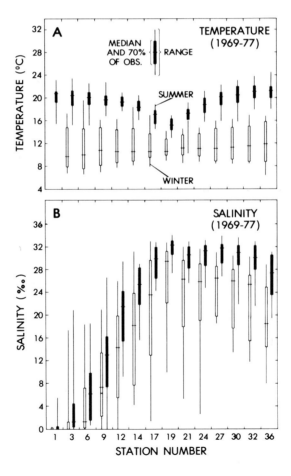

Fig. 5. Longitudinal distribution of water temperature (A) and salinity (B) at ≤ 2 m during winter (December through April) and summer (July through October) at near-monthly intervals (1969 through 1977) at hydrographic stations (Fig. 1).

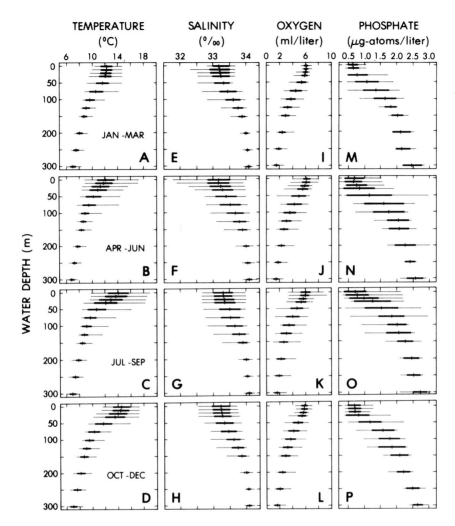

Fig. 6. Temperature (A-D), salinity (E-H), dissolved oxygen (I-L) and orthophosphate (M-P) distributions as a function of depth in the ocean adjacent to San Francisco Bay. The data were collected from 1920 through 1970 in an area bounded by 36-38°N, 121-126°W and summarized by Churgin and Halminski (1974).

PROPERTIES AND MODULATING PROCESSES

Suspended Particulate Matter

Particulate matter is contributed by rivers, the ocean, sewage effluents, is resuspended from the substrate, and produced *in situ* primarily by biological processes. Knowledge of the composition, distribution, and processes affecting suspended particles is important as they adsorb and concentrate trace contaminants (such as trace metals and synthetic organic compounds), are substrates for bacteria, and are food for planktonic and benthic filter-feeders. Furthermore, together with dissolved material, they attenuate light and thereby limit photosynthetic activity.

The particulate matter, based on composition, is divided into two groups: lithogenous particles, which are inorganic and mostly are crystalline products of weathering and erosion of rocks;

SAN FRANCISCO BAY

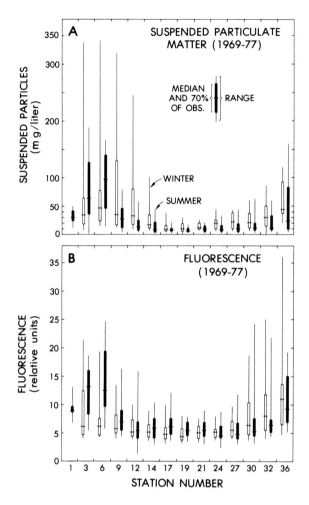

Fig. 7. Longitudinal distribution of suspended particulate matter (A) and *in vivo* fluorescence (B) at ≤2 m during winter (December through April) and summer (July through October) at near-monthly intervals (1969 through 1977) at hydrographic stations (Fig. 1).

and biogenous particles, which are organic and may be either living (phytoplankton, zooplankton, and bacteria) or nonliving (organic detritus). The gross composition of the lithogenous fraction has been discussed elsewhere (Knebel et al. 1977). The biogenous fraction, discussed in more detail in this volume (see for example, Spiker and Schemel [1979]), is important to this discussion of water properties. We report particle dry weight per unit volume as an index of concentration (Fig. 7A). Light transmission data (Fig. 8) provide an estimate of the relative attenuation of light caused by both particulate and dissolved matter in the water column. The biogenous particles are represented by *in vivo* chlorophyll *a* fluorescence (Fig. 7B), an indicator of phytoplankton abundance, and particulate organic carbon (Fig. 9), a measure of total particulate organic matter abundance.

Sources and Distribution of Suspended Particles. Rivers contribute the bulk of the lithogenous particles and some biogenous matter, particularly plant fragments (detritus) and fresh-water phytoplankton, mainly into the northern reach (Porterfield et al. 1961; Conomos and Peterson

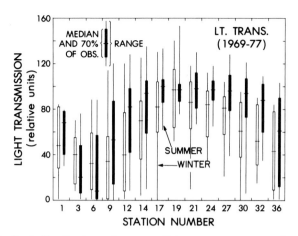

Fig. 8. Longitudinal distribution of percent light transmission at ≤2 m during winter (December through April) and summer (July through October) at near-monthly intervals (1969 through 1977) at hydrographic stations (Fig. 1).

1977; Krone 1979; Arthur and Ball 1979; Spiker and Schemel 1979; Peterson 1979; Cloern 1979; Fig. 7). The ocean contributes marine phytoplankton, mostly during the spring and summer coastal upwelling season (Cloern 1979).

In the shallow areas of the northern reach and in much of the southern reach the primary source of suspended particles is the substrate, having been distributed by wind waves and tidal currents (Conomos and Peterson 1977). Although most of the particles are lithogenous, large quantities of benthic diatoms are present (Nichols 1979). During winter, Delta-derived suspended particles intrude into the southern reach with the stratified water (Carlson and McCulloch 1974). Throughout the Bay, sewage effluents contribute detrital particulate matter (Table 3). Sewage detritus may be relatively more important during summer in the southern part of the southern reach, where the sewage water discharges exceed those from the local streams (Fig. 10), water-residence time is longest, and the ratio of receiving-water volume to sewage-water inflow is low (Table 3).

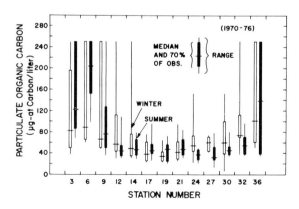

Fig. 9. Longitudinal distribution of particulate organic carbon at ≤2 m during winter (December through April) and summer (July through October) at near-monthly intervals (1970 through 1976) at hydrographic stations (Fig. 1). Problems with analytical sensitivity prohibited detection of concentrations greater than 250 μg-atoms·liter^{-1}.

SAN FRANCISCO BAY

TABLE 3. POINT-SOURCE WASTE-WATER LOADINGS (METRIC TONNES PER DAY) IN SAN FRANCISCO BAY, 1975[a]

Segment (USGS Sta. No.)[b]	A Receiving-water volume[c] (km^3)	B Waste-water inflow (10^{-3} $km \cdot d^{-1}$)	A:B (Daily ratio)	Waste-water loadings ($t \cdot d^{-1}$)						Total phosphorus	Biological oxygen demand
				Suspended Solids	Organic nitrogen	Ammonia	Nitrate	Nitrite			
San Jose-Dumb. Br. (36)	0.09	0.55	154	17.5	3.0	10.7	0.58	0.03		11.0	16.8
Dumb. Br.-San Mateo Br. (30, 32)	0.34	0.15	2250	9.7	1.7	2.5	0.11	0.22		2.4	15.8
San Mateo Br.-Bay Br. (21, 24, 27)	2.11	2.88	732	51.1	6.9	10.3	1.05	0.11		8.1	79.4
Bay Br.-Pt. Richmond (17, 19)	2.01	0.31	6400	13.7	2.5	4.5	0.33	0.02		3.1	26.9
Pt. Richmond-Carq. Br. (12, 14)	1.52	0.79	1930	12.1	0.8	3.4	0.51	0.03		1.6	13.4
Carq. Br.-Benicia Br. (9)	0.17	0.02	9100	0.7	0.1	0.1	0.02	0.002		0.5	0.6
Benicia Br.-Antioch (3, 6)	0.42[d]	0.42	995	16.4	0.7	3.3	0.19	0.04		2.8	14.6
TOTAL	6.65	5.13	1300	121.0	15.6	34.8	2.79	0.45		28.9	165.8

[a] Unpublished waste-water data courtesy of California State Water Quality Control Board, San Francisco Bay Region.
[b] Station number locations indicated on Fig. 1.
[c] Data from Selleck et al. 1966; at mean tide level.
[d] This volume is for the segment between Benicia Bridge and Pittsburg.

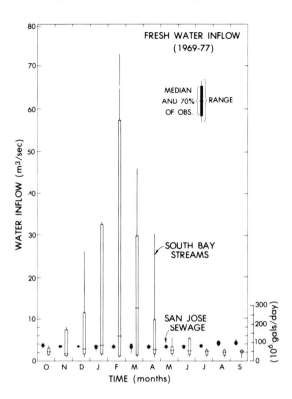

Fig. 10. Monthly means of southern-reach stream (Fig. 1) and San Jose-Santa Clara sewage effluent inflows, 1969 through 1977. Stream discharge from U. S. Geological Survey; sewage data courtesy of San Jose-Santa Clara Water Pollution Control Facility.

Most particulate matter in Bay waters is permanently deposited on the Bay bottom or is flushed into the ocean. Some biogenous particles are destroyed by bacterial decomposition (mineralization) and by ingestion by herbivorous zooplankton and filter-feeding benthic organisms.

The gross distributions of near-surface suspended particle concentrations, in the main channels, estimated by several methods (Figs. 7, 8, 9), are similar and show distinct differences between reaches. A suspended-particle (turbidity) maximum is present in the northern reach, a minimum occurs at the Golden Gate, and a concentration gradient is observed increasing southward in the southern reach.

Seasonal differences in distributions between reaches also occur. Typical concentrations of total suspended particles (Fig. 7A) and the organic fraction (Figs. 7B, 9) in the turbidity maximum during summer and higher than in winter, although the amount of suspended particles entering the northern reach through the Delta is much less than winter. The converse is true in the southern reach where all indices of particulate matter are higher during winter. In both reaches the ratio of biogenous to total particle concentrations is typically greater during summer than winter (Fig. 11), since the quantity of lithogenous matter diluting the biogenous particles is probably greater during winter.

Processes Affecting Particle Composition and Distribution. Delta-derived suspended-particle distributions in the deeper channels of the northern reach are controlled primarily by the estuarine circulation cell generated by Delta outflow (Conomos 1979). A null zone and an associated turbidity maximum are present as a consequence of the tidally averaged water movements of this circulation cell (Peterson et al. 1975a; Arthur and Ball 1979; Conomos 1979; Festa and Hansen 1978). Advective transport traps suspended particles by the convergence of near-bottom currents at the null zone and retains them there, forming a particle-concentration maximum (see Cloern 1979, Fig. 9).

The abundance and composition of the suspended particles in the turbidity maximum change seasonally in response to relative changes in Delta outflow and biological activity (Conomos and Peterson 1974; Arthur and Ball 1979; Cloern 1979; see also Peterson et al. 1975b, Fig. 9). During winter when Delta outflow and its suspended particle load are high, the turbidity maximum is not as pronounced as during summer, apparently because much of the sediment is held in suspension in the well-stratified upper water layers and advected seaward to the shoal areas in San Pablo and Central bays where it is deposited (Conomos and Peterson 1977). Some particles settle out into the lower water layers and are returned landward to the null zone by estuarine circulation. Furthermore, in winter there is little or no development of a biogenous-particle component to enhance the maximum. There is a low standing stock of living matter (plankton) as the water residence time is quite short, and insolation (Fig. 3) and the water temperatures (Fig. 5A) are low. The riverborne biogenous fraction (mainly detrital) is advected seaward like the lithogenous matter. Probably few of these biogenous particles are deposited, for the settling rates are somewhat slower than lithogenous matter (Conomos and Gross 1972).

During summer, with the decline of winter high Delta outflow, the concentration of riverborne suspended particles supplied to the null zone decreases, but the particle concentration of the turbidity maximum increases. This is apparently caused by the greater volumes of resuspended sediment transported landward by the near-bottom nontidal currents (Conomos and Peterson

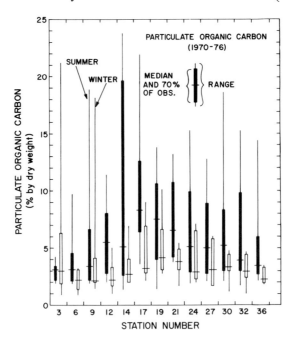

Fig. 11. Longitudinal distribution of the ratio (in percent) of particulate organic matter to total suspended particulate matter (by dry weight) at ≤2 m winter (December through April) and summer (July through October) at near-monthly intervals (1970 through 1976) at hydrographic stations (Fig. 1).

1977) and by the greater production of organic matter (phytoplankton) in the null zone by phytoplankton (Cloern 1979; Peterson 1979; Arthur and Ball 1979). The high influx of resuspended material to the turbidity maximum is made possible by greater wave-induced resuspension present during summer (see Conomos 1979, Figs. 6, 7, 8). The higher standing stock of phytoplankton is maintained by estuarine circulation, the higher production of phytoplankton associated with the null zone (Peterson et al. 1975a, b), and the greater amounts of biogenous matter exported from the surrounding marshes (Arthur and Ball 1979; Ball and Arthur 1979).

There is no estuarine circulation cell in the southern reach because little river inflow discharges directly into the southern boundary (Conomos 1979). The high particle concentrations at the southern boundary are contributed by local streams and major sewage discharges (Fig. 7A). The concentrations of both the lithogenous and biogenous fractions are high because the receiving-water volume is small relative to these inputs (Fig. 10; Table 3) and advective transport is negligible. Furthermore, the effects of wave-induced resuspension are increasingly important in a southeastward direction because the water becomes progressively shallower and the wind fetch increases. Although the water of northern boundary of the southern reach is a sink because of dilution with less turbid ocean water, turbid Delta-derived water intrusions may sometimes act as secondary sources of suspended particles during winter. Considering the sluggish circulation of this reach (Conomos 1979) the gradient implies that, overall, particles introduced at the southern boundary are diffused northward by tidal-current and wind mixing. From the limited data available (Smith 1965), it appears that not only does the southern reach not accumulate much new sediment, but in the last several decades it could be losing sediment to the northern reach (Fig. 4).

Seasonal differences in particle concentrations of the southern reach are not modulated only by Delta outflow. The higher winter concentrations of suspended particles throughout the gradient (Fig. 7A) are also caused during these months by the higher inputs of local streams and storm overflow drains from the south.

The higher concentrations of the particulate organic fraction (Figs. 7B, 9) in the southern reach during winter are apparently caused by an increase of the living matter, as sewage inputs (mainly detrital) are relatively constant throughout the year. The relative contribution of sewage-derived detritus should be lower because it is diluted by increased stream inflow during these winter months. This increase is probably caused by the photosynthetic activity (spring bloom) occurring during the latter months of the winter period (February through April) (Cloern 1979; Schemel and Dedini 1979b). The fluorescence data, replotted so that winter (December through April) is subdivided into winter (November through February) and spring (March through June) show that the spring chlorophyll *a* concentrations are higher than those during summer or winter (Fig. 12A). This has been previously demonstrated with southern-reach particulate organic carbon data (Schemel and Dedini 1979b).

We believe that production occurs *in situ* because the mixing characteristics of this reach (Conomos 1979) prevent advection from great distances. High production and the abundance of phytoplankton and zooplankton associated with the leading edges of the low-salinity Delta-derived water masses (Cloern 1979) are not usually found this far south. Our limited knowledge of the phytoplankton (Cloern 1979) and benthic diatom (Nichols 1979) distributions, suggests that the algae in the channels near the southern boundary grow *in situ* or grow in the river channels, sloughs, shoal areas, and adjacent marshes during the spring months and then are flushed into the channels. These seemingly anomalous higher winter concentrations of living organic particles suggest that photosynthetic activity, higher production of oxygen, and higher utilization of plant nutrients occur during winter than summer in the southern reach. This is opposite to the seasonal trend observed in Suisun Bay.

The higher production noted during the latter part of the winter season rather than during

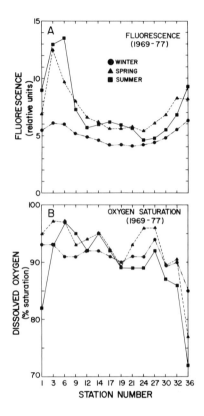

Fig. 12. Longitudinal distributions of *in vivo* chlorophyll *a* fluorescence (A) and percent oxygen saturation (B) at ≤2 m during winter (November through February), spring (March through June) and summer (July through October) at near-monthly intervals (1969 through 1977) at hydrographic stations (Fig. 1).

summer may reflect the relative availability of light and the stability of the water column. Because nutrients are abundant throughout the Bay (particularly at the southern boundary of the southern reach; Figs. 14, 15, 16), light probably limits primary production (Peterson 1979; Cloern 1979; Arthur and Ball 1979; Ball and Arthur 1979). In areas of constant water depth, three major factors determine light levels available to algae: amount of daily insolation, relative turbidity of the water, and relative stability of the water column. During the months we chose as winter, the insolation is 20% less than that of summer (Fig. 3) and the waters are, on the average, more turbid (Figs. 7A, 8). Considering these factors alone, primary production should be greater during summer. However, the increased freshwater inflows during winter cause salinity-induced vertical density stratification as well as contribute to turbidity. This stratification, which tends to maintain the algal cells in the photic zone and thus expose them to more light during the latter part of winter than during summer, is maintained during long periods (weeks) of relative quiescence because persistent wind-generated waves of summer are absent (Conomos 1979, Fig. 8). The high winter (March-April) production in the northern part of the southern reach, and probably the southern part as well, is a result (Cloern 1979).

Dissolved Oxygen

The distribution of oxygen differs from that of conservative properties such as salinity and temperature in that it is biologically reactive: it is closely associated with changes in carbon and plant-nutrient concentrations. Dissolved oxygen is influenced by a variety of important processes: (1) exchange of oxygen across the water surface through atmospheric invasion (gain) and outgassing (loss); (2) photosynthesis; (3) respiration by plants and animals, decomposition of organic matter by bacteria and chemical oxidation; and (4) advection and diffusion.

Dissolved oxygen concentrations in the near-surface waters show marked spatial and seasonal patterns. The winter oxygen concentrations throughout the Bay are notably higher than those during summer (Fig. 13A), promoted in part by the greater solubility of oxygen in colder water (Fig. 5A). The temperature control of oxygen solubility is also evident at the Golden Gate (Station 19) during summer, where higher concentrations are present in the colder ocean water. Oxygen concentrations generally decrease from north (Delta) to south (Station 36) throughout the year.

Peterson (1979) indicates that the atmosphere, Delta outflow and oxygen production by photosynthesis are oxygen sources, whereas the ocean and respiration (both in the water column and on the substrate) are oxygen sinks. Of these oxygen sources, atmospheric exchange rates are much greater than the mixing and advective inputs from Delta outflow. Although oxygen invasion from the atmosphere is significant, it is more than offset by oxygen losses to the substrate (benthic respiration). Production of oxygen by photosynthesis in the photic zone of the water column outweighs respiration below the photic zone. Oxygen loss by the oxidation of sewage waste is apparently minor, and the gain of oxygen by Delta outflows is in approximate balance with the loss to the ocean. The northern reach, then, is an oxygen sink to Delta outflow and atmospheric invasion, but is a source of oxygen to the adjacent ocean and to the southern reach.

Although no similar budget has been made for the southern reach, this reach also appears to be an oxygen sink. The rates of the processes utilizing oxygen (i.e. uptake by the substrate, bacterial decomposition of the copious quantities of organic wastes, respiration of plants) seem greater

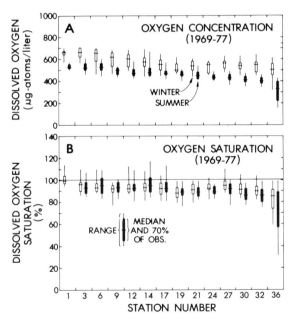

Fig. 13. Longitudinal distributions of dissolved oxygen concentration (A) and percent oxygen saturation (B) at ≤2 m during winter (December through April) and summer (July through October) at near-monthly intervals (1969 through 1977) at hydrographic stations (Fig. 1).

than those contributing oxygen (i.e. photosynthesis, atmospheric invasion, intrusion of oxygenated Delta outflow and ocean water).

Seasonal differences in the rates of the processes determining dissolved oxygen concentrations are not well understood. Oxygen saturation is greater during summer than winter in the northern reach, whereas in the southern reach, the converse is true (Fig. 13B). The high photosynthetic rate during summer in the northern reach apparently causes the higher oxygen saturation values since advection of Delta-derived oxygen is less than in winter. Oxygen utilization by benthic respiration of organic-matter decomposition is probably higher because of the higher water temperatures.[3]

Even less is known about the relative rates of oxygen production and utilization in the southern reach. The higher winter saturation values are apparently caused by the high photosynthetic activity during the latter part of the winter season (Fig. 12B) and the lower demand for the oxidation (decomposition) of sewage because of increased flushing and dilution and colder ambient water temperatures (Fig. 5A).

Plant Nutrients

The plant nutrients, like dissolved oxygen, are nonconservative and vary considerably with both space and time. These substances, in their simpler dissolved inorganic forms, are required for the nutrition of primary producers such as phytoplankton, benthic algae, and vascular marsh plants. Because these nutrients are taken up (during photosynthesis) and released (regenerated by mineralization of organic matter) in relatively predictable ratios (Redfield et al. 1963) it is necessary to consider both their spatial and seasonal distributions as well as their relation to one another.

Nutrient Sources and Sinks. River inflow, the ocean, sewage inputs, and marsh drainage are the primary sources of nutrients in the Bay. Inputs from the ocean and sewage effluents are relatively constant throughout the year, whereas the inputs from the rivers and marshes show considerable seasonal variability. A secondary source is mineralization in the water column and on the substrate. Sinks include the ocean, the substrate, and perhaps the marshes.

Temporal differences in nutrient concentrations, free of the effects of large tidal variations, are best represented geographically when data are correlated with salinity rather than distance (Fig. 14). The Bay-wide covariance of the nutrient concentration with salinity during any survey period forms a continuous array of data points that form a "v"-shaped distribution curve. Such a distribution suggests a three-component system with sources of markedly differing nutrient concentrations and relative ratios. These major sources are river water (Delta outflow), ocean water (Golden Gate exchange), and local stream inflow and sewage effluent (ambient waters of the southern boundary of the southern reach). Although these vary seasonally in response to changes in supply and removal rates, they remain in a similar pattern from year to year. On the basis of Bay-wide distributions, the southern boundary of the southern reach is believed to be the major source of phosphate, ammonia and nitrate+nitrite to the southern reach whereas Delta outflow is the major source of silicate to the northern reach. "Ocean water" contains moderate concentrations of phosphate, ammonia and nitrate+nitrite. Because of mixing with Bay water these values are somewhat higher than "true" ambient coastal-ocean water (Conomos and Peterson 1975), but they do generally reflect the seasonal fluctuations of ocean nutrient supply controlled by coastal upwelling (Fig. 6).

In the northern reach, nutrient supply is dominated by Delta outflow and exchange with the ocean (Conomos and Peterson 1975; Peterson 1979). The winter input of all nutrients from the

---

[3] We tacitly assume that the atmospheric invasion rate of oxygen is relatively constant between seasons.

Delta is at least ten times greater than during summer. The relatively constant ocean and sewage inputs are usually insignificant compared to Delta inputs. Supply from the mudflats and marshes is possibly important but is as yet not measured. Significant amounts of ammonia and phosphate are contributed by sewage effluents, whereas the ocean and Delta outflow are the major sources of nitrate+nitrite and silicate. The nutrients in the southern reach are primarily contributed by sewage effluents although exchange from the substrate can also be important: during summer, more silicate can be contributed from the substrate (D. Hammond pers. comm.) than from sewage inflow.

Despite the large man-derived additions of nutrients to the Bay, the fact that the Bay is

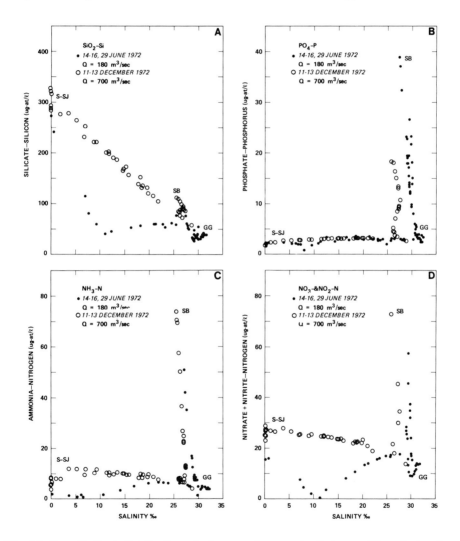

Fig. 14. Longitudinal distribution of plant nutrients and salinity at ≤2 m during June and December 1972 in the San Francisco Bay. Geographic end members are S-SJ—Sacramento-San Joaquin (Station 1); GG—Golden Gate (Station 19); SB—South Bay (Station 36). Mean monthly Delta outflow (Q) data are calculated by the U. S. Bureau of Reclamation. From Conomos and Peterson 1975.

naturally nutrient-rich precludes detection of changes in the biota resulting from these additions.

Processes Controlling Nutrient Distributions. The nutrient concentrations of the water sources have been used, together with estimates of the relative fluxes to estimate the major sources of the nutrients in the northern reach. The relative linearity of the nutrient-salinity relations (Fig. 14) has allowed us to estimate the amount, location, and timing of apparent net nutrient utilization or regeneration in the northern reach (Conomos and Peterson 1975; Peterson et al. 1975a, 1978).

Because near-surface salinity covaries in a general way with distance in the northern reach (see Peterson et al. 1975a, Fig. 9), this relation can be extended to summaries of nutrient concentrations along a longitudinal section (Figs. 15, 16). The winter silicate distribution in the northern reach is relatively linear (Fig. 15A), suggesting that the contrast between utilization and mineralization is minor in comparison to supply by Delta outflow. During summer, however, utilization by

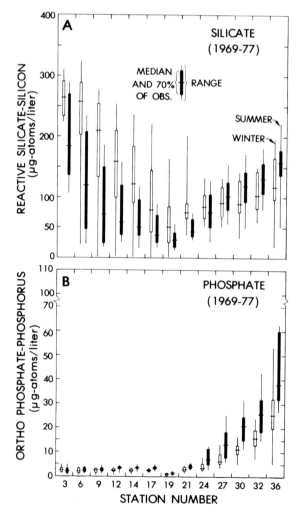

Fig. 15. Longitudinal distribution of reactive silicate (A) and orthophosphate (B) at ≤2 m during winter (December through April) and summer (July through October) at near-monthly intervals (1969 through 1977) at hydrographic stations (Fig. 1).

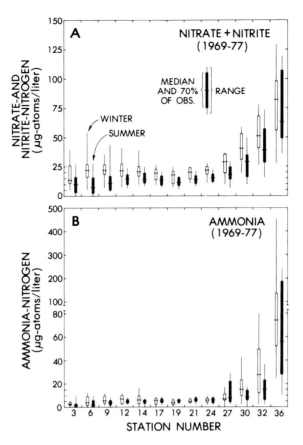

Fig. 16. Longitudinal distribution of nitrate+nitrite (A) and ammonia (B) at ≤2 m during winter (December through April) and summer (July through October) at near-monthly intervals (through 1977) at hydrographic stations (Fig. 1).

algae may exceed Delta input and median values show a marked departure from linearity. The maximum apparent utilization occurs in the null zone (see also Conomos and Peterson 1974, 1975; Peterson et al. 1975a).

Relative seasonal concentration differences are reversed in the southern reach, however, as summer silicate concentrations exceed those of winter (Fig. 15A). The seasonal variations in silicate are controlled by stream inputs, dilution, flushing and nutrient uptake by phytoplankton.[4] Sewage and the substrate inputs are relatively constant throughout the year. In the absence of silicate removal by phytoplankton, the stream inputs should be greater in winter than summer.

We suggest that the ambient silicate concentrations are lower during winter because the rate of silicate removal, caused by phytoplankton utilization during the latter part of winter (Fig. 17A) is apparently greater than during summer. This interpretation is consistent with the chlorophyll *a* (Figs. 7B, 12A) and oxygen-saturation (Figs. 12B, 13B) distributions discussed above. Furthermore, it appears that silicate diffuses upward from the substrate at a rate exceeding

---

[4] Our interpretation of the relative importance of processes occurring here is speculative because there are no nutrient data from most important southern-reach streams.

the rate of input from local streams and sewage inputs. The concentrations are lower during winter because the water is flushed more rapidly.

Seasonal differences in phosphorus, nitrate+nitrite and ammonia concentrations in the northern reach are not large (Figs. 15B, 16A, B). The processes and rates which control supply and removal, like silicate, are seasonally modulated by Delta outflow. Summer utilization rates are greater in comparison to supply rates at the null zone because of high phytoplankton production (Peterson et al. 1975b; Peterson 1979).

During winter, rates of phytoplankton production decrease and shift the utilization-regeneration balance toward regeneration. The biogenous fraction of the suspended matter (viable

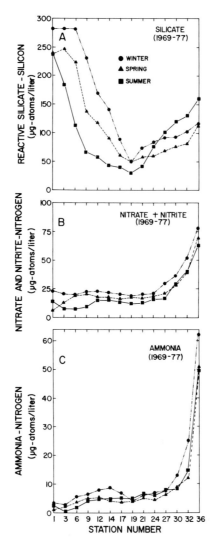

Fig. 17. Longitudinal distribution of reactive silicate (A), nitrate+nitrite (B), and ammonia (C) at ≤2 m during winter (November through February), spring (March through June) and summer (July through October) at near-monthly intervals (1969 through 1977) at hydrographic stations (Fig. 1).

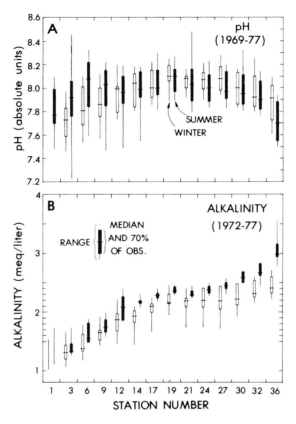

Fig. 18. Longitudinal distribution of hydrogen-ion activity—pH (A) and alkalinity (B) at ≤2 m during winter (December through April) and summer (July through October) at hydrographic stations (Fig. 1). pH was measured from 1969 through 1977; alkalinity was measured from 1972 through 1977.

plankton as well as detritus) comprising the turbidity maximum and accumulating in the null zone is a reservoir of particulate nutrients. Organic nitrogenous compounds are converted into ammonia and thence to nitrite, and nitrate, and would create the winter maximum (Fig. 16). Organic forms of phosphate are similarly transformed into inorganic phosphate. These winter regeneration rates, apparently in excess of utilization rates, may be the cause of the mid-estuary nitrate+nitrite and ammonia maxima often found during winter.

In the southern reach, sewage effluents contribute virtually all of the phosphate (Fig. 15B) and much of the nitrogen (Fig. 16) at a relatively constant rate throughout the year. The local streams probably contribute high concentrations of nitrate+nitrite but little phosphate during winter (V. Kennedy pers. comm.). These large inputs at the southern boundary create a pronounced gradient northward. Their concentrations, at least an order of magnitude greater than those found in the northern reach, show distinct seasonal patterns. Like silicate, summer phosphorus concentrations are greater than winter, whereas winter nitrate+nitrite and ammonia concentrations are greater than those measured during summer.

Algal utilization does not demonstrably depress phosphorus concentrations. In contrast, although inputs of nitrate+nitrite and ammonia from the southern boundary of southern reach are high (Figs. 16, 17B, C) utilization by phytoplankton lowers these concentrations noticeably. Such production can be expected to remove about 16 times more nitrate (by atoms) than phosphate. Winter concentrations of these nutrients are higher than those of summer perhaps because the supply of additional nitrate+nitrite and ammonia by local stream inflow is greater than the removal by algae. Ambient phosphorus is physically removed during winter by the flushing action of the Delta-derived low-salinity, low-phosphate water (McCulloch et al. 1970; McCulloch 1972) and by dilution and limited flushing by local streams. Seasonal differences in vertical fluxes of phosphate and nitrogen between the substrate and water column have not been defined.

Dissolved Inorganic Carbon

Like oxygen, carbon dioxide (CO_2) plays a fundamental role in the metabolism of organisms. Photosynthesis, which combines CO_2 with other substances to form algal biomass and oxygen, and mineralization and respiration, which release CO_2 to the water, are basic biological processes which affect the chemistry of aqueous systems. Estuarine waters contain dissolved inorganic carbon as CO_2, carbonic acid, and bicarbonate and carbonate ions. These ion forms contribute the largest fraction of the alkalinity (see Spiker and Schemel 1979). A net effect of aquatic primary production is removal of CO_2 from the water. Since dissolved CO_2 is in equilibrium with bicarbonate, carbonate, and hydrogen ions, removal of CO_2 also decreases the hydrogen-ion concentration, thus increasing pH. It follows then that observed distributions of pH (and alkalinity to a lesser extent) in estuaries can show the net effects of biologic processes (photosynthesis, mineralization, and respiration) and may demonstrate sources and sinks of inorganic carbon.

In the northern reach, mixing of river and ocean waters determines major features of the pH and alkalinity distributions (Fig. 18A, B). Alkalinity in the northern reach typically increases near-linearly with salinity from river values of about half that of ocean water. The lower alkalinities during winter probably reflect generally lower salinities (Fig. 5) and consequently the smaller fraction of higher-alkalinity ocean water. However, summer values apparently can increase due to higher river alkalinity (Spiker and Schemel 1979). In contrast to the northern reach, the alkalinity of the southern reach typically increases with decreasing salinity in a southward direction reflecting the influence of waste inputs. During winter, ambient alkalinities are lowered because of the intrusion of lower-alkalinity Delta-derived water from the northern reach. During summer, these intrusions diminish and waste inputs increase alkalinity. The substrate is probably an additional alkalinity source.

Observed distributions of pH in the Bay can be complex because temperature, salinity, alkalinity, and processes which supply or remove CO_2 all significantly affect pH. In the northern reach, mixing of river and ocean waters alone would result in a non-linear pH-salinity distribution (Mook and Koene 1975). This assumes a typical northern-reach alkalinity distribution (Fig. 18B) and that Delta outflow, which is typically over-saturated with CO_2 by a factor of two or three, is mixing with ocean water which is in near-equilibrium with the atmosphere (Spiker and Schemel 1979). Lateral variations in water temperature would further complicate the distribution because CO_2 solubility is temperature dependent.

The pH distributions (Fig. 18A) are consistent with our understanding of processes affecting dissolved CO_2 in the Bay. In general, pH in the northern reach appears primarily to be the result of mixing river and ocean waters during winter. Since pH is temperature dependent, a winter to summer temperature increase of 10°C would result in a decrease in pH of about 0.1 unit, if all other factors (salinity, alkalinity, etc.) were constant. However, median values during summer are higher than those in winter near the null zone. In that region high values together with high chlorophyll *a* fluorescence (Fig. 7B), suggest that the high photosynthetic activity found here (Cloern

1979; Peterson 1979) is perhaps most significant in affecting the pH distribution during summer.

Considering the winter to summer water-temperature increase (Fig. 5B), median pH values in the southern reach do not indicate major differences in the factors affecting the distributions. Waste-water is effective in lowering pH to the south[5]. The lower median pH value at Station 36 during summer is possibly caused by the greater effectiveness of waste inputs and higher rates of mineralization at the warmer summer water temperatures. The range of pH is greater during summer, suggesting that photosynthetic activity, respiration, and mineralization may each dominate at particular times. However, tidal effects and the temperature range may account for much of the summer pH range.

RESEARCH NEEDS

Development of a long-term (long period of record) data base is necessary to distinguish yearly variations and trends from important short-term events such as storms and freshets. Further, the data presented herein need to be extended into three dimensions by mapping the vast shallow areas of the Bay. The major external sources of dissolved and particulate substances (from sewage outfalls, rivers, and "nonpoint sources" such as street runoff, marshes and sloughs) need to be monitored quantitatively. Such mapping and monitoring tasks, while not simple, are relatively straightforward. Equally important are the exchange rates of these parameters at the boundaries: the atmosphere and substrate (Hammond and Fuller 1979) and ocean (Conomos 1979) need to be measured. And, of paramount importance are measurements of the rates of biological processes within the water column and on the substrate, such as photosynthesis by benthic and planktonic algae and regeneration of particulate organic matter by bacteria and zooplankton (Cloern 1979; Nichols 1979; Peterson 1979). These measurements are considerably more difficult to make.

Finally, we must better understand hydrodynamic processes such as transport and mixing of the water masses because these processes largely determine the biological processes in the Bay. Until such information is available, we must depend on the measured seasonal features of the river hydrograph for making a preliminary separation of the physical processes involved.

ACKNOWLEDGMENTS

We thank J. E. Cloern, D. E. Hammond, S. N. Luoma and F. H. Nichols for reading the manuscript and for making helpful suggestions that improved our discussion. We are indebted to many other persons as well, particularly A. E. Alpine, B. E. Cole, L. A. Dedini, R. E. Herndon, D. D. Harmon, Anne Hutchinson, S. M. Wienke, and R. L. J. Wong, for able and enthusiastic laboratory and field assistance and for valuable comments offered during the course of our continuing studies.

LITERATURE CITED

Atlas, E. L., S. W. Hager, L. I. Gordon, and P. K. Park. 1971. A practical manual for use of the Technicon AutoAnalyzer in seawater nutrient analyses, revised. Department of Oceanography, Oregon State University, Corvallis, Ore. Ref. 71-22.

---

[5] Many processes related to waste-waters can contribute to the observed distribution and levels of pH. Waste water itself may be greatly over-saturated with CO_2, and mineralization of the high organic-carbon content of sewage effluent produces CO_2 in the southern reach. The high total CO_2 content of waste water (see Spiker and Schemel 1979) and our data (Fig. 18B) suggest that alkalinity is high in the effluent itself, a factor which would affect the pH distribution.

SAN FRANCISCO BAY

Arthur, J. F., and M. D. Ball. 1979. Factors influencing the entrapment of suspended material in the San Francisco Bay-Delta estuary. Pages 143-174 *in* T. J. Conomos, ed. San Francisco Bay: The Urbanized Estuary. Pacific Division, Amer. Assoc. Advance. Sci., San Francisco, Calif.

Ball, M. D., and J. F. Arthur. 1979. Planktonic chlorophyll dynamics in the northern San Francisco Bay and Delta. Pages 265-285 *in* T. J. Conomos, ed. San Francisco Bay: The Urbanized Estuary. Pacific Division, Amer. Assoc. Advance. Sci., San Francisco, Calif.

Carlson, P. R., and D. S. McCulloch. 1974. Aerial observations of suspended sediment plumes in San Francisco Bay and the adjacent Pacific Ocean. J. Res. U. S. Geol. Surv. 2(5):519-526.

Carpenter, J. H. 1965. The Chesapeake Bay Institute technique for the Winkler dissolved oxygen method. Limnol. Oceanogr. 10:141-143.

Churgin, J., and S. J. Halminski. 1974. Key to oceanographic records documentation No. 2. Temperature, salinity, oxygen, and phosphate in waters off the United States. Vol. III. Eastern North Pacific. National Oceanic and Atmospheric Administration, U. S. Department of Commerce, Washington, D. C. 259 pp.

Cloern, J. E. 1979. Phytoplankton ecology of the San Francisco Bay system: The status of our current understanding. Pages 247-264 *in* T. J. Conomos, ed. San Francisco Bay: The Urbanized Estuary. Pacific Division, Amer. Assoc. Advance. Sci., San Francisco, Calif.

Conomos, T. J. 1979. Properties and circulation of San Francisco Bay waters. Pages 47-84 *in* T. J. Conomos, ed. San Francisco Bay: The Urbanized Estuary. Pacific Division, Amer. Assoc. Advance. Sci., San Francisco, Calif.

Conomos, T. J., and M. G. Gross. 1972. River-ocean suspended particulate matter relations in summer. Pages 176-202 *in* A. T. Pruter and D. L. Alverson, eds. Columbia River Estuary and Adjacent Ocean Waters: Bioenvironmental Studies. Chap. 8. University of Washington Press, Seattle, Wash.

Conomos, T. J., and D. H. Peterson. 1974. Biological and chemical aspects of the San Francisco Bay turbidity maximum. Mem. Inst. du Bassin d'Aquitaine 7:45-52.

Conomos, T. J., and D. H. Peterson. 1975. Longitudinal distribution of selected micronutrients in northern San Francisco Bay during 1972. Pages 103-126 *in* R. L. Brown, ed. Proceedings of a Workshop on Algal Nutrient Relationships in the San Francisco Bay and Delta (8-10 November 1973, Clear Lake, California). San Francisco Bay and Estuarine Assoc.

Conomos, T. J., and D. H. Peterson. 1977. Suspended particle transport and circulation in San Francisco Bay: An overview. Pages 82-97 *in* M. Wiley, ed. Estuarine Processes. Vol. 2. Academic Press, New York.

Conomos, T. J., F. H. Nichols, R. T. Cheng, and D. H. Peterson. 1978. Field and modeling studies of San Francisco Bay. Pages 1917-1927 *in* Vol. 3, Coastal Zone '78. Symp. on Technical, Environmental, Socioeconomic and Regulatory Aspects of Coastal Zone Management, March 14-16, San Francisco, Calif.

Culberson, C., R. M. Pytkowicz, and J. E. Hawley. 1969. Seawater alkalinity determination by the pH method. J. Mar. Res. 28:15-21.

Festa, J. F., and D. V. Hansen. 1978. Turbidity maxima in partially mixed estuaries: A two-dimensional numerical model. Estuarine Coastal Mar. Sci. 7:347-359.

Hammond, D. E., and C. Fuller. 1979. The use of Radon-222 to estimate benthic exchange and atmospheric exchange rates in the San Francisco Bay. Pages 213-230 *in* T. J. Conomos, ed. San Francisco Bay: The Urbanized Estuary. Pacific Division, Amer. Assoc. Advance. Sci., San Francisco, Calif.

Harris, H. S., D. L. Feuerstein, and E. A. Pearson. A pilot study of physical, chemical and biological characteristics of waters and sediments of south San Francisco Bay (south of Dumbarton Bridge). SERL Report. University of California, Berkeley, Calif.

Kahn, L., and F. T. Brezenski. 1967. Determination of nitrate in estuarine waters. Environmental Science Technology 6:488-491.

Knebel, H. J., T. J. Conomos and J. A. Commeau. 1977. Clay-mineral variability in the suspended sediments of the San Francisco Bay system, California. J. Sed. Petrol. 47:229-236.

Krone, R. B. 1979. Sedimentation in the San Francisco Bay system. Pages 85-96 *in* T. J. Conomos, ed. San Francisco Bay: The Urbanized Estuary. Pacific Division, Amer. Assoc. Advance. Sci., San Francisco, Calif.

Lorenzen, C. J. 1966. A method for the continuous measurement of *in vivo* chlorophyll concentration. Deep-Sea Res. 13:223-227.

McCarty, J. C., R. A. Wagner, M. Macomber, et al. 1962. An investigation of water and sediment quality and pollutional characteristics of three areas in San Francisco Bay 1960-1961. Sanitary Engineering Research Laboratory, University of California, Berkeley, Calif. 571 pp.

McCulloch, D. S. 1972. Seasonal flushing of South San Francisco Bay: 1969-1972. Pages 39-46 *in* V. A. Frizzell, ed. Progress Report on U. S. Geological Survey Quaternary Studies in the San Francisco Bay Area. Guidebook for Friends of the Pleistocene.

McCulloch, D. S., D. H. Peterson, P. R. Carlson, and T. J. Conomos. 1970. Some effects of freshwater inflow on the flushing of South San Francisco Bay: A preliminary report. U. S. Geol. Surv. Circ. 637A. 27 pp.

Miller, R. C., W. D. Ramage, and E. L. Lazier. 1928. A study of physical and chemical conditions in San Francisco Bay especially in relation to the tides. Univ. Calif. Publ. Zool. 31:201-267.

Mook, W. G., and B. K. S. Koene. 1975. Chemistry of dissolved inorganic carbon in estuarine and coastal brackish waters. Estuarine Coastal Mar. Sci. 3:325-336.

Nichols, F. H. 1979. Natural and anthropogenic influences on benthic community structure in San Francisco Bay. Pages 409-426 *in* T. J. Conomos, ed. San Francisco Bay: The Urbanized Estuary. Pacific Division, Amer. Assoc. Advance. Sci., San Francisco, Calif.

Pearson, E. A., P. N. Storrs, and R. E. Selleck. 1967. Some physical parameters and their significance in marine waste disposal. Pages 297-315 *in* T. A. Olson and F. J. Burgess, eds. Pollution and Marine Ecology. Interscience-Wiley & Sons, New York.

Pearson, E. A., P. N. Storrs, and R. E. Selleck. 1969. A comprehensive study of San Francisco Bay. Final report. Vol. III. Waste discharges and loadings. SERL Rep. 67-3. 98 pp.

Peterson, D. H. 1979. Sources and sinks of biologically reactive oxygen, carbon, nitrogen, and silica in northern San Francisco Bay. Pages 175-193 *in* T. J. Conomos, ed. San Francisco Bay: The Urbanized Estuary. Pacific Division, Amer. Assoc. Advance. Sci., San Francisco, Calif.

Peterson, D. H., T. J. Conomos, W. W. Broenkow, and P. C. Doherty. 1975a. Location of the nontidal current null zone in northern San Francisco Bay. Estuarine Coastal Mar. Sci. 3:1-11.

Peterson, D. H., T. J. Conomos, W. W. Broenkow, and E. P. Scrivani. 1975b. Processes controlling the dissolved silica distribution in San Francisco Bay. Pages 153-187 *in* L. E. Cronin, ed. Estuarine Research. Vol. 1. Chemistry and Biology. Academic Press, New York.

Peterson, D. H., J. F. Festa, and T. J. Conomos. 1978. Numerical simulation of dissolved silica in the San Francisco Bay. Estuarine Coastal Mar. Sci. 7:99-116.

Porterfield, G., N. L. Hawley, and C. A. Dunnam. 1961. Fluvial sediments transported by streams tributary to the San Francisco Bay area. U. S. Geol. Surv. Open-File Rep. 70 pp.

Redfield, A. C., B. H. Ketchum, and F. A. Richards. 1963. The influence of organisms on the composition of sea water. Pages 26-77 *in* M. N. Hill, ed. The Sea. Vol. 2. Interscience, New York.

Regional Water Quality Control Board. 1975. Water quality control plan, San Francisco Bay Basin (2). Part 1. State Water Resources Control Board, San Francisco Bay region (2). 135 pp.

Schemel, L. E., and L. A. Dedini. 1979a. A continuous water-sampling and multiparameter-measurement system for estuaries. U. S. Geol. Surv. Open-File Rep. 79-273. 92 pp.

Schemel, L. E., and L. A. Dedini. 1979b. Particulate organic carbon in San Francisco Bay, California 1971-1977. U. S. Geol. Surv. Open-File Rep. 79-512. 30 pp.

Selleck, R. E., B. Glenne, and E. A. Pearson. 1966a. A comprehensive study of San Francisco Bay. Final report. Vol. VII. A model of mixing and dispersion in San Francisco Bay. SERL Rep. 67-1. 111 pp.

Selleck, R. E., E. A. Pearson, B. Glenne, and P. N. Storrs. 1966b. A comprehensive study of San Francisco Bay. Final report. Vol. IV. Physical and hydrological characteristics of San Francisco

Bay. SERL Rep. 65-10. 99 pp.

Smith, B. J. 1965. Sedimentation in the San Francisco Bay system. Pages 675-708 *in* Proc. Federal Interagency Sedimentation Conf. 1963. U. S. Dep. Agr. Misc. Pub. 970. 933 pp.

Smith, R. E., R. E. Herndon, and D. D. Harmon. 1979. Physical and chemical properties of San Francisco Bay waters, 1969-1976. U. S. Geol. Surv. Open-File Rep. 79-511. 630 pp.

Solorzano, L. 1969. Determination of ammonia in natural waters by the phenolhypochlorite method. Limnol. Oceanogr. 14:799-801.

Spiker, E. C., and L. E. Schemel. 1979. Distribution and stable-isotope composition of carbon in San Francisco Bay. Pages 195-212 *in* T. J. Conomos, ed. San Francisco Bay: The Urbanized Estuary. Pacific Division, Amer. Assoc. Advance. Sci., San Francisco, Calif.

Storrs, P. N., E. A. Pearson, and R. E. Selleck. 1966. A comprehensive study of San Francisco Bay. Final report. Vol. V. Summary of physical, chemical and biological water and sediment data. SERL Rep. 67-2. 140 pp.

Storrs, P. N., E. A. Pearson, H. F. Ludwig, R. Walsh, and E. J. Stann. 1968. Estuarine water quality and biological population indices. J. International Assoc. Water Pollution Res. 2(1):128-129.

Storrs, P. N., E. A. Pearson, and R. E. Selleck. 1969. A comprehensive study of San Francisco Bay. Final report. Vol. VI. Water and sediment quality and waste discharge relationships. SERL Rep. 67-4. 80 pp.

Sumner, F. B., G. D. Louderback, W. L. Schmitt, and E. C. Johnston. 1914. A report upon the physical conditions in San Francisco Bay, based upon the operations of the United States Fisheries steamer *Albatross* during the years 1912 and 1913. Univ. Calif. Publ. Zool. 14(1):1-198.

Strickland, J. D. H., and T. R. Parsons. 1972. A practical handbook of seawater analysis. Can. Fish. Res. Bd. Bull. 167, 2nd ed., pp. 21-28.

Technicon Corporation. 1973. Nitrate and nitrite in water and waste-water. Technicon AutoAnalyzer. Industrial Method No. 100-70W. Released Sept. 1973.

Technicon Corporation. 1976. Silicates in waste and wastewater. Technicon AutoAnalyzer II. Industrial Method. No. 105-71WB. Released Feb. 1973/revised Jan. 1976.

FACTORS INFLUENCING THE ENTRAPMENT OF SUSPENDED MATERIAL IN THE SAN FRANCISCO BAY-DELTA ESTUARY

JAMES F. ARTHUR AND MELVIN D. BALL
U. S. Bureau of Reclamation, 2800 Cottage Way, Sacramento, CA 95825

Inorganic suspended particulate matter, turbidity, particulate nutrients, phytoplankton, *Neomysis mercedis* (Holmes), certain other zooplankton, and juvenile striped bass (young-of-the-year) accumulate in an entrapment zone located in the upper San Francisco Bay-Delta estuary (Sacramento-San Joaquin River System). The location of this entrapment zone is regulated by the magnitude and the pattern of river inflow, as well as the tidal excursion. At Delta outflow indices of 20 $m^3 \cdot s^{-1}$, the zone was located 40-45 km upstream of its location at 1,800 $m^3 \cdot s^{-1}$; tidal movement of the zone is from 3 to 10 km, depending on tidal phase and height. The location of the zone is related to, and can be approximated from, specific conductance values of 2 to 10 millimho·cm^{-1} (1-6 °/oo salinity). The concentration of constituents in the zone varied directly with Delta outflow, water depth, and tidal velocity. Depending on the constituent and environmental conditions at the time of measurement, the suspended-material concentration varied from as little as twice to as much as several hundred times the upstream or downstream concentration. The most significant environmental aspect of the entrapment zone may be that the quantity of phytoplankton and certain other estuarine biota appear to be enhanced when the zone is located in upper Suisun Bay.

Bureau of Reclamation (USBR) and the California Department of Water Resources (DWR) plans call for large pumped diversions from the southern portion of the Sacramento-San Joaquin Delta and possible construction of a drain (for removal of saline subsurface agricultural water from the San Joaquin Valley) which may discharge in the general vicinity of Suisun Bay.

The USBR is cooperating with the U. S. Fish and Wildlife Service (USFWS), the California Department of Fish and Game (DFG), and DWR in conducting environmental studies ("Interagency Ecological Study Program") to evaluate the potential impact of these projects on the estuary. This chapter describes one aspect of this program: the determination of how changes in Delta outflow and flow patterns, attributable to the operation of the federal and state water projects, might influence the distribution and abundance of estuarine phytoplankton and other particulate material (Ball 1977; Arthur and Ball 1978). Among the factors evaluated thus far, the entrapment zone appears to be a major feature regulating the phytoplankton standing crop in Suisun Bay (Arthur 1975; Arthur and Ball 1978).

BACKGROUND

Phytoplankton are important to the estuarine environment as primary producers, with certain species forming the base of the food web. However, in many aquatic environments, excessive concentrations of phytoplankton cause eutrophication (i.e. reductions in dissolved oxygen concentrations to a point detrimental to higher aquatic organisms), create taste and odor problems in municipal water supplies, clog filters in water treatment plants and/or are aesthetically undesirable for recreationists. However, phytoplankton problems presently appear minor and the maximum desirable concentration and species composition of phytoplankton has yet to be determined

(Arthur and Ball 1978) in the study area (Fig. 1).

The quantity of freshwater flowing through the estuary is important to phytoplankton growth because it regulates nutrient concentration, determines riverborne sediment inflow and influences suspended-particle transport which in turn affects light-penetration (required for algal growth), determines phytoplankton residence time, and directly regulates salinity intrusion and

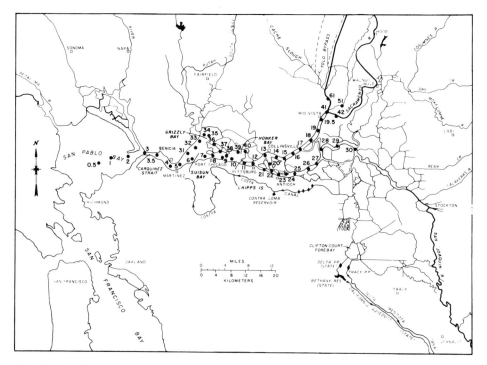

Fig. 1. Sampling sites of entrapment zone study.

the location of the entrapment zone. These and other factors interact to determine the amount and type of phytoplankton in the estuary (Arthur and Ball 1978). Ball (1977) and Ball and Arthur (1979) have evaluated factors influencing the temporal and spatial distribution and abundance of phytoplankton throughout the San Francisco Bay-Delta estuary.

Krone (1966, 1979), among others, has speculated that under projected low flow conditions resulting from water development projects the sediment input to the estuary would be greatly reduced, resulting in a greater photic depth in Suisun Bay. This increase in photic depth could potentially increase the phytoplankton standing crop to undesirable concentrations.

In evaluating the probable effects of subsurface agricultural drain discharge to the estuary, Bain (1968) concluded that this discharge would about double the concentration of nitrogen in Suisun Bay which could result in severe algal blooms accompanied by depressions in dissolved oxygen as the blooms decline. As a result, methods were studied for removing nitrogen from drainage water (Brown 1975) and studies were conducted on the potential impact of drain water on the Delta environment (USBR 1972).

In reviewing water quality data and factors controlling phytoplankton growth during the 1968-74 period (Arthur 1975; Ball 1975, 1977), long-term averages of phytoplankton, chlorophyll, particulate organic nitrogen and particulate phosphate, turbidity, and inorganic suspended solids were found to be at higher concentrations in Suisun Bay than in the adjacent upstream or downstream areas. Since phytoplankton concentrations were highest in Suisun Bay, while light penetration was lowest and water temperatures and nutrient concentrations were generally favorable, some other mechanism(s) apparently was responsible for the high phytoplankton concentrations.

Further evaluation of other historical water quality data and review of suspended-materials distribution studies for the area (for example, Simmons 1955; Einstein and Krone 1961; Meade 1972; Peterson and Charnell 1969; Conomos and Peterson 1974, 1977; Peterson et al. 1975a,b) and for other estuaries (Wiley 1977) has led us (Arthur 1975; Ball 1977; Arthur and Ball 1978) to conclude that suspended materials are entrapped and accumulate in the estuary at the upstream end of the fresh-water—salt-water mixing zone. We theorize that the causes of this entrapment are the increased flocculation, aggregation, and/or settling of suspended materials at specific conductances above 1 millimho/cm (0.6 °/oo salinity) and the effects of net two-layered estuarine circulation flow (California DFG et al. 1975, 1976). Terms used by others to describe the area of maximum concentration of suspended materials are the "turbidity maximum," "critical zone," "nutrient trap," "sediment trap" and "null zone" (Arthur and Ball, 1978). We prefer the more descriptive "entrapment zone" (Arthur 1975).

Studies in the San Francisco Bay-Delta Estuary

As early as 1931, Grimm stated there were net upstream bottom currents in the San Francisco Bay Estuary. Since then, studies (Simmons 1955; U. S. Army Corps of Engineers 1967, 1977; Smith 1966; McCulloch et al. 1970; Conomos 1975, 1979; Conomos et al. 1970, 1971; Conomos and Peterson 1974; Peterson et al. 1975a) have demonstrated that a net two-layered flow circulation pattern exists throughout much of the northern reach of the Bay system. This generalized flow is believed to be significantly modified by "trapping" and "pumping" (two forms of tidal dispersion) and wind dispersion (Fischer 1976).

The location of the entrapment zone, the effects of riverflow on its location, and seasonal changes in the abundance and composition of suspended matter in the zone have been described (Conomos and Peterson 1974; Peterson et al. 1975a, b; Arthur 1975; Ball 1977; Arthur and Ball 1978). These and other studies produced a reasonably good understanding of how two-layered flow influences sediment transport in this and other estuaries.

In contrast, very little is known about the effects of two-layered flow on the estuarine biota. Although no specific studies were conducted on the effects of two-layered flow on the plankton and benthos in the entrapment zone, an early conclusion (Kelley 1966) was that of the environmental factors studied, chlorinity (salinity) was most responsible for species distribution of zooplankton and zoobenthos. Recent evaluations (Arthur 1975; Arthur and Ball 1978; Siegfried et al. 1978; Orsi and Knutsen 1979) indicate that zooplankton entrapment occurs. Riverflow and salinity were considered the dominant factors controlling longitudinal distribution of a number of species of fish in the estuary (Turner and Kelley 1966). Furthermore, the maximum concentration of juvenile striped bass (young-of-the-year) are known to occur within specific salinity ranges (Turner and Chadwick 1972; Stevens 1979).

The summer phytoplankton and zooplankton maxima were observed in the entrapment zone (Conomos and Peterson 1974; Peterson et al. 1975a,b; Arthur 1975; Ball 1977; Arthur and Ball 1978).

ESTUARINE HYDRODYNAMICS AND SUSPENDED MATERIAL TRANSPORT

The study area is considered an estuary characterized by two-layered flow with appreciable vertical mixing during most of the year (Conomos 1979). According to Bowden (1967), estuaries having such flow and mixing are generally shallow. The tidal currents are of increasing amplitude and extend throughout the depth mixing the fresher water downwards and the more saline water upwards. Although vertical mixing occurs, there are still two layers of net flow separated by a plane of no-net-motion which is generally above mid-depth (Fig. 2). The salinity continuously increases from the water surface to the bottom with the maximum salt gradient occurring at the plane of no-net-motion.

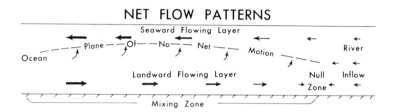

Fig. 2. Theoretical net flow patterns in a two-layered flow with vertical mixing estuary.

A wide range in water stratification exists in this type of estuary and is dependent on the ratio of the amplitude of tidal currents to the riverflow and depth. The increase in salinity from surface to bottom may vary from 1 to 10 °/oo (specific conductance of approximately 1.5 to 15 millimho·cm^{-1}). The net seaward flow of the upper layer may be several times the river inflow (Bowden 1967).

The primary driving force causing the net upstream flow in the lower layer of water is the salt-induced density difference between the surface and bottom waters (Fig. 3). Because of this density difference, freshwater entering the estuary with a greater hydraulic head tends to flow over the denser, more saline water (Simmons 1955; Schultz and Simmons 1957; Helliwell and Bossanyi 1975; Krone 1972). The greater the river inflow, the greater the hydraulic head or vertical gradient and, consequently, the greater the seaward-driving force. High river flows drive the mixing zone of freshwater and seawater farther seaward, increase salinity stratification, and compress the mixing zone (Arthur and Ball 1978; Conomos 1979). The turbulent forces of tides and winds tend to destroy vertical salinity stratifications (Nichols and Poor 1967; Conomos 1979).

The two-layered flow theoretically influences the maximum tidal-current velocities of each layer. Because there is a net downstream flow in the surface layer, surface velocities are greater

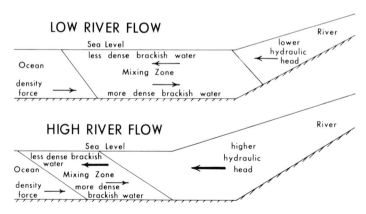

Fig. 3. The primary driving forces controlling two-layered flow circulation in the estuary.

during ebb tides than during flood tides, while in the lower layer, the reverse occurs. Higher velocities occur during flood tides than during ebb tides and increase the net upstream transport of materials along the bottom (Postma 1967).

Several factors influence the transport of suspended materials (Fig. 4). In our laboratory studies we have demonstrated that increasing the salinity of Delta water (starting at about 1.0 millimho/cm specific conductance [0.6 °/oo]) enhances flocculation of the suspended inorganic particles (primarily in the 2- to 10-μ size range) into aggregates. These aggregates settle at rates

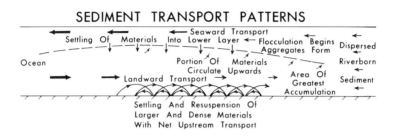

Fig. 4. Theoretical transport patterns of suspended materials in a two-layered flow with vercally mixing estuary.

greater than the unaggregated materials and are transported downstream out of the estuary or settle into the lower layer and are returned upstream where they concentrate in the entrapment zone (Simmons 1955; Krone 1966; Meade 1972; Conomos and Peterson 1974, 1977). Larger and denser materials may settle out near slack tides and then be resuspended as tidal velocity increases. The less dense and smallest suspended materials tend to be carried into the upper layer as a result of the net upward vertical flow and are transported seaward. A portion of the suspended material is transported laterally into shallow areas and may be deposited in shoals. Some of the sedimented materials may be resuspended by tidal or wind action and transported back to the channel. Suspended materials in the lower layer may be transported upstream to the entrapment zone where the areas of maximum concentration and maximum water residence time occur. Theoretically the entrapment zone occurs slightly downstream of where the net vertical water velocities are thought to be the greatest. As the aggregates move into the fresher water, partial disaggregation may occur. The materials that enter the upper layer are again transported seaward and theoretically can be recirculated numerous times. Under low riverflows, suspended sediment settles into the lower layer farther upstream in the estuary than during high flows. Conversely, during high riverflows, a larger portion of the fine suspended sediment is transported to the ocean.

FIELD OBSERVATIONS

Studies conducted from 1973 through 1977 (Arthur and Ball 1978) and summarized in this chapter were designed to characterize the distribution of suspended materials in the entrapment zone over a range of river discharge in order to determine how the zone influences the water quality and biota (primarily the phytoplankton).

River Discharge

Delta outflow (river discharge past Chipps Island, site 11) was the main variable in the study.

SAN FRANCISCO BAY

Daily Delta Outflow indices (DOI), calculated by the USBR and DWR, were used in this study. The DOI consists of the Sacramento River discharged at Sacramento plus the San Joaquin River discharge at Vernalis, less the pumped Delta export and the estimated Delta consumptive use (see Conomos 1979, Fig. 11). The consumptive use coefficient estimate varies seasonally but is constant between years. The coefficient is as high as 130 $m^3 \cdot s^{-1}$ in midsummer. Consequently, since crop and weather patterns change between years, under very low flows the DOI lacks precision. This error during typical summer outflow conditions may be as great as ±30 to 60 $m^3 \cdot s^{-1}$. Furthermore, the Yolo Bypass (which has tidally influenced discharge that would be hard to measure) and other peripheral streams also contribute significant discharges to the Delta outflow especially during periods of high runoff (over 1,400 $m^3 \cdot s^{-1}$). Measurements of these additional stream discharges are not included in the DOI but have been incorporated into another calculated outflow (average monthly historical Delta outflow) which still utilizes the consumptive use estimate. The historical Delta outflow, although only a monthly average, is the more accurate of the two for total discharges from the Delta (Fig. 5). Since the DOI is the only daily calculation readily available, the index is usually used when referring to Delta outflow even though it is an underestimate of high flow.

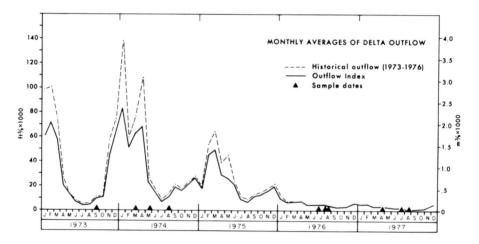

Fig. 5. Comparison of the Delta outflow index and the historical Delta outflow during the study period.

The Delta outflow during the 1973 through 1975 period was above normal, while during 1976 and 1977 it was the lowest since completion of Shasta Reservoir (the main water storage reservoir of the Central Valley Water Project).

Salinity Intrusion

The 2 millimho/cm specific conductance (1 ‰ salinity) isocontour shifted nearly 45 km over the range of DOI studied (23-1,800 $m^3 \cdot s^{-1}$) (Fig. 6).

In addition to the quantity of the riverflow, the pattern of flow also appears to influence the salinity distribution. For example, although the September 1973 and August 1974 surveys were conducted at near identical Delta outflows, there was greater compression of the 2-25 millimho·cm^{-1} (1 to 15 ‰ salinity) water mass in 1973. There had been several months of low flow (the average DOI for July and August was 130 $m^3 \cdot s^{-1}$) prior to September 1973; while prior to the

August 1974 survey the DOI was nearly twice as large (Fig. 5).

Salinity stratification increased with increasing Delta outflow. The isoconductivity contours in March 1974 demonstrated greater vertical stratifications than during the low outflow of August 1977 (Fig. 6). The degree of stratification apparently also influences the distribution patterns of suspended materials.

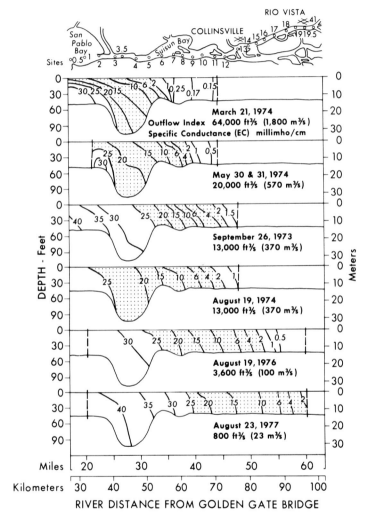

Fig. 6. Isoconductivity (salinity) contours measured on high slack tides at various Delta outflows (the 2-25 millimho/cm EC range has been arbitrarily shaded).

Typical variations in tidal excursion that occur in the study area are demonstrated for the 19-21 August 1974 and the 23 August 1977 data (Figs. 7, 8). The 1974 data were collected on three consecutive days with DOI = 370 m$^3 \cdot$s^{-1}, while the 1977 data were collected on a single day at DOI = 23 m$^3 \cdot$s^{-1}. The tidal excursion measured for the August 1974 run was nearly 10 km and occurred on a greater flood, close to a spring tide (with relatively high tidal velocities). Conversely, the tidal excursion for the August 1977 observations was only about 3 km. The reduced excursion

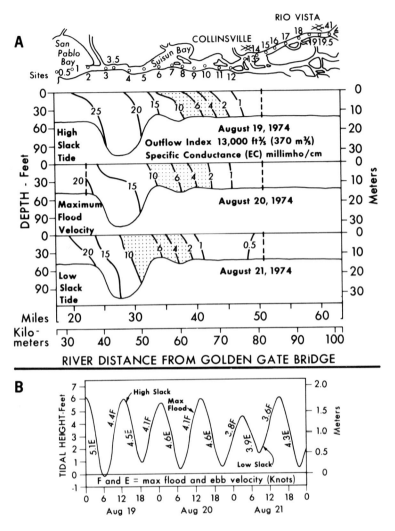

Fig. 7. A. Isoconductivity (salinity) contours measured on three consecutive days during different tidal phases during August 1974. B. Calculated Golden Gate Bridge tidal heights and maximum flood and ebb (F and E) velocities in knots.

resulted from the low tidal velocities and the slight difference in tidal heights occurring on the lesser ebb near a neap tidal period.

Suspended Material Distributions

The distribution patterns of suspended particulate matter and dissolved constituents were characterized in the upper estuary at selected DOI ranging from 23 to 1,800 $m^3 \cdot s^{-1}$ between September 1973 and September 1977 (Arthur and Ball 1978).

Total suspended solids (TSS) correlated well with turbidity, the latter of which was measured more extensively. Areas of maximum turbidity at various Delta outflow were typically located where the surface water was in the 2-10 millimho·cm^{-1} specific conductance (1-6 °/oo salinity) range in both the Sacramento and San Joaquin rivers (Figs. 9 and 10). Because suspended materials accumulated in this salinity range, the 2-10 millimho·cm^{-1} surface specific conductance contour (SUR EC) was added to the illustrations as a reference.

The maximum turbidity in the entrapment zone varied from 2 to 40 times the upstream and downstream levels and increased up to 20 times with depth. The maximum turbidity, over 800 Formazin turbidity units (FTU; USEPA 1971) was centered in Carquinez Strait during the highest Delta outflow studied (1,800 m^3·s^{-1}). In contrast, during 1977 (one of the lowest river discharge years on record), maximum turbidities of about 60 FTU were measured at DOI = 23 m^3·s^{-1} and the entrapment zone was centered about 40 km upstream of Carquinez Strait.

Volatile suspended solids (VSS) also peaked in the entrapment zone and were approximately 10-20% of the TSS.

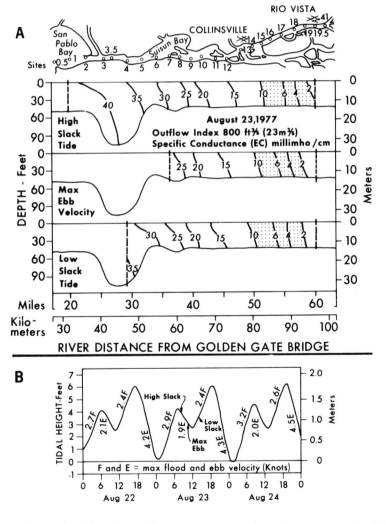

Fig. 8. A. Isoconductivity (salinity) contours measured on three consecutive tidal phases on 23 August 1977. B. Calculated Golden Gate Bridge tidal heights and maximum flood and ebb (F and E) velocities in knots.

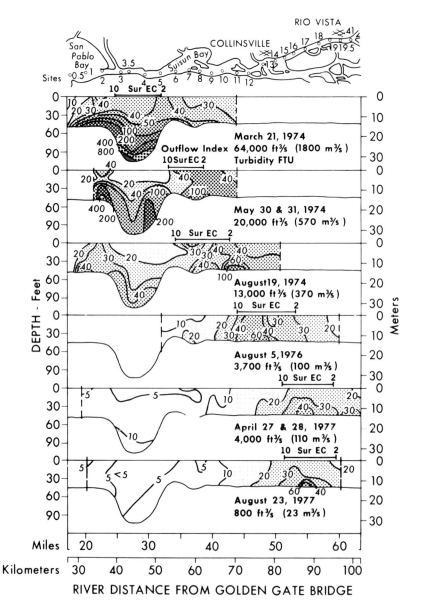

Fig. 9. Turbidity distribution relative to salinity on high slack tides at various Delta outflows.

Differences in the amount of resuspension and settling were observed between the greater and lesser flood and ebb tides (Fig. 11). The greatest resuspension of materials (between slack and maximum tidal velocity) was observed when tidal height differences and maximum velocities were high (Fig. 7) as opposed to when they were low (Fig. 8).

The maximum concentration of particulate organic nitrogen and phosphorus also typically occurred in the same general area as the maximum turbidity.

Distribution of Dissolved Materials

Dissolved constituents, of course, are not subject to entrapment by two-layered flow circulation. The concentration of nitrate+nitrite (Fig. 12), ammonia and orthophosphate generally increased with water depth and peaked downstream of the entrapment zone (see also Peterson 1979; Conomos et al. 1979).

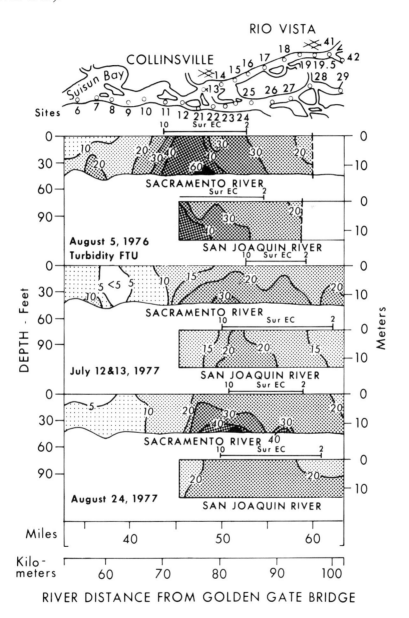

Fig. 10. Turbidity distribution in the Sacramento and San Joaquin rivers relative to salinity on high slack tides during low Delta outflow in 1976 and 1977.

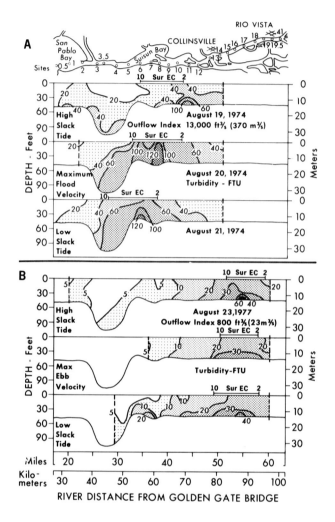

Fig. 11. A. Turbidity distribution relative to salinity measured on three consecutive days during different tidal phases in August 1974. B. Turbidity distribution relative to salinity measured on three consecutive tidal phases on 23 August 1977.

Distribution of Estuarine Biota

The same estuarine-circulation forces that influence the accumulation of suspended solids and particulate nutrients in the entrapment zone also appear to determine the distribution patterns of phytoplankton, certain zooplankton, and juvenile striped bass (young-of-the-year).

The chlorophyll *a* concentration, over a range of Delta outflows (Fig. 13), typically peaked in the entrapment zone at all Delta outflows studied. The peak concentrations in 1976 and 1977 (the two low-flow years) were the lowest ever recorded. The distribution of chlorophyll *a* and the dominant phytoplankton genera were similar throughout the study area and peaked in the 2-10 millimho·cm^{-1} specific conductance (1 to 6 °/oo salinity) range (Fig. 14). The maximum concentration on the surface generally occurred downstream of the maximum concentration on the bottom during bloom periods (see also Ball and Arthur 1979; Conomos et al. 1979).

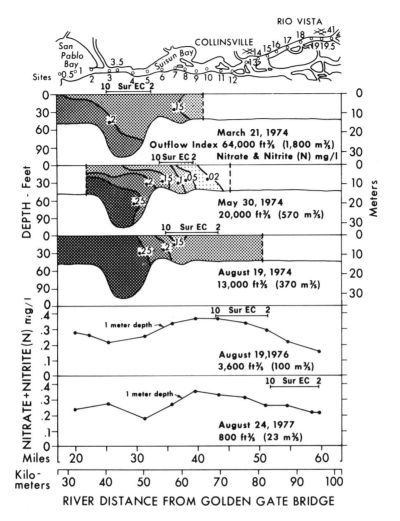

Fig. 12. Nitrate+nitrite distribution relative to salinity during high slack tides at various Delta outflows.

The various factors influencing the distribution of *Neomysis mercedis* and other zooplankton are discussed by Orsi and Knutsen (1979). The maximum abundance of *N. mercedis* and certain other zooplankton occurred in the 2-10 millimho·cm^{-1} specific conductance (1 to 6 ‰ salinity) range. Their distribution pattern relative to salinity (Fig. 15) was similar to that of the other constituents.

The copepod distribution indicated two peaks of abundance (Fig. 16). One peak, composed of *Eurytemora hirundoides*, was centered in the approximate location of maximum suspended solids concentration. The other peak, dominated by *Acartia clausi*, was farther downstream.

The distribution of 50-mm juvenile striped bass (young-of-the-year collected in July, 1973, 1974, 1976, and 1977; Fig. 17) also appears to be related to the distribution of other suspended constituents. The peak concentrations are also related to the surface 2-10 specific conductance (1 to 6 ‰ salinity) range. Similar distribution patterns were noted for other study periods.

SAN FRANCISCO BAY

DISCUSSION

Entrapment of suspended materials and certain estuarine biota were evident at the entire range of outflows studied in both the Sacramento and San Joaquin rivers. Since the peak concentrations of constituents typically occurred where the surface specific conductivity was approximately in the 2-10 millimho·cm^{-1} (1-6 °/oo salinity) range, this salinity range was selected to estimate the location of the entrapment zone.

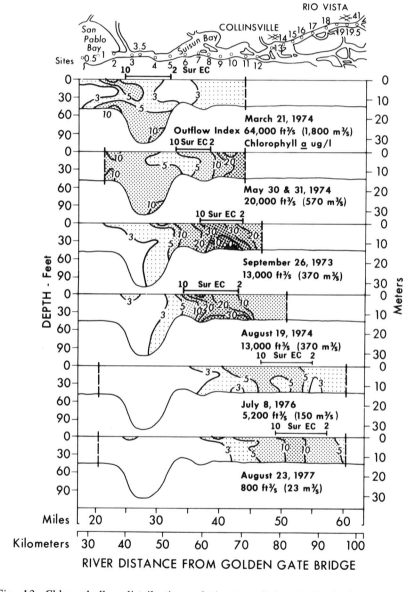

Fig. 13. Chlorophyll *a* distribution relative to salinity during high slack tides at various Delta outflows.

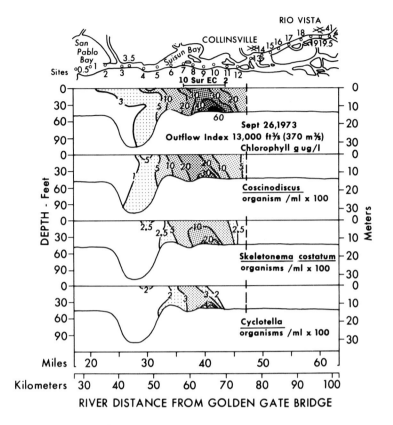

Fig. 14. Distribution of chlorophyll *a* and dominant phytoplankton genera relative to salinity during high slack tides on 26 September 1973.

Factors Influencing Suspended Materials Entrapment

Factors thought to influence the quantity of suspended materials in the entrapment zone include the riverborne suspended sediment load; flocculation, aggregation, and settling rates of particles; tidal- and wind-induced resuspension; bathymetry; dredging activities, and seasonal growth patterns of biota.

High suspended-sediment concentrations and loads to the estuary typically occur with winter floods and to a lesser extent in the late fall and early spring and increase the concentration of suspended materials observed in the entrapment zone.

In recent years, reservoir regulation of riverflows has reduced winter and spring riverflows and increased riverflows throughout the summer and early fall. Releases and drainage return flows have increased suspended sediment loads during the summer. However, flow regulation has resulted in an overall reduction of the total suspended sediment load as a result of settling that occurs in the reservoirs and sediments lost to export.

The Sacramento and San Joaquin rivers are the two main systems discharging suspended sediment to the Delta (Fig. 18). The Sacramento River (including the Yolo Bypass) contributes most (80%) of the total. The combined discharge is an estimate of the total suspended sediment load; however, during the flooding and very high outflows, suspended sediment discharge to the Delta from the Yolo Bypass may be equal to or even greater than that from the Sacramento River.

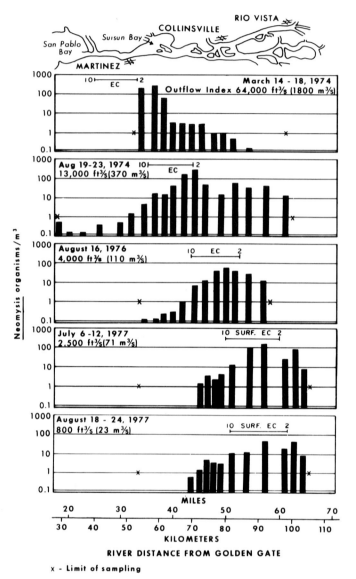

Fig. 15. *Neomysis mercedis* distribution relative to salinity on high slack tides at various Delta outflows (data collected by DFG).

Since the discharge from the Yolo Bypass is not measured, the total discharge to the Delta is often grossly underestimated.

The entrapment zone was located further seaward and with higher suspended-solids concentrations during periods of high suspended sediment discharge as compared to periods of low suspended sediment discharge (Fig. 9). These data support Postma's (1967) belief that the magnitude of the turbidity maximum (entrapment zone) is a direct function of the amount of suspended matter in the river or sea and the strength of the estuarine current.

The maximum suspended solids occurred in higher salinity water during high outflows as compared to low outflows (Fig. 9). This variation may have resulted from seasonal differences in water

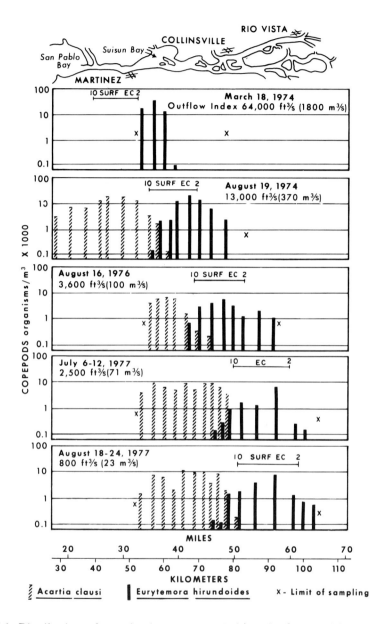

Fig. 16. Distribution of two dominant copepods (*Acartia clausi* and *Eurytemora hirundoides*) relative to salinity on high slack tides at various Delta outflows (data collected by DFG).

velocity or water temperatures. The greater net downstream velocity in the upper layer during high flows may carry the suspended materials further downstream and into more saline water before flocculation, aggregation, and settling of particles occurs. Alternatively, the settling velocity of particles could be decreased during winter by the colder water temperatures increasing the water viscosity.

There are different opinions as to what will happen to the water transparency in Suisun Bay as the amounts of riverborne sediment are decreased by future river diversions. One opinion is the water transparency is inversely correlated to the sediment load entering the estuary during any

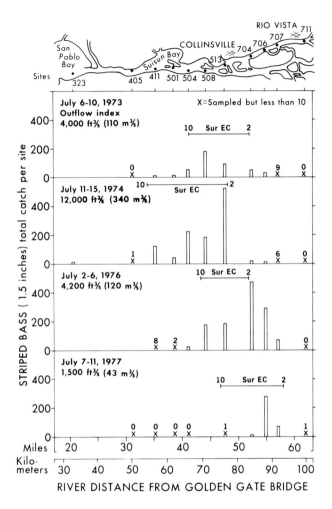

Fig. 17. Distribution of juvenile striped bass (young-of-the-year) relative to salinity on high slack tides during July 1973, 1974, 1976, and 1977 (data collected by DFG).

given year, and therefore, transparencies would increase with decreasing sediment loads. A second opinion is that winds and tidal currents along with tidal dispersion will resuspend large quantities of estuarine sediment and will maintain fairly constant transparency for many years of low river inflow.

Summer Secchi-disc measurements (made monthly during high-slack-tide from 1968-71 and twice monthly from 1972-77 by the DFG), as well as our turbidity measurements, have demonstrated a pronounced increase in water transparencies in Suisun Bay during 1976 and 1977, our two lowest outflow years. An inverse relationship between Suisun Bay water transparency and the summer Delta outflow (Fig. 19) suggests that the summer water transparency in Suisun Bay is strongly influenced by the Delta outflow. This outflow also regulates the entrapment zone location, with the zone moving upstream with the salinity intrusion and the waters of Suisun Bay becoming more transparent. Even though summer wind and tidal resuspension forces were considered to be about equal each summer, considerable transparency variation each year occurred between 1968 and 1977.

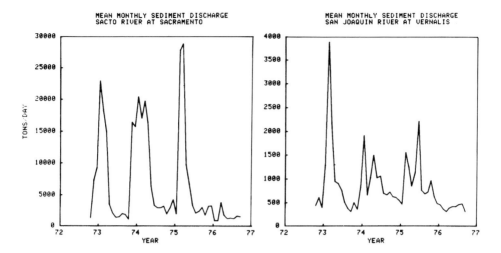

Fig. 18. Suspended sediment loads to the Delta. A. Sacramento River at Sacramento. B. San Joaquin River at Vernalis. Tons·day^{-1} scale differs for the two rivers.

In addition to outflow, both winter and summer sediment loads, the summer inflow sediment concentration and the location of the entrapment zone relative to shallow bays have been thought to influence the summer variation in transparency between 1968 and 1977.

To evaluate the first three factors, the routine Secchi-disc measurements at 14 channel sites between Rio Vista and Martinez (when occurring in water of 2-10 millimho·cm^{-1}) were averaged

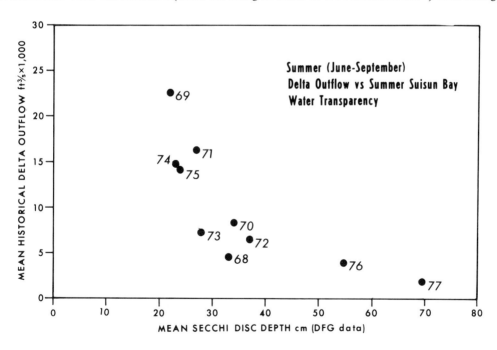

Fig. 19. Summer Suisun Bay water transparency versus summer historical Delta outflow. The 1977 outflow value was calculated as the Delta outflow index.

for each summer (June-September) and compared with the winter suspended sediment load (Fig. 20), summer suspended sediment load (Fig. 21), and summer suspended sediment concentration (Fig. 22). Although there appears to be a slight inverse relationship between the summer water transparency in the entrapment zone and the above factors, the relationships are not conclusive. Furthermore, the summer suspended sediment load as well as concentration were related to the winter load, as summer outflows that followed high outflow winters were usually also high. Since there is also so much variation in water transparency due to wind and tides one must use those evaluations with caution.

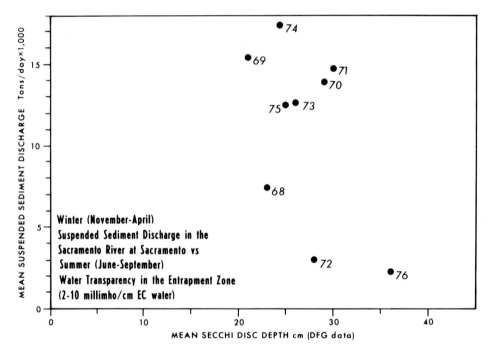

Fig. 20. Summer water transparency in the entrapment zone (2-10 millimho/cm EC range) versus winter suspended sediment load at Sacramento.

Arthur (1975), using Sacramento River water in which the salinity was adjusted with concentrated seawater brine from San Francisco Bay, demonstrated that flocculation, aggregation, and/or settling rates of suspended material increased as the specific conductance of the water samples was increased above 1 millimho·cm^{-1} (0.6 ‰ salinity) (Fig. 23).

We initiated field measurements in 1975 to obtain particle settling-rate data for verification of a suspended-solids model (O'Connor and Lung 1977) used by our study program (Arthur and Ball 1978). Particle settling rates were compared using two sampling methods. Samples from the entrapment zone were pumped into the first set of settling chambers, while the second set (special sampling-settling chambers designed by R. Krone, Univ. Calif. Davis) was lowered to the depth of sampling, the ends closed, and the settling chamber returned to the surface. Settling rates for both sets were determined by changes in turbidity at various heights in the chambers. The settling rates of particles collected by the submersible pump were several times less than those collected in settling chambers. These data suggest that the high turbulence caused by pumping disaggregates particles and imply that the particles were flocculated and/or aggregated before collection (USBR

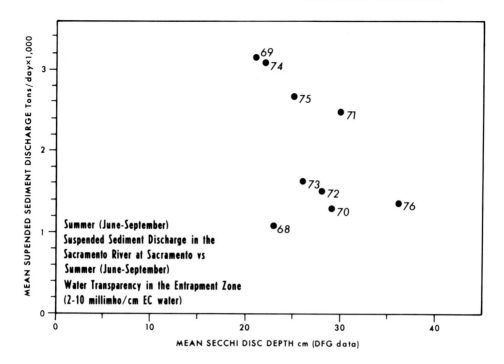

Fig. 21. Summer water transparency in the entrapment zone (2-10 millimho/cm EC range) versus summer suspended sediment load at Sacramento.

unpublished data).

The increased surface-water transparencies with distance downstream of the entrapment zone may be caused by the removal of suspended material with settling velocities greater than the upward vertical water velocity and by increasing dilution with low-turbidity ocean water. These combined effects have not been quantified.

We do not know the extent to which flocculation increases the settling rates of suspended materials and the quantity of suspended materials in the entrapment zone. Our limited data agree with Postma (1967) who suggested that flocculation is an important factor influencing the spatial distribution and entrapment of suspended materials.

Resuspension induced by wind, tide, and dredging activities results in the continual relocation of a portion of the deposited sediments. The TSS concentrations and turbidity in the shallow areas of Suisun and San Pablo bays more than double following periods of high wind (Rumboltz et al. 1976). Increasing tidal current velocities also increase the rate of sediment resuspension, with differences in the amount of resuspension and settling observed between greater and lesser flood or ebb tides. The greatest resuspension was observed when tidal height differences and maximum velocities were highest. During calm days we have often observed highly turbid water masses a few meters in diameter to come billowing to the surface with increasing tidal-current velocities.

Dredging also tends to relocate as well as resuspend sediments. The most intense dredging occurs near Mare Island adjacent to Carquinez Strait, and when the spoils are deposited in San Pablo Bay they increase water turbidity.

The effect of estuarine circulation on suspended sediment distribution in the study area is greatly influenced by bathymetry.

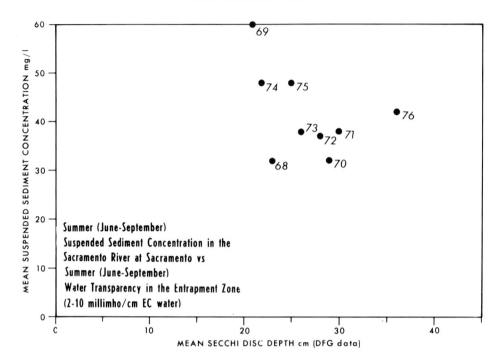

Fig. 22. Summer water transparency in the entrapment zone (2-10 millimho/cm EC range) versus summer suspended sediment concentration at Sacramento.

Distribution of Dissolved Constituents

Dissolved constituents are not directly affected by the entrapment zone. The general increase in nitrate+nitrite (Fig. 12), ammonia, and orthophosphate with depth and distance downstream of the entrapment zone was apparently caused by numerous municipal and industrial waste discharges. Depressions in inorganic nitrogen and dissolved silica concentrations were observed when high phytoplankton standing crops accumulated in the entrapment zone (Arthur and Ball 1978; Peterson et al. 1975b; Peterson 1979).

Dissolved oxygen concentrations (at 1-m depth) in the western Delta-Suisun Bay area were always near saturation values (USBR 1972; Macy 1976) even when chlorophyll a concentrations were relatively high (50-100 μg·liter^{-1}.) Oxygen concentrations one meter from the bottom were generally a few tenths of a mg·liter^{-1} lower than near the surface (these near-bottom oxygen measurements, although made during 1976-77, do not cover periods when high phytoplankton standing crops were present). Presumably, mixing by tidal currents and wind are adequate to maintain near-saturation levels at the present level of eutrophication (Arthur and Ball 1978).

Effects of Entrapment on the Phytoplankton Standing Crop

The location of the entrapment zone adjacent to the Honker Bay area is one of several factors which appears to greatly stimulate phytoplankton growth in the western Delta-Suisun Bay area (Arthur and Ball 1978). In the initial years of our studies (1968-75) when "typical" Delta outflows were present, the standing crop of phytoplankton tended to be highest in the years with the greatest water transparency (Ball 1977; Ball and Arthur 1979).

The unusually low phytoplankton standing crop in Suisun Bay during the recent drought

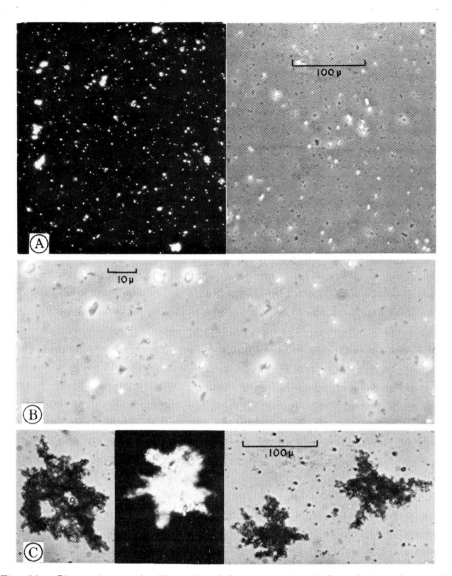

Fig. 23. Photomicrographs illustrating laboratory-induced flocculation of suspended sediments collected from the Sacramento River during flooding conditions on 25 March 1975. A. Control (0.116 millimho/cm EC). B. Control (0.116 millimho/cm EC) (enlarged). C. After addition of concentrated sea brine (2,500 micromhos/cm EC in beaker) and 8 hr of stirring at 30 r·min^{-1}.

(summer of 1976 and throughout 1977) was contrary to predictions based on the 1968-75 data period. We conducted a number of field and laboratory studies during 1977 to study the low phytoplankton standing crop associated with low outflow conditions. We evaluated water transparency, water temperature, solar radiation, salinity, nutrient limitation, toxicity, parasitism, zooplankton grazing, filter feeding of benthic organisms, and the location of the entrapment zone and compared them to our previous (1968-75) observations.

SAN FRANCISCO BAY

The 1976-1977 data indicated that water transparencies in Suisun Bay were approximately double (Fig. 19) that of the previous years of high standing crop while solar radiation (insolation), water temperatures, and algal macro nutrients were within the normal range. Furthermore, the phytoplankton standing crop in the northern and southern Delta during 1976 and 1977 were the highest recorded although climatical conditions in these areas were similar to Suisun Bay.

A number of algal growth potential (AGP) and phytoplankton productivity studies were conducted during 1977 to determine if nutrient depletion, increased salinity, or toxicity might have been responsible for the low phytoplankton standing crop.

The AGP-test results demonstrated that the growth rates of the endemic phytoplankton tend to increase with increasing salinity and suggested that salinity intrusion into Suisun Bay during the low flow years did not directly inhibit the algal growth rates. Furthermore, because the concentration of phytoplankton in the unaltered water of the AGP tests peaked several times higher than in the field, it appeared that neither toxicity nor low concentrations of macro or micro nutrients were limiting algal growth. The primary productivity test results (DO method) in 1977 also supported this contention as the dissolved oxygen production per unit chlorophyll was equal to or higher than that of previous years.

Zooplankton concentrations were lower than normal in 1977, suggesting that grazing rates on phytoplankton should also have been lower than normal.

Although there may have been some movement of marine benthic organisms into Suisun Bay during 1976-77, it was impossible to draw any definite conclusions because there is little previous benthic data with which to compare. Comparison of 1976-77 data with future years of high phytoplankton standing crops may provide further insight into the possible significance of filter feeding of benthic organisms.

Comparison of chlorophyll a data in Suisun Bay with Delta outflows (Fig. 24a, b) shows that moderate to high chlorophyll a concentrations (above 20 μg·liter^{-1}) were present when Delta outflows ranged from 110 to 700 m^3·s^{-1}. When the outflows were below 110 m^3·s^{-1}, the standing phytoplankton crop either declined or remained low. This outflow range places the tidally averaged location of the entrapment zone at various positions adjacent to the Suisun-Honker Bay area. The highest chlorophyll concentrations were measured when the outflow varied from 140 to 200 m^3·s^{-1} in August 1970, 1972, and 1973, and September 1968 when the averaged tidal location of the entrapment zone (based on the 1-6 °/oo salinity range) was adjacent to the Suisun-Honker Bay area. In February 1976, a substantial algal bloom developed as the entrapment zone moved upstream into the Suisun-Honker Bay area earlier than normal as a result of low river flow. This bloom occurred earliest of any year. Significantly, during the bloom water temperatures were only about 12°C and the photoperiod was short (although water transparencies were high). This bloom declined in March as the water transparency decreased. A second bloom developed in April 1976 and declined as the entrapment zone moved further upstream in June 1976 with decreasing riverflow.

When the entrapment zone was upstream of Honker Bay under low (30 to 110 m^3·s^{-1}) Delta outflows (such as occurred in July and August 1966, July 1970, and June-December 1976), chlorophyll concentrations either remained low or were declining. As the 1976 drought continued into 1977 and Delta outflows remained low, the entrapment zone remained several kilometers upstream of Honker Bay for the entire year. Significantly, 1977 was the first year on record when a phytoplankton bloom did not develop in Suisun Bay. The chlorophyll a concentration in Suisun Bay was generally less than 5 μg·liter^{-1} with an occasional value of about 10 μg·liter^{-1} (see Ball and Arthur 1979).

The highest chlorophyll a concentrations (nearly 20 μg·liter^{-1}) measured west of Antioch during 1977 were in the entrapment zone (at 1 to 6 °/oo salinity) at locations above Collinsville on the Sacramento River and near Antioch on the San Joaquin River. Summer chlorophyll a

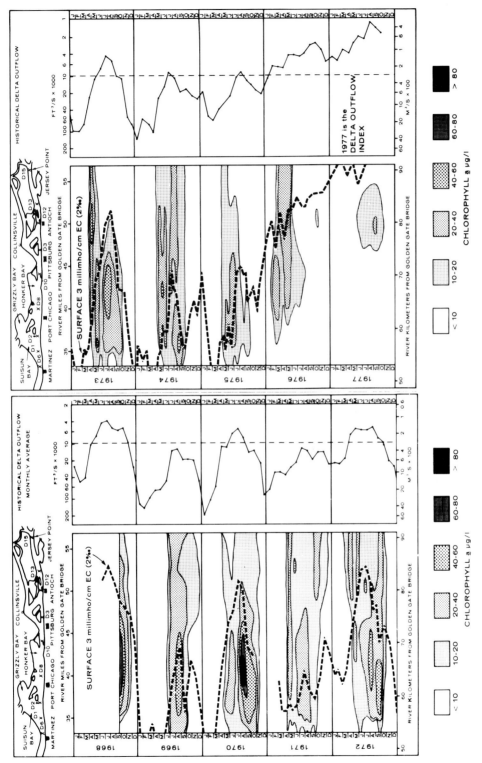

Fig. 24A and B. Chlorophyll *a* distribution on high slack tides from 1968-1977, between Jersey Point and Martinez, as related to salinity intrusion and Delta outflow (from Ball and Arthur 1979). (The 3 millimho/cm EC contour represents the upstream location of the entrapment zone on high slack tides.)

concentrations at the sites where the entrapment zone occurred in 1977, were in the same range as for previous years (1969-75) when the entrapment zone was farther downstream of these sites. Water transparencies at these sites in 1977 were lower than normal, suggesting that the higher phytoplankton standing crop was maintained by entrapment.

An important factor in evaluating algal growth is the residence time of algae in any given location. Whereas the residence time in any stretch of a river can be estimated by knowing the volume of water and the rate of flow, in an estuary where two-layered flow and tidal exchange occur, the residence time of algae (and other suspended materials) can either be greatly increased or reduced over that of the net downstream flow of water. The residence time of phytoplankton in two-layered flow circulation has not been directly measured. In theory, phytoplankton tend to be carried seaward if their settling velocity is less than the net vertical velocity, tend to be recirculated to and about the entrapment zone if their settling velocity is nearly equal to the net vertical velocity, or become entrapped and remain near the bottom if their settling velocity is much greater than the net vertical (upward) water velocity.

Certain algal species of the genus *Coscinodiscus* are consistently associated with the entrapment zone (Arthur and Ball 1978). These organisms have since been identified as belonging to the species *Coscinodiscus decipiens* which is synonymous with *Thalossiosira excentricus*. The organisms have thick cell walls, generally have inorganic particles attached to their exterior, and have been seen to settle rapidly in counting chambers. Their settling velocity relative to the net vertical water velocity presently being studied may provide these organisms with an ecological advantage which allows accumulation in the entrapment zone. In contrast, certain species of the genus *Chaetoceros* have cells much smaller in size which settle very slowly, have high growth rates, and at times become very dominant in the AGP test. *Chaetoceros* probably do not become dominant in the entrapment zone because their settling rates are so low; however, they often are the dominant form downstream of the entrapment zone. In addition to entrapment, the most important aspect of algal residence time related to the algal standing crop is the percent of time algal cells reside in the photic zone.

A substantial phytoplankton bloom (chlorophyll *a* >700 $\mu g \cdot liter^{-1}$ at water surface) occurred in the summer of 1977 in the McAvoy marina (south side of Honker Bay) which consisted almost entirely of *Exuviella,* a motile dinoflagellate. The intensity of the bloom gave the water a reddish-brown cast. This organism was also observed at very low concentrations in Suisun Bay during 1977. Apparently, such areas, although physically connected to the main channel, are isolated from the effects of wind, tidal current mixing and river flushing. The most logical explanation seems to be that the residence time of the algae is longer in these isolated areas than in the main channel and their mobility can maintain them near the water surface.

We do not know exactly how reduced Delta outflow and the location of the entrapment zone influence the phytoplankton standing crop in the Suisun Bay area. We offer, however, several hypotheses which when considered either singularly or in some combination may explain how the upstream movement of the zone could have caused a reduction in the Suisun Bay phytoplankton standing crop during the drought of summer 1976 and throughout 1977 (Arthur and Ball 1978):

1. *Decreased phytoplankton residence time in the Suisun Bay area when the entrapment was located upstream.* The residence time of suspended materials in rivers increases as river flow decreases. The record high phytoplankton crop in 1976 and 1977 in the northern and southern Delta (upstream of the study area) may be attributed to the increase in phytoplankton residence time resulting from lower river flows (Ball and Arthur 1979). However, in the fresh/saltwater mixing zone the water flow and mixing processes are much more complex. The longer residence time in the entrapment zone, relative to the immediate upstream and downstream areas,

may be a major factor regulating the phytoplankton standing crop. We postulate that when the entrapment zone moved upstream in 1976 and 1977, the residence time influencing the phytoplankton standing crop in Suisun Bay (both the shoals and the channel) decreased and resulted in the low phytoplankton standing crop in that area.

2. *Upstream movement of the area of maximum flocculation-aggregation-settling.* Suspended materials are in relatively low concentrations in San Francisco Bay and in the ocean. When Delta outflow were low during 1976 and 1977, the percent of ocean water nearly doubled in Suisun Bay over that of more typical years (1969-75). Furthermore, chlorophyll *a* levels during 1977 in Suisun Bay were similar to those observed in Central San Pablo Bay during the higher flow years.

We are uncertain why phytoplankton standing crops observed in the field were low in high salinity water (over 25 millimho/cm EC water) yet growth rates were highest at similar salinities in our field and laboratory growth rate tests. Perhaps the phytoplankton standing crop is characteristically low in high salinity water in the field because increased flocculation, aggregation, and/or settling of suspended particles occurs in the area downstream of the entrapment zone (the area where the net upward vertical water velocities are assumed to decrease). Phytoplankton may be affected by the increased particle settling and thus are unable to maintain themselves in the photic zone downstream of the entrapment zone. Consequently, as the entrapment zone moved upstream throughout 1976-77, greater settling rates may have occurred in Suisun Bay.

3. *Decreased phytoplankton residence time in the photic zone.* Phytoplankton are concentrated where the entrapment zone is located, with their growth rate directly proportional to the length of time they spend in the photic zone. When the entrapment zone is adjacent to the shallow bays, the average water depth present at the zone is much less than when the zone is located a dozen kilometers upstream in the more confined channels (assuming tidal exchange of the phytoplankton between the channel and the adjacent shallow bays). When the entrapment zone was located upstream in 1977, the contained phytoplankton spent less average time in the photic zone as compared to a downstream location. This hypothesis assumes complete vertical mixing of the water column.

4. *Increased vertical mixing with reduced salinity stratification.* During the low Delta outflows of 1977 the salinity stratification was less and the vertical mixing of the water column was apparently greater than during moderate to high summer outflows. The greater salinity stratification during the higher summer outflows could maintain the algae nearer the water surface and in the photic zone to a greater extent than during low outflows. Consequently, during low outflow, the reduced water stratification results in increased mixing which lowers the growth rate and standing crop of phytoplankton.

5. *Intrusion of marine benthic filter feeders.* We are uncertain whether the upstream movement of marine filter-feeding benthic organisms influenced the phytoplankton crop in 1976 and 1977.

We offer the following hypotheses that may account for the lower suspended materials concentrations observed in the entrapment zone during periods of low flow (as compared to high outflow), but do not know if or how these hypotheses may explain the low phytoplankton standing crop in Suisun Bay during 1976-77.

1. *Reduction of two-layered flow circulation.* The intensity of two-layered flow circulation should decrease as riverflow to the estuary decreases. This reduced circulation could increase the residence time of suspended materials in the entrapment zone while simultaneously reducing the quantity of suspended materials circulated through the zone. The interactions of these factors are unknown, however.

2. *Reduced aggregation and settling.* High concentrations of river-borne suspended materials increase the chances of particle aggregation in the estuary which in turn increases the settling rates of the suspended materials (R. Krone, pers. comm.). This factor may increase the quantity of material entrapped. Conversely, the quantity of suspended material entrapped decreases as the suspended-particle concentration decreases. The suspended-particle concentration entering the estuary usually varies directly with riverflow.

Factors Influencing Entrapment of Zooplankton and Striped Bass

The results of this and other studies suggest that the maximum abundance of *Neomysis mercedis* (Fig. 15) and certain copepods (Fig. 16) relative to salinity is primarily influenced by the interaction of two-layered flow circulation on their instinctive vertical swimming behavior. Cronin and Mansueti (1971), Heubach (1969), and Siegfried et al. (1978) state that certain species of zooplankton migrate upward during the night and downward during the day. In a two-layered flow estuary this movement translates into downstream transport at night and upstream transport during the day, resulting in a roughly circular motion that retains the species near its optimal salinity range. High tidal-current velocities also result in their upstream movement (Heubach 1969; and Siegfried et al. 1978).

The different distribution patterns of *Eurytemora hirundoides* and *Acartia clausi* (Fig. 16) are attributed to differences in the optimal salinity range for these genera (Kelley 1966). The mechanism responsible could be differences in vertical swimming behavior between the two species.

We partially attributed the decrease in the total zooplankton standing crop during 1976 and 1977 to the fact that the center of the populations shifted upstream with movement of the entrapment zone into an area occupied by a smaller surface area and volume of water (Arthur and Ball 1978). The DFG has suggested that *Neomysis* and certain other zooplankton concentrations are directly related to the concentration of phytoplankton in the entrapment zone (see also Orsi and Knutson 1979). It is interesting to note that *Neomysis* (Fig. 15) and zooplankton (Fig. 16) concentrations were relatively high in March of 1974—prior to the development of a phytoplankton bloom. Unfortunately, routine sampling did not extend downstream of Martinez to characterize the distribution of both zooplankton and phytoplankton during higher Delta outflows.

The relatively high concentrations of juvenile striped bass (young-of-the-year) present in the entrapment zone may be caused by (1) the bass tending to swim to where the food supply peaks, or (2) the juvenile bass are concentrated by two-layered flow circulation in the essentially plankton stage in their early life cycles. The latter explanation seems more reasonable. Cronin and Mansueti (1971) have found that the larval forms of many Atlantic Coast fish species that spawn both in freshwater and at the entrance to estuaries are carried to the plankton-rich low salinity area (entrapment zone) where zooplankton are abundant. Stevens (1979) further discusses the factors influencing the striped bass population.

Predicting the Entrapment Zone Location

Evaluation of salinity and suspended materials data over the past 10 years indicates that the location of the entrapment zone can be predicted from salinity gradients and occurs in the upstream portion of the mixing zone where the surface specific conductance is approximately in the 2-10 millimho/cm (1 to 6 °/oo salinity) range. A plot of geographic location of this salinity range versus the DOI (at high slack tide) could be used to estimate the location of the entrapment zone at future outflows within the outflow range presented (Fig. 25). Although the overall relationship between the location of the entrapment zone and the DOI is good, it is less precise at low outflows. This may be due to the lack of precision in calculating the DOI at low outflows, and that the location of the zone is also dependent upon both the history (variation and magnitude) of the previous outflow and on changes in tidal elevation.

Environmental Significance

The most significant environmental aspect of the entrapment zone, other than influencing the location of maximum shoaling (sediment deposition), may be that the quantity of phytoplankton and certain other estuarine biota are enhanced when the zone is located in upper Suisun Bay. The lowest levels of phytoplankton and certain zooplankton recorded in the Suisun Bay area occurred during 1976 and 1977 when the Delta outflow was low and the entrapment zone was located several kilometers upstream of Honker Bay. However, we do not yet know the significance of a long-term low Delta outflow on total estuarine productivity.

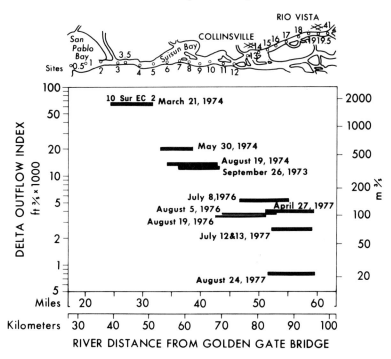

Fig. 25. Estimated high slack tide locations of the entrapment zone, based on the 2-10 millimho/cm EC (1-6 °/oo) range at various Delta outflows.

ACKNOWLEDGMENTS

We thank scientists and engineers of research and educational institutions and of other governmental agencies who kindly furnished field and laboratory support, biological data, and technical advice.

LITERATURE CITED

Arthur, J. F. 1975. Preliminary studies of the entrapment of suspended materials in Suisun Bay,

San Francisco Bay-Delta estuary. Pages 17-36 in R. L. Brown, ed. Proceedings of a Workshop on Algal Nutrient Relationships in the San Francisco Bay and Delta, November 1973. Calif. Dep. Water Resources, Sacramento, Calif.

Arthur, J. F., and M. D. Ball. 1978. Entrapment of suspended materials in the San Francisco Bay-Delta estuary. U. S. Bureau of Reclamation, Sacramento, Calif. 106 pp.

Bain, R. C. 1968. San Joaquin master drain--effects on water quality of San Francisco Bay and Delta. Appendix C. Nutrients and Biological Response. FWPCA, S. W. Region.

Ball, M. D. 1975. Chlorophyll levels in the Sacramento-San Joaquin Delta to San Pablo Bay. Pages 54-102 in R. L. Brown, ed. Proceedings of a Workshop on Algal Nutrient Relationships in the San Francisco Bay and Delta, November 1973. Calif. Dep. Water Resources, Sacramento.

Ball, M. D. 1977. Phytoplankton growth and chlorophyll levels in the Sacramento-San Joaquin Delta through San Pablo Bay. U. S. Bureau of Reclamation, Sacramento, Calif. 96 pp.

Ball, M. D., and J. F. Arthur. 1979. Planktonic chlorophyll dynamics in the northern San Francisco Bay and Delta. Pages 265-285 in T. J. Conomos, ed. San Francisco Bay: The Urbanized Estuary. Pacific Division, Amer. Assoc. Advance. Sci., San Francisco, Calif.

Bowden, K. F. 1967. Circulation and diffusion. Pages 15-36 in G. H. Lauff, ed. Estuaries. Amer. Assoc. Advance. Sci. Publ. No. 83. Washington, D. C.

Brown, R. L. 1975. The occurrence and removal of nitrogen in subsurface agricultural drainage from the San Joaquin Valley, California. Water Resources 9:529-546.

California Department of Fish and Game, California Department of Water Resources, U.S. Fish and Wildlife Service, and U. S. Bureau of Reclamation. 1975. Interagency Ecological Study Program for the Sacramento-San Joaquin Estuary. 4th Annual Report. State of California, Sacramento, Calif.

California Department of Fish and Game, California Department of Water Resources, U. S. Fish and Wildlife Service, and U. S. Bureau of Reclamation. 1976. Interagency Ecological Study Program for the Sacramento-San Joaquin Estuary. 5th Annual Report. State of California, Sacramento, Calif.

Conomos, T. J. 1975. Movement of spilled oil as predicted by estuarine nontidal drift. Limnol. Oceanogr. 20:159-173.

Conomos, T. J. 1979. Properties and circulation of San Francisco Bay waters. Pages 47-84 in T. J. Conomos, ed. San Francisco Bay: The Urbanized Estuary. Pacific Division, Amer. Assoc. Advance. Sci., San Francisco, Calif.

Conomos, T. J., D. S. McCulloch, D. H. Peterson, and P. R. Carlson. 1971. Drift of surface and near-bottom waters of the San Francisco Bay system. March 1970 through April 1971. U. S. Geol. Surv. Open-File Map.

Conomos, T. J., and D. H. Peterson. 1974. Biological and chemical aspects of the San Francisco Bay turbidity maximum. Symposium International Relations Sedimentaires entre Estuaires et Plateaux Continentaux. Institut de Geologie du Bassin d'Aquitaine. 15 pp.

Conomos, T. J., and D. H. Peterson. 1977. Suspended particle transport and circulation in San Francisco Bay: An overview. Pages 82-97 in L. E. Cronin, ed. Estuarine Processes. Vol. 2. Academic Press, New York.

Conomos, T. J., D. H. Peterson, P. R. Carlson, and D. S. McCulloch. 1970. Movement of seabed drifters in the San Francisco Bay estuary and the adjacent Pacific Ocean. U. S. Geol. Surv. Circ. 637B. 7 pp.

Conomos, T. J., R. E. Smith, D. H. Peterson, S. W. Hager, and L. P. Schemel. 1979. Processes affecting seasonal distributions of water properties in the San Francisco Bay estuarine system. Pages 115-141 in T. J. Conomos, ed. San Francisco Bay: The Urbanized Estuary. Pacific Division, Amer. Assoc. Advance. Sci., San Francisco, Calif.

Cronin, L. E., and A. J. Mansueti. 1971. The biology of the estuary. Pages 14-39 in P. Douglas and R. Stroud, eds. A Symposium on the Biological Significance of Estuaries. Sport Fishing Institute, Washington, D. C.

Einstein, H. A., and R. B. Krone. 1961. Estuarial sediment transport patterns. Amer. Soc. Civil

Eng. Proc., J. Hydraulics Div. 87(HY2):51-59.
Fisher, H. B. 1976. Mixing and dispersion in estuaries. Annu. Rev. Fluid Mech. 8:107-132.
Grimm, C. I. 1931. Study of tidal currents and silt movements in the San Francisco Bay area with particular reference to the effect of a saltwater barrier upon them. Unpublished report, U. S. Army Corps of Engineers. Appendix D to House Document 191, 71st Congress, Third Session.
Helliwell, P. R., and J. Bossanyi. 1975. Pollution criteria for estuaries. J. Wiley & Sons, New New York. 102 pp.
Heubach, W. 1969. *Neomysis awatschensis* in the Sacramento-San Joaquin River estuary. Limnol. Oceanogr. 14:533-546.
Kelley, D. W. 1966. Ecological studies of the Sacramento-San Joaquin estuary. Part I. Calif. Dep. Fish Game Fish Bull. No. 133. 133 pp.
Krone, R. B. 1966. Predicted suspended sediment inflows to the San Francisco Bay system. Central Pacific River Basins Project. Fed. Water Pollution Control Admin. 33 pp.
Krone, R. B. 1972. A field study of flocculation as a factor in estuarial shoaling processes. U. S. Army Corps Eng. Tech. Bull. No. 19.
Krone, R. B. 1979. Sedimentation in the San Francisco Bay system. Pages 85-96 *in* T. J. Conomos, ed. San Francisco Bay: The Urbanized Estuary. Pacific Division, Amer. Assoc. Advance. Sci., San Francisco, Calif.
Macy, T. L. 1976. A report of water quality in the Sacramento-San Joaquin estuary during the low flow year, 1976. A cooperative study between the U.S. Bureau of Reclamation, M. P. Region and the Environmental Protection Agency, IX Region. Contract No. EPA-1A6-D6-F078. 227 pp.
McCulloch, D. S., D. H. Peterson, P. R. Carlson, and T. J. Conomos. 1970. Some effects of freshwater inflow on the flushing of South San Francisco Bay. U. S. Geol. Surv. Circ. 637A. 27 pp.
Meade, R. H. 1972. Transport and deposition of sediments in estuaries. Pages 91-120 *in* B. W. Nelson, ed. Environmental Framework of Coastal Plain Estuaries. Geol. Soc. Amer. Memoir 133.
Nichols, M. M., and G. Poor. 1967. Sediment transport in a coastal plain estuary. Proceedings of the American Society of Civil Engineers of the Waterway and Harbors 4:83-94.
O'Connor, D. J., and W. Lung. 1977. Preliminary report on two-dimensional hydrodynamic, salinity transport and suspended solids distribution in western Delta. Hydroscience, Inc., Westwood, N.J. 66 pp.
Orsi, J. J., and A. C. Knutson, Jr. 1979. The role of mysid shrimp in the Sacramento-San Joaquin estuary and factors affecting their abundance and distribution. Pages 401-408 *in* T. J. Conomos, ed. San Francisco Bay: The Urbanized Estuary. Pacific Division, Amer. Assoc. Advance. Sci., San Francisco, Calif.
Peterson, D. H. 1979. Sources and sinks of biologically reactive oxygen, carbon, nitrogen and silica in northern San Francisco Bay. Pages 175-193 *in* T. J. Conomos, ed. San Francisco Bay: The Urbanized Estuary. Pacific Division Amer. Assoc. Advance. Sci., San Francisco, Calif.
Peterson, D. H., T. J. Conomos, W. W. Broenkow, and P. C. Doherty. 1975a. Location of the non-tidal current null zone in northern San Francisco Bay. Estuarine Coastal Mar. Sci. 3:1-11.
Peterson, D H., T. J. Conomos, W. W. Broenkow, and E. P. Scrivani. 1975b. Processes controlling the dissolved silica distribution in San Francisco Bay. Pages 153-187 *in* L. E. Cronin, ed. Estuarine Research. Vol. 1. Academic Press, New York.
Peterson, D. H., and R. Charnell. 1969. Seasonal changes in suspended particle composition of the turbidity maximum, San Francisco Bay, California. Amer. Geophys. Union Trans. 50(11):629. (Abstr.)
Postma, H. 1967. Sediment transport and sedimentation in the estuarine environment. Pages 158-179 *in* G. H. Lauff, ed. Estuaries. Amer. Assoc. Advance. Sci. Publ. No. 83. Washington, D. C.
Rumboltz, M., J. F. Arthur, and M. D. Ball. 1976. Sediment transport characteristics of the upper San Francisco Bay-Delta estuary. Pages 12-37 *in* Proceedings of Third Fed. Interagency Sed. Conf., Denver, Colo. Prepared by Sedimentation Comm. Water Resources Council.

Schultz, E. A., and H. B. Simmons. 1957. Fresh-water/salt-water density currents, a major cause of siltation in estuaries. Comm. on Tidal Hydrol. Tech. Bull. No. 2. U. S. Army Corps of Engineers, Vicksburg, Miss. 28 pp.

Siegfried, C. A., A. W. Knight, and M. E. Kopache. 1978. Ecological studies in the western Sacramento-San Joaquin Delta during a dry year. Water Sci. Eng. Paper No. 4506. University of California, Davis, Calif. 121 pp.

Simmons, H. B. 1955. Some effects of upland discharge on estuarine hydraulics. Amer. Soc. Civil Eng. Proc. 181:792-812.

Smith, B. 1966. The tides of San Francisco. Prepared for the San Francisco Bay Conservation and Development Commission, San Francisco, Calif. 42 pp.

Stevens, D. E. 1979. Environmental factors affecting striped bass (*Morone saxatilis*) in the Sacramento-San Joaquin estuary. Pages 469-478 *in* T. J. Conomos, ed. San Francisco Bay: The Urbanized Estuary. Pacific Division, Amer. Assoc. Advance. Sci., San Francisco, Calif.

Turner, J. L., and D. W. Kelley. 1966. Ecological studies of the Sacramento-San Joaquin Delta. Part II. Calif. Dep. Fish Game Fish Bull. 136. 168 pp.

Turner, J. L., and H. K. Chadwick. 1972. Distribution and abundance of young-of-the-year striped bass, *Morone saxatilis,* in relation to riverflow in the Sacramento-San Joaquin estuary. Trans. Amer. Fish Soc. 101:442-452.

U. S. Army Corps of Engineers. 1967. Report on survey of San Francisco Bay and tributaries, California. Appendix V. Sedimentation and Shoaling and Model Test. Army Eng. District, San Francisco, Calif. 176 pp.

U. S. Army Corps of Engineers. 1977. Dredge disposal study. San Francisco Bay and estuary. Appendix E. Army Eng. District, San Francisco, Calif. 623 pp.

U. S. Bureau of Reclamation. 1972. Delta-Suisun Bay surveillance program. A progress report on the Delta San Luis Drain surveillance portion of the program. U. S. Bureau of Reclamation, Sacramento, Calif. 78 pp.

U. S. Environmental Protection Agency. 1971. Methods for chemical analysis of water and wastes. Natl. Environmental Res. Center, Cincinnati, Ohio. 312 pp.

Wiley, M. 1977. Estuarine Processes. Vol. 2. Circulation, Sediments, and Transfer of Material in the Estuary. Academic Press, New York. 428 pp.

SOURCES AND SINKS OF BIOLOGICALLY REACTIVE OXYGEN, CARBON, NITROGEN, AND SILICA IN NORTHERN SAN FRANCISCO BAY

DAVID H. PETERSON
U.S. Geological Survey, 345 Middlefield Road, Menlo Park, CA 94025

The distributions of biologically reactive dissolved oxygen, carbon, nitrogen, and silicon (OCNSi) in the main channels of northern San Francisco Bay appear to be related to winter and summer variations in the dynamics of the estuary. At moderate or higher (>500 m$^3 \cdot$s^{-1}) river flow, OCSi distributions in the estuary frequently are nearly conservative. Thus, during high river discharge periods, the relative effects of additional estuarine sources and sinks (waste inputs, phytoplankton production and remineralization, or atmospheric- and benthic-exchange processes) appear to be minimal. At such river flows replacement time for estuarine water is on the order of weeks, whereas the OCNSi replacement (turnover) times due to additional sources and sinks are longer. The turnover time of NH$_3$-N, however, is shorter. The river and ocean are probably not major sources of NH$_3$ to the estuary.

Marked departures from near-conservative OCNSi distributions occur during low river flow (<200 m$^3 \cdot$s^{-1}) when the magnitudes of the local sources and sinks may exceed river and ocean inputs. As an overview, however, several processes seem to control these distributions at comparable rates and no one factor dominates: dissolved oxygen is typically 5 to 10% below saturation concentrations; dissolved carbon dioxide is 150-200% above saturation concentrations and in approximate balance with oxygen consumption; phytoplankton production keeps pace with waste inputs of nitrogen; and dissolved silica is maintained above concentrations that would be limiting for phytoplankton growth.

Knowledge of estuarine hydrodynamics and of the appropriate sources and sinks is needed to predict micronutrient and dissolved-gas distributions in an estuary. This chapter presents a series of inferences about the processes which control oxygen, carbon, nitrogen, and silica (OCNSi), based on their observed distributions. These elements were studied because an understanding of their behavior is basic to our knowledge of natural water chemistry in an estuary. The discussion is limited to the northern part of the San Francisco Bay estuary between the Golden Gate and Rio Vista herein termed North Bay (Fig. 1). The southern reach (South Bay), from the Golden Gate to San Jose, has only a small freshwater inflow and is not discussed here (see Conomos 1979; Conomos et al. 1979).

Under certain assumptions and with appropriate rate measurements we can estimate sources and sinks of these elements throughout North Bay. To some extent the magnitudes and positions of these sources and sinks are, of course, always shifting and changing. Thus, to put the sources and sinks into perspective and to illustrate how they might interrelate with one another, a simple conceptual model of North Bay is used. The model, which has fixed dimensions and receives seasonally varying runoff and insolation, is used for discussion purposes with the understanding it can provide only gross budgets.

Copyright © 1979, Pacific Division, AAAS.

SAN FRANCISCO BAY

Fig. 1. San Francisco Bay and environs. The study area, North Bay, is between the Golden Gate, the entrance to the ocean, and Rio Vista. All field data presented herein were collected at indicated main-channel stations.

PREVIOUS WORK

Most of the interest in processes controlling water chemistry in the Bay concerns understanding the effects of increasing municipal and agricultural waste loads coupled with a decreasing freshwater inflow. These problems, related to urbanization, seem to have been largely ignored until the first surveys of the dissolved oxygen in the seaward regime of the Bay (Miller et al. 1928). It was not until the early 1960's that such studies were extended from river to ocean (McCarty et al. 1962; Storrs et al. 1963, 1964). Subsequent studies treated near-surface oxygen distributions (Bain and McCarty 1965) and processes controlling oxygen in the landward estuary and lower river (Bailey 1967; California Departments of Water Resources and Fish and Game 1972). The distribution of dissolved CO_2 in the Bay has only recently been investigated (Spiker and Schemel 1979).

During spring and summer, 1964, the abundance and distribution of inorganic nitrogen and phosphorus in North Bay indicated light as the prime factor limiting phytoplankton productivity (Bain and McCarty 1965). It appeared, however, that inorganic nitrogen may be depleted to growth-rate limiting concentrations in the fall. As an extension of these results the potential impact of massive nitrogen inputs from agricultural waste was assessed by Brown (1975). Although the significance of nitrogen to phytoplankton production is not fully understood, it is considered more important than phosphorus in setting an upper limit on phytoplankton biomass. Dissolved silica is different because its principal source is natural runoff and its seasonal patterns are related to river supply and to phytoplankton removal (Peterson et al. 1975). Although a quantitative synthesis of OCNSi distributions has not been attempted for the entire Bay system, distributions of these constituents have been simulated with a numerical model of phytoplankton dynamics in

REVIEW OF WINTER AND SUMMER FEATURES

The data base is from our near-monthly surveys of hydrographic properties (Smith et al. 1979) at 20 stations spaced nearly equally between Rio Vista and Golden Gate (Fig. 1). Sampling and analytical procedures for micro nutrients and dissolved gases have been described elsewhere (Peterson et al. 1975; Peterson et al. unpublished). Analytical methods are summarized in Table 1 of this chapter. The methodology for and typical observations of particulate nitrogen and dissolved organic nitrogen are given in Peterson et al. (unpublished) and of particulate carbon in Spiker and Schemel (1979).

Distributions of biologically reactive substances are determined by circulation patterns and rates of both the water and biochemical constituents. Sacramento-San Joaquin river (Delta) inflow supplies more than 90% of the freshwater to the estuary. It averages 600 $m^3 \cdot s^{-1}$ but ranges from 2,000 $m^3 \cdot s^{-1}$ in winter to less than 200 $m^3 \cdot s^{-1}$ in summer. River inflow is the prime seasonal regulator of water residence time in the northern part of the estuary (Conomos 1979).

Strong winds in the Bay can move water at speeds exceeding other nontidal components (>10 $cm \cdot s^{-1}$, cf. Hansen 1967, Elliot 1976 and Weisberg 1976); such effects, however, have not yet been studied quantitatively for this estuary (Conomos 1979). Mean monthly wind speeds range seasonally between 2 $m \cdot s^{-1}$ in December and 6 $m \cdot s^{-1}$ in June, and generally persist in a landward direction opposing the natural nontidal surface flow (Conomos 1979). Wind stress on water is also important for increasing gas-exchange rates between atmosphere and water, for resuspending bottom material (which decreases light), and, perhaps, for increasing mixing rates between shoal and channel waters.

The salt field is a driving component of gravitational circulation in the estuary and influences the solubility of dissolved O_2 and CO_2. Sea salt may penetrate to Rio Vista (Fig. 1) in summer, depending on the river inflow (Conomos 1979, Fig. 18). Surface-to-bottom differences, about 3 °/oo, typify a partially to well-mixed estuary, except during winter floods. A salt-driven estuarine circulation pattern has been identified for the main channels, but seasonal and spatial details of this circulation are unknown.

Phytoplankton abundance is correlated with variations in water circulation in a general way except during extended periods of extremely low river inflow. Phytoplankton biomass, estimated with fluorometric chlorophyll-a observations, decreases in winter (Fig. 2) and increases in summer, particularly in the landward portions of the estuary (Fig. 3). These observations are plotted against salinity, rather than against location, to minimize variability due to tides (10 to 15 $km \cdot cycle$). Although short-term events such as spring tides, wind storms, and cloudy days may alter the OCNSi distributions, data suggest that effects of common short-term (tidal and diurnal) events do not exceed seasonal effects.

Mean-monthly sunlight ranges from 37 $Einsteins \cdot m^{-2} \cdot d^{-1}$ (about 200 $langleys \cdot d^{-1}$) in December to 110 in June (Conomos et al. 1979, Fig. 3). Light-saturation levels (10 to 30 $Einsteins \cdot m^{-2} \cdot d^{-1}$, Peterson et al. unpublished) probably continue year-round at the water surface with some local interruptions caused by dense fog. Photic depth, which ranges from about 6 m near the ocean to 0.5 m inland, is determined by the concentrations of suspended particles. The near-surface concentrations average about 10 $mg \cdot liter^{-1}$ near the ocean to 50 $mg \cdot liter^{-1}$ inland in winter and 5 $mg \cdot liter^{-1}$ near the ocean, and about 100 $mg \cdot liter^{-1}$ inland in summer (Conomos and Peterson 1977; Conomos et al. 1979).

The relative importance of conservative and nonconservative processes can be inferred from

SAN FRANCISCO BAY

TABLE 1. ANALYTICAL METHODS

Property	Method	Principal Analysts
S (°/oo), T(°C)	Salinity-temperature meter (Schemel and Dedini 1979).	L. E. Schemel and L. A. Dedini
O_2 (μg-atoms·liter^{-1})	Modified Winkler titration (Carpenter 1965 a, b); for calculation of % saturation see Weiss (1970).	A. Hutchinson and R. E. Smith
CO_2 (μg-atoms·liter^{-1})	Calculated from equations of Smith and Broenkow (1978) using electrometric observations of pH (Schemel and Dedini 1979) and alkalinity (Strickland and Parsons 1968); also, see Spiker and Schemel (1979) for comparison of observed and calculated pCO_2.	L. E. Schemel and L. A. Dedini
NH_3 (μg-atoms·liter^{-1})	AutoAnalyzer: adaptation of Solorzano (1969) similar to Head (1971); more details are in Table III of Peterson et al. (1978).	S. W. Hager and D. D. Harmon
NO_2, NO_3 (μg-atoms·liter^{-1})	Technicon method number AAIII 100-70W (Technicon Corp., Terrytown, N. Y.).	S. W. Hager and D. D. Harmon
SiO_2 (μg-atoms·liter^{-1})	Technicon method number AA105-71W.	S. W. Hager and D. D. Harmon
Chlorophyll *a* (fluorescence units)	Spectrophotometric (Strickland and Parsons 1972).	B. E. Cole and A. E. Alpine

micronutrient distributions in the estuary. In summer, when phytoplankton biomass increases, micronutrient concentrations are lower than those resulting from a mixture of river and ocean waters. Nevertheless their concentrations are generally well above rate-limiting levels. Consequently, a summer minimum in phytoplankton abundance typical of ocean waters at temperate latitudes is not found in the Bay.

A linear relation between a biologically reactive substance and salinity suggests the relative importance of conservative and nonconservative processes to its distribution. For extreme examples, dissolved silica always exhibits near-linear salinity correlations in winter, whereas during a low river-flow period in summer 1961 (see McCarty et al. 1962) silica was observed below rate-limiting concentrations (10 μg-atoms·liter^{-1}, cf.: Paasche 1973 a, b; Goering et al. 1973; Davis et al. 1973). Other micronutrients with more complex histories exhibit less consistent winter-linearity and summer non-linearity with salinity. At times during winter even fluorometric chlorophyll-*a* distributions seem to be near-linear with respect to salinity (Fig. 2).

The complexity of nitrogen sources and sinks may best be understood when their seasonal three-dimensional distributions are known. Present two-dimensional data demonstrate that NO_3

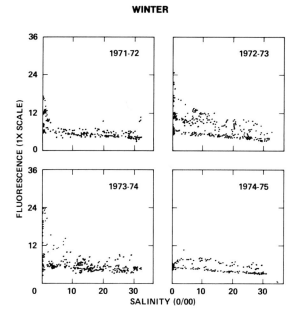

Fig. 2. Winter (October through April) fluorescence (mostly chlorophyll *a*) distributions with salinity during the years indicated in each panel. Data from near-monthly surveys (Smith et al. 1979).

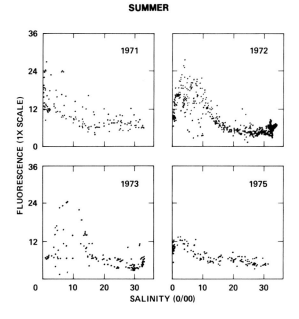

Fig. 3. Summer (May through September) fluorescence (mostly chlorophyll *a*) distributions with salinity during the years indicated in each panel. Data from near-monthly surveys (Smith et al. 1979).

concentrations are exceedingly variable in the river, whereas NH_3 concentrations are lower in the river than in the estuary (Figs. 4, 5). Phytoplankton productivity experiments show that NH_3 utilization is preferred before NO_3, if ambient NH_3 concentrations remain above 2 μg-atoms·liter^{-1} (Peterson et al. unpublished). By late summer (August and September) both these inorganic nitrogen species may be depressed below rate-limiting concentrations (Fig. 5). And, because total N:P is generally less than 16:1 by atoms, nitrogen is considered to be the growth-limiting micronutrient.

The seasonal distributions of dissolved O_2 and CO_2 are less well defined than the micronutrient distributions. Oxygen is generally slightly below saturation concentrations and CO_2 is generally well above saturation concentrations throughout the year (an exception is discussed in the last section).

SOURCES AND SINKS

To compute annual OCNSi sources and sinks (Table 2) I first approximate the atmospheric exchange rates of O_2 and CO_2 from their average or typical concentrations in the Bay, and then superimpose the photosynthetic activities of the phytoplankton on effects of atmospheric exchange processes to establish their influence on micronutrient circulation and dissolved O_2 and CO_2 distributions. Next, exchange rates between the substrate and overlying water are included and compared with the contributions from river inflow and waste. Finally, I discuss the problem of defining exchange at the ocean boundary.

Atmosphere

Oxygen is a useful parameter for initial calculations because its seasonal and spatial variations are small. Average near-surface oxygen concentrations along the main channel are 94% of saturation with respect to the atmosphere (Fig. 6). Thus the Bay waters act as an oxygen sink, and the oxygen-invasion rate from the atmosphere can be estimated using the stagnant boundary layer model (cf. Broecker and Peng 1974):

$$\text{Oxygen flux} = D\frac{C_S - \alpha P}{\mu}$$

where D is the molecular diffusivity of gas, μ is the film thickness, C_S is the gas concentration at the bottom of the film, α is the gas solubility, and P is the partial pressure of gas in the air.

Typical summer and winter dissolved oxygen concentrations range from 500 to 700 μg-atoms·liter^{-1} depending on seasonally varying water temperature and salinity (Fig. 7). Using this range of values and assuming that near-surface O_2 is at 94% of saturation concentrations, $C_S - \alpha P$ ranges from -32 to -45 μg-atoms O_2:liter^{-1} (a value of -40 μg-atoms·liter^{-1} is used in the calculation, Table 3). Other assumptions include: average temperature is 10°C in winter and 20°C in summer (this requires an accompanying change in D); μ is estimated from wind speeds (Emerson 1975) of 6 m·s^{-1} in summer and 2 m·s^{-1} in winter. An annual mean oxygen surface-exchange rate is made from the average winter and summer exchange rates using the North Bay volume (4.12 x 10^9 m^3) and area (6 x 10^8 m^2; Selleck et al. 1966).

Dissolved CO_2 in the estuary is generally 125 to 250% above saturation concentrations (Spiker and Schemel 1979). Saturation concentration is about 12 μg-atoms·liter^{-1} depending primarily on alkalinity, temperature, and salinity (see Skirrow 1975). Thus, the CO_2 evasion rate from the estuary is estimated following a similar calculation as above for O_2. $C_S - \alpha P$ is assumed to be 12 μg-atoms·liter^{-1} in both winter and summer (Table 3), or about 200% saturation.

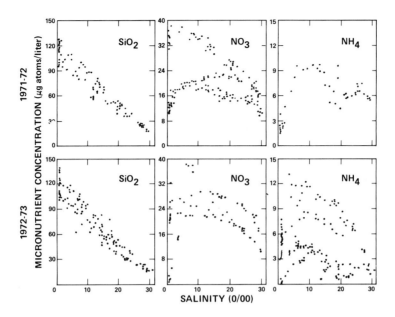

Fig. 4. Winter (October through April) micronutrient distributions with salinity during years indicated in each panel. Data from near-monthly surveys (Smith et al. 1979).

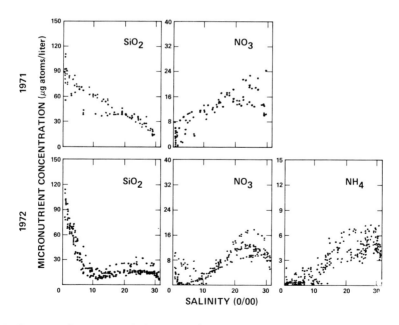

Fig. 5. Summer (May through September) micronutrient distributions with salinity during years indicated in each panel. Data from near-monthly surveys (Smith et al. 1979).

TABLE 2. ORDER-OF-MAGNITUDE ESTIMATES OF OCNSi SOURCES AND SINKS (REAL AND POTENTIAL) FOR NORTHERN SAN FRANCISCO BAY ESTUARY.[a]

SOURCES/SINKS[b]	OXYGEN (1)	OXYGEN (2)	OXYGEN (3)	CARBON (1)	CARBON (2)	CARBON (3)	NITROGEN (1)	NITROGEN (2)	NITROGEN (3)	SILICON (1)	SILICON (2)	SILICON (3)
BOUNDARIES												
Atmosphere	+540	+56	+8.2	-94	-13	-1.9					0	
Substrate	-640	-67	-9.8	+170	+24	+3.5	+34	+4	+0.60	+85	+5[d]	+0.8
Ocean[c]	-580	-60	-8.7		?			?			?	
River												
dissolved inorganic	+580	+60	+8.7	+780	+110	+16	+26	+3	+0.45	+290	+17	+2.5
particulate				+270	+38	+5.5	+27	+3.2	+0.47			
WATER COLUMN												
photic depth[e]	+430	+45	+13	-180	-25	-7.2	-34	-4	-1.2	-170	-10	-1.5
aphotic depth[f]	-330	-34	-10	+100	+14	+4	+9	+1	+0.28			
Total Depth	+100	+11	+1.6	-80	-11	-1.6	-25	-3	-0.4	-170	-10	-1.5
ANTHROPOGENIC												
Waste	-86	-8.9	-1.3	+29	+4	+0.6	+21	+2.5	+0.36			

[a] Note that these rates illustrate mean-annual approximations and that seasonal variations may be large (see text). Values are reported as (1) metric tonnes per day; (2) milligram-atoms per meter^2 per day; and (3) microgram-atoms per liter per day. Two figures are generally given for ease in following the calculations and do not indicate precision.

[b] Sources (+) mean the element is introduced to the water column as a dissolved species (with the exception of river-borne particles); sinks (-) mean the dissolved species is converted to a particulate phase or leaves the system. See the text for a discussion of assumptions. Volume = 4.12×10^9 m^3, area = 6×10^8 m^2, depth = 7 m.

[c] See text for discussion of uncertainties.

[d] From Hammond and Fuller (1979).

[e] One-half total volume.

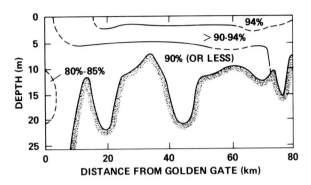

Fig. 6. Longitudinal and depth distribution of dissolved oxygen % saturation (median value) from near-monthly surveys over the period from Fall 1971 to 1974.

Phytoplankton

The photosynthetic rates utilized in the calculations are from observations made in 1976 and 1977, when phytoplankton biomass and associated activity in the inner estuary were probably below typical levels (Peterson et al. unpublished). Net 24-hour oxygen production and carbon assimilation, averaged over the photic depth, ranged from 43 to 100 and 35 to 87 mg-atoms·m^{-2}·d^{-1} respectively, in the outer estuary, and from 16 to 72 and from 13 to 27 mg-atoms·m^{-2}·d^{-1}, respectively, in the turbid landward regime for "sunny" spring and fall days (75 Einsteins·m^{-2}·d^{-1}). Photic depth ranged from 3.4 to 7.3 m in the seaward portion of the estuary and from 0.5 to 2.3 m landward.

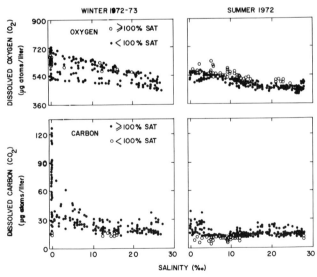

Fig. 7. Winter and summer (1972) dissolved oxygen and carbon dioxide distribution with salinity. Note the different symbol for above and below 100% saturation with respect to atmosphere and compare with summer (1972) fluorescence (Fig. 3) and micronutrient (Fig. 5) distributions.

TABLE 3. ASSUMED PARAMETERS AND ESTIMATES OF AVERAGE NET GAS-EXCHANGE RATES BETWEEN THE ATMOSPHERE AND SAN FRANCISCO BAY[a]

Parameter[b]	O_2 Winter	O_2 Summer	CO_2 Winter	CO_2 Summer
$C_S - \alpha P$ (μg-atoms·liter^{-1})	-40	-40	+12	+12
D (10^{-5} cm^2·s^{-1})	1.57	2.06	1.25	1.64
μ (10^{-6} m)	240	80	240	80
flux (mg-atoms·m^{-2}·d^{-1})	-23	-89	+5.4	+21

[a] (- indicates a sink, + a source).
[b] cf. Broecker and Peng 1974.

Rates of dark-bottle biochemical mineralization, O_2 consumption, and CO_2 production were of the same order as the analytical precision and probably estimated an upper limit to the real rates in the aphotic water column. Seventy percent of the values were between 4.8 and 17 μg-atoms·liter^{-1}·d^{-1} for O_2 consumption and between -1.2 and 4.8 μg-atoms·liter^{-1}·d^{-1} for CO_2 production. Ten and 4 μg-atoms·liter^{-1}:d^{-1} respectively are the values used in Table 2.

In the oxygen-production and carbon-assimilation model, I assume that photic processes occur in half the North Bay volume and aphotic processes in the other half. The average North Bay depth, 7 m, is used. I also assume that photic depth-integrated rates represent the upper half-volume, and that dark-bottle experimental rates represent the lower half-volume. Photic depth productivity selected for oxygen and carbon are well within the observed ranges (Table 2) but may seem low compared to values inferred in the literature. An upper limiting value for the Bay might be the annual average of 80 mg-atoms C·m^{-2}·d^{-1} as observed in the New York Bight (Malone 1976). Ambient chlorophyll-*a* levels are similar in the two regions but average specific productivity is higher in New York Bight and the photic depth there is deeper (see also Cloern 1979).

Photosynthetic nitrogen assimilation is six times carbon assimilation (by atoms), and average dark bottle (aphotic) nitrogen mineralization is estimated to be 0.28 μg-atoms·liter^{-1}·d^{-1} (ranging from 0.02 to 0.4). These values were also applied to "photic and aphotic" North Bay volumes (Table 2). The estimate of silica utilization by phytoplankton was made using a numerical model (Peterson et al. 1978) and it appears consistent with the other values if a major fraction of the phytoplankton consists of diatoms.

Substrate

The processes of atmospheric-oxygen invasion and carbon evasion at the surface and of net oxygen production and carbon assimilation within the water do not balance. For the present it seems attractive to assume that these processes are balanced by exchange with the substrate (Table 2). Experimental observations of benthic oxygen consumption and NH_3 release made during winter indicate similar rates for O_2 and NH_3 as we have estimated (Hammond and Fuller 1979).

Sacramento-San Joaquin River (Delta Outflow)

Transport of river-borne nitrogen is estimated by multiplying the river concentration of dissolved inorganic nitrogen species (NH_3, NO_3, NO_2) by mean-monthly discharge and, for about half the months when no observations were available, by using a linear interpolation of concentration. Because concentrations are highest during high river discharge, monthly nitrogen income varies by several orders of magnitude from winter to summer (220 metric tonnes per day [$t \cdot d^{-1}$] in February, 1.5 in July, 1973, 80 $t \cdot d^{-1}$ in January and 1.6 in July, 1974). The annual average dissolved inorganic nitrogen transport was 31 $t \cdot d^{-1}$ in 1973 and 21 $t \cdot d^{-1}$ in 1974. Mean flows for these years were 720 and 1,100 $m^3 \cdot s^{-1}$, respectively, whereas the mean annual discharge is 600 $m^3 \cdot s^{-1}$. From the few observations available it appears that dissolved organic nitrogen is about equal in abundance to inorganic forms (Peterson et al. unpublished). Thus, measuring inorganic nitrogen underestimates the total dissolved nitrogen supply by a factor of two.

River-borne particulate nitrogen and carbon is estimated assuming: (1) a mean annual sediment load (Conomos and Peterson 1977); (2) that the sediment is 3% organic carbon (unpublished data); and (3), that sediment contains 0.3% organic nitrogen by weight (see Storrs et al. 1966).

Dissolved inorganic carbon is estimated from alkalinity. The month-to-month variations in alkalinity in the river, typically 0.8 to 1.5 $meq \cdot liter^{-1}$, are not well understood. We assume that a value of 1.25 $meq \cdot liter^{-1}$ is representative of average flow (600 $m^3 \cdot s^{-1}$) and that both dissolved inorganic carbon and alkalinity are mostly the bicarbonate species (1.2 mg-atoms $C \cdot liter^{-1}$). In the river, pCO_2 frequently exceeds three times the saturation level and is probably an important source of dissolved CO_2 in the estuary (Fig. 7). However, because the bicarbonate species represents most of the inorganic carbon supplied to the estuary, dissolved CO_2 levels in the river can be ignored here.

Waste

Municipal and industrial waste sources (Table 4) are averaged over the region from below Rio Vista to the San Mateo Bridge (Fig. 1), which includes the upper reaches of South Bay, because a substantial portion of the waste is released in this segment. To some unknown extent this waste is mixed with waters of the central region and may be available for transport into North Bay. Estimated loads have been averaged by volume over the input region (6.23 x 10^9 m^3) but estimates are given only for the volume of the North Bay, 4.12 x 10^9 m^3 (Table 3). This may overestimate the waste contributions from point sources to North Bay. Waste from non-point sources, such as urban runoff into local streams, although believed to be significant, is unknown and is not estimated.

The Sacramento and San Joaquin rivers are not solely a natural source of substances to the estuary; much of their water is used and reused before arriving at the Bay. The relative magnitude of anthropogenic contributions to the enormous winter runoff of nitrogen is unknown. Streams draining agricultural land have high nitrogen concentrations (Omernick 1976), however, and it is likely that fertilizer is an important source of nitrogen brought to the Bay. Livestock waste is another likely source. Nitrogen reported as fertilizer for the 1975 agricultural consumption total for the State of California is 1,300 $t \cdot d^{-1}$ (California Department of Food and Agriculture 1976). While this total should be corrected to reflect only that portion used within the Sacramento-San Joaquin drainage basin, this high value illustrates that only a small leakage could supply substantial amounts of nitrogen to the Bay.

An exemplary reference model documenting waste sources has been made for the New York Bight (Gross 1976; Mueller 1976). This region is of interest here even though its features are not the same (e.g. dredging, barging of waste, and atmospheric washout of nitrogen are important there) because the bulk of the wastes (from New York) are discharged near the mouth of the estuary as in San Francisco Bay (from San Francisco to Oakland).

SAN FRANCISCO BAY

TABLE 4. POINT SOURCE WASTE LOADS (METRIC TONNES PER DAY) FOR CENTRAL AND NORTH SAN FRANCISCO BAY[a]

Location[b]	Property			
	BOD	Organic Nitrogen	NH_3	NO_3
South Central (San Mateo to Bay Bridge)	79	6.9	10	1.0
Central-North (Bay Bridge to Pt. Richmond)	25	2.5	2.2	0.3
San Pablo Bay (Pt. Richmond to Carquinez)	13	0.8	3.4	0.5
Carquinez Straits (Carquinez to Benicia Bridge)	0.6	0.09	0.09	0.02
Suisun Bay (Benicia Bridge to Antioch)	15	0.7	3.3	0.2
TOTAL	130	11	19	2

[a] Unpublished data for 1975, supplied by California State Water Quality Control Board, San Francisco Bay Region.

[b] Location areas indicated in Fig. 1.

Ocean

The direction of net exchange at the river, atmosphere and bottom boundaries is usually apparent (e.g. from river to estuary). At the ocean boundary, however, it is difficult to determine the direction of net exchange without knowledge of circulation. In the Hudson River estuary, for example, the net transfer of nutrients is clearly from estuary to ocean (Garside et al. 1976; Simpson et al. 1977). Because average water replacement time there is short (a few days), *in situ* production is not the only major supply of phytoplankton found in the estuary. Rather, incoming near-bottom water has been identified as a dominant source of phytoplankton (Malone 1977). Establishing the net exchange of phytoplankton between the San Francisco Bay estuary and the coastal ocean is not so simple. Several difficulties in qualitatively describing the exchange process for the Bay are apparent.

If, rather than attempting an independent evaluation of exchange processes at the ocean boundary, we assume a steady-state budget, then dissolved-oxygen exchange between the ocean and Bay equals the supply of O_2 from river, atmosphere, and net photosynthesis in the water column minus O_2 consumption by the bottom. This predicts a net O_2 loss from the Bay to the ocean (580 t·d^{-1}; see Table 2).

The average distribution of dissolved O_2 suggests that the above calculation is qualitatively correct (as % saturation, Fig. 6). Incoming waters at the bottom tend to be low in dissolved O_2 because they often originate from wind- or estuarine-water-induced upwelling of low-oxygen deep water offshore (Conomos 1979). The central part of the Bay also receives a large quantity of sewage. Effects of the latter process are not clearly separated from those of the former but the relation of decreasing O_2 with decreasing NH_3, when the NH_3 concentrations are below maximum natural levels (about 3 μg-atoms·liter^{-1}, suggest upwelling as the primary low-oxygen source near the mouth of the Bay.

Incoming low-oxygen near-bottom waters may cause part of the decrease in oxygen concentration with increasing depth in the interior of the estuary (lowering our estimates of benthic O_2

consumption). Of course, a more obvious cause is simply that the contributing atmospheric, phytoplankton, and riverine sources are near-surface phenomena, while benthic consumption is a near-bottom phenomenon. In either case the landward-flowing near-bottom waters tend to be lower in O_2 than the seaward-flowing surface waters, thus maintaining an advective loss of O_2 from the Bay.

The key to understanding estuarine-transport processes is, however, in the nonadvective flux terms (Kinsman et al. 1977). The few observations made in San Francisco Bay (restricted to the main channel) suggest that horizontal eddy diffusion contributes about 70% of the sea-salt to the estuary (Conomos 1979, Fig. 25). The eddy exchange coefficients for sea salt may also apply to dissolved O_2. However, the net exchange of oxygen by eddy diffusivity remains qualitatively unknown because the horizontal oxygen gradients are small and not well-described, especially seaward of the Golden Gate. If we assume the converse, or a gain of O_2 from the ocean to Bay, then, O_2 consumption by the substrate becomes even more important (unless we also reverse the net O_2 flows as described).

Carbon exchange is more complicated than oxygen exchange because both particulate and dissolved phases are involved. Also, the inorganic reactions of dissolved inorganic carbon must be considered (cf. Skirrow 1975). Ignoring such complexities for dissolved CO_2, making the same calculation as for dissolved O_2 shows a loss of CO_2 from Bay to ocean (equal to river input).

We do not know if this calculated loss is reasonable because the average depth distributions of dissolved CO_2 near the ocean boundary have not been estimated. In the Bay pCO_2 generally exceeds atmospheric equilibrium values, whereas pCO_2 in surface ocean waters near the Bay is probably lower. This suggests some loss of CO_2 from the Bay to ocean, although this process represents a very small portion of the total exchange. The bulk of the inorganic carbon is in the form of dissolved bicarbonate. Its concentration in river water is about half that of seawater, whereas the concentration of particulate organic carbon is usually several times greater in the river than in the adjacent ocean (see also Spiker and Schemel 1979). Preliminary observations indicate that the concentration of dissolved organic carbon is about 80 μg-atoms·liter^{-1} for oceanic waters and about two times greater for the Sacramento River. Thus, dissolved organic carbon is quantitatively important, but its participation in the carbon circulation of the estuary is unknown.

The supply of river-borne particulate organic carbon is very large because the supply of sediment to the Bay is also very large. The supply of river-borne particulate organic carbon may rival *in situ* net phytoplankton production averaged over the photic depth (Table 2). It follows that, because the sediment supply is large, subtracting the accumulation rate of carbon or nitrogen in bottom sediments from their supply rates would not provide a reliable estimate of their loss to the ocean. Expected uncertainties in either the river supply or bottom accumulation rates of carbon or nitrogen, for example, are of the same magnitude as their waste-input rates (note that this large sediment dilution effect is also manifest in the general absence of high concentrations of anthropogenic trace elements in the bottom sediments [McCulloch et al. 1971; Peterson et al. 1972].)

The near-surface plume of NH_3-enriched water which commonly extends outside the Golden Gate reveals that NH_3 is almost continuously released from the Bay. Nitrate, however, appears to be transferred from the ocean to Bay or vice versa (Fig. 4).

The net exchange of dissolved silica is almost invariably from the Bay to ocean, as illustrated by higher silica concentrations in outflowing (surface) than inflowing (near-bottom) waters (Fig. 8, Panels A, B and C). In instances when silica concentrations in the Bay are depressed below ocean concentrations, the concentrations at the Golden Gate are lower at the surface than in near-bottom water, and net exchange is probably reversed from the ocean to the Bay (Fig. 8, Panel D).

Biogenic silica, primarily diatom remains, accumulates in bottom sediment. Dissolution supplies silica to the interstitial water and, ultimately, to the overlying water column. Hammond and

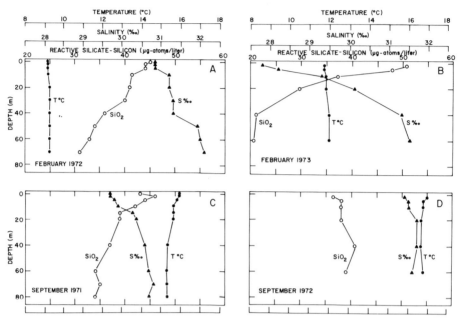

Fig. 8. Depth distribution of salinity, temperature and dissolved silica at the Golden Gate. A, B, winter; C, summer of high river discharge; D, summer of low river discharge.

Fuller (1979) have measured this rate *in situ* to be about 5 mg atoms·m^{-2}·d^{-1}. This measurement provides a lower limit to an estimate of incoming sedimentation of biogenic silica, assuming abiogenic silica dissolution to be negligible, because complete dissolution is not expected. In this sense both the estimated silica utilization in the water and Hammond and Fuller's observed supply from the bottom seem reasonable.

There is, then, a large export of dissolved silica from the Bay to the ocean, but this export is probably less than the river supply because there must be some loss associated with accumulation of bottom sediment. The average oxygen distribution at the Golden Gate also indicates a net loss of O_2 from the Bay to ocean, opposite to what one might intuitively expect. For example, average landward nontidal flow is 5 km·d^{-1} and seaward flow is 6 km·d^{-1} (Conomos 1975). Assuming that these velocities characterize half the cross-sectional area at the Golden Gate (44 x 10^3 m^2), that incoming waters are at 85% saturation concentrations (510 μg-atoms·liter^{-1}), and that outflowing waters are at 95% of saturation (570 μg-atoms·liter^{-1}), advective net transfer is 190 t·d^{-1} (about one-third the river supply). Perhaps there is also a net loss of nitrogen from the Bay; furthermore, atmospheric evasion along with net exchange to the ocean seems to be a mechanism for loss of dissolved CO_2 from the Bay. Also, there is probably a net transport of dissolved organic carbon from the Bay, especially if the river supply does not participate in biochemical processes in the estuary. Describing the direction of net exchange for carbon, however, will require a better knowledge of its distribution at the ocean boundary and in sediments.

SEASONAL VARIATIONS

Winter and summer distributions of gases and nutrients (Figs. 2, 3, 4, 5, 7) illustrate the effects of water circulation and of nonconservative sources and sinks. When water circulation is strong (when river inflow is high), the effects of sources and sinks not directly associated with the river or ocean are minimized and the overall distributions tend to be linearly related to salinity. When circulation weakens (when river flow is low), effects of these sources and sinks become increasingly apparent. Winter (November through April) is typically characterized by river flows greater than 500 $m^3 \cdot s^{-1}$, and summer by flows of 200 to 300 $m^3 \cdot s^{-1}$ or less. The lower flows are probably the main reason why the distributions tend to be nonlinear in summer. To contrast seasonal effects we will approximate the water-replacement time in the estuary at 500 $m^3 \cdot s^{-1}$ as 20 days and at 100 $m^3 \cdot s^{-1}$ as 70 days (Peterson et al. 1978).

Winter

River inflow is clearly a major source of substances to the estuary and, generally, must increase in importance with increasing winter discharge. During winter, photosynthetic processes are minimized because the standing crop of phytoplankton is less than in summer (Cloern 1979) and sunlight is about one-third that of summer values; atmospheric exchange may weaken because wind speeds average one-third those of summer values (Conomos 1979); and benthic activity is assumed to be subdued because of lower water temperatures. Thus, source or sink values are possibly lower than average while the river supply is about the same or greater. In a simplified model a reduction by a factor of two seems justified in the winter rates of net water column photosynthesis and atmospheric and benthic exchanges.

With regard to the level of dissolved O_2 in winter, net photosynthetic processes produce less O_2 in winter than in summer, atmospheric oxygen invasion is less in winter, and benthic oxygen consumption is also less. Therefore, the dissolved oxygen concentration in the estuary is controlled more by its concentration at the river and ocean boundaries than internal sources or sinks. If, for example, dissolved O_2 is near 100% saturation concentration in the river in winter, then the estuary would also tend to approach 100% saturation concentrations (depending on the O_2 concentrations at the ocean boundary). Similar reasoning applies to the other elements.

The relative importance of water circulation and of nonconservative sources and sinks is demonstrated by comparing replacement times of various elements in the water column, for each rate of supply or removal, with the water-replacement time, assuming the winter rates discussed above (Table 5). The dissolved-OCNSi concentrations used are based on the distributions presented herein; representative winter particulate carbon and nitrogen concentrations were chosen to be 60 and 6 μg-atoms·liter^{-1} respectively. If the water replacement time is clearly less than the substance-turnover time, then that source or sink is relatively ineffective.

The winter ratios of water-replacement time to OCSi (sources/sinks) turnover time are all less than one. Thus, near-conservative OCSi distributions are predicted to be a common occurrence in winter. This is indeed observed.

Near conservative winter distributions, then, are a consistent pattern for dissolved silica and to a lesser extent for chlorophyll-*a* fluorescence (or phytoplankton) and for dissolved O_2. Most of the NO_3 and, as might be expected from its low concentration in river and sea water, all of the NH_3 distributions appear nonconservative. It follows that the NH_3 sources may be relatively strong, even in winter, as is suggested by the distributions and calculations.

Summer

Nonconservative effects during summer are illustrated by increasing the rates of atmospheric and benthic exchange and net photosynthesis in Table 2 by 50% and decreasing river inflow from 600 $m^3 \cdot s^{-1}$ to 100 $m^3 \cdot s^{-1}$, a relatively low but not unusual river flow. Total inorganic carbon, primarily in the form of bicarbonate, is predicted to maintain conservative or near-conservative distributions in summer (Table 5).

TABLE 5. RATIO OF WATER REPLACEMENT TIME TO SUBSTANCE REPLACEMENT TIME DURING WINTER (W) AND SUMMER (S)[b]

Sources and Sinks	Oxygen		ΣCO$_2$[c]		POC[d]		NO$_3$		NH$_3$		PON[e]		Silicon	
	W	S	W	S	W	S	W	S	W	S	W	S	W	S
Atmosphere	0.13	1.4	0.01	0.11										
Substrate	0.16	1.7	0.02	0.21										
River Dissolved-inorganic	0.29	0.17	0.18	0.39			0.45	0.08	1	16			0.7	0.5
Particulate					1.8	0.71								
Water Column Net photosynthesis	0.03	0.28	0.26	1.8							0.66	4.7	0.2	2.6
Waste	0.04	0.15	0.007						0.66	3.5				
(See note[a])	600	600	1,700	1,700	60	90	20	10	6	4	6	9	20	60

[a] Concentrations (μg-atoms·liter^{-1} at/ℓ) in estuary.
[b] High value means water replacement is less important than source or sink in influencing substance distribution; water replacement is assumed to be 20 days in winter, 70 days in summer.
[c] Total dissolved inorganic carbon primarily as bicarbonate.
[d] Particulate organic carbon.
[e] Particulate organic nitrogen.

SAN FRANCISCO BAY

In summer 1972 the distributions were complex, but they largely reflected the effects of photosynthetic processes in the inner estuary, where the salinity was less than 15 °/oo. In this instance the usual distributions of dissolved O_2 and CO_2 were reversed: oxygen was above saturation concentrations with respect to the atmosphere and carbon dioxide was depressed below saturation concentrations. In that case photosynthetic rates must balance or must exceed the carbon supply associated with both benthic oxygen consumption and atmospheric invasion. The approximation of net photosynthetic effects is now probably low, especially for this section of the estuary. The selected value can be easily revised upwards, however, in part because we do not know very well what the real net photosynthetic rate is over the entire water column.

As noted, the similar magnitudes of many of the estimated sources and sinks emphasize this dynamic nature of the system. In summer 1972 (Fig. 7) photosynthetic O_2 production needed only about 50 μg-atoms·liter^{-1} to achieve equilibrium concentrations, making net atmospheric exchange zero. Assuming a daily increase in net photosynthetic dissolved oxygen equal to the precision of the method, or, about 10 μg-atoms·liter^{-1}·d^{-1}, we conclude that it would take only 5 days to make up this deficit! It should also be noted that photosynthetic production:mineralization processes disturb the dissolved CO_2 equilibration with the atmosphere disproportionately, as compared with O_2 equilibration, primarily because the ratio of O_2 to CO_2 in the atmosphere is very small (O:C = 630:1 by atoms).

The last example illustrating the dynamic nature of the system is suggested by comparing the summer of 1971 with 1972. In 1971 river inflow was stronger, averaging 400 m^3·s^{-1}, as compared with 200 in 1972, and photosynthetic effects in 1971 appear relatively weaker (Figs. 3, 5). Thus, it is apparent from the observations that the calculations as made have considerable freedom for adjustment up or down and will still appear realistic.

During periods of high river flow, then, many nonconservative substances appear to be passing through the estuary. During low river flow, however, the magnitudes of local sources and sinks may exceed river and ocean inputs. Also, their estimated magnitudes indicate that several processes control the oxygen, carbon, nitrogen and silicon distributions at comparable rates. With more observations, future refinement in the above estimates of sources and sinks and water replacement times is expected. It is important that such refinement be accompanied by observations from the main channels into the "unknown" waters overlying the shoals of the Bay and into the shelf waters entering the Bay.

ACKNOWLEDGMENTS

I thank W. W. Broenkow and D. E. Hammond for their critical reviews, and my colleagues at the U. S. Geological Survey for their comments and for the pleasure of working with them in the field.

LITERATURE CITED

Bailey, T. E. 1967. Estuarine oxygen resources - photosynthesis and reaeration. Pages 310-334 *in* Natl. Symp. on Estuarine Pollution Proc. Stanford University, Stanford, Calif.

Bain, R. C., and J. C. McCarty. 1965. Nutrient-productivity studies in San Francisco Bay. U. S. Public Health Service, Central Pacific Basins Water Pollution Control Admin. 116 pp.

Broecker, W. S., and T. H. Peng. 1974. Gas exchange rates between air and sea. Tellus 24:21-35.

Brown, R. L. 1975. The occurrence and removal of nitrogen in subsurface agricultural drainage from San Joaquin Valley, California. Water Res. 9:529-546.

California Department of Water Resources and Fish and Game. 1972. Dissolved oxygen dynamics Sacramento-San Joaquin Delta and Suisun Bay. State of California, The Resources Agency. 129 pp.

California Department of Food and Agriculture. 1976. Fertilizing materials, tonnage report 1975, ACF 58-018. 172 pp.

Carpenter, J. H. 1965a. The accuracy of the Winkler method for dissolved oxygen analysis. Limnol. Oceanogr. 10:135-140.

Carpenter, J. H. 1965b. The Chesapeake Bay Institute Technique for the Winkler dissolved oxygen method. Limnol. Oceanogr. 10:141-143.

Cloern, J. E. 1979. Phytoplankton ecology of the San Francisco Bay system: The status of our current understanding. Pages 247-264 in T. J. Conomos, ed. San Francisco Bay: The Urbanized Estuary. Pacific Division, Amer. Assoc. Advance. Sci., San Francisco, Calif.

Conomos, T. J. 1975. Movement of spilled oil as predicted by estuarine nontidal drift. Limnol. Oceanogr. 20(2):159-173.

Conomos, T. J. 1979. Properties and circulation of San Francisco Bay waters. Pages 47-84 in T. J. Conomos, ed. San Francisco Bay: The Urbanized Estuary. Pacific Division, Amer. Assoc. Advance. Sci., San Francisco, Calif.

Conomos, T. J., and D. H. Peterson. 1977. Suspended-particle transport and circulation in San Francisco Bay: An overview. Pages 82-97 in L. E. Cronin, ed. Estuarine Processes. Vol. 2. Academic Press, New York.

Conomos, T. J., R. E. Smith, D. H. Peterson, S. W. Hager, and L. E. Schemel. 1979. Processes affecting seasonal distributions of water properties in the San Francisco Bay estuarine system. Pages 115-141 in T. J. Conomos, ed. San Francisco Bay: The Urbanized Estuary. Pacific Division, Amer. Assoc. Advance. Sci., San Francisco, Calif.

Davis, C. O., P. J. Harrison, and R. C. Dugdale. 1973. Continuous culture of marine diatoms under silicate limitation, I. Synchronized life cycle of *Skeletonema costatum*. J. Phycology 9:175-180.

DiToro, D. M., R. V. Thomann, D. J. O'Connor, and J. L. Mancini. 1977. Estuarine phytoplankton biomass models—verification analyses and preliminary applications. Pages 969-1020 in E. D. Goldberg et al., eds. The Sea, Ideas and Observations of Progress in the Study of the Seas. Vol. 6. J. Wiley & Sons, New York.

Elliott, A. J. 1976. A study of the effect of meteorological forcing on the circulation in the Potomac estuary. Chesapeake Bay Institute, The Johns Hopkins University, Spec. Rep. 56. 66 pp.

Emerson, S. R. 1975. Gas exchange rates in small Canadian shield lakes. Limnol. Oceanogr. 20:754-761.

Garside, C., T. C. Malone, O. A. Roels, and B. C. Sharfstein. 1976. An evaluation of sewage derived nutrients and their influence on the Hudson estuary and New York Bight. Estuarine Coastal Mar. Sci. 4:281-290.

Goering, J. J., D. M. Nelson, and J. A. Carter. 1973. Silicic acid uptake by natural populations of marine phytoplankton. Deep-Sea Res. 20:777-789.

Gross, M. G. 1976. Sources of urban wastes. Amer. Soc. Limnol. Oceanogr. Spec. Symp. 2:150-161.

Hammond, D. E., and C. Fuller. 1979. The use of Radon-222 to estimate benthic exchange and atmospheric exchange rates in San Francisco Bay. Pages 213-230 in T. J. Conomos, ed. San Francisco Bay: The Urbanized Estuary. Pacific Division, Amer. Assoc. Advance. Sci., San Francisco, Calif.

Hansen, D. H. 1967. Salt balance and circulation in partially mixed estuaries. Pages 45-51 in G. H. Lauff, ed. Estuaries. Amer. Assoc. Advance. Sci. Pub. No. 83.

Head, P. C. 1971. An automated phenolhypochlorite method for the determination of ammonia in sea water. Deep-Sea Res. 18:531-532.

Kinsman, B., et al. 1977. Transport processes in estuaries: recommendations for research. Marine Sciences Research Center, State University of New York, Stony Brook, N. Y. Spec. Rep. 6, Reference 77-2. 21 pp.

Malone, T. C. 1976. Phytoplankton productivity in the apex of the New York Bight: Environmental regulation of productivity/chlorophyll *a*. Pages 260-272 in M. G. Gross, ed. The Middle Atlantic Shelf and New York Bight. Amer. Soc. Limnol. Oceanogr. Spec. Symp. No. 2.

Malone, T. C. 1977. Environmental regulation of phytoplankton productivity in the lower Hudson estuary. Estuarine Coastal Mar. Sci. 5:157-171.

McCarty, J. C., et al. 1962. An investigation of water and sediment quality and pollutional characteristics of three areas in San Francisco Bay 1960-61. Sanitary Eng. Res. Lab., University of

California, Berkeley, Calif. 571 pp.

McCulloch, D. S., T. J. Conomos, D. H. Peterson, and K. W. Leong. 1971. Distribution of mercury in surface sediments in San Francisco Bay estuary, California. U. S. Geol. Surv. Open-File Rep.

Miller, R. C., W. D. Ramage, and E. L. Lazier. 1928. A study of physical and chemical conditions in San Francisco Bay, especially in relation to tides. Univ. Calif. Pub. Zool. 31:201-267.

Mueller, J. A., A. R. Anderson, and J. S. Jeris. 1976. Contaminants entering the New York Bight: sources, mass loads, significants. Pages 162-170 in M. G. Gross, ed. The Middle Atlantic Shelf and New York Bight. Amer. Soc. Limnol. Oceanogr. Spec. Symp. No. 2.

Omernick, J. M. 1976. The influence of land use on stream nutrient levels. EPA (Environmental Protection Agency)-600/3-76-014. 106 pp.

Paasche, E. 1973a. Silicon and the ecology of marine plankton diatoms. I. *Thalassiosira pseudonana (Cyclotella nana)* grown in a chemostat with silicata as limiting nutrient. Mar. Biol. 19:117-126.

Paasche, E. 1973b. Silicon and the ecology of marine plankton diatoms. II. Silicate-uptake kinetics in five diatom species. Mar. Biol. 19:262-269.

Peterson, D. H., D. S. McCulloch, T. J. Conomos, and P. R. Carlson. 1972. Distribution of lead and copper in surface sediments in the San Francisco Bay estuary, California. U. S. Geol. Surv. Misc. Field Studies Map MF-323.

Peterson, D. H., T. J. Conomos, W. W. Broenkow, and E. P. Scrivani. 1975. Processes controlling the dissolved silica distribution in San Francisco Bay. Pages 153-187 in L. E. Cronin, ed. Estuarine Research. Vol. 1. Chemistry and Biology. Academic Press, New York.

Peterson, D. H., J. F. Festa, and T. J. Conomos. 1978. Numerical simulation of dissolved silica in the San Francisco Bay. Estuarine Coastal Mar. Sci. 7:99-116.

Peterson, D. H., et al. 1978. Estimates of phytoplankton productivity using O_2, C, and N, San Francisco Bay estuary. (Unpublished.)

Schemel, L. E., and L. A. Dedini. 1978. A continuous water sampling and multi-parameter analysis system. U. S. Geol. Surv. Open-File Rep. 200 pp.

Selleck, R. E., E. A. Pearson, B. Glenne, and P. N. Storrs. 1966. A comprehensive study of San Francisco Bay, final report. IV. Physical and hydrological characteristics of San Francisco University of California, Berkeley, Sanitary Eng. Res. Lab. Rep. 65(10):1-99.

Simpson, H. J., S. C. Williams, C. R. Olsen, and D. E. Hammond. 1977. Estuaries, geophysics, and the environment. National Academy of Science. pp. 94-103.

Skirrow, G. 1975. The dissolved gases-carbon dioxide. Pages 1-192 in J. P. Riley and G. Skirrow, eds. Chemical Oceanography. Vol. 2. Academic Press, New York.

Smith, R. E., and W. W. Broenkow. 1978. Equations for dissociation constants of carbonic and boric acids in estuaries. (Unpublished.)

Smith, R. E., R. E. Herndon, and D. D. Harmon. 1979. Physical and chemical properties of San Francisco Bay waters, 1969-1976. U. S. Geol. Surv. Open-File Rep.

Solorzano, L. 1969. Determination of ammonia in natural waters by the phenolhypochlorite method. Limnol. Oceanogr. 14:799-801.

Spiker, E., and L. E. Schemel. 1979. Distribution and stable-isotope composition of carbon in San Francisco Bay. Pages 195-212 in T. J. Conomos, ed. San Francisco Bay: The Urbanized Estuary. Pacific Division, Amer. Assoc. Advance. Sci., San Francisco, Calif.

Storrs, P. N., R. E. Selleck, and E. A. Pearson. 1963. A comprehensive study of San Francisco Bay, 1961-62: Second annual report. Univ. Calif. Sanitary Eng. Res. Lab Rep. No. 63-4. 323 pp.

Storrs, P. M., R. E. Selleck, and E. A. Pearson. 1964. A comprehensive study of San Francisco Bay, 1962-63: third annual report. Univ. Calif. Sanitary Eng. Res. Lab. Rep. No. 64-3. 227 pp.

Storrs, P. N., E. A. Pearson, and R. E. Selleck. 1966. A comprehensive study of San Francisco Bay, final report; V. Summary of physical, chemical and biological water and sediment data. Univ. Calif. Sanitary Eng. Res. Lab. Rep., vol. 67, No. 2. 1940 pp.

Strickland, J. D. H., and T. R. Parsons. 1968. A manual of sea water analysis. Can. Fish. Res. Bd. Bull. No. 125, 2nd ed. 203 pp.

Strickland, J. D. H., and T. R. Parsons. 1972. A practical handbook of seawater analysis. Can. Fish. Res. Bd. Bull. 167 (2nd ed.):21-28.

Weisberg, R. H. 1976. A note on estuarine mean flow estimation. J. Mar. Res. pp. 387-394.

Weiss, R. F. 1970. The solubility of nitrogen, oxygen, and argon in water and seawater. Deep-Sea Res. 17:721-736.

DISTRIBUTION AND STABLE-ISOTOPE COMPOSITION OF CARBON IN SAN FRANCISCO BAY

ELLIOTT C. SPIKER AND LAURENCE E. SCHEMEL

U. S. Geological Survey National Center, Reston, VA 22092
U. S. Geological Survey, 345 Middlefield Road, Menlo Park, CA 94025

Observed distributions of alkalinity, pCO_2 and $\delta^{13}C(\Sigma CO_2)$ indicate that dissolved inorganic carbon (DIC) was primarily supplied to San Francisco Bay by ocean, Delta, and municipal waste waters during the low Delta-outflow period from March 1976 to March 1977. Delta-derived alkalinity was typically about half that of ocean water and increased slightly with time. The pCO_2 values were highest (2 to 3 times the atmospheric value of approximately 325 ppm) in the Sacramento River and southern boundary of South Bay and decreased to near-atmospheric values seaward of the Golden Gate. The $\delta^{13}C(\Sigma CO_2)$ was lowest in the Sacramento River (approx. -10 $^{\circ}/oo$), increasing to marine values in the Gulf of the Farallones (approx. +2 $^{\circ}/oo$). Golden Gate values were approximately 2 $^{\circ}/oo$ less than those seaward, indicating that at least 10% of the ΣCO_2 was biogenic and is apparently the product of respiration and the mineralization of organic matter in the Bay. South Bay alkalinity and pCO_2 levels increased southward, whereas $\delta^{13}C(\Sigma CO_2)$ and salinity decreased. Municipal waste discharged into South Bay is the most probable source of the excess biogenic CO_2.

Distributions of particulate organic carbon (POC) in North Bay were influenced by *in situ* phytoplankton production and seaward dilution of riverine and estuarine POC. Apparent depletions of pCO_2 in North Bay coincide with chlorophyll *a*, POC, and $\delta^{13}C(\Sigma CO_2)$ increases. The $\delta^{13}C(POC)$ values during March 1977 approached those predicted for *in situ* algal production, suggesting that about 80 to 90% of the POC was produced in the seaward part of the estuary. *In situ* algal production was an important source of POC in the river. However, in the null-zone associated turbidity maximum, less than two-thirds of the POC appears to be riverborne, the remaining one-third being produced in the estuary or associated with resuspended bottom sediment. South Bay suspended POC appears to be a mixture of resuspended bottom sediments, *in situ* produced POC and land-derived organic carbon. Based on $\delta^{13}C$ data, *Spartina* salt-marsh grass does not appear to be a significant source of detritus in the Bay.

The $\delta^{13}C$ of sediment total organic carbon (TOC) indicates that riverine carbon from the Delta is diluted in the Bay by estuarine and marine carbon. The suspended POC and sediment TOC $\delta^{13}C$-measurements approached marine values seaward of the Golden Gate.

The importance of carbon in estuaries is fundamental because estuaries are valuable environments for carbon fixation by aquatic plants and this estuarine production is essential in maintaining some fisheries (Woodwell et al. 1973). The San Francisco Bay estuarine system (Fig. 1) receives carbon from natural sources as well as large amounts from the surrounding urban area, primarily in the form of municipal wastes. In this chapter, we discuss the sources and dynamics of carbon in San Francisco Bay waters by presenting distributions of some important organic and inorganic forms and their stable carbon isotope compositions.

SAN FRANCISCO BAY

Carbon in Estuaries

Dissolved inorganic carbon (DIC) in estuarine waters comprises dissolved carbon dioxide (CO_2) and the ionic species of dissociated carbonic acid:

$$H_2O + CO_2 \overset{K_1}{\rightleftharpoons} H^+ + HCO_3^- \overset{K_2}{\rightleftharpoons} 2H^+ + CO_3^{2-} \quad \text{(eq. 1)}.$$

Concentrations of these species depend on constants, K_1 and K_2, which are temperature and salinity dependent. A detailed discussion of DIC in estuaries is presented by Mook and Koene (1975).

DIC is most commonly described in terms of four measurable quantities: alkalinity (A), total CO_2 (ΣCO_2), negative logarithm of the hydrogen-ion activity (pH), and partial pressure of the CO_2 dissolved in the water (pCO_2). Alkalinity is the amount of hydrogen ion, in milliequivalents per liter (meq·liter^{-1}), required to convert carbonic and weaker acids to their undissociated forms:

$$A = C_{HCO_3^-} + 2C_{CO_3^{2-}} + C_{B(OH)_4^-} + C_{(OH^- \cdot H^+)} + C_{ma} \quad \text{(eq. 2)}.$$

The boric acid concentration ($C_{B(OH)_4^-}$) is low in river water and it increases to account for about 4 to 5% of the alkalinity in sea water (Skirrow 1975). The excess hydroxide concentration, $C_{(OH^- \cdot H^+)}$, is very low in the pH range naturally occurring in estuaries and can be ignored. The alkalinity due to other miscellaneous weak acids and bases, C_{ma}, can be important but, in general, over 95% of the alkalinity is attributable to carbonate and bicarbonate ions, which comprise all but a few percent of the total DIC in estuarine waters (eq. 1).

Although ΣCO_2, the total amount of CO_2 which can be removed from a solution after acidification, can be measured, in practice A is more easily determined to a higher degree of accuracy (Skirrow 1975). Furthermore, A is perhaps a more useful measurement because it is not affected by the addition or removal of molecular CO_2, as occurs during photosynthesis. In an estuary with a single large freshwater inflow, mixing with ocean water can result in a conservative (linear) alkalinity-salinity distribution, if the river and ocean compositions remain constant over the time necessary for water to mix through the estuary (water-replacement time). When internal sources or sinks are significant, a nonlinear distribution occurs and the degree of nonlinearity is related to their relative importance (Liss 1976). Nonconservative alkalinity distributions can result from precipitation or dissolution of carbonate minerals and from processes which release acids or bases. For example, utilization of ammonium or nitrate ions by phytoplankton during photosynthesis and other biochemical processes can produce measurable effects (Brewer and Goldman 1976).

The pH and pCO_2 of estuarine water will change if molecular CO_2 is added or removed. For example, CO_2 respired by organisms dissociates (eq. 1), increasing the hydrogen-ion activity and thus lowering the pH. Similarly, pCO_2 will increase because it is directly related to the concentration of dissolved CO_2. From freshwater to seawater, the magnitude of such a pH or pCO_2 change will vary primarily because of the salinity and temperature dependence of the dissociation constants and variations in the chemical composition of the water. The measurable quantities of the CO_2 system can be estimated if salinity, temperature, and any two of these four variables are known (Skirrow 1975). Using two salinity-alkalinity couples typical of the Bay, calculated changes in pH and ΣCO_2 are presented in Table 1 for increases in pCO_2 of 100 ppm (parts per million of 1 atmosphere). The decreases in pH correspond to relatively small increases in ΣCO_2. At lower salinity and alkalinity the changes in pH or pCO_2 represent smaller changes in CO_2.

The high sensitivities of pH and pCO_2 to changes in CO_2 make them excellent choices for studying photosynthetic processes in estuaries (Park 1969). Considering the precisions achievable for these measurements, very small changes are detectable. For the examples in Table 1, ±0.003

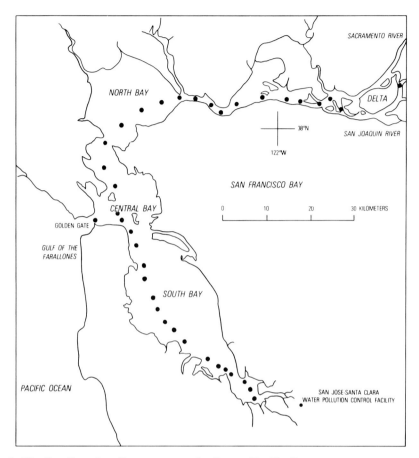

Fig. 1. The San Francisco Bay system and adjacent Pacific Ocean.

pH is equivalent to ±0.02 to 0.6 $\mu M \cdot liter^{-1}$ CO_2 and ±3.0 ppm pCO_2 corresponds to ±0.2 to ±1.0 $\mu M \cdot liter^{-1}$ CO_2. Analytically, pCO_2 offers some advantages in estuaries. Uncertainties in electrometric pH measurements can arise from variation of ionic strength, which is salinity and temperature dependent. Measurements of pCO_2 are not directly affected by this and analytical methods are adaptable to automated and continuous measurement (Schemel and Dedini 1979a).

Observed pCO_2 levels in rivers and coastal waters indicate that estuaries are typically oversaturated with CO_2. For example, pCO_2 was reported in excess of 1000 ppm in the Columbia River (Park et al. 1969) and 2000 ppm in the Hudson River (Hammond 1975). Values exceeding the atmospheric level (approx. 325 ppm) in coastal waters are often the result of mixing with river water (Kelley and Hood 1971; Gordon et al. 1973). The high pCO_2 in river and estuarine waters can be attributed to the mineralization of organic carbon, which is more abundant in these environments than in the ocean (Head 1976).

Following the approach taken by Mook and Koene (1975) in discussing pH distributions in estuaries, conservative pCO_2 distributions can be calculated, assuming that A and ΣCO_2 are conserved. Although the biological reactivity and other properties of CO_2 are such that conservative distributions might rarely be observed, these estimates can be useful in evaluating the relative importance of processes involving CO_2 in estuaries. The family of curves in Figure 2 illustrates the results of conservatively mixing ocean and river waters at constant temperature. Longitudinal

TABLE 1. CALCULATED CHANGES IN pH AND ΣCO_2 CORRESPONDING TO pCO_2 INCREASES FROM 300 TO 400 PPM AND 900 to 1000 PPM AT TWO SALINITIES AND TEMPERATURES.[a]

Salinity (°/oo)	Temperature °C	Alkalinity (meq·liter^{-1})	pCO_2 300 to 400 ppm		pCO_2 900 to 1000 ppm	
			pH change	ΣCO_2 change (mM·liter^{-1})	pH change	ΣCO_2 change (mM·liter^{-1})
6	10	1.48	-0.118	+0.018	-0.044	+0.008
6	20	1.48	-0.114	+0.020	-0.045	+0.008
24	10	1.96	-0.112	+0.036	-0.044	+0.012
24	20	1.96	-0.107	+0.042	-0.042	+0.011

[a] Calculations were made using expressions for the dissociation constants of carbonic and boric acids and the solubility of CO_2 based on the data of Buch (1951), Lyman (1956), and Weiss (1974).

temperature variations will, of course, influence the shapes of the curves. Net removal of CO_2 by processes operating in the estuary would cause the observed levels to be lower relative to the mixing curve, while CO_2 sources, such as respiration and mineralization of organic matter, would increase the pCO_2 levels.

Organic carbon is an important but difficult component to fully evaluate in estuarine waters. Measurements of organic carbon are normally made on two fractions, dissolved organic carbon (DOC) and particulate organic carbon (POC). DOC and POC are separated by filtration; POC is the fraction greater than about 0.5 to 1.0 μm in diameter (Parsons 1975). The concentrations of DOC and POC are similar in estuaries (Head 1976). POC is composed of both living organisms, such as plankton, and detrital (nonliving) material. The phytoplankton POC fraction is often estimated as the chlorophyll *a* concentration multiplied by an empirical factor, the carbon to chlorophyll *a* ratio. Because this ratio often varies (Banse 1974), for practical purposes, the chlorophyll *a* concentration alone is used as an indicator of relative phytoplankton POC abundance.

Riverine POC concentrations are typically higher than those in the marine environment (Head 1976). Although POC abundance generally decreases between the river and ocean because of dilution and deposition, resuspension of bottom sediments, primarily by wind-induced turbulence, may also influence local POC abundance in the Bay. Phytoplankton production can also be a major factor contributing to the seasonal abundance of POC in estuaries (Biggs and Flemer 1972). Seasonal POC data (Schemel and Dedini 1979b) indicate that this is important in the Bay.

Stable Carbon Isotopes in Estuaries

Differences in the thermodynamic and kinetic properties of the stable carbon isotopes, ^{12}C and ^{13}C, result in measurable isotope-composition variations in natural substances. For example, carbonate minerals contain more ^{13}C than organic matter produced by photosynthesis. The stable carbon isotope composition is expressed as the difference in parts per thousand (°/oo) between the substance and the PDB reference standard (Craig 1957):

$$\delta^{13}C(°/oo) = \frac{(R_s - R_{st}) \times 10^3}{R_{st}} \quad \text{(eq. 3)}$$

R_s and R_{st} are the $^{13}C/^{12}C$ ratio of the substance and standard, respectively. A δ-value of -10 °/oo means that the $^{13}C/^{12}C$ ratio of the substance is 10 °/oo lower than that of the standard.

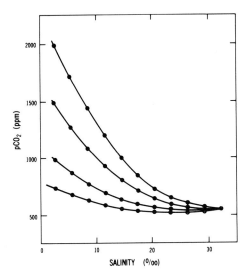

Fig. 2. Calculated conservative pCO_2 distributions for conditions typical of San Francisco Bay. Values of pCO_2 are estimated assuming conservation of ΣCO_2 and alkalinity (A = 1.174 + 0.034 [S ‰]) at 10°C, using expressions for the dissociation constants of carbonic and boric acids and the solubility of CO_2 based on the data of Buch (1951), Lyman (1956), and Weiss (1974).

Many processes operating in aqueous systems produce changes in the stable carbon isotope composition of DIC and POC. An observed $\delta^{13}C$ of the total dissolved inorganic carbon, $\delta^{13}C(\Sigma CO_2)$, is the product of the sources and sinks of carbon and the result of isotope fractionation occurring between solid, dissolved, and gaseous phases (Mook et al. 1974; Wigley et al. 1978). Major sources are biogenic carbon from the mineralization and respiration of organic matter (approx. -20 to -30 ‰), atmospheric carbon dioxide (approx. -7 ‰), and carbon derived from the dissolution of carbonate minerals (approx. 0 ‰). The effect of adding carbon from an identifiable source can be estimated by a mass balance calculation,

$$\delta_p = \delta_i f + \delta_x (1-f), \quad (eq.\ 4)$$

where δ_i, δ_p and δ_x are the $\delta^{13}C$ of the initial solution, the product, and the input carbon, respectively, and f and (1-f) are the fractions of the initial and input carbon sources.

The isotopic effects of processes involving fractionating outputs, such as photosynthetic utilization of CO_2 or outgassing, can be described by Rayleigh distillation equations for equilibrium (see Hendy 1971; Claypool and Kaplan 1974; Katz et al. 1977). For example, preferential utilization of ^{12}C during photosynthesis results in a ^{13}C enrichment of the residual solution, thus increasing the $\delta^{13}C(\Sigma CO_2)$. The isotopic composition resulting from a single output is estimated by:

$$\delta_r = [\delta_i + 1000\,(1-f)\,(1/a_{so}-1)] - 1000 \quad (eq.\ 5)$$

where δ_r and δ_i are the $\delta^{13}C$ of the residual and initial solution; f is the fraction of carbon removed from the solution, a_{so} is the fractionation factor between solution(s) and the output species(o): $a_{so} = R_s/R_o$, where $R = {^{13}C}/{^{12}C}$ of each. The composite effect of multiple sources and sinks in aqueous systems is discussed by Wigley et al. (1978).

The $\delta^{13}C(\Sigma CO_2)$ of fresh water ranges from zero to as low as -25 ‰. Dissolution of carbonate minerals produces a solution with an intermediate $\delta^{13}C$ value of about -11 ‰ (Fig. 3).

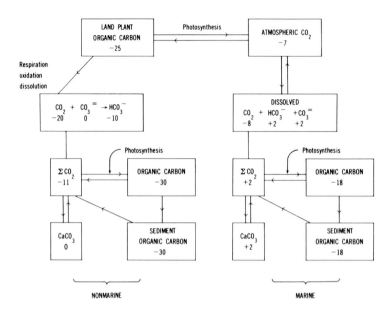

Fig. 3. Major carbon cycle oxidation and reduction processes and their attendant isotope fractionation effects.

More negative values are due mostly to the addition of biogenic CO_2, and more positive values may be a result of photosynthetic activity and atmospheric exchange (Broecker and Walton 1959; Oana and Deevey 1960; Parker and Calder 1970; Mook 1970; Rau 1978). In lakes with sufficient water-residence time, atmospheric exchange may result in a ^{13}C enrichment towards equilibrium $\delta^{13}C$ values (approx. +2 ‰), as in the surface ocean (Kroopnick 1974). Thus, seasonal trends are controlled by variations in biologic processes, atmospheric exchange, and the surface and groundwater base runoffs.

The $\delta^{13}C(\Sigma CO_2)$-salinity distribution has been observed to be near-conservative in the Hudson and Mississippi estuaries (Sackett and Moore 1966), and in the Western Scheldt estuary (Mook 1970). The forms of $\delta^{13}C$ conservative mixing curves in estuaries depend on the concentrations of carbon, C, and the $\delta^{13}C$, δ, in the river and ocean waters, $C_r\delta_r$ and $C_o\delta_o$, respectively. The $\delta^{13}C$ of the conservative mixture is:

$$\delta_{mix} = \frac{C_r\delta_r f + C_o\delta_o(1-f)}{C_{mix}} \quad (eq.\ 6)$$

The fraction of river (fresh), f, water is calculated from the salinity.

The $\delta^{13}C$ of estuarine and near-shore surface-sediment total organic carbon, $\delta^{13}C$ (TOC), and suspended particulate organic carbon, $\delta^{13}C$ (POC), increases from the river (approx. -27 ‰) to ocean (approx. -19 ‰). This is generally attributed to the mixing of carbon from land plants and freshwater algae (approx. -25 to -30 ‰) with estuarine and marine algae (approx. -12 to -25 ‰; Sackett and Thompson 1963; Hunt 1970; Shultz and Calder 1976; Gearing et al. 1977). The $\delta^{13}C$ of algae produced in the estuary is primarily related to the isotopic composition of the CO_2 source and the isotopic fractionation resulting from photosynthesis. The observed fractionation between ΣCO_2 and cell carbon may have a wide range (approx. -13 to -33 ‰), because of growth-rate differences, temperature, and species (see Wong and Sackett 1978). In order to approximate the importance of estuarine primary production, we assume an apparent photosynthetic fractionation effect of -19 ‰. Thus, if the $\delta^{13}C(\Sigma CO_2)$ varies (river to ocean) from -10 to +2 ‰, then, the $\delta^{13}C$ of algae produced *in situ* would vary from -29 to -17 ‰. Alternatively, the $\delta^{13}C(POC)$ distribution may be related to conservative dilution (eq. 6). Predicted $\delta^{13}C(POC)$ distributions based on conservative dilution and algal production are a reference for estimating the importance of these processes as well as identifying inputs from point sources and resuspended bottom sediments. Terrestrial organic carbon (approx. -25 ‰), and *Spartina* salt-marsh grasses (approx. -13 ‰) are two point sources of detritus that can be identified (Haines 1977; Fry et al. 1977).

METHODS

Measurements were made in San Francisco Bay during March and September 1976 and March 1977 at locations established in the channels (Fig. 1). Additional measurements were made in the Gulf of the Farallones during September 1976 and in North Bay during January 1976. Near-surface sediments were collected with a van Veen sampler in August 1973. Specific sampling locations, numerical values for the measurements, and a more detailed discussion of the analytical methods are found in Schemel et al. (1978).

Water samples were taken with a submersible pump or pumped through a hull fitting near the bow of the vessel at a 2-m depth. Salinity, temperature, and pCO_2 of the pumped water were continuously measured (Schemel and Dedini 1979a; Table 2). Salinity was measured with an induction salinometer on the vessel. *In situ* temperatures were measured at the submersible pump and at the hull-fitting intake.

The pCO_2 was estimated by continuously equilibrating a circulating volume of air with the sample-water flow and measuring the pCO_2 of the dried air with an infrared analyzer (Broecker and Takahashi 1966; Gordon and Park 1972). The accuracy of the method was tested by comparing measured values with values calculated from discrete pH and alkalinity measurements. Results compare within 6% (Table 3).

Discrete samples were collected for chlorophyll *a*, *A*, POC, and $\delta^{13}C$ analyses. Chlorophyll *a* was determined by the spectrophotometric method described by Strickland and Parsons (1972). *A* was determined by measuring the pH of a 50-ml filtered sample after addition of 0.010N HCL (Strickland and Parson 1972). POC samples were collected on precombusted glass-fiber filters and

TABLE 2. PRECISION AND ESTIMATED ACCURACY OF MEASUREMENTS.

Parameter	Precision	Estimated Accuracy
Salinity	±0.01 ‰	±0.05 ‰
Temperature	±0.1°C	±0.2°C
pCO_2	±0.3% of full scale	±5% of value
Chlorophyll *a*	±8% of value	not estimated
Particulate organic carbon	±5 μM (approx. ±20 μM C·liter^{-1})	not estimated
Alkalinity	±0.02 meq·liter^{-1}	not estimated
$\delta^{13}C$	±0.2 ‰	not estimated

SAN FRANCISCO BAY

TABLE 3. COMPARISON OF MEASURED AND CALCULATED pCO_2.[a]

Date	Salinity (°/oo)	Calculated pCO_2 (ppm)	Measured pCO_2 (ppm)	(% of calculated)
04Mar76	26.3	502	532	106
09Mar76	29.8	432	434	100
10Mar76	29.0	445	442	99
30Mar76	28.5	514	536	104
18Mar76	13.2	596	617	104
23Mar76	4.5	703	672	96
25Mar76	4.4	651	645	99
24Aug76	30.0	730	689	94
26Aug76	30.1	639	655	103
31Aug76	4.4	948	910	96
02Sep76	5.1	818	831	102
21Sep76	30.5	1218	1212	100
23Sep76	32.5	757	753	99

[a] The pCO_2 was calculated from discrete pH and alkalinity measurements using expressions for the dissociation constants of carbonic and boric acids and the solubility of CO_2 based on the data of Buch (1951), Lyman (1956), and Weiss (1974).

analyzed by a modification of the DOC method of Menzel and Vaccaro (1964; Schemel and Dedini 1979b).

Water samples for $\delta^{13}C(\Sigma CO_2)$ measurements were pressure-filtered (5 psi nitrogen) through Whatman GF/C glass-fiber filters directly into 500-ml bottles, poisoned with 2 ml saturated $HgCl_2$ solution, sealed, and refrigerated. CO_2 was extracted from acidified samples on a vacuum line. POC for $\delta^{13}C$ analyses was collected on precombusted GF/C glass-fiber filters and frozen. Near-surface sediment samples for $\delta^{13}C(TOC)$ analyses were stored in plastic cups, refrigerated, then dried at 50°C. Carbonate was removed from POC and sediment samples by acidification with HCl. Samples were combusted in a vacuum line, and the CO_2 was purified before isotopic analysis (method after Craig 1953; Degens 1969). The $^{13}C/^{12}C$ ratio of the CO_2 was measured with a 15-cm, 60°-sector ratio mass spectrometer.

RESULTS AND DISCUSSION

Delta outflow was abnormally low during most of the study period (Fig. 4; Conomos 1979). As a result, salinity was higher in the Bay, large segments of the Delta contained brackish water, and the null zone (Peterson et al. 1975a; Arthur and Ball 1979; Conomos 1979) moved landward. Water replacement time in North Bay was probably about 6 to 12 weeks, except in January 1976, a period of higher river flow, when water replacement time was about 3 weeks or less (Peterson et al. 1978). Because removal (flushing) of accumulated dissolved constituents from South Bay depends, in part, on the winter increase in Delta outflow (Conomos 1979), this flushing was probably minimal between September 1976 and March 1977.

Our observations from North Bay are related to salinity; the inner estuary extends from the confluence of the Sacramento and San Joaquin rivers to the region of mid-salinity (approx. 15

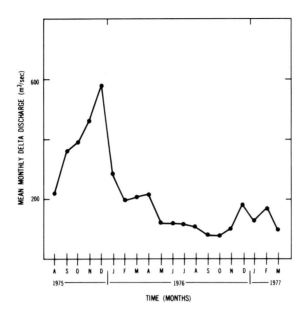

Fig. 4. Mean monthly Delta outflow ($m^3 \cdot s^{-1}$) from August 1975 through March 1977. Data from U. S. Bureau of Reclamation.

‰), and the outer estuary extends from the mid-salinity region to the Golden Gate. Upstream measurements (termed river) refer to Sacramento River unless otherwise noted. Observations in South Bay are related to distance from the Golden Gate.

Distributions of Dissolved Inorganic Carbon in North Bay Waters

Although variable, the alkalinity of Delta outflow was about half that of sea water (Fig. 5). Delta-outflow alkalinity was higher in March 1977 than March 1976, indicating a subtle but general increase in river-water alkalinity over the extended period of low outflow. The alkalinity distribution in North Bay was near-conservative in March 1976 and 1977 and September 1977, which suggests that the alkalinities of Delta outflow and ocean water were relatively invariant for at least 6 to 12 weeks, the time necessary for river water to mix through the estuary. Thus, for the purposes of this discussion, we make the general assumption that Delta outflow and ocean water are the major contributors to the DIC distribution in North Bay and that the distribution would be conservative if influenced by mixing alone. The net effects of other supply and removal processes operating in North Bay might then be indicated by pCO_2 or $\delta^{13}C(\Sigma CO_2)$ anomalies relative to conservative mixing of river and ocean waters.

While samples for $\delta^{13}C(\Sigma CO_2)$ analysis were taken in San Francisco Bay during March 1976 and 1977 and September 1976 (Figs. 6, 7, 8), the only Gulf of the Farallones sampling occurred during September 1976 (Fig. 7). Marine values were observed in the Gulf of the Farallones, but values were significantly lower at the Golden Gate even though salinity was only 1.4 ‰ lower in the near-surface water. In addition, pCO_2 was over 200 ppm higher at the Golden Gate than seaward. Golden Gate $\delta^{13}C(\Sigma CO_2)$ values (2- and 30-m depth) are about 2 ‰ less than predicted by conservative mixing of river and ocean waters, and suggest that about 10% of the ΣCO_2 is biogenic (eq. 4). Thus, processes which supply biogenic CO_2, such as respiration and the mineralization of organic matter, are effective in influencing the DIC in even high-salinity water at the Golden Gate.

SAN FRANCISCO BAY

The $\delta^{13}C(\Sigma CO_2)$ values at the Golden Gate were also lower than marine values in March 1976 and 1977; this may be typical of low Delta-outflow conditions. In normal years, when river flow is higher and water replacement time is shorter, the isotopic distribution may indicate a smaller biogenic fraction or may even appear conservative. Although the dynamic nature of the sources and sinks are not illustrated by the near-conservative ΣCO_2 distribution (Peterson 1979), isotopic data suggest that biogenic CO_2 produced within the estuary is exchanging with the DIC before it escapes to the atmosphere. If isotopic equilibrium exists between the DIC species, then there are two opposing effects to consider. Outgassing CO_2 is isotopically lighter (more negative $\delta^{13}C$) than the parent solution by about 9 ‰ (Mook et al. 1974), thus the $\delta^{13}C$ of the residual solution is increased (eq. 5). However, the input of biogenic CO_2, which is much lighter (approx. -20 to 25 ‰) will more effectively decrease the $\delta^{13}C(\Sigma CO_2)$. Values at the Golden Gate indicate that the supply of biogenic CO_2 is significant relative to the effect of outgassing. In addition, if isotopic

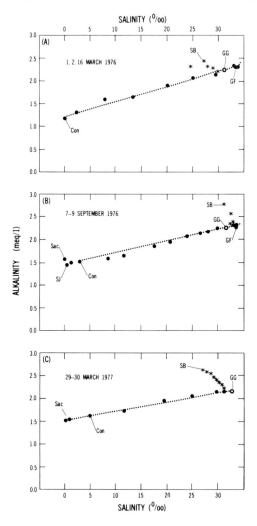

Fig. 5. Alkalinity distributions in San Francisco Bay and the Gulf of the Farallones; March 1976 (A), September 1976 (B), and March 1977 (C). Symbols: Sac, Sacramento River; Con, confluence of Sacramento and San Joaquin rivers; GG, Golden Gate; SB, South Bay; GF, Gulf of the Farallones; SJ, San Joaquin River.

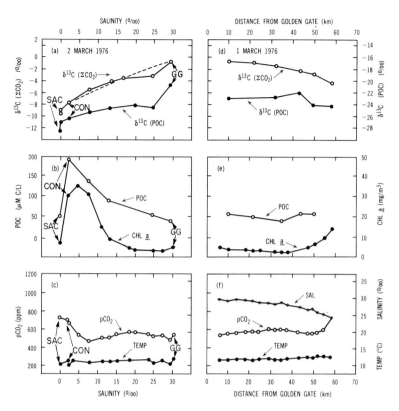

Fig. 6. Temperature (TEMP), salinity (SAL), chlorophyll a (CHL a), pCO_2, particulate organic carbon (POC), $\delta^{13}C(\Sigma CO_2)$, and $\delta^{13}C$(POC) in North Bay, 2 March 1976 (a, b, c) and South Bay, 1 March 1976 (d, e, f). Estimated conservative mix distribution is indicated by dashed line. Symbols: Sac R, Sacramento River; Con, confluence of Sacramento and San Joaquin rivers; GG, Golden Gate.

exchange with the atmosphere is significant, an estimate of the biogenic carbon input would be larger (see Wigley et al. 1978).

Estuarine CO_2 sources may also affect North-Bay DIC during times of higher Delta outflow. January is typically a period of low mean wind speed, low water temperature and high river flow (Conomos 1979). During January 1976, pCO_2 levels slightly exceeded that predicted by conservative mixing (Fig. 9), suggesting that CO_2 sources exceeded CO_2 sinks in North Bay.

Biological processes, particularly photosynthesis, can produce large pCO_2 variations in only a few days, which is considerably less than the water replacement time in North Bay during our observations. Phytoplankton biomass, as indicated by chlorophyll a concentration, was relatively high in the estuary in March 1976 (Fig. 6). Within the 2-wk period between the January and March samplings, pCO_2 levels became depleted by 200 to 400 ppm near the chlorophyll a maximum. This depression corresponds to an uptake of about 15 to 30 $\mu M \cdot liter^{-1}$ CO_2. Similarly, CO_2 uptake estimated from the $\delta^{13}C(\Sigma CO_2)$ anomaly is about 45 ± 20 $\mu M \cdot liter^{-1}$ CO_2 (eq. 5), assuming that the $\delta^{13}C$ of the biogenic (utilized) CO_2 is -26 ‰.

During March 1977, chlorophyll a was abundant in the outer estuary (Fig. 8). In this region, pCO_2 was depleted about 250 ppm below that estimated by conservative mixing, corresponding to about 75 $\mu M \cdot liter^{-1}$ CO_2. Similarly, the high $\delta^{13}C(\Sigma CO_2)$ values indicate a net utilization of about

SAN FRANCISCO BAY

70 ± 20 μM·liter^{-1} CO_2, if the $\delta^{13}C$ of the biogenic carbon is -20 $°/oo$ (eq. 5).

Chlorophyll a was less abundant and the effects of net photosynthesis were less apparent in North Bay during September 1976 (Fig. 7).

Distributions of Dissolved Inorganic Carbon in South Bay

Salinity in South Bay normally decreases southward because of dilution by municipal waste and local streams (Conomos 1979; Conomos et al. 1979). The effectiveness of these sources in influencing South-Bay water increases southward because the sectional volume of the basin decreases southward and the degree of mixing with North Bay water probably decreases southward.

In contrast to North Bay, alkalinity typically increased with decreasing salinity in South Bay (Fig. 5). South-Bay pCO_2 was about two times the atmospheric level during March 1976 and 1977 and over three times the atmospheric level in the southernmost reach during September 1976 (Figs. 6, 7, 8). The $\delta^{13}C(\Sigma CO_2)$ typically decreased southward, indicating that the carbon source is biogenic. Analysis of waste water from the San Jose-Santa Clara water pollution control facility (Fig. 1), the major single source of municipal waste in South Bay, was 4.0 mM·liter^{-1} ΣCO_2 with a $\delta^{13}C$ value of -13 $°/oo$ (sampled November 1977). Thus, it appears that the abundance of DIC in South Bay primarily relates to the increasing waste-enrichment of waters to the south.

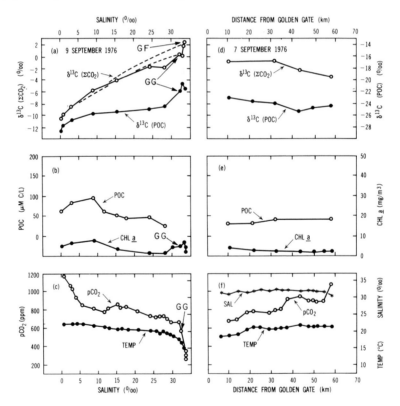

Fig. 7. Temperature (TEMP), salinity (SAL), chlorophyll a (CHL a), pCO_2, particulate organic carbon (POC), $\delta^{13}C(\Sigma CO_2)$, and $\delta^{13}C$(POC) in North Bay, and the Gulf of the Farallones, 8, 9 September 1976 (a, b, c) and South Bay, 7 September 1976 (d, e, f). Symbols: GG, Golden Gate; GF, Gulf of the Farallones. Estimated conservative mix distributions are indicated by dashed lines.

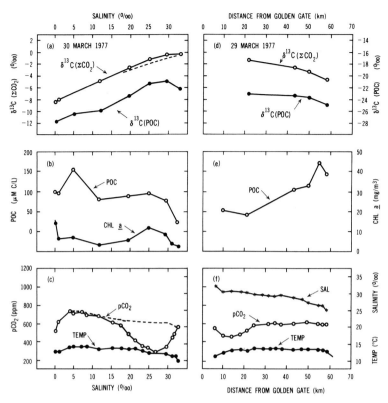

Fig. 8. Temperature (TEMP), salinity (SAL), chlorophyll *a* (CHL *a*), pCO_2, particulate organic carbon (POC), $\delta^{13}C(\Sigma CO_2)$, and $\delta^{13}C(POC)$ in North Bay, 30 March 1977 (a, b, c), and in South Bay, 19 March 1977 (d, e, f). Estimated conservative mix distributions are indicated by dashed lines.

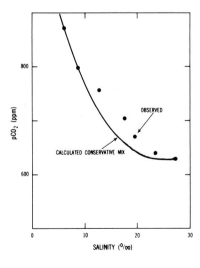

Fig. 9. Near-surface (2m) pCO_2 in North Bay, 27 January 1976. Conservative mix distribution was estimated using the alkalinity and pCO_2 values near the confluence of the Sacramento and San Joaquin rivers and at a salinity of 27.3 °/oo.

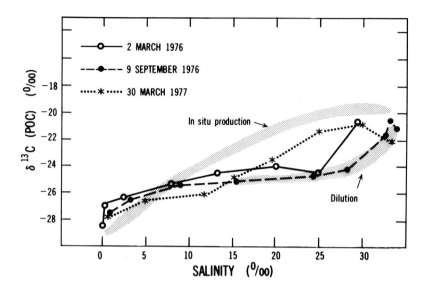

Fig. 10. $\delta^{13}C(POC)$ with respect to salinity in North Bay, 2 March 1976 (−o−), 9 September 1976 (--●--), and 30 March 1977 (·····*·····). Estimated distributions are based on *in situ* production and dilution models.

Suspended Particulate Carbon and Sediment Organic Carbon

In North Bay, an estuarine circulation cell maintains a turbidity (suspended particle) maximum in the null zone (Arthur and Ball 1979; Conomos 1979). The phytoplankton maximum associated with the null zone is maintained by advective transport and *in situ* phytoplankton production in this region of longer advective water-replacement time (Peterson et al. 1975a, b).

In general, POC abundance in North and South bays correlates well with chlorophyll *a* abundance (Figs. 6, 7, 8) indicating that phytoplankton production is an important source. Seaward of the river or apparent phytoplankton-POC sources in North Bay, concentrations decrease, indicating dilution and possibly deposition of POC. The $\delta^{13}C(POC)$ is particularly useful in evaluating the relative importance of processes and identifying major POC sources. Observed $\delta^{13}C(POC)$ distributions in North Bay are related to theoretical distributions based on conservative dilution seaward of the turbidity maximum and production by phytoplankton (Fig. 10). The distribution resulting from dilution is a concave-upward curve (eq. 6) because of the high POC concentration at low salinity. The distribution which would result from *in situ* algal production alone is estimated from the $\delta^{13}C(\Sigma CO_2)$ distribution and an average photosynthetic fractionation of -19 ‰. The $\delta^{13}C(\Sigma CO_2)$ varies from -10 to +2 ‰ over the salinity range 0 to 35 ‰, thus the estimated $\delta^{13}C(POC)$ will vary from -29 to -17 ‰.

In the outer estuary during March 1977, the $\delta^{13}C(POC)$ maximum, coincident with pCO_2 depletion and increased chlorophyll *a* and POC abundance, approached values predicted for *in situ* production (Fig. 10). An estimated 80 to 90% of the POC was produced locally (eq. 4). This was not observed when phytoplankton were less abundant in September and March 1976. Then, the distributions appeared to be primarily the result of dilution, with possible contributions from locally-resuspended bottom sediment.

The $\delta^{13}C(POC)$ values in the river were 2 to 4 ‰ less than those of land plants (approx. -25 ‰), indicating algal production (approx. -29 ‰) was an important source during all our samplings. The values in the turbidity maximum were several ‰ higher than the values upstream, but were similar to the $\delta^{13}C(TOC)$ in the bottom sediment (Fig. 11). The observed values indicate less than two-thirds of the POC in the turbidity maximum was riverborne (eq. 4) and, therefore, that the remainder was produced in the estuary or derived from resuspended sediments.

The spatial distribution of sediment $\delta^{13}C(TOC)$ is consistent with known sources of carbon in the Bay. Riverine carbon dominates the region landward of the null zone, and seaward there is a near-linear δ^{13} increase with distance to marine carbon values in the Gulf of the Farallones (Fig. 11). The $\delta^{13}C(TOC)$ in bottom sediments is more negative than the $\delta^{13}C(POC)$ observed in the water column in both North and South bays. This can be interpreted as a larger fraction of land-plant carbon in the sediments, reflecting deposition predominantly during periods of high Delta outflow. Alternatively, selective preservation and isotopic fractionation effects related to decomposition could account for the low sediment values (Degens et al. 1968; Eadie and Jeffrey 1973).

The value in sediment 16 km seaward of Golden Gate (-21.2 ‰) is within the range of mid-latitude marine POC, indicating that less than 10% of the carbon in these sediments is land derived. If this offshore carbon is considered as a mixture of three sources (marine, land-derived, and estuarine phytoplankton), then the fraction of land-derived carbon is probably even smaller.

In South Bay, local sources of riverborne sediments are small and suspended particle concentrations are probably controlled by wind-driven resuspension of bottom sediment and phytoplankton during most of the year (Conomos and Peterson 1977; Schemel and Dedini 1979b). During winter, Delta-derived low-salinity water, a source of suspended particulate matter, often penetrates South Bay (Conomos 1979). The apparent sources of suspended particulate matter predict that South Bay POC is a time-variable mixture of these carbon sources. In general, its carbon-isotope composition (-22 to -25 ‰) is close to that predicted for *in situ* phytoplankton production (-18 to -24 ‰) and resuspended bottom sediment (about -24 ‰; Fig. 11). In March 1976 mid-Bay values were highest, possibly because of phytoplankton production. In September 1976 levels were lowest, perhaps indicating a component of land-derived POC. The March 1977 values were similar to those for bottom sediment and were between those of the two previous surveys, and suggest that the increased POC abundance observed southward during March 1977 (Fig. 8) was the result of bottom-sediment resuspension rather than *in situ* production.

Bottom-sediment values are apparently time and source integrated. In spite of abundant South Bay tidal and marsh areas, detritus originating from *Spartina* marsh grass ($\delta^{13}C$, approx. -13 ‰) was not identifiable as an important carbon source.

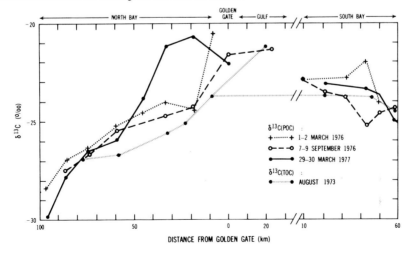

Fig. 11. $\delta^{13}C(POC)$ and $\delta^{13}C(TOC)$ in San Francisco Bay and the Gulf of the Farallones as a function of distance from the Golden Gate.

SAN FRANCISCO BAY

SUMMARY

Although our results are insufficient to define seasonal variations or describe details of the carbon cycle and fluxes through the estuary, they do substantiate the importance of some processes, sources, and sinks in the Bay. Respiration and mineralization in the water and sediments are important sources of inorganic carbon, and plant production and atmospheric outgassing remove significant amounts of inorganic carbon. Municipal waste effluent is an apparent source of biogenic carbon in South Bay. The Bay is a trap for riverine POC, whereas marine algae constitute most of the sediment-associated TOC seaward of the Golden Gate.

ACKNOWLEDGMENTS

We thank T. B. Coplen, F. J. Pearson, Jr., D. H. Peterson, D. S. McCulloch, and T. J. Conomos for review and helpful criticism of the manuscript. We also thank L. A. Dedini, S. M. Wienke, B. E. Cole, and A. E. Alpine for analytical assistance and T. B. Coplen for providing the mass spectrometer.

LITERATURE CITED

Arthur, J. F., and M. D. Ball. 1979. Factors influencing the entrapment of suspended material in the San Francisco Bay-Delta estuary. Pages 14?-174 *in* T. J. Conomos, ed. San Francisco Bay: The Urbanized Estuary. Pacific Division, Amer. Assoc. Advance. Sci., San Francisco, Calif.

Banse, K. 1977. Determining the carbon-to-chlorophyll ratio of natural phytoplankton. Mar. Biol. 41:199-212.

Brewer, P. G., and J. C. Goldman. 1976. Alkalinity changes generated by phytoplankton growth. Limnol. Oceanogr. 21:108-117.

Briggs, R. B., and D. A. Flemer. 1972. The flux of particulate carbon in an estuary. Mar. Biol. 12:11-17.

Broecker, W. S., and T. Takahashi. 1966. Calcium carbonate precipitation on the Bahama Banks. J. Geophys. Res. 71(6):1575-1602.

Broecker, W. S., and A. Walton. 1959. The geochemistry of C^{14} in fresh-water systems. Geochim. Cosmochim. Acta 16:15-38.

Buch, K. 1951. Das Kohlensaure Gleichgewichts System in Meerwasser. Hausforskingsinstitutets Skrift Helsingfors, No. 151, Helsingfors. 18 pp.

Claypool, G. E., and I. R. Kaplan. 1974. The origin and distribution of methane in marine sediments. Pages 99-139 *in* I. R. Kaplan, ed. Natural Gases in Marine Sediments. Plenum Publishing Corp., New York.

Conomos, T. J. 1979. Properties and circulation of San Francisco Bay waters. Pages 47-84 *in* T. J. Conomos, ed. San Francisco Bay: The Urbanized Estuary. Pacific Division, Amer. Assoc. Advance. Sci., San Francisco, Calif.

Conomos, T. J., and D. H. Peterson. 1977. Suspended-particle transport and circulation in San Francisco Bay: An overview. Pages 82-97 *in* M. Wiley, ed. Estuarine Processes. Vol. 1. Academic Press, New York.

Conomos, T. J., R. E. Smith, D. H. Peterson, S. W. Hager, and L. E. Schemel. 1979. Processes affecting seasonal distributions of water properties in the San Francisco Bay estuarine system. Pages 115-142 *in* T. J. Conomos, ed. San Francisco Bay: The Urbanized Estuary. Pacific Division, Amer. Assoc. Advance. Sci., San Francisco, Calif.

Craig, H. 1953. The geochemistry of the stable carbon isotopes. Geochim. Cosmochim. Acta 3:53-92.

Craig, H. 1957. Isotopic standards for carbon and oxygen and correction factors for mass spectrometric analysis of carbon dioxide. Geochim. Cosmochim. Acta 12:133-149.

Degens, E. T. 1969. Biogeochemistry of stable carbon isotopes. Pages 304-329 *in* E. Eglington and M. T. J. Murphy, eds. Organic Geochemistry. Springer-Verlag, Berlin.

Degens, E. T., M. Behrendt, B. Gotthardt, and E. Reppmann. 1968. Metabolic fractionation of carbon isotopes in marine plankton-II. Data on samples collected off the coasts of Peru and Ecuador. Deep-Sea Res. 15:11-20.

Eadie, B. J., and L. M. Jeffrey. 1973. $\delta^{13}C$ analysis of oceanic particulate organic matter. Mar. Chem. 1:199-209.

Fry, B., R. S. Scalon, and P. L. Parker. 1977. Stable carbon isotope evidence for two sources of organic matter in coastal sediments: seagrasses and plankton. Geochim. Cosmochim. Acta 14:1875-1877.

Gearing, P., F. E. Plunker, and P. L. Parker. 1977. Organic carbon stable isotope ratios of continental margin sediments. Mar. Chem. 4:251-266.

Gordon, L. I., and P. K. Park. 1972. A continuous pCO_2 measurement system. Tech. Rept. 240, Oregon State Univ., Corvallis, Ore. 77 pp.

Gordon, L. I., P. K. Park, J. J. Kelley, and D. W. Hood. 1973. Carbon dioxide partial pressures in North Pacific surface waters, pt. 2. General late summer distribution. Mar. Chem. 1:191-198.

Haines, E. B. 1977. The origins of detritus in Georgia salt marsh estuaries. Oikos 29:254-260.

Hammond, D. E. 1975. Dissolved gases and kinetic processes in the Hudson River estuary. Ph.D. Thesis. Columbia University, New York. 161 pp.

Head, P. C. 1976. Organic processes in estuaries. Pages 53-91 *in* J. D. Burton and P. S. Liss, eds. Estuarine Chemistry. Academic Press, New York.

Hendy, C. H. 1971. The isotopic geochemistry of speleotherms, pt. 1. The calculation of the effects of different modes of formation on the isotopic composition of speleotherms. Geochim. Cosmochim. Acta 35:801-824.

Hunt, J. M. 1970. The significance of carbon isotope variations in marine sediments. Pages 27-35 *in* G. D. Hobson and G. C. Spears, eds. Advances in Organic Geochemistry. Pergamon, Oxford.

Katz, A., Y. Kolodny, and A. Nissenbaum. 1977. The geochemical evolution of the Pleistocene Lake Lisan—Dead Sea system. Geochim. Cosmochim. Acta 41:1609-1626.

Kelley, J. J., and D. W. Hood. 1971. Carbon dioxide in the Pacific Ocean and Bering Sea: upwelling and mixing. J. Geophys. Res. 76(3):745-752.

Kroopnick, P. 1974. Correlations between ^{13}C and ΣCO_2 in surface waters and atmospheric CO_2. Earth and Planetary Sci. Letters 22:397-403.

Liss, P. S. 1976. Conservative and non-conservative behavior of dissolved constituents during estuarine mixing. Pages 93-130 *in* J. D. Burton and P. S. Liss, eds. Estuarine Chemistry. Academic Press, New York.

Lyman, J. 1956. Buffer mechanism of sea water. Ph. D. Thesis. University of California, Los Angeles, Calif. 196 pp.

Menzel, D. W., and R. F. Vaccaro. 1964. The measurement of dissolved organic and particulate carbon in seawater. Limnol. Oceanogr. 9:138-142.

Mook, W. G. 1970. Stable carbon and oxygen isotopes of natural waters in the Netherlands. Pages 163-190 *in* Isotope Hydrology 1970. Int. Atomic Energy Agency, Vienna.

Mook, W. G., J. C. Bommerson, and W. H. Staverman. 1974. Carbon isotope fractionation between dissolved bicarbonate and gaseous carbon dioxide. Earth and Planetary Sci. Letters 22:169-176.

Mook, W. G., and B. K. S. Koene. 1975. Chemistry of dissolved inorganic carbon in estuarine and coastal brackish waters. Estuarine Coastal Mar. Sci. 3:325-336.

Oana, S., and E. S. Deevey. 1960. Carbon-13 in lake waters and its possible bearing on paleolimnology. Amer. J. Sci. 258-A:253-272.

Park, P. K. 1969. Oceanic CO_2 system: an evaluation of ten methods of investigation. Limnol. Oceanogr. 14(2):179-186.

Park, P. K., L. I. Gordon, S. W. Hager, and M. C. Cissel. 1969. Carbon dioxide partial pressure in the Columbia River. Science 166:867-868.

Parker, P. L., and J. A. Calder. 1970. Stable carbon isotope ratio variations in biological systems. Pages 107-127 *in* D. W. Hood, ed. Organic Matter in Natural Waters. Institute of Marine Science, Occasional Publication No. 1. University of Alaska, College, Alaska.

Parsons, T. R. 1975. Particulate organic carbon in the sea. Pages 365-383 *in* J. P. Riley and G. Skirrow, eds. Chemical Oceanography. Vol. 2. Academic Press, New York.

Peterson, D. H. 1979. Sources and sinks of biologically reactive oxygen, carbon, nitrogen, and silica in northern San Francisco Bay. Pages 175-193 *in* T. J. Conomos, ed. San Francisco Bay: The Urbanized Estuary. Pacific Division, Amer. Assoc. Advance. Sci., San Francisco, Calif.

Peterson, D. H., T. J. Conomos, W. W. Broenkow, and P. C. Doherty. 1975a. Location of the non-tidal current null zone in northern San Francisco Bay. Estuarine Coastal Mar. Sci. 3:1-11.

Peterson, D. H., T. J. Conomos, W. W. Broenkow, and E. P. Scrivani. 1975b. Processes controlling the dissolved silica distribution in San Francisco Bay. Pages 153-187 *in* L. E. Cronin, ed. Estuarine Research. Vol. 1. Academic Press, New York.

Peterson, D. H., J. F. Festa, and T. J. Conomos. 1978. Numerical simulation of dissolved silica in the San Francisco Bay. Estuarine Coastal Mar. Sci. 7:99-116.

Rau, G. 1978. Carbon-13 depletion in a subalpine lake: Carbon flow implications. Science 201:901-902.

Sackett, W. M., and W. S. Moore. 1966. Isotopic variations of dissolved inorganic carbon. Chem. Geol. 1:323-328.

Sackett, W. M., and R. R. Thompson. 1963. Isotopic organic composition of recent continental derived clastic sediments of eastern Gulf Coast, Gulf of Mexico. Bull. Amer. Assoc. Petrol. Geol. 47:525-531.

Schemel, L. E., A. E. Alpine, B. E. Cole, L. A. Dedini, and E. C. Spiker. 1978. Water and sediment measurements in San Francisco Bay. U. S. Geol. Surv. Open-File Rept. 78-973. 28 pp.

Schemel, L. E., and L. A. Dedini. 1979a. A continuous water-sampling and multiparameter-measurement system for estuaries. U. S. Geol. Surv. Open-File Rept. 79-273. 92 pp.

Schemel, L. E., and L. A. Dedini. 1979b. Particulate organic carbon in San Francisco Bay, California, 1971-1977. U. S. Geol. Surv. Open-File Rept. 79-512. 30 pp.

Shultz, D. J., and J. A. Calder. 1976. Organic carbon $^{13}C/^{12}C$ variations in estuarine sediments. Geochim. Cosmochim. Acta 40:481-485.

Skirrow, G. 1975. The dissolved gases-carbon dioxide. Pages 1-192 *in* J. P. Riley and G. Skirrow, eds. Chemical Oceanography. Vol. 2. Academic Press, New York.

Strickland, J. D. H., and T. R. Parsons. 1972. A practical handbook of seawater analysis. Fish. Res. Bd. Can. Bull. 167. 311 pp.

Weiss, R. F. 1974. Carbon dioxide in water and seawater: The solubility of non-ideal gas. Mar. Chem. 2:203-215.

Wigley, T. M. L., L. N. Plummer, F. J. Pearson, Jr. 1978. Mass transfer and carbon isotope evolution in natural water systems. Geochim. Cosmochim. Acta 42:1117-1139.

Wong, W. W., and W. M. Sackett. 1978. Fractionation of stable carbon isotopes by marine phytoplankton. Geochim. Cosmochim. Acta 42:1809-1815.

Woodwell, G. M., P. H. Rich, and C. A. Hall. 1973. Carbon in estuaries. Pages 221-240 *in* G. M. Woodwell and E. V. Pecner, eds. Carbon in the Biosphere. NTIS:CONF-720510.

THE USE OF RADON-222 TO ESTIMATE BENTHIC EXCHANGE AND ATMOSPHERIC EXCHANGE RATES IN SAN FRANCISCO BAY

DOUGLAS E. HAMMOND AND CHRISTOPHER FULLER
Department of Geological Sciences, University of Southern California
Los Angeles, California 90007

Using a slurry technique, measurements of radon-222/radium-226 in San Francisco Bay sediments range from 0.3 to equilibrium. Radon deficiency generally decreases with increasing depth. A small deficiency may exist as deep as 40 cm in some cases. This deficiency is attributed primarily to irrigation of sediments by polychaete worms. If irrigation is modeled as an advective process, an irrigation rate of 3 cm·d^{-1} is calculated as a lower limit at a station in South Bay in August 1976. Using this model and nutrient measurements in interstitial waters, fluxes across the sediment-water interface for ΣCO_2, NH_4^+, PO_4^{-3}, and SiO_2 are calculated to be 40, 4, 0.03, and 6 mmol·m^{-2}·d^{-1} at this station. The flux of radon across the sediment-water interface is estimated to be 200±70 atoms·m^{-2}·s^{-1} on the basis of integrated radon deficiencies and benthic flux chamber measurements.

Simultaneous measurements of radon and wind speed in July 1977 suggest that flow-induced turbulence, rather than wind speed, is the primary factor controlling gas exchange across the air-water interface. From a radon mass balance, the mass transfer coefficient for radon across this interface is calculated to be 1.0±0.5 m·d^{-1}. Using this information, a vertical mixing coefficient in the water column of South Bay is calculated to be greater than 1 x 10^{-2} cm·s^{-1}, indicating the water column mixes more rapidly than once in 12 h. The volume transport of sand bedforms in Central Bay is estimated to be 8 x 10^6 m^3·d^{-1}.

The distribution of chemical and biological properties in estuarine waters and sediments is strongly influenced by physical processes, including turbulent mixing and exchanges across the sediment-water and air-water interfaces. Unfortunately, the rates of these processes are often difficult to estimate. Extrapolation of laboratory experiments to field conditions may introduce major scaling errors, and field measurements using introduced tracers are difficult to do in large systems. One solution to these problems is to model the distribution of naturally-occurring tracers in terms of these physical processes. This paper discusses preliminary results of the use of naturally-occurring radon-222 to estimate the rate of vertical mixing in the water column and the rate of exchange across the sediment-water and air-water interfaces in San Francisco Bay.

The use of this isotope in marine systems was first proposed by Broecker (1965), and it has recently been used in estuarine systems to study exchange across the sediment-water interface (Hammond et al. 1977).

Radon is a noble gas with a 4-day half-life. It is produced primarily in sediments by the decay of radium-226 (Fig. 1). A fraction of the radon which is produced in sediments will escape to the overlying water column, leaving a deficiency in sediments, so that the activity ratio of radon to radium is less than one. This deficiency is a measure of the rate of exchange across the sediment-water interface. Once radon is in the water column, it is mixed vertically and will either decay there or escape to the atmosphere. In a steady state system, the depth-integrated radon deficiency in sediments (in activity units) must equal the radon flux (in atoms per area time) across the

SAN FRANCISCO BAY

ESTUARINE RADON

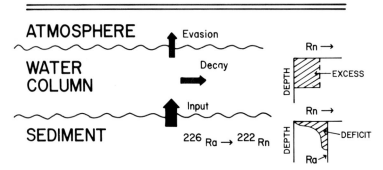

Fig. 1. Radon transport in estuaries. Vertical profiles of radon in the water column and in sediments are shown schematically. The shaded areas are integrated excess in the water column and integrated deficiency in sediments.

sediment-water interface. It must also equal the depth-integrated decay rate of excess (unsupported by decay of dissolved or suspended radium) radon in the water column plus the flux across the air-water interface.

ANALYTICAL TECHNIQUES

Measurements of radon in the water column were made using techniques described by Broecker (1965) and Mathieu (1977). Briefly, this involves collection of a 20-liter water sample in an evacuated bottle, extraction of radon onto activated charcoal at dry-ice temperatures with a helium carrier, and (alpha) counting radon and its two polonium daughters in a phosphored chamber. Measurements on samples from flux chambers were done on 2-liter samples. Analytical precision is about 5%.

Measurements in sediments were made by collecting gravity cores (5-cm ID), extruding mud sections, adding these to 100 ml of estuary water to create a slurry, and purging to measure radon as described by Hammond et al. (1977). These analyses were nearly always completed within 24 hours of sample collection. The supported radon in sediments (referred to as radium) was measured by storing the slurry for 1-4 weeks and extracting the new crop of radon which had been produced. Radium analyses were repeated until the standard deviation in this parameter was less than 5%. Key et al. (1977) have noted that there may be a problem with using the slurry technique to measure supported radon in deep sea sediments. They find that measurements of supported radon at depth are consistently 10-20% greater than the initial measurements. They attribute this to the physical process of creating a slurry but do not elucidate the mechanism. Our laboratory has observed a similar effect for marine sediments from basins in the southern California borderlands, but with muds from San Francisco Bay (Table 1) and from the Hudson River Estuary (Hammond et al. 1977), equilibrium ratios of radon and supported radon have been observed at depth.

It is possible that a problem may arise due to the 0.086 MeV recoil energy received by a radon atom as radium-226 decays. Extrapolating range-energy relations for fission fragments in aluminum (Friedlander et al. 1964:100) to low energy suggests the range should be about 0.02 mg·cm^{-2}, equivalent to 800Å in a silicate phase or 2000Å in water. Assuming sediment grains are cubic with an edge of length r, the distance between grains arranged in a primitive cubic lattice is

TABLE 1. RADON/RADIUM IN SAN FRANCISCO BAY SEDIMENT CORES

Station Number	Date	Interval (cm)	Radium-226 dpm·g^{-1}	Radium-226 dpm·cm^{-3}	Rn/Ra	Integrated Deficiency[a]
28C	8-09-76	0- 2	0.23	0.17	0.33	
		2- 4	0.17	0.22	0.25	
		4- 6	0.24	0.18	0.71	
		6- 8	0.30	0.19	0.66	
		12-14	0.18	0.19	0.83	†163
	3-07-77	0- 2	0.24	0.22	0.31	
		2- 3	(0.51)	0.33	0.30	
		3- 4	0.31	0.32	0.43	
		4- 6	0.28	0.22	0.66	
		6- 8	0.26	0.24	0.61	†175
	7-12-77	0- 3	0.20	0.15	0.25	
		3- 6	0.25	0.18	0.49	
		6- 9	0.28	0.23	0.42	
		9-12	0.14	0.20	0.79	†188
	10-19-78	0- 3	0.23	0.20	0.46	
		3- 6	0.24	0.20	0.85	
		6- 9	0.25	0.20	0.89	
		12-15	0.23	0.20	0.82	
		21-24	0.24	0.21	0.86	†138
27		0- 1.5	0.31	0.22	0.37	
		1.5- 3	0.30	0.23	0.78	
		3- 4.5	0.27	0.23	0.75	
		4.5- 6	0.30	0.26	0.81	
		6- 8	0.24	0.26	0.84	† 92
28	10-22-77	0- 3	0.29	0.22	0.58	
		3- 6	0.30	0.21	0.33	
		6- 9	0.30	0.20	0.59	
		12-15	0.25	0.19	0.83	
		17-20	0.30	0.18	0.92	†209
	12-12-77	0- 3		0.19	0.52	
		3- 6		0.22	0.88	
		6- 9		0.22	1.12	
		18-21		0.24	0.97	
		51-54		0.27	1.15	
	12-12-77	0- 3		0.20	0.59	
		6- 9		0.20	0.96	
		18-21		0.26	0.87	
		27-30		0.25	0.98	
		42-45		0.23	1.38	
		51-54		0.28	1.00	
	12-14-77	0- 3		0.23	0.60	
		3- 6		0.23	0.92	
		6- 9		0.22	0.98	
		24-27		0.22	0.93	
		39-42		0.23	0.93	
	12-14-77	15-18		0.20	0.82	
		24-27		0.23	1.00	
		33-36		0.24	0.90	
		42-45		0.25	1.10	
		60-63		0.24	1.03	
		78-81		0.25	1.00	177

[a] Integrated deficiency of radon (atoms·m^{-2}·s^{-1}).

$R = 2r[(1-\phi)^{-1/3} -1]$ where ϕ is sediment porosity. Typical values for r and ϕ in San Francisco Bay are 6 μm and 0.7, so that R = 6 μm or 30 times the recoil range. In deep sea sediments we might choose 3 μm and 0.5 so that R = 1.5 μm or 8 times the recoil range. To rigorously calculate the change in the probability of recoil from one grain to another would require data on the position of radium in sediment grains and the statistical variation in grain spacing before and after the slurry is created. The calculation above shows that radon recoil may cause problems in measuring emanation from fine-grained, low porosity sediments, but should not be a problem for estuarine sediments. Thus the slurry technique should be satisfactory for measuring supported radon in these sediments.

On 9 August 1976, duplicate gravity cores were collected at station 28C. One core was sectioned for radon, the second was sectioned and squeezed to obtain interstitial water. Sections were squeezed at room temperature (about 5°C above *in situ*) in Reeburgh (1967) squeezers with water passing through a fiber filter (Whatman #42) and an 0.45 μm Nucleopore filter. Since only two squeezers were available and sediment permeability was low, squeezing was not completed until 2 days after the core was collected. The core was stored in its liner during this operation and a fresh horizon was exposed prior to loading each sample. The initial 4-5 cm^3 of water was discarded, then 3 cm^3 was taken for nutrient analysis, 1 cm^3 for ΣCO_2, and a second 5-cm^3 aliquot for nutrient analysis. ΣCO_2 analyses were performed on 12 August 1976 using a Swinnerton stripper and gas chromatograph. The precision of these analyses is 3%. Nutrient samples were refrigerated for one week and diluted aliquots were analyzed with a Technicon AutoAnalyzer. Despite this storage time, analyses of the two aliquots for nutrients generally agreed within 5% for SiO_2 and NH_4^+, and within 10% for PO_4^{-3}.

Fluxes of radon and nutrients from sediments were measured directly by using benthic-flux chambers. These chambers were inverted boxes made of plexiglass (25 cm x 25 cm x 15 cm). They were deployed by divers so that the lower 5 cm was below the sediment surface, leaving 10 cm above the surface. Boxes were sampled immediately and at 1-day intervals after deployment through a nylon tube extending to the surface. Occasionally they were sampled 4-6 h after deployment, but changes in nutrients and radon over this period were generally too small to measure accurately. Fresh water was introduced during the sampling by inflow through a check valve as the sample was withdrawn. Measurements on samples drawn from the box were corrected for this inflow by assuming that the inflowing water mixes rapidly with the box water. Chambers were equipped with several types of devices intended to transmit turbulence mechanically from tidal currents to the enclosed water, but observation by divers indicated that these were not successful. Oxygen analyses on a separate aliquot were done by Winkler titration (Carpenter 1965).

RESULTS AND DISCUSSION

Radon in Sediments

To demonstrate the validity of the slurry technique, duplicate cores were collected and sectioned on each of 2 days at station 28 in South San Francisco Bay (Fig. 2). The activity ratios of radon to radium observed in surface samples of these cores are quite consistent and the ratio approaches equilibrium in the 9-12 cm interval (Fig. 3). Below that interval there seems to be a deficiency of a few percent down to about 40 cm, and below that, samples are in equilibrium. The high value at 42-45 cm is unexplained. This radon analysis was done 2 days after collection and any analytical error in radon would be increased by 20% in the decay corrections. It is clear, however, that the deficiency is largest in surface sediments, that it may extend well below the sediment-water interface, and that equilibrium values can be found with this technique.

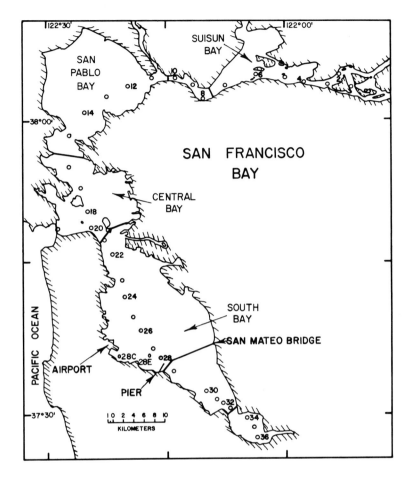

Fig. 2. Location map of San Francisco Bay showing standard U.S.G.S. stations and other stations described in the text.

The data from station 28C (Fig. 2) in a shallow-water (mean depth = 2 m) area in South Bay, show some variation in the supported radon as a function of depth (Fig. 4). The largest source of this error was probably the length of the extruded section. To remove this variation, the radon/radium activity ratio was multiplied by the average radium to yield the values plotted in the figure. The deficiency generally decreases with increasing depth, although some deep minima are observed. The depth of this deficient zone is uncertain because equilibrium samples were not obtained. Data obtained through the year at station 28C and 27 or 28 are listed in Table 1. Integrated deficiencies are also listed, although these are usually lower limits.

Nutrients in Interstitial Waters

Interstitial water data from the duplicate core collected in August 1976 show that all nutrient species have small concentration gradients in the upper 6 cm and much larger gradients below this (Fig. 5). ΣCO_2 shows a small but significant minimum at 3-6 cm. The upper zone correlates well with the zone which is quite deficient in radon. Similar profiles have been observed in cores from this site during other seasons and at other locations (Korosec in press).

SAN FRANCISCO BAY

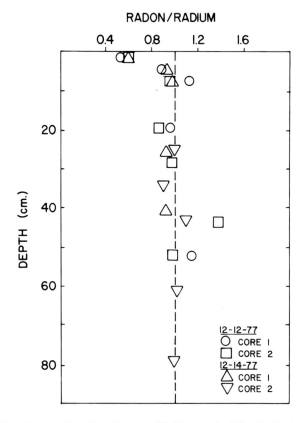

Fig. 3. Radon/Radium vs. Depth at Station 28. The dashed line indicates secular equilibrium. Data to the left of the line indicate radon deficiency.

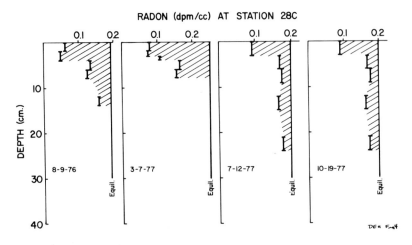

Fig. 4. Radon vs. Depth at Station 28C. Solid lines represent secular equilibrium and dashed areas show deficient zone. This zone may extend below the deepest samples.

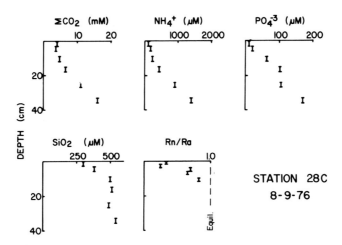

Fig. 5. Interstitial water chemistry at Station 28C. See text for explanation.

Processes which transport dissolved species through interstitial waters and across the sediment-water interface include molecular diffusion (Hammond et al. 1977), irrigation of sediments by benthic organisms (Aller and Yingst 1978) and physical stirring of sediments by wind waves and tidal currents (Krone 1979). Molecular diffusion alone would create a radon deficiency in sediments which decreases exponentially with increasing depth (Hammond et al. 1977). The half-distance of this profile would be about 2 cm. It is clear from the data (Figs. 3, 4, 5) that at least one other mechanism must be important. Physical stirring would create a radon deficiency which decreases monotonically with depth, yet profiles (Figs. 3, 4) are not always monotonic. This suggests that irrigation may be important.

South Bay sediments are inhabited by a number of benthic species which can irrigate sediments as they construct or move about in their burrows. Cores collected at station 28C were always found to contain live specimens of *Asychis elongata*, a large polychaete which builds thick-walled tubes ($\sim$ 3-5 mm ID [see Nichols 1979]); and cores from stations 27 and 28 were always found to contain live specimens of *Heteromastis filiformis*, a smaller polychaete which builds narrower ($\sim$ 2-4 mm ID) soft-walled burrows. These burrows were observed at the sediment-water interface and were often found during sectioning to be open to 40-50 cm in the core. Live worms have also been found close to these depths. Radiographs of cores show worm burrow densities which are typically 0.5-1 burrows·cm^{-2} of sediment surface (Korosec in press). Irrigation can enhance the flux of dissolved species across the sediment-water interface and can also produce minima in profiles of dissolved nutrients in interstitial water (Goldhaber et al. 1977). The observations above suggest that irrigation may be of major importance in these sediments and that the polychaete worms are the primary perpetrators, although other unidentified species may also be important. Physical stirring of surface sediments by currents may be occurring, but the surface of these sediments is fairly cohesive and this process is probably not important below the upper 2 cm, if at all. Since any effects of current stirring cannot be distinguished from infaunal irrigation with the techniques discussed here, any effects of this process will be attributed to irrigation.

Past work on interstitial water chemistry has usually treated irrigation as a diffusive process (see Goldhaber et al. 1977). A coefficient of eddy diffusion can be calculated for a substance in the irrigated zone by multiplying the molecular diffusivity times the ratio of the gradient in a deep quiescent zone to the gradient in the irrigated zone. The coefficient determined for one substance can be applied to other substances.

This type of model would not produce a minimum in ΣCO_2, however. A more appropriate model might be an advective pumping model in which worm tubes are a conduit for overlying water to be pumped into the sediments. A minimum in ΣCO_2 or radon could be produced by a localized input. A similar model has been proposed by McCaffrey et al. (in prep.).

In this model (Fig. 6), the sediments can be divided into irrigation zones defined by the stepwise structure of the radon profile. Each zone is assumed to be well mixed. The upper zone

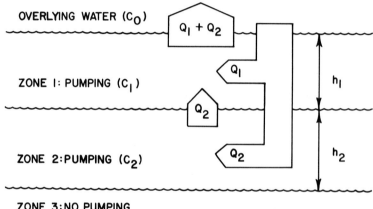

Fig. 6. Worm Pumping Model. The advective flux across the interface is $(Q_1 + Q_2)(C_1 - C_0)$ where Q is the flow rate and C is the concentration in each zone.

receives flow Q_1 directly from the overlying water and Q_2 from the underlying zone. Assuming radon in the water column to be negligible, the radon balance for zone 2 is:

$$\text{Production} = \text{Loss}$$
$$Ph_2 A = Q_2 C_2 + \lambda h_2 A C_2 + J_2 A$$

where C_2 = radon concentration in Zone 2 (atoms·vol^{-1})

h_2 = thickness of Zone 2

A = area

P = production rate of radon from radium per unit volume

λ = decay constant for radon

J_2 = net loss by molecular diffusion from Zone 2 per unit area

The balance for Zone 1 is:

$$Ph_1 A + Q_2 C_2 = (Q_1 + Q_2) C_1 + \lambda h_1 A C_1 + J_1 A$$

where the subscripted symbols are analogous to those above. These equations can be rearranged so that:

$$v C_1 = D - J_D$$

where $D = P(h_1 + h_2) - \lambda(h_2 C_2 + h_1 C_1)$
  = integrated radon deficiency

$J_D = J_1 + J_2$

$v = (Q_1 + Q_2)/A$
  = irrigation velocity in upper zone

Thus irrigation velocities can be calculated from radon, and fluxes of nutrients due to irrigation can be estimated.

Two problems are encountered in using this approach with our data (Fig. 5). The lower limit of radon deficiency was not reached, thus the calculation will yield a lower limit for v. Also, the net diffusive flux cannot be directly evaluated from these data. If the sample immediately below those collected had an equilibrium amount of radon, the diffusive flux into the irrigated zone from below would approximately equal the diffusive flux across the interface. Thus J_D would be nearly zero because the gradients would be quite similar. If the sample immediately below was deficient, J_D would increase, but so would D. Thus, the major error will arise in the value used for D, and v will be a lower limit. Defining 0-4 cm as Zone 1 on the basis of radon, taking $D \simeq 163$ atoms·m^{-2}·s^{-1}, $J_D \simeq 0$, and $C_1 \simeq 5 \times 10^8$ atoms·m^{-3} (Table 1) thus yields $v = 3 \times 10^{-7}$ m·s^{-1} $\simeq 3$ cm·d^{-1}. This is the same order of magnitude as the value of 0.7 ± 0.4 cm·d^{-1} observed by McCaffrey et al (in prep.) in laboratory measurements on Narrangansett Bay cores.

The fluxes of dissolved nutrients due to irrigation can now be estimated from the expression:

$$\text{Flux} = v(C_1 - C_0)$$

where C_0 is the concentration in the overlying water column. These fluxes are listed in Table 2 and are an order of magnitude greater than fluxes which can be attributed to molecular diffusion (Korosec in press). An additional source of uncertainty is introduced in these calculations because water column measurements were not made simultaneously with the core collection. This could be

TABLE 2. CALCULATION OF NUTRIENT FLUXES
FROM INTERSTITIAL WATERS[a]

Dissolved Nutrient	$C_1 (\mu M)$[b]	$C_0 (\mu M)$[c]	$F(\text{mmol·m}^{-2}\cdot\text{d}^{-1})$[d]
ΣCO_2	3840	2400	40
NH_4^+	129	6	4
PO_4^{-3}	13.5	12.5	0.03
SiO_2	315	110	6

[a] Data collected at station 28C on 9 August 1976
[b] Weighted average of analyses from 0-4 cm
[c] Values typical for this season
[d] Lower limit, assumes $v = 3$ cm·d^{-1}

a major problem for PO_4^{-3}, but not other species. Despite the uncertainties, these estimates should be accurate within a factor of two. The importance of these fluxes in the nutrient balance of the estuary is discussed by Korosec (in press), Peterson (1979) and Conomos et al. (1979).

SAN FRANCISCO BAY

Flux Chambers

The depth-integrated radon deficiency in sediments must equal the flux across the sediment-water interface. This flux was directly measured by placing inverted plexiglass boxes on the sediments and measuring the change in radon. Some difficulties were encountered in sampling as it was difficult to avoid pulling the boxes up by the tubing while sampling on choppy and windy days. About 50% of the deployments were successful and these results are listed in Table 3. Each flux represents a one-day experiment. In one case (box 1, October) the box was left in place for a second day. Substantial variability exists within the experiments at a single station and thus the apparent differences between July and October and between 28C and 28E may not be significant. These measurements can be criticized for two reasons. The failure of the stirring devices to transmit turbulence effectively may allow gradients to build near the interface, reducing fluxes. Also, oxygen in the boxes dropped to nearly zero after one day and this may have reduced irrigation. In either case, these results should represent a lower limit. It is interesting to note that they are within a factor of two of the integrated deficiencies in Table 1. The box experiments and the integrated

TABLE 3. RADON FLUXES ACROSS THE SEDIMENT-WATER INTERFACE MEASURED *IN SITU* WITH BENTHIC CHAMBERS

Station	Sampling period	Box	(atoms·m^{-2}·s^{-1})
28C	July 1977	1	143
		2	123
28C	October 1977	1	198[a]
			280
		3	167
		4	303
		1	296
28E	October 1977	5	127

[a] Mean value of the five observations for station 28C during October, 1977: 249 ± 66.

deficiencies suggest that the flux of radon from South Bay sediments is about 200 ± 70 atoms·m^{-2}·s^{-1}. Molecular diffusion alone would supply radon at a rate:

$$J = \sqrt{\lambda D_s P}$$

where D_s is the effective diffusivity (Hammond et al. 1977). This would be equal to about 80 atoms·m^{-2}·s^{-1} at 15°C, or 40% of the observed flux.

Water Column Analyses

If the rate at which radon escapes to the atmosphere could be predicted, the flux of radon across the sediment-water interface could be obtained by constructing a mass balance for radon in the water column. Thee transects have been made along the Bay axis from station 30 to Rio Vista to collect samples from the water column. All three were similar and one of these is plotted (Fig. 7) to illustrate the uniformity of the distribution. To a first approximation, radon is well-mixed vertically and along the estuary axis. While some variations appear, these are not consistent

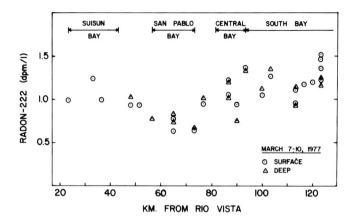

Fig. 7. Radon transect in March 1977. Surface-water samples were collected at 1-m depth and deep-water samples were usually 6-12 m in depth.

spatially or temporally. A few transects have been made perpendicular to the bay axis. These measurements have been similar to channel measurements, although samples collected in shallow (<1 m) water within 200-400 m of shore may have concentrations twice those in channel samples. In the absence of significant horizontal gradients, a box model for the water column can be constructed which ignores horizontal transport. Ground-water can also be ignored because the ratio of water area to recharge area is large. The radon budget for South Bay is computed in Table 4. The decay rate is found by multiplying the average water column concentration by the mean depth. Estimates of radon flux to the atmosphere (evasion) are based on the stagnant film model for gas exchange (Broecker and Peng 1974). This model assumes that film thickness controls exchange rates and an empirical relation between exchange rates and wind speed developed by Emerson (1975) can be used to estimate air-water exchange rates. Radium-226 accounts for the input from dissolved and suspended radium. The input from sediments is used to balance the budget.

If the wind speed-film thickness model is correct, a substantially larger radon flux from sediments is required during periods of high wind than during periods of low wind. Conomos and Peterson (1977) have shown that high winds create waves which resuspend surficial sediments, but this is equivalent to a sediment thickness of only a few millimeters. About 5 cm must be disturbed to supply the extra flux, and the presence of stable benthic communities in surface sediments suggests this does not occur. It is also unlikely that the flux difference in Table 4 would be due to a difference in irrigation because March and January have similar temperatures and are likely to have similar benthic activity. It seems more likely that the calculation of exchange rates from wind speed alone may not be satisfactory.

To test this problem, a sampling station was set up on the end of the San Mateo fishing pier, on the edge of the deep channel in South Bay, about 1.5 km from shore. Surface samples were collected at approximately 1.5-h intervals over a 2-day period (Fig. 8). The average ($\pm 1\sigma$) of 29 analyses was 1.17 ± 0.10 dpm·liter^{-1}. Wind speed was measured at the San Francisco International Airport by the National Weather Service and also with a hand-held integrating anemometer (Taylor Inst. Co.) at the end of the pier about 5 m above the water surface. The wind measurements are in fair agreement, although the pier measurements are a little lower during the first day. This may be due to a failure of the pier sitters to orient the anemometer properly or to keep it well away from obstructions on the pier. At high wind speeds, gusts made it difficult to measure speed accurately at the pier. The solid line in Figure 8 was chosen to represent the data. The large diurnal variation

SAN FRANCISCO BAY

TABLE 4. RADON BUDGET AND ENVIRONMENTAL PARAMETERS FOR SOUTH BAY WATER COLUMN

Date	29-31 Jan 1976	8-11 Aug 1976	7-11 Mar 1976
Environmental Parameters			
T (°C)	10	24	15
Wind (m·s^{-1})	1.5	5	4.5
Film thickness (μ)	500	75	100
Avg. radon (dpm·liter^{-1})	1.49±0.32	1.27±0.22	1.29±0.18
No. samples	10	18	25
Budget (atoms·m^{-2}·s^{-1})			
Losses			
Decay	99±21	85±15	86±12
Evasion	24±12	187±93	122±61
Inputs			
^{226}Ra decay	7±3	7±3	7±3
From sediment[a]	116±24	265±94	201±62

[a] Input from sediment required to balance inputs and losses.

in wind speed during the summer is normal (Conomos 1979). If the flux of radon from sediments is uniform, vertical and horizontal mixing are rapid, and wind speed is the only factor controlling gas exchange, radon in the water column should reflect the diurnal wind variation. In this case, concentration should follow the solid line labeled "model" (Fig. 8). The major features of the model curve are the first minimum and the second maximum. These features should be present because of the large contrast in exchange rates between high wind speed and low wind speed in the model, but neither feature is apparent in the data. Thus, gas exchange in estuarine waters does not seem to be controlled by wind speed alone.

The gas exchange rate per unit area is characterized by a mass transfer coefficient k which has units of velocity:

$$k = J_{atm}/(C_W - C_{atm})$$

where J_{atm} = gas flux/area to the atmosphere

C_W = concentration in the water column

C_{atm} = concentration in water when equilibrated with the atmosphere.

O'Connor and Dobbins (1958) have suggested that flow-generated turbulence may be important in controlling gas exchange rates. On the basis of turbulent velocity fluctuations and scale sizes of vertical eddies, they proposed that the mass transfer coefficient for gas exchange in streams and rivers is approximately:

$$k = (D_m \bar{v}/h)^{1/2}$$

where D_m = molecular diffusivity

$\bar{v}$ = average current velocity

h = water depth

TABLE 5. CALCULATION OF SOUTH BAY GAS EXCHANGE RATE
IN OCTOBER 1977

Radon Budget in Water Column (atoms·m^{-2}·s^{-1})
    Average Input   200
    Excess Decay   94
    Evasion[a]   106

Mass Transfer Coefficient (m·d^{-1})
    Radon   1.0±0.5
    O'Connor-Dobbins   1.3

[a] Flux to atmosphere required to balance input and decay.

Applying this model to South San Francisco Bay, we obtain $k = 1.3$ m·d^{-1}.

The average rate of exchange in October can be calculated by constructing a mass balance for radon in the water column (Table 5). The influx across the sediment-water interface was previously shown to be about 200±70 atoms·m^{-2}·s^{-1}. The depth-integrated decay rate is calculated

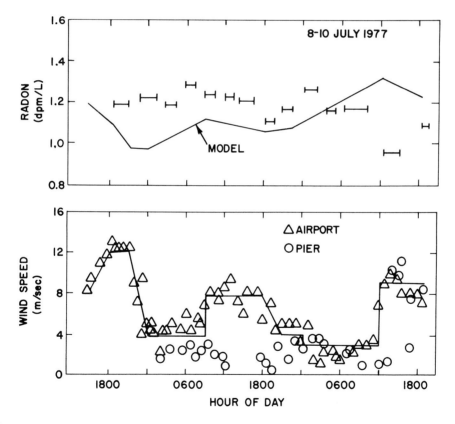

Fig. 8. Radon and wind speed. Data were obtained at the end of the San Mateo fishing pier (Fig. 1). The lower solid line is the average wind speed, the upper solid line is radon expected in the water column. The model curve was calculated assuming a constant decay rate in the water column, a constant input from sediments, and a gas exchange rate which depends on wind speed. Bars show the average of sequential radon analyses.

from the measured concentration in the water column and the mean depth (4 m). The budget is balanced by evasion to the atmosphere which requires $k = 1.0 \pm 0.5$ m·d^{-1}, remarkably close to the O'Connor-Dobbins prediction.

The diurnal field data (Fig. 8) indicate that the gas exchange rate is not closely tied to wind speed, but it is interesting to note that if an average wind speed of 4-6 m·s^{-1} at 10 m above the water surface (the October conditions) is used with the Emerson (1975) wind speed-film thickness relation, the mass transfer coefficient would be 1.5-2.0 m·d^{-1}, fortuitously close to the estimate from the radon mass balance. While the data presented here suggested that gas exchange in estuarine waters is not controlled by wind speed alone, it may play some role. The O'Connor-Dobbins model predicts an exchange rate close to the radon mass balance estimate, but since gas exchange has not been measured over a range of current speeds, the validity of this model cannot be properly assessed from these data. It is clear that further work must be done to elucidate the mechanisms which control gas exchange.

Taking the average mass transfer coefficient to be 1.0 m·d^{-1}, radon budgets can be calculated for different areas of the Bay (Table 6). The difference between fluxes from South Bay and San Pablo Bay sediments may reflect a difference in benthic irrigation. San Pablo Bay has lower densities of these species of deep dwelling polychaetes than South Bay, perhaps because salinity variations are much larger.

The large flux from Central Bay sediments may be due to physical stirring of the sandy sediments by bed-form migration. This does not occur in the fine-grained sediments of South Bay and San Pablo Bay. The remainder of these sediments are medium to coarse sands and would need to supply 445 atoms·m^{-2}·s^{-1} to balance the water column budget. Unfortunately, it is difficult to collect cores in sand and no measurements of radon deficiency have been made in these. Rubin and McCulloch (1979) have shown that sand waves migrate on the Bay floor. Sandy sediments produce radon at a rate of about 1000 atoms·m^{-3}·s^{-1}. Thus, about 40 cm of sand must be continuously stirred to supply the required flux. Alternatively, it may be erosion of bedforms which supplies radon. Assuming sandy sediments to be in secular equilibrium, containing 5×10^8 atoms·m^{-3}, the rate of erosion must be about 8 cm·d^{-1} averaged over the sandy portion. Taking this area to be 100 km^2, the volume transport of bedforms should be about 8×10^6 m^3·d^{-1}.

TABLE 6. RADON BUDGETS FOR REGIONS IN SAN FRANCISCO BAY (WATER COLUMN)[a]

	South Bay	Central Bay	San Pablo Bay
Mean±1σ(dpm·liter^{-1})	1.32±.24	1.07±.19	0.90±.26
Number of samples	52	22	17
Mean depth (m)	4	14	4
Losses (atoms·m^{-2}·s^{-1})			
Decay	88	250	60
Evasion	123	100	84
Inputs			
Ra^{226} Decay	7	25	7
From Sediments[b]	205	325	137

[a] Based on January, March, and August transects. South Bay includes USGS Sta No. 21-30, Central Bay includes Sta No. 17-20, San Pablo Bay includes Sta No. 11-15 (Fig. 2).

[b] Required to balance inputs and losses.

Vertical Mixing in the Water Column

Surface and deep samples have been collected at 13 stations. The median of top to bottom radon concentration ratios is 1.01, demonstrating that vertical gradients are small. The absence of these vertical gradients can be used to estimate a lower limit for the rate of vertical mixing in South Bay. These estimates can be made by constructing a two-box model (Fig. 9) which divides the water column into a well-mixed surface box of mean thickness h_1 which exchanges with the atmosphere and with a fraction of the sediments, and a well-mixed lower box of mean thickness h_2 which exchanges with the remainder of the sediments. Exchange between the two boxes is characterized by a mixing coefficient K_{12} and exchange between the upper box and the atmosphere is characterized by a mass transfer coefficient K_a. Both K_{12} and K_a have units of velocity.

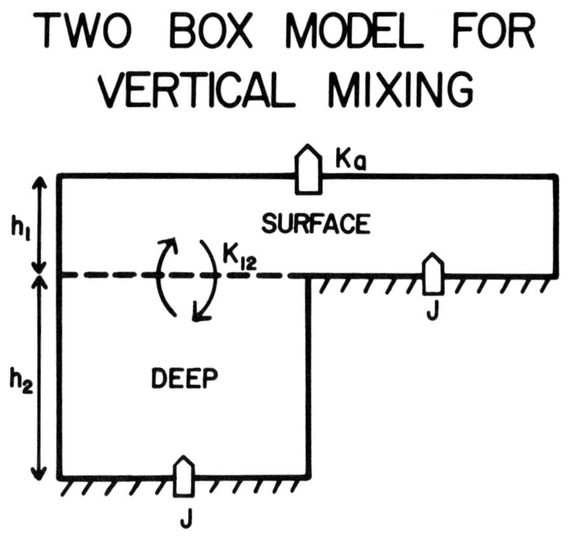

Fig. 9. Two-box model for vertical mixing. See text for explanation.

Input of radon from a unit area of sediment (J) is taken to be constant. The areal fraction of the estuary which is less deep than h_1 is f_1, the fraction which is deeper than h_1 is $1-f_1$. Assuming a steady state, the mass balance for radon over a unit area in the upper box requires that:

$$Jf_1 + K_{12}(1-f_1)(C_2 - C_1) = K_a C_1 + \lambda h_1 C_1$$

In the lower box:
$$J = K_{12}(C_2 - C_1) + \lambda h_2 C_2$$

These equations can be solved to calculate the ratio C_1/C_2 in terms of the geometrical characteristics of the system (h_1, h_2, f_1), and the transport coefficients (K_a, K_{12}). A hypsographic curve for South Bay shows that about half the area is quite shallow (<2 m at mid tide) and half the area is a relatively deep channel (Conomos and Peterson 1977). The mean depth of South Bay is 4 m. Taking $h_1=2$ m, $h_2=4$ m, $f_1=0.50$, and $K_a=0.8$ m·d^{-1}, the ratio of C_1/C_2 can be calculated in terms of K_{12} (Fig. 10). Field data indicate that C_1/C_2 is certainly greater than 0.9 so K_{12} must be greater than 10^{-2} cm·s^{-1}. The model residence time of water in the lower box before it enters the surface box is $h_2/K_{12} < 4 \times 10^4$ or less than 12 h. If the two boxes were separated by a halocline

SAN FRANCISCO BAY

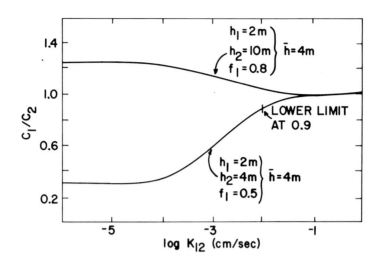

Fig. 10. Surface/deep radon vs. vertical mixing coefficient. If the ratio is greater than 0.9, K_{12} must be greater than 1×10^{-2} cm·s^{-1}. The lower curve is drawn for a geometry approximating South Bay; the upper curve is drawn for a hypothetical geometry.

1 m thick, this mixing coefficient would be equivalent to a vertical eddy diffusivity of greater than 1 cm^2·s^{-1}.

It is interesting to note that in an estuary with different geometry, the ratio C_1/C_2 would be quite different. For example, if the mean depth was 4 m and $f_1 = 0.80$, the concentration ratio could be as great as 1.25 (Fig. 10). Thus, estuaries with very large shoal areas and narrow, deep channels could have concentration ratios well above 1 if vertical mixing was slow.

CONCLUSIONS

(1) Irrigation of sediments by benthic organisms, probably by one or two species of deep-dwelling polychaete worms, creates radon deficiencies in sediments and zones in which concentrations of nutrients in interstitial waters are nearly uniform.
(2) Using a model which treats irrigation as an advective process, fluxes of nutrients across the sediment-water interface due to irrigation can be calculated from radon deficiencies and nutrient measurements. At one station in South Bay, the rate of this transport is an order of magnitude greater than the transport rate which molecular diffusion could accomplish.
(3) Direct measurement of radon fluxes using benthic chambers agrees with fluxes calculated from sediment deficiencies within a factor of two.
(4) The rate of gas exchange across the air-water interface is equivalent to a mass transfer coefficient of 1.0 ± 0.5 m·d^{-1}, in reasonable agreement with the rate predicted by the O'Connor-Dobbins (1958) model. This rate seems to be primarily controlled by flow-generated turbulence rather than by wind-generated turbulence.
(5) A lower limit can be placed on the rate of vertical mixing in South Bay. This limit indicates that vertical mixing is complete in less than 12 hours.
(6) A large flux of radon from the sediments occurs in central San Francisco Bay. If this flux is attributed to the continuous stirring of sandy sediments, a thickness of 40 cm must turn over rapidly. From observations of bedforms, Rubin (pers. comm.) suggests that a thickness of 3 to

several tens of centimeters should turn over on a short time scale, which supports the conclusions from radon measurements.

ACKNOWLEDGMENTS

The authors wish to acknowledge the assistance and company of Blayne Hartman, Michael Korosec, and Lawrence Miller in field work. T. John Conomos, David H. Peterson, and Frederic H. Nichols of the U. S. Geological Survey have provided shiptime and logistical support as well as invaluable discussion. Stephen W. Hager of the U. S. Geological Survey kindly performed nutrient analyses in the interstitial waters. Richard Ku and T. John Conomos reviewed the manuscript and provided valuable comments.

This research has been supported by National Science Foundation Grant No. OCE76-81154 and NOAA, Office of Sea Grant, Department of Commerce Grant No. 04-7-158-44113.

LITERATURE CITED

Aller, R. C., and J. Y. Yingst. 1978. Biogeochemistry of tube-dwellings: A study of the sedentary polychaete *Amphitrite ornata* (Leidy). J. Mar. Res. 36:201-254.

Broecker, W. S. 1965. The application of natural radon to problems in ocean circulation. Pages 116-145 *in* T. Ichiye, ed. Symposium on Diffusion in Oceans and Fresh Waters. Lamont-Doherty Geological Observatory, Palisades, N.Y.

Broecker, W. S., and T-H. Peng. 1974. Gas exchange rates between sea and air. Tellus 26:21-35.

Carpenter, J. H. 1965. The Chesapeake Bay Institute technique for the Winkler dissolved oxygen method. Limnol. Oceanogr. 10:141-143.

Conomos, T. J. 1979. Properties and circulation of San Francisco Bay waters. Pages 47-84 *in* T. J. Conomos, ed. San Francisco Bay: The Urbanized Estuary. Pacific Division, Amer. Assoc. Advance. Sci., San Francisco, Calif.

Conomos, T. J., and D. H. Peterson. 1977. Suspended particle transport and circulation in San Francisco Bay: An overview. Pages 82-97 *in* L. E. Cronin, ed. Estuarine Processes. Vol. 2. Academic Press, New York.

Emerson, S. R. 1975. Gas exchange rates in small Canadian shield lakes. Limnol. Oceanogr. 20:754-761.

Friedlander, G., J. W. Kennedy, and J. M. Miller. 1964. Nuclear and radiochemistry, 2nd ed. J. Wiley & Sons, New York. 585 pp.

Goldhaber, M. B. et al. 1977. Sulfate reduction, diffusion, and bioturbation in Long Island Sound sediments. Report of the FOAM Group. Amer. J. Sci. 277: 193-237.

Hammond, D. E., H. J. Simpson, and G. Mathieu. 1977. ^{222}Radon distribution and transport across the sediment-water interface in the Hudson River estuary. J. Geophys. Res. 82: 3913-3920.

Key, R. M., N. L. Guinasso, and D. R. Schink. 1977. The release of radon from sediments. Trans. Amer. Geophys. Union 58:421. (Abstr.)

Korosec, M. In press. The effects of biological activity on transport of dissolved species across the sediment-water interface in San Francisco Bay. M. S. Thesis. University of Southern California.

Krone, R. B. 1979. Sedimentation in the San Francisco Bay system. Pages 85-96 *in* T. J. Conomos, ed. San Francisco Bay: The Urbanized Estuary. Pacific Division, Amer. Assoc. Advance. Sci., San Francisco, Calif.

Mathieu, G. 1977. Radon-222/radium-226 technique of analysis. Appendix I *in* P. Biscaye, Annual Report to ERDA, Transport and Transfer Rates in the Waters of the Continental Shelf. Contract EY76-S-02-2185. 30 pp.

McCaffrey, R. J., A. C. Myers, E. Davey, G. Morrison, M. Bender, N. Luedtke, D. Cullen,

SAN FRANCISCO BAY

P. Froehlich, G. Klinkhammer. 1979. Benthic fluxes of nutrients and manganese in Narragansett Bay, Rhode Island. (Paper submitted to Limnol. Oceanog.)

Nichols, F. H. 1979. Natural and anthropogenic influences on benthic community structure in San Francisco Bay. Pages 409-426 *in* T. J. Conomos, ed. San Francisco Bay: The Urbanized Estuary. Pacific Division, Amer. Assoc. Advance. Sci., San Francisco, Calif.

O'Connor, D. J., and W. E. Dobbins. 1958. Mechanism of reaeration in natural streams. Trans. Amer. Soc. Civil. Eng. 123: 641-666.

Peterson, D. H. 1979. Sources and sinks of biologically reactive oxygen, carbon, nitrogen, and silica in northern San Francisco Bay. Pages 175-193 *in* T. J. Conomos, ed. San Francisco Bay: The Urbanized Estuary. Pacific Division, Amer. Assoc. Advance. Sci., San Francisco, Calif.

Reeburgh, W. S. 1967. An improved interstitial water sampler. Limnol. Oceanogr. 12:163-165.

Rubin, D. M., and D. S. McCulloch. 1979. The movement and equilibrium of bedforms in central San Francisco Bay. Pages 97-113 *in* T. J. Conomos, ed. San Francisco Bay: The Urbanized Estuary. Pacific Division, Amer. Assoc. Advance. Sci., San Francisco, Calif.

FLUCTUATIONS OF COPPER, ZINC, AND SILVER IN TELLENID CLAMS AS RELATED TO FRESHWATER DISCHARGE—SOUTH SAN FRANCISCO BAY

SAMUEL N. LUOMA AND DANIEL J. CAIN
U. S. Geological Survey, 345 Middlefield Road, Menlo Park, CA 94025

Significant contamination of the tellenid clam *Macoma balthica* with Cu and Ag was observed at stations in southern San Francisco Bay. The degree of contamination appeared to be greatly influenced by the discharge of fresh water into South Bay. Local runoff appeared to provide an important source of the contaminants, especially in the summer and fall. Fresh-water discharge, either from local sources or from the Sacramento-San Joaquin Delta, also provided the force that flushed biologically available Cu and Ag from South Bay, and the degree of this flushing force appeared to determine the magnitude of the annual peak in Cu and Ag concentrations of the clam. A metal discharge index, combining an indirect estimate of annual metal loading (derived from cumulative rainfall) with the inverse of fresh-water discharge at the Delta, was found to explain 60-80% of the temporal variance in the Ag and Cu concentrations of *M. balthica*. The index represents a first step toward quantitatively predicting the effect of any reduction in fresh-water discharge into the Bay on Ag and Cu contamination in South Bay. Significant differences between temporal variations in Zn concentrations in the clams and the variations in Cu and Ag concentrations suggest all contaminants do not behave similarly in South Bay.

Trace metal contamination is often associated with the type of industrial/urban development which surrounds San Francisco Bay. South San Francisco Bay may be especially vulnerable to trace contaminant effects because the residence times of South Bay waters are long during most of the year. McCulloch et al. (1970), Imberger et al. (1977) and Conomos (1979) suggest that South Bay is well flushed only during periods of high fresh-water discharge from the Sacramento-San Joaquin Delta. Girvin et al. (1975) suggest that pollutants may accumulate in South Bay waters, sediments, and biota during periods of restricted flushing.

The discharge of fresh water into South Bay may affect the concentration of trace metals available to organisms in several ways: (1) River, stream and sewer discharge may carry elevated concentrations of solute and particulate-bound metals into the estuary during the rainy season. Urban storm runoff is characterized by high concentrations of many trace metals (Pitt and Amy 1973). (2) Terrigenous (land-derived) sediments are carried into the estuary primarily during the winter (Conomos and Peterson 1977). The physicochemical characteristics of the terrigenous sediments (especially those in urban runoff) may differ from the characteristics of the estuarine sediments. These differences may affect both the partitioning of metals between particulates and solution and the ability of organisms which ingest particulates to accumulate metals (Luoma and Jenne 1977; Luoma 1977a). (3) The facilitation of flushing by fresh-water discharge into South Bay may reduce concentrations of solute trace metals which accumulated in the water and/or change the chemistry of the sediments (again, affecting the availability of the metals to organisms) and (4) decreases in salinity of ambient waters, associated with fresh-water influx, may directly affect metal uptake by organisms (Phillips 1977a).

Any or all of the above effects should be reflected in temporal changes in the metal concentrations of organisms in South Bay. Temporal variability in the metal concentrations in bivalves (Anderlini et al. 1975) and seston (Flegal 1977) were reported in North San Francisco Bay, but the causes of the variation were not discussed. Careful analyses of the causes of such variability have been useful in studying biological, physical and chemical influences on metal cycling in other estuaries (Luoma 1977a; Frazier 1975). In this chapter we present data from South San Francisco Bay on temporal changes in the concentration of silver (Ag), copper (Cu) and zinc (Zn) in the soft tissues of the tellenid clam *Macoma balthica*. The data extend from March 1975 to February 1978 and include a 2-yr period of severe drought, which significantly affected physical and chemical processes relevant to San Francisco Bay. We show that increases in fresh-water discharge enhance the biologically available concentrations of all three metals in South Bay, but also appear to modulate the removal of Cu and Ag from the estuary.

Macoma balthica was chosen for study because it is a deposit feeder (ingests sediment and associated organic matter for food), it concentrates a number of metals to a greater extent than the two other clams (*Tapes japonica* and *Mya arenaria*) common in South Bay (Luoma unpublished data) and it is widespread on intertidal mudflats throughout the Bay (see also Nichols 1979 and Carlton 1979). Silver (Ag), Cu and Zn were chosen because all are potentially toxic to estuarine organisms and are potentially important contaminants in South Bay.

Methods and Materials

Clams and sediments were collected at eight intertidal stations and one subtidal station in South Bay (Fig. 1). Our discussion will focus primarily on two of these stations (1 and 5) which have been sampled periodically since early 1975. Sediments were scraped from the surface oxidized layer, sieved through 250μm polyethylene mesh and extracted within 24 h of collection with either hydroxylamine hydrochloride in 0.01N nitric acid, 25% acetic acid, 0.1N sodium hydroxide or a mixture of concentrated nitric and sulfuric acids (for "total" metal). Specific extraction methodology is described elsewhere (Luoma in press).

The sediment extractions were used to assess the effects of fresh-water discharge on sediment chemistry, and to identify periods of terrigenous sediment movement in South Bay. Sodium hydroxide extractions were used to estimate the concentration of humic materials in the sediments. Humic acid concentrations in the extract were measured by absorbance at 480 nm. Humic materials originate primarily from bacterial metabolites in terrestrial soils and occur at higher concentrations when terrigenous input into sediments increases (Luoma and Bryan unpublished data). Hydroxylamine and acetic acid extractions were used to estimate the proportion of freshly precipitated iron in the sediments. In oxidized sediments, Fe occurs as a hydrated oxide primarily associated with the surface of particles (Jenne 1968, 1977). Iron oxides are highly amorphous when they precipitate but gradually become more crystalline with age. The solubility of Fe in hydroxylamine hydrochloride declines more rapidly as iron oxides age (crystallize) than does the solubility of Fe in acetic acid (Luoma in press). Thus, the ratio of Fe extracted from sediments by hydroxylamine relative to that extracted by acetic acid is an index of the proportion of freshly formed iron oxide. Freshly formed iron oxide enters the estuary and its tributaries primarily during periods of high runoff (Elder et al. 1976; Tefrey and Presley 1976). When terrigenous input of Fe is low, the oxides should increase in crystallinity as they age or mix with the "older" Fe in marine sediments. These seasonal changes should be reflected in the relative extraction of Fe by the two extractants.

Fifteen to 30 clams were collected at each sampling. The animals were kept for 24 h after collection in clean seawater to clear their gut of undigested sediments. Soft tissues were dissected from the shells, weighed, digested in a mixture of concentrated nitric and sulfuric acids (2:1, with the addition of excess HNO_3 where necessary), then evaporated to dryness and reconstituted in

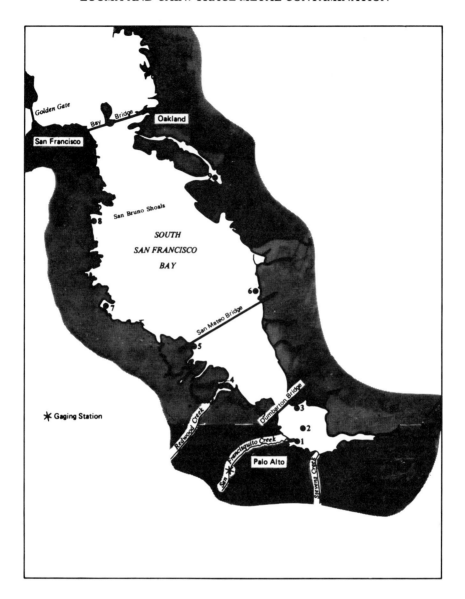

Fig. 1. Sampling stations in South San Francisco Bay.

25% HCl. (Reconstitution in HCl was essential to prevent precipitation of Ag in the samples). Analyses were conducted on either samples of pooled animals (in 1975 and early 1976) or on the tissues of individual animals by atomic absorption spectrophotometry with background correction where necessary. In a few instances significant correlations between metal concentration and the weight of individual animals were observed. Metal concentrations calculated for a median-sized animal (100 mg) from the regression equation for the concentration-weight relationship were used instead of mean concentration in all such samples.

River flow data were obtained from calculations of a Sacramento River-San Joaquin Delta

SAN FRANCISCO BAY

Outflow Index (see Arthur and Ball 1979); stream flow data were from the U. S. Geological Survey (USGS) gauging station on San Francisquito Creek (Fig. 1); and rainfall data were from National Weather Service data for Palo Alto. Estuarine water salinities were determined with a salinometer from shallow pools on the surface of the mudflats.

RESULTS

Physical and Chemical Environment

Physical variables. Delta discharge provides the major source of fresh-water influx into the entire San Francisco Bay system (Conomos 1979). San Bruno shoals and the narrows at the Dumbarton Bridge may impede the penetration of this Delta water into the southern reaches of the Bay (Imberger et al. 1977). Thus, local streams and sewers may be a very important source of fresh water in South Bay despite their relatively low rates of discharge. Local stream discharge was highly irregular during our study, closely reflecting the pattern of local rainfall (Fig. 2). Discharge from most local streams was negligible during the drought summers of 1976 and 1977. The discharge of the Sacramento River also declined substantially as the drought progressed (Fig. 2).

Chemical variables. Salinities at stations 1 and 5 were lower in the spring and winter of both drought years than in the summer, reflecting some significant seasonal increase in fresh-water inflow in South Bay despite the drought (Figs. 3, 4). Local stream discharge appeared to be a significant source of this fresh water. The minimum salinities observed at station 1 in 1976 (late March), and in 1977 (January, April, November) all followed the largest pulses of discharge from San Francisquito Creek observed in those years. Reduced salinities at station 5 in March 1976, and April 1977 also followed large pulses of stream discharge. The minimum salinity observed at station 5 (5 May 1977), however, was preceded by more than 30 days of relatively low stream discharge, suggesting Delta discharge may have penetrated to station 5 in May 1977. A surprising salinity minimum occurred in late August 1977 at stations throughout the South Bay (1, 3, 5, 6, 7, 8). This followed 90 days of no rainfall and zero discharge from all local streams (USGS Data Report in preparation; Santa Clara Valley Water District unpublished data) and coincided with a period of very low Delta discharge.

The humic acid concentrations measured in 1977-78 indicated little terrigenous input of sediment at station 5 until November (following the second storm of the year, but the first pulse of stream runoff measured at the USGS gauging station) and at station 1 until January (Figs. 3, 4). The proportion of fresh iron oxide in the sediments at station 5 followed the expected seasonal pattern with minor peaks following the large pulse of stream discharge in April 1977 and the first storm of the fall in September 1977 (Fig. 4). The proportion of freshly precipitated iron oxide at station 1 was consistently higher and less variable than at station 5 (Fig. 3). This is consistent with the strong influence of a high, relatively constant input of sewage into the southernmost reach (Imberger et al. 1977). Peaks in the proportion of fresh iron oxide in April 1976 and in February and April 1977, following the largest stream discharges of those years, suggested runoff also contributed Fe to the sediments at station 1. The winter maximum and the end of the summer minimum in the proportion of fresh iron oxide occurred earlier at station 5 than at station 1 in both years, as did the beginning of the winter increase in humic acid concentration. At both stations the anomalous salinity minimum in August 1977 was accompanied by a small increase in the proportion of fresh iron oxide, and the winter increase in humic acid concentrations preceded the increase in fresh iron.

Trace metal concentrations. To facilitate a meaningful perspective, the concentrations of Cu,

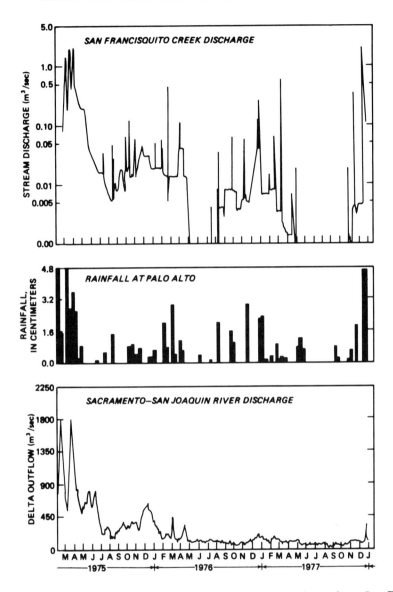

Fig. 2. Rainfall at Palo Alto (near station 1), fresh-water discharge from San Francisquito Creek (near station 1), and fresh-water discharge through the Sacramento-San Joaquin Delta between February 1975, and January, 1978.

Ag and Zn observed in South Bay sediments and clams (Table 1) were compared to similar observations from 17 English estuaries which ranged from pristine in nature to some of the most polluted estuaries in the world (Luoma and Bryan unpublished data). The tellenid clam *Scrobicularia plana*, from the English estuaries, is ecologically, morphologically and behaviorally quite similar to *M. balthica*. Where the two species co-occur in England their Ag, Cu and Zn concentrations are comparable (Bryan and Hummerstone 1977). Median Ag concentrations in clams from station 1 were nearly 100 times greater than concentrations in clams from more pristine environments (Table 1). Moreover, Ag enrichment in both sediments and animals was as great at station 1 in South Bay as in any estuary in the English survey (Fig. 5) which included several estuaries with silver mines in their drainage basin. The Ag enrichment appeared to originate from a point source near station 1 on the western shoal, below Dumbarton Bridge, and was rapidly diluted north and east of station 1 (Table 1).

The concentrations of Cu and Zn in South Bay sediments were low relative to concentrations observed in metalliferous, or industrially enriched areas (Fig. 5; Table 1). Concentrations of

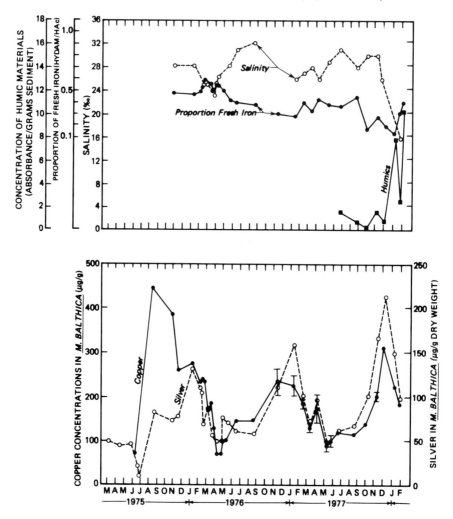

Fig. 3. Salinity, the proportion of freshly precipitated iron oxide in sediments, humic acid concentrations in sediments, and concentrations of copper (Cu) and silver (Ag) in *Macoma balthica* observed at station 1 between March 1975, and February 1978.

Zn in *Macoma balthica* were also low for this family of bivalves (Fig. 5). In contrast, the concentrations of Cu periodically observed in clams at stations 1 and 4 in South Bay were 30 times background (Luoma, unpublished data) and the highest Cu levels were as high as any observed in the English clams. The high degree of biological Cu enrichment in some parts of South Bay suggested the benthic community of this portion of the estuary was especially vulnerable to Cu input, probably due to undefined physicochemical characteristics of the system.

Silver and copper dynamics of M. balthica. Between March 1975 and February 1978, concentrations of Ag in clams at station 1 varied by over 30 times (from 7 to 220 $\mu g \cdot g^{-1}$) and concentrations of Cu by six times (74 to 440 $\mu g \cdot g^{-1}$, Fig. 3). Copper concentrations in the clams varied by five times during the same period at station 5 (Fig. 4). Concentrations of Ag were too low at station 5 for assessments of the range of variation to be meaningful. The variation of Cu and Ag concentrations in the animals did not follow the variations of concentrations in the sediments.

The highest concentration of Cu observed in *M. balthica* occurred at station 1 in August 1975 (Fig. 3). This peak coincided with a period of severe anoxia on the mudflat (due to an intense bloom of the benthic macro-algae *Polysiphonia* sp.) which destroyed all but a few isolated patches of the benthic infauna (see also Nichols 1979). Mobilization of biologically available Cu

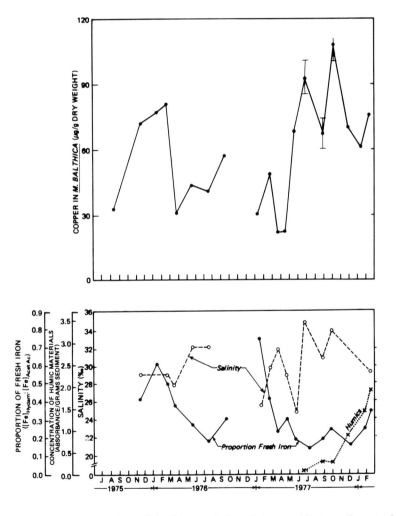

Fig. 4. Salinity, the proportion of freshly precipitated iron oxide in sediments, humic acid concentrations in sediments, and concentrations of copper (Cu) in *Macoma balthica* observed at station 5 between August 1975 and February 1978.

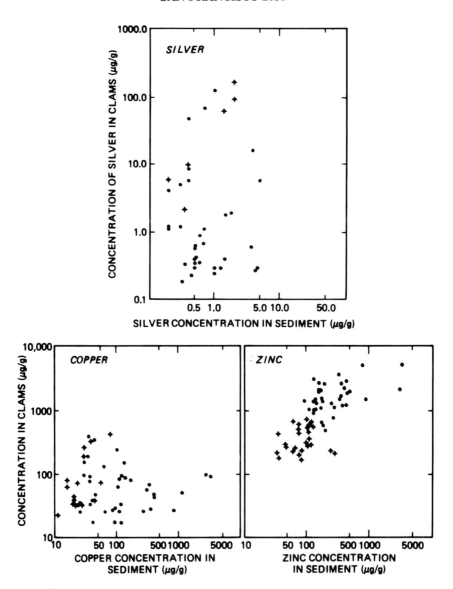

Fig. 5. Concentrations of zinc (Zn), copper (Cu) and silver (Ag) in *Macoma balthica* (✦) from San Francisco Bay and *Scrobicularia plana* (•) from southwest England, compared to concentrations of the metals in sediments from the two locations.

has been observed in other estuaries under similar types of anaerobic conditions (Luoma and Bryan in prep.); although the mechanism involved has not been explained.

With the exception of the Cu peak in late 1975, the dynamics of Ag and Cu in *M. balthica* at station 1 followed a similar pattern (Fig. 3). The lowest concentrations of both metals were observed in early summer. Concentrations began to increase with the onset of infrequent rainfall in the late summer and fall. Concentrations peaked in the early winter, followed by a relatively rapid decline, which coincided with the period of maximum rainfall. Temporal variations in Cu concentrations in *M. balthica* at station 5 followed a pattern similar to that observed at station 1 (Fig. 4).

TABLE 1. CONCENTRATIONS OF AG AND CU IN SEDIMENTS AND CLAMS FROM SOUTH SAN FRANCISCO BAY, COMPARED TO CONCENTRATIONS OBSERVED IN A RELATIVELY PRISTINE HARBOR ON THE PACIFIC COAST.[a]

Station	Silver Concentrations ($\mu g \cdot g^{-1}$)		Copper Concentrations ($\mu g \cdot g^{-1}$)	
	Sediment (mean)	*M. balthica* (median)	Sediment (mean)	*M. balthica* (median)
1	1.8	104	52 ± 9	252
2		14	no data	45
3	0.4	7	36 ± 12	83
4		13	100 ± 44	231
5	0.1	6	29 ± 9	65
6		7	22 ± 10	48
7		4	33 ± 13	46
8	0.3	2	42 ± 6	38
9		3.5	92	25
Princeton Harbor		1.2	15	8

[a] Silver analyses of sediments were conducted by Girvin, et al., using Zeeman spectroscopy.

The autumn buildup of Cu and Ag in the clams was apparently caused by the influx of local runoff into South Bay. In both 1976 and 1977 the onset of metal accumulation at both stations followed early storms in the watershed of South Bay. Moreover, the concentration of Ag and Cu in the animals, between the time of minimum and maximum concentration in each year, was a function of cumulative rainfall during that period of the year (Fig. 6). If Ag and Cu enter South Bay primarily in runoff, then cumulative rainfall should be an indirect measure of the sum of the metal discharge into South Bay at a given point during the year. The concentration of Cu and Ag in the clams at a given quantity of rainfall (i.e. quantity of metal discharged) was significantly greater in 1977 than in 1976, and (for Ag) slightly higher in 1976 than in 1975. Fresh-water discharge from all sources declined between 1975 and the end of 1977 (Fig. 2) suggesting an inverse relationship between the rate of fresh-water discharge and the accumulation of Cu and Ag in the clams per unit metal discharge in runoff. To quantify this relationship, a metal discharge index, M_d, was calculated for each sampling date between the time of minimum and maximum Ag and Cu concentrations in the clams where

$$M_d = (\Sigma R)(1/FW_d)$$

with ΣR = cumulative rainfall (cm) over the stated period and FW_d = the discharge rate of fresh water ($m^3 \cdot s^{-1}$) from a specified source. The concentrations of Ag in the clams from station 1, as observed during the period of accumulation in all three years of the study, fell into a single, highly significant regression with the metal discharge index (Fig. 7) when FW_d was determined from Delta discharge 10 days before the sampling date (the 10-day lag was chosen somewhat arbitrarily, but made little difference in the discharge value chosen). The correlation with Ag concentrations in the clams was insignificant when FW_d was determined from San Francisquito Creek discharge within the week prior to sampling, or as the mean discharge rate of the creek for either the 10 or 20 days before the sampling date. Copper concentrations at both station 1 and station 5 also showed highly significant correlations with the metal discharge index when Delta discharge was used for FW_d

SAN FRANCISCO BAY

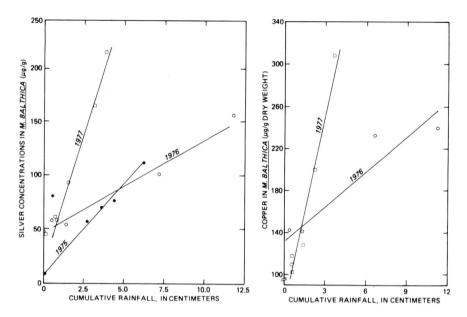

Fig. 6. Correlation of silver (Ag) and copper (Cu) concentrations in *Macoma balthica* from station 1 with cumulative rainfall in 1975, 1976 and 1977. Data are taken from the time between minimum metal concentrations and maximum metal concentrations in the animals in each year. Data for Cu are not presented from 1975 due to the effects of anoxia on Cu concentrations at station 1 in that year.

(Fig. 7). There was no significant difference between the slopes of the relationships at the two stations; however, Cu concentrations at station 1 were consistently higher than those at station 5.

Zinc dynamics. The dynamics of Zn in *M. balthica* at stations 1 and 5 differed from the dynamics of Ag and Cu in several ways (Fig. 8). (1) The period of low discharge was characterized by declining or stable Zn concentrations in the clams. Zinc accumulation in the animals did not begin until after the period of maximum fresh-water discharge, and was not related to cumulative rainfall. (2) The magnitude of the winter peaks in Zn concentration declined as fresh-water discharge declined over the course of the drought. (3) Whereas Cu and Ag concentrations in the clams showed no consistent relationship with salinity, temporal fluctuations in Zn concentrations were consistently the inverse of temporal fluctuations in salinity. (4) Zn concentrations in *M. balthica* were more similar at stations 1 and 5 throughout the study period than were Cu and Ag concentrations.

DISCUSSION

Biologically significant points of Ag and Cu enrichment occur in South Bay. The Ag enrichment is largely the result of Ag discharge from a point source on the western shoal of the southernmost reaches of the Bay. Copper enrichment appears to result from undefined physicochemical conditions in the Bay, which make the system especially susceptible to the relatively small discharges of Cu. Zinc does not appear to be a contaminant of great significance but its behavior is of interest in that it differs significantly from Ag and Cu. The concentrations of Cu and Ag observed at station 1 in November 1977, were as high as any ever reported for tellenid clams. Within two months of this peak, adult *M. balthica* essentially disappeared from the mudflat at this station (but not from station 3 where metal concentrations are lower). Proving that trace

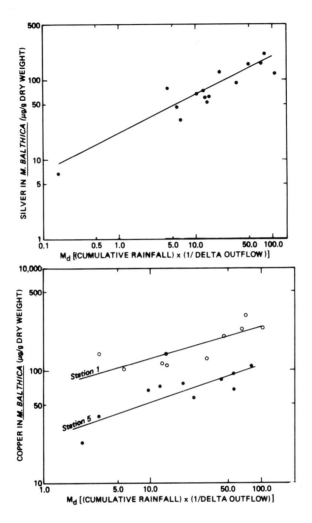

Fig. 7. Correlation of silver (Ag) concentrations at station 1 and copper (Cu) concentrations at stations 1 and 5 with the metal discharge index, M_d. Regression statistics: for AG, $r = 0.91$ and $\log y = 24 + 0.45 \log x$; for Cu at station 1, $r = 0.79$ and $\log y = 69 + 0.26 \log x$; for Cu at station 5, $r = 0.87$ and $\log y = 25 + 0.31 \log x$. All correlations are significant ($p<0.01$).

contaminants are affecting organisms in a natural system is nearly an impossible task (Nichols 1979; Luoma 1977b). However, the coincident occurrence of high tissue concentrations of two potent toxicants, and the disappearance of a population suggests the Cu and Ag enrichment in South Bay is worth further investigation.

Fresh-water discharge rates were very important in determining the degree of biological Ag and Cu contamination at least at stations 1 and 5. The accumulation of Cu and Ag in the clams as rainfall increased in frequency through the summer and fall, and the annual correlation between metal concentrations in the clams and cumulative rainfall, suggested local runoff provided the primary input of the two metals. This was substantiated by the observation by Girvin et al. (unpublished data) of elevated concentrations of Ag in solution at a location offshore from our station 1 (where our data indicate biological contamination is diluted relative to station 1) in

SAN FRANCISCO BAY

September 1976, following a rainstorm. Three days later the high concentrations of Ag had disappeared. In a diel study at the same station they also observed increasing concentrations of Ag in solution with the onset of rainfall.

The rate of fresh-water discharge also had an inhibitory effect on biologically available Cu and Ag during the summer and fall period. The year of lowest fresh-water discharge was the year when metal accumulation in the clams per unit rainfall was greatest. Moreover, 60-80% of the temporal variance in the summer-fall concentrations of Ag and Cu in the clams was explained by our metal discharge index, which included the inverse of Delta discharge as one term. The correlation with the metal discharge index suggested concentrations of Cu and Ag available to *M. balthica* in South Bay between 1975 and 1978 were controlled during the summer-fall period by a dynamic balance between metal input rates from local runoff and a slow rate of metal loss, modulated by fresh-water discharge rates. (Of course, the input and loss functions need not represent metal loadings alone, but could also occur through chemical changes in the sediments or water column that enhance or decrease the biological availability of the metals.) The magnitude of the fresh-water discharge during this period of relatively long water residence times in South Bay, appeared to be the crucial factor determining the size of the annual peak in Cu and Ag concentrations.

The rapid decline of Cu and Ag concentrations in the clam tissues during winter and spring occurred as significant quantities of fresh-water and terrigenous sediment entered South Bay. The

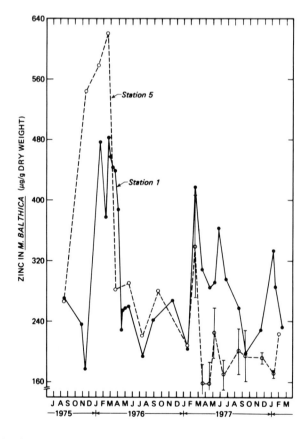

Fig. 8. Zinc (Zn) concentrations in *Macoma balthica* at stations 1 and 5 between August 1975, and February 1978.

onset of the decline in Cu concentrations at station 5 coincided with the minimum salinity observed at that station in 1975-76 and with the onset of declining salinity and increasing humic acid concentrations in the sediments in 1977-78. At station 1 the onset of the 1975-76 decline coincided with a salinity decline and a gradual increase in the proportion of freshly precipitated iron oxide in the sediment. In 1977-78 the decline did not occur until salinities fell below 26 °/oo, but coincided with a sharp increase in the humic acid content of the sediments. The 1976-77 decline at station 1 did not coincide with a salinity decline, but the proportion of freshly precipitated iron in the sediment did rise as Ag and Cu levels in the clam began to drop.

There are at least two possible explanations for the decline in Ag and Cu concentrations in the animals. First, at some point in the early winter, the flushing action of fresh-water discharge is sufficiently strong to remove biologically available Cu and Ag from South Bay as rapidly as it enters in the runoff. By limiting the residence time of the contaminated water and/or sediment, the exposure of the clams to Ag and Cu is limited. The second possibility is that as the rainy season progresses the chemical and/or physical nature of the terrigenous sediment in the runoff changes (V. Kennedy pers. comm.). Early runoff should be dominated by sediment from the urban watershed of South Bay. Not only will the initial storms wash urban streets containing several months' accumulation of contaminants, but also the initial rainfall of the year in the less urbanized areas will largely be converted to groundwater, rather than running off, until the soils are saturated. An example of the latter was observed after the first storm of the year in September 1977, when no stream flow was detected in either San Francisquito Creek or Redwood Creek (USGS Data Report, in prep.) at gauging stations upstream from most of the urban area. However, a significant pulse of flow was observed at an urban gauging station in Stevens Creek (Santa Clara Valley Water District unpublished data). Later in the rainy season the urban runoff will carry less sediment, but more important, the stream discharge will also carry a heavy load of less contaminated sediment from the more rural parts of the watershed. Humic acids should be more concentrated in the rural runoff than in the urban runoff; the close correspondence of declines in Ag and Cu concentrations in 1977-78 with increased humic acid concentrations could reflect a change in the dominant sediment type in the runoff. These hypotheses are, of course, not mutually exclusive since any flushing modulated by fresh-water discharge would also be important in removing urban runoff from the estuary.

The metal discharge index represents a first step toward quantifying the degree of Cu and Ag enrichment we might expect in South Bay organisms as fresh-water discharge into the estuary is reduced. Unfortunately, the index will not be useful in predicting environmental impact until we understand more about the relative importance in flushing of South Bay by local stream discharge versus Delta discharge. Initial model studies of hydrodynamics in South Bay suggest significant quantities of fresh water from the Delta should not have penetrated beyond San Bruno shoals during the summer or fall in either 1976 or 1977 (Imberger et al. 1977). If so, local stream discharge is the most likely source modulating the different rates at which biologically available Cu and Ag were removed from South Bay in these two years. The winter decline in concentrations of Cu also clearly followed major pulses of local runoff more closely than Delta discharge at station 5 in March 1976, and November 1977, and at station 1 in January, 1977 (although the chemical nature of the sediment in the runoff may also explain these effects). Several lines of evidence raise questions about the dominance of local stream discharge as a cleansing force in South Bay, however: (1) The lowest Ag concentrations ever observed at station 1 followed a period of zero rainfall between 1 May and 20 June 1975. The Ag minimum suggests a negligible input of Ag combined with a rapid flushing of available Ag at this time. Local stream flow declined to relatively low levels by June 1975; but Delta discharge peaked at a sufficient rate to penetrate San Bruno shoals shortly before the Ag minimum. At least in the spring of 1975, Delta discharge was the most likely force driving the removal of biologically available Ag at station 1. (2) The salinity anomaly in August 1977, was clearly not the result of local runoff. This salinity minimum coincided with a small influx of terrigenous sediment (reflected by an increase in the proportion of freshly precipitated iron) at both stations 1 and 5, but, more importantly, with a sharp reduction in Cu concentrations in clams at station 5. A similar reduction in Cu concentrations at station 8 was observed at this time, but not at stations 1, 3 or 6 (Luoma and Cain unpublished data). (3) When the metal discharge index was calculated from the inverse of local stream flow, no correlation with the three years of data on Cu and Ag levels in the clams was observed. It is possible that we did not adequately quantify crucial aspects of the highly irregular stream flows in the relationship; but it is also possible that Delta discharge is a more important flushing force on the shoals of the South Bay than predicted from the initial studies of Imberger et al. (1977) in the main channel of the estuary.

The differences between Zn dynamics and the dynamics of Cu and Ag in South Bay demonstrate that all trace metals do not behave similarly in the system. The differences between the metals may reflect their source, or the source of the chemical factors that control their availability. In 1977-78, Zn concentrations in the clams were lower at both stations than in 1975-76, and, like Cu, concentrations at station 1 exceeded those at station 5 approximately twofold. The distribution of Zn and Cu in 1977-78 suggest factors controlling the availability of the metals originated from similar local sources, probably throughout South Bay (since the northward dilution of Ag greatly exceeded that of Cu and Zn). The south-to-north gradient in Cu and Zn concentrations in 1977-78 could reflect differences in discharge rates, dilution by an increasing volume of water toward the north, or more efficient flushing in the northern reaches. The similarity of the slopes of the regressions between Cu in the clams versus M_d at stations 1 and 5 suggests one of the latter two suggestions is most likely.

In contrast to Cu, the south-to-north gradient for Zn was not observed in 1975-76 when Delta discharge was certainly sufficient to flow southward of San Bruno shoals. Zinc concentrations at station 5 exceeded those at station 1 in 1975-76, and peak concentrations at both stations were much greater than in the following years of low fresh-water discharge. Although further study is required, it appears that Delta-derived water may control the biological availability of Zn in years of normal rainfall. This hypothesis is consistent with our earlier suggestion that Zn is not discharged in sufficient quantities into South Bay to be a biologically significant contaminant.

Factors such as temperature, salinity (Wolfe 1971; Phillips 1977a), and biological phenomena (tissue growth; seasonal variations in physiology–Frazier 1975; Betzer and Pilson 1975; Phillips 1977b), have also been cited to explain temporal variations in the metal concentrations of benthic organisms within estuaries. In South Bay, temperature was not found to affect Zn, Ag or Cu concentrations in *M. balthica* directly. The temporal pattern of variation in metal concentrations was roughly the inverse of the seasonal temperature cycle. Seasonal contrasts in the relationship with Cu and Ag concentrations also suggested salinity was not a variable to which these metals were directly responding. Concentrations of Cu and Ag in the clam varied inversely with salinity at station 1 in the spring (March, April 1976; April 1977) and positively with salinity at both stations 1 and 5 in the winter. Zinc concentrations in the clam varied inversely with salinity in a consistent manner; thus, the possibility that salinity variations may have had some direct effect on Zn uptake rates by *M. balthica* cannot be discounted.

The winter decline in Cu and Ag concentrations in the clams coincided with the period of maximum growth of the organism (Nichols in prep.; Cloern and Nichols 1978). Two arguments suggest the effects of tissue growth on metal concentrations were minimal, however: (1) temporal changes in the content (μg) of Ag and Cu in the animals closely followed changes in concentration ($\mu g \cdot g^{-1}$), thus eliminating the possibility that the seasonal differences in metal concentrations were the result of simply a changing tissue mass, rather than metal fluxes; and (2) sharp changes in the Cu concentrations of *M. balthica* were not observed during the spring at stations 6 and 8 in 1977 (Luoma and Cain unpublished data). If undefined physiological processes were responsible for the temporal changes observed, those processes were specific to stations 1 and 5 - a highly unlikely prospect.

Our evidence strongly suggests that the interaction of hydrodynamic processes, local weather, and, quite possibly, the chemical characteristics of sediments entering South San Francisco Bay in local runoff, play the major roles in controlling the contamination of organisms south of the San Mateo Bridge with Cu and Ag. The rate of fresh-water discharge into the Bay is a primary factor mitigating the effects of those interactions, and appears to control the amplitude of the annual peak in contaminant concentrations.

LITERATURE CITED

Anderlini, V. C. et al. 1975. Dredge disposal study San Francisco Bay and estuary: pollutant uptake study. Lawrence Berkeley Laboratory, University of California Rep. UCID-3666. 152 pp.

Arthur, J. F., and M. D. Ball. 1979. Factors influencing the entrapment of suspended material in the San Francisco Bay-Delta estuary. Pages 143-174 *in* T. J. Conomos, ed. San Francisco Bay: The Urbanized Estuary. Pacific Division, Amer. Assoc. Advance. Sci., San Francisco, Calif.

cisco Bay: The Urbanized Estuary. Pacific Division, Amer. Assoc. Advance. Sci., San Francisco,

Betzer, S. B., and M. E. Q. Pilson. 1975. Copper uptake and excretion of *Busycon canaliculatum* L. Biol. Bull. 148:1-15.

Bryan, G. W., and L. G. Hummerstone. 1977. Indicators of heavy metal contamination in the Looe Estuary (Cornwall) with particular regard to silver and lead. J. Mar. Biol. Assoc. U. K. 57:75-92.

Carlton, J. T. 1979. Introduced invertebrates of San Francisco Bay. Pages 427-444 *in* T. J. Conomos, ed. San Francisco Bay: The Urbanized Estuary. Pacific Division, Amer. Assoc. Advance. Sci., San Francisco, Calif.

Cloern, J. E. 1979. Phytoplankton ecology of the San Francisco Bay sytem: the status of our current understanding. Pages 247-264 *in* T. J. Conomos, ed. San Francisco Bay: The Urbanized Estuary. Pacific Division, Amer. Assoc. Advance. Sci., San Francisco, Calif.

Cloern, J. E., and F. H. Nichols. 1978. A von Bertalanffy growth model with a seasonally varying coefficient. J. Fish. Res. Bd. Can. 35:1479-1482.

Conomos, T. J. 1979. Properties and circulation of San Francisco Bay waters. Pages 247-264 *in* T. J. Conomos, ed. San Francisco Bay: The Urbanized Estuary. Pacific Division, Amer. Assoc. Advance. Sci., San Francisco, Calif.

Conomos, T. J., and D. H. Peterson. 1977. Suspended-particle transport and circulation in San Francisco Bay: an overview. Pages 82-97 *in* L. E. Cronin, ed. Estuarine Processes. Vol. 2. Academic Press, New York.

Elder, J. F., K. E. Osborn and C. R. Goldman. 1976. Iron transport in a Lake Tahoe tributary and its potential influence upon phytoplankton growth. Water Res. 10:783-789.

Flegal, A. R. 1977. Mercury in the seston of the San Francisco Bay estuary. Bull. Environ. Contam. Toxicol. 17:733-738.

Frazier, T. M. 1975. The dynamics of metals in the American Oyster, *Crassostrea virginica* I. Seasonal effects. Chesapeake Sci. 16:162-171.

Girvin, D. C., A. T. Hodgson, and M. H. Panietz. 1975. Assessment of trace metal and chlorinated hydrocarbon contamination in selected San Francisco Bay estuary shellfish. Lawrence Berkeley Laboratory, University of California Rep. UC10-3778. 80 pp.

Imberger, J., W. B. Kirkland, and H. B. Fischer. 1977. The effect of delta outflow on the density stratification in San Francisco Bay. ABAG (Assoc. San Francisco Bay Area Govern.) Rep.

HBF-77/02. 109 pp.

Jenne, E. A. 1968. Controls on Mn, Fe, Co, Ni, Cu and Zn concentrations in soil and water: The significant role of hydrous Mn and Fe oxides. Pages 337-387 *in* R. F. Gould, ed. Trace Inorganics in Water. Amer. Chem. Soc.

Jenne, E. A. 1977. Trace element sorption by sediments and soils—sites and processes. Pages 425-553 *in* W. Chappell and K. Peterson, eds. Symposium on Molybdenum in the Environment, vol. 2. M. Dekker, Inc., New York.

Luoma, S. N. In press. Trace metal bioavailability: modeling biological and chemical interactions of sediment-bound zinc. *In* E. A. Jenne, ed. Chemical Modeling-Speciation, Sorption, Solubility and Kinetics in Aqueous Systems. Amer. Chem. Soc.

Luoma, S. N. 1977a. The dynamics of biologically available mercury in a small estuary. Estuarine Coastal Mar. Sci. 5:643-652.

Luoma, S. N. 1977b. Detection of trace contaminant effects in aquatic ecosystems. J. Fish. Res. Bd. Can. 37:437-439.

Luoma, S. N., and E. A. Jenne. 1977. The availability of sediment-bound cobalt, silver and zinc to a deposit feeding clam. Pages 213-230 *in* H. Drucker and R. E. Wildung, eds. Biological Implications of Metals in the Environment. U. S. NTIS-CONF-750929, Springfield, Va.

McCulloch, D. S., D. H. Peterson, P. R. Carlson, and T. J. Conomos. 1970. Some effects of freshwater inflow on the flushing of south San Francisco Bay—a preliminary report. U. S. Geol. Surv. Circ 637A. 27 pp.

Nichols, F. H. 1979. Natural and anthropogenic influences on benthic community structure in San Francisco Bay. Pages 409-426 *in* T. J. Conomos, ed. San Francisco Bay: The Urbanized Estuary. Pacific Division, Amer. Assoc. Advance. Sci., San Francisco, Calif.

Phillips, D. J. H. 1977a. The common mussel *Mytilus edulis* as an indicator of pollution by zinc, cadmium, lead and copper. I. Effects of environmental variables on uptake of metals. Mar. Biol. 38:59-69.

Phillips, D. J. H. 1977b. The use of biological indicator organisms to monitor trace metal pollution in marine and estuarine environments—A review. Environ. Poll. 13:281-319.

Pitt, R. E., and G. A. Amy. 1973. Toxic materials analysis of street surface contaminants. Environ. Protection Technol. Ser. EPA-R2-73-283, Washington, D. C. 135 pp.

Tefrey, J. H., and B. J. Presley. 1976. Heavy metal transport from the Mississippi River to the Gulf of Mexico. Pages 39-76 *in* H. L. Windom and R. Duce, eds. Marine Pollutant Transfer. D. C. Health and Co., Lexington, Mass.

Wolfe, D. A. 1971. Fallout ^{137}Cs in clams (*Rangia cuneata*) from the Neuse River estuary, North Carolina. Limnol. Oceanogr. 16:797-803.

PHYTOPLANKTON ECOLOGY OF THE SAN FRANCISCO BAY SYSTEM: THE STATUS OF OUR CURRENT UNDERSTANDING

JAMES E. CLOERN
U. S. Geological Survey, 345 Middlefield Road, Menlo Park, CA 94025

Although past studies of phytoplankton dynamics in the San Francisco Bay system are limited in number and scope, they have provided sufficient information to define gross spatial and temporal patterns. Annual changes in the density and composition of phytoplankton populations differ among major geographic areas within the system, and recent studies suggest that phytoplankton dynamics in each major portion of San Francisco Bay are governed by a unique set of environmental factors. The annual maximum abundance of phytoplankton in central San Francisco Bay during spring may be a direct consequence of diatom blooms that occur in coastal waters during the upwelling season. The spring maximum of phytoplankton abundance in South Bay may also result from the dispersion of neritic diatoms from offshore during some years, although the 1978 spring maximum resulted from rapid *in situ* growth of microflagellate populations. Apparently, stratification of the South Bay water column (initiated by movement of Delta-derived low-density water from the northern reach) creates a shallow surface layer where flagellates are given sufficient solar irradiation to maintain rapid growth rates. Phytoplankton populations in the northern reach of San Francisco Bay apparently are most strongly regulated by the physical accumulation of suspended particulates by gravitational circulation, the rapid growth of planktonic algae over shoals, and phytoplankton dynamics in coastal waters and/or tributaries.

Because few research efforts have been implemented to define environmental factors that regulate phytoplankton dynamics, basic unanswered (or unasked) questions remain. There is need (1) to define those functional groups of planktonic algae responsible for fixing inorganic carbon and energy, and then to follow pathways of energy and material transfer from the phytoplankton to other trophic levels, (2) to define the relationships between the physics of water movement and phytoplankton dynamics, and (3) to identify those physical-chemical-biological factors most responsible for regulating phytoplankton population size and composition, and then to quantify the response of algal population growth to changes in these important environmental factors.

The quality of estuarine waters is reflected in and a consequence of the phytoplankton community because the density (i.e. abundance) and composition of plankton populations both respond to environmental stress and can, in turn, cause environmental stress. The species composition and population density of phytoplankton are sensitive to environmental changes, and continual documentation of phytoplankton population dynamics can provide an invaluable record of water quality, can signal if radical changes occur within an estuarine system, and can offer clues to the causes of changes when they do occur. Planktonic algae affect concentrations of dissolved gases (oxygen and carbon dioxide), concentrations of dissolved inorganic and organic substances, and affect pH. Finally, the photosynthetic fixation of inorganic carbon by phytoplankton offers a source of organic carbon and energy for higher trophic levels and ultimately determines the success of fisheries, including those having commercial and recreational value. The understanding of phytoplankton dynamics (i.e. changes in population abundance, composition and distribution, and rates of physiological processes) is, therefore, central to the understanding of how estuarine ecosystems work and how they respond to stresses imposed by man and nature.

Despite the size and economic importance of San Francisco Bay, the biological components of its water column have been meagerly studied. Virtually no baywide phytoplankton studies were done until the University of California's Sanitary Engineering Research Laboratory (SERL) undertook a five-year study in the early 1960's (Storrs et al. 1966). The U. S. Geological Survey initiated a study of water properties of the Bay system in 1969 (Conomos et al. 1978); these surveys included measurements of relative phytoplankton abundance and distribution. In 1970 four agencies (U. S. Bureau of Reclamation [USBR], U. S. Fish and Wildlife Service, and California Departments of Water Resources and Fish and Game) pooled their resources into a cooperative ecological study of the Sacramento-San Joaquin Delta and parts of northern San Francisco Bay. These three major efforts all included descriptive studies of phytoplankton abundance or composition, but with few exceptions (e.g. Peterson et al. 1975a; and Delta studies of Arthur and Ball 1979 and Ball and Arthur 1979) there has been no research effort to define and quantify those environmental factors that regulate phytoplankton dynamics throughout San Francisco Bay. Until this is done, our capability to forecast impacts of proposed perturbations (such as construction of the Peripheral Canal and San Luis Drain and the deepening of the Baldwin Ship Channel; see, e.g. Seckler 1971), and our understanding of factors that determine the success of important fisheries, such as the striped bass (*Morone saxatilis*) and dungeness crab (*Cancer magister*), will continue to be restricted.

Although our knowledge of dynamic processes affecting the plankton is limited, past studies have provided sufficient information to define gross spatial and temporal patterns of changing phytoplankton abundance that occur over an annual cycle. In this chapter I integrate the results of previous studies into a generalized description of these patterns, then present the results of a recent (1977-78) study that was designed to offer preliminary hypotheses concerning mechanisms that cause these observed patterns. Finally, I point out critical new directions for future research and significant questions that must be addressed before our understanding of the water column of San Francisco Bay will allow for intelligent management decisions in the future.

TEMPORAL AND SPATIAL PATTERNS

Storrs et al. (1966), who first described the seasonal changes of phytoplankton abundance in different portions of the San Francisco Bay system, observed that Suisun Bay (Fig. 1) typically has a maximum phytoplankton standing stock during summer (Fig. 2), whereas all other portions of the Bay system have annual maxima during spring, and that the population maximum in Suisun Bay is at least 10 times greater than the maxima seen elsewhere. Although year-to-year variations exist, subsequent studies (Peterson 1979; Peterson et al. 1975a; Arthur and Ball 1979; Ball and Arthur 1979; Conomos et al. 1979) have confirmed this general pattern. The similarity in both the magnitude and timing of population maxima in San Pablo Bay, Central Bay, and South Bay (Fig. 2) superficially suggests that phytoplankton dynamics in all three embayments are regulated by the same set of environmental factors. However, our recent studies demonstrate that this is not the case. It is appropriate, therefore, that each geographic section of the Bay system be discussed separately.

Central Bay

In the earliest quantitative study of phytoplankton populations in San Francisco Bay, Whedon

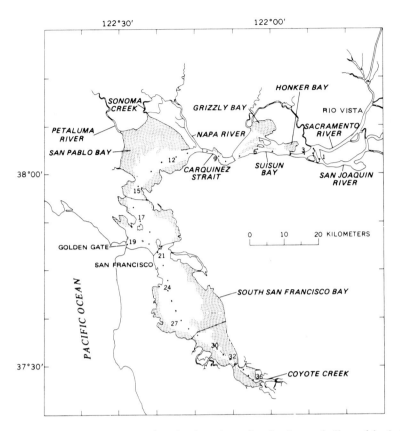

Fig. 1. San Francisco Bay system showing location of major bays, shallows (shaded area with a mean tide depth ≤ 2 m), central channel (dashed line represents 10-m isobath) and location of sampling stations.

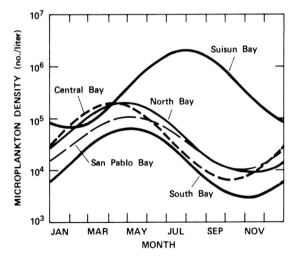

Fig. 2. Seasonal variation of total microplankton density in different parts of the Bay system (redrawn from Storrs et al. 1966).

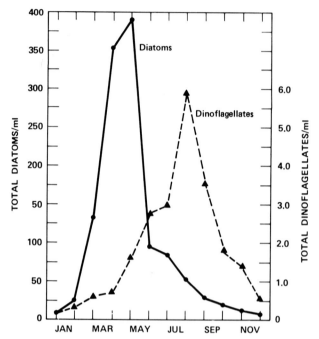

Fig. 3. Monthly mean densities of diatoms and dinoflagellates in surface waters taken from a San Francisco pier, 1933-1935 (data from Whedon 1939).

(1939) found that the Central Bay (bordered by stations 17, 19 and 21 - Fig. 1) has its maximum standing stocks between March and June (Fig. 3), consistent with the observations of Storrs et al. (1966) 30 years later. Results of both studies showed that the spring maximum is a consequence of increased numbers of neritic (i.e. coastal marine) diatoms. From March to as late as September, northerly winds along the California coast generate periods of upwelling that produce episodic blooms of netplankton (single cells or chains larger than about 20 μ) offshore (Bolin and Abbott 1963; Malone 1971). Since these blooms typically are dominated by the same species (*Chaetoceros* spp., *Nitzschia* spp., *Rhizosolenia* spp. and *Skeletonema costatum*) as reported by Whedon (1939) and Storrs et al. (1966), it is likely that the spring maximum in Central Bay results from the dispersion of planktonic diatoms into San Francisco Bay from these offshore blooms during the upwelling season.

We did not see a dramatic spring increase in phytoplankton abundance (measured as chlorophyll *a* concentration) at the Golden Gate during 1978 (Fig. 4), although the modest increases we did measure in the spring resulted from increased numbers of neritic diatoms. Causes of the increased chlorophyll *a* concentration at the Golden Gate during early December 1977 (Fig. 4) are not known, but microscopic enumeration revealed that this pulse was dominated by the neritic diatom *Nitzschia seriata* (R. Wong pers. comm.). Conceivably, anomalous meteorological conditions created a winter upwelling event that allowed for rapid population growth of this species offshore, but this hypothesis cannot be tested, and the proposed relationship between coastal upwelling and phytoplankton dynamics in San Francisco Bay will remain speculation until simultaneous studies are done both inside and outside the Golden Gate.

South Bay

Storrs et al. (1963) observed that phytoplankton populations (measured as cell density) are relatively small in South Bay (Fig. 2), and that South Bay, like Central Bay, has its annual maximum during spring (Fig. 5). They also reported a population composition that was dominated by neritic diatoms; this again suggests that the spring maximum results from the importation of marine diatoms from offshore. But our study gave very different results in 1977-78: although phytoplankton density (measured both as chlorophyll a concentration and cell density) was again highest in March (Fig. 4), the partitioning of chlorophyll a into three size classes demonstrated that over 90% of this phytoplankton maximum comprised ultraplankton (very small cells capable of passing through the 5-μ pores of a membrane filter). Microscopic examination of preserved samples confirmed

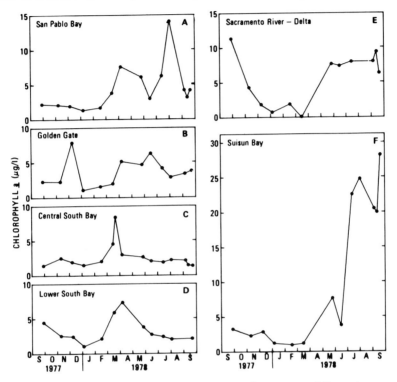

Fig. 4. Mean concentration of chlorophyll a in near-surface waters of the major geographic areas comprising the San Francisco Bay system, between Sept. 1977-Sept. 1978. (San Pablo Bay encompasses the area between stations 9-15; Golden Gate is station 19; Central South Bay lies between stations 24-30; Lower South Bay includes stations 32-36; Suisun Bay lies between stations 4-8.)

that the South Bay water mass had a large number of microflagellates (primarily cryptophytes, haptophytes and naked dinoflagellates) and relatively few large diatoms (R. Wong pers. comm.). Since offshore blooms are dominated by the large netplankton (Malone 1971; Garrison 1976), the South Bay maximum of March 1978 clearly did not originate from coastal netplankton blooms.

Vertical profiles of salinity and chlorophyll a in mid-South Bay (Fig. 6) suggest the cause of rapid phytoplankton population growth during March 1978. On 14 March, vertical distributions of both salinity and chlorophyll a were fairly homogeneous at stations 27 and 30, suggesting that the water column was well mixed; data from station 24 showed pronounced salinity stratification. On 20 March, the entire South Bay showed some degree of salinity stratification, and concentrations of chlorophyll a had increased dramatically in the surface layer (Fig. 6). Apparently, the spring maximum of 1978 occurred under conditions of a salinity stratification which created a shallow surface

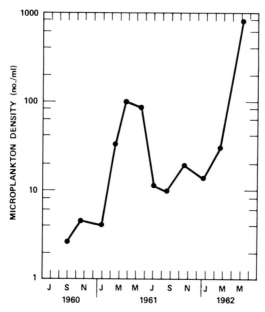

Fig. 5. Average microplankton density in surface waters of South Bay, September 1960 to May 1962 (data from Storrs et al. 1963).

layer where phytoplankton were given sufficient sunlight for rapid growth. When the water column is vertically mixed, planktonic algae spend a majority of time in the lower aphotic (dark) zone of the water column and do not receive sufficient solar irradiation to maintain rapid growth rates.

McCulloch et al. (1970), Imberger et al. (1977) and Conomos (1979) found that pronounced

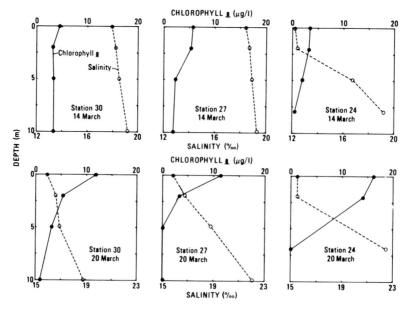

Fig. 6. Vertical distributions of chlorophyll a (solid line) and salinity (dashed line) at three South Bay stations on 14 March 1978 (top frames) and 20 March 1978 (bottom frames).

salinity stratification of South Bay results from the penetration of low-salinity water from the northern reach during periods of rapid freshwater discharge through the Sacramento-San Joaquin Delta. This mechanism explains the salinity stratification of South Bay during March 1978, since Delta outflow increased dramatically between 3-9 March (Fig. 7). Stratification was observed throughout South Bay 16 days after the start of this flood; this is consistent with Imberger et al. (1977) and Conomos (1979) who inferred a time delay of about 1-2 weeks before the South Bay responds to a large flood through the Delta.

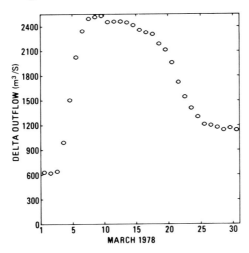

Fig. 7. Net Delta Outflow of the Sacramento-San Joaquin Delta during March 1978 (from USBR, Sacramento, Calif.).

Results of Storrs et al. (1963) study are consistent with the hypothesis that phytoplankton dynamics in coastal waters dominate seasonal patterns in South Bay, and that diatoms predominate. Our results in 1978 demonstrate that spring increases are caused by stratification of the water column during periods of rapid Delta outflow, and that microflagellates dominate the spring increase in phytoplankton biomass. The incongruity between these two studies may reflect the fact that: (1) the methods of Storrs et al. (1963) may have given a severe underestimate of the biomass of microflagellates in South Bay during the 1960's; (2) meteorological conditions in 1978 were not conducive to strong or frequent coastal upwelling events, hence minimizing the importance of neritic netplankton in South Bay during the spring of 1978; or (3) real changes have occurred in the composition of South Bay phytoplankton assemblages since the 1960's. Answers to these basic questions will only come from long term studies that include efforts to determine the relationship between plankton dynamics both in coastal waters and in San Francisco Bay, and studies that provide information about species composition as well as biomass of phytoplankton populations.

Results of previous studies have demonstrated that vertical and longitudinal (north-south) variations in phytoplankton abundance are generally small, implying that South Bay is a fairly homogenous embayment. But examination of the composition of phytoplankton populations in South Bay reveals a different picture. The longitudinal distribution of chlorophyll a on 24 May 1978, for example, was fairly uniform in surface waters of South Bay (Fig. 8a). The relative contribution of ultraplankton (i.e. microflagellates), however, increased rapidly from the Golden Gate to station 24, then declined in mid-South Bay (station 27), increased rapidly again near station 30, and declined again at the southern extremity (Fig. 8b). Reasons for this heterogeneity of phytoplankton composition are not yet known but must be related to the complex bathymetry of South

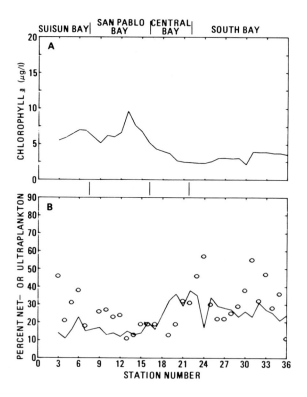

Fig. 8. Longitudinal profiles of near-surface chlorophyll a (A) and size distribution of phytoplankton (B) in the San Francisco Bay system, 24-25 May 1978. Percent netplankton (solid line) is the fraction of chlorophyll a retained by Nitex screens having 22-μ mesh, and percent ultraplankton (0) is the fraction passing a Nuclepore filter having 5-μ pores.

Bay, influences from local discharges and surface runoff, and the formation of surface lenses of low-density water as Delta-derived water penetrates South Bay on each tidal cycle (Imberger et al. 1977; Conomos 1979). Our initial perception of South Bay as a longitudinally homogeneous embayment changes upon examination of plankton size composition and species composition. Extension of sampling over the lateral shallow areas of South Bay should also offer new insights into spatial patterns observed in the channel.

San Pablo Bay

The northern reach of the San Francisco Bay system (defined here as the area between stations 17 and 1 - Fig. 1) is a partially to well-mixed estuary comprising a central deep channel and two isolated shallow embayments, San Pablo Bay and Grizzly-Honker bays, that contain phytoplankton communities with very different dynamics. Seasonal patterns in the density and composition of phytoplankton populations in San Pablo Bay can vary dramatically from year to year, but Storrs et al. (1966), USBR et al. (1977) and Conomos et al. (1979) all observed annual phytoplankton maxima during spring. Several mechanisms may contribute to the growth of phytoplankton populations in San Pablo Bay during spring, the simplest being the movement of coastal plankton blooms into San Pablo Bay during spring. This hypothesis is consistent with the observation that increased concentrations of chlorophyll a between March and May 1978 (Fig. 4) were accompanied by large increases in the population density of *Skeletonema costatum,* a neritic diatom that is dominant in netplankton blooms along the California coast (Bolin and Abbott 1963; Malone 1971). However, the simple dispersion of coastal plankton into San Pablo Bay is not sufficient to explain observed spatial patterns. We consistently measured higher phytoplankton densities in San Pablo Bay than in Central Bay during spring of 1978 (see, for example, Fig. 8a). Therefore, if the spring maximum originates outside the Golden Gate, some additional mechanism causes an accumulation or growth of these coastal-derived populations within San Pablo Bay.

Conomos et al. (1971) demonstrated that surface waters of the northern reach have net (tidally-averaged) advective movement toward the Pacific Ocean, while bottom waters have a net landward movement (see also Conomos 1979). This two-layer flow, caused by longitudinal and vertical salinity gradients characteristic of partially and well-mixed estuaries (e.g. Bowden 1967; Dyer 1973), has a profound influence on the distribution of phytoplankton and other suspended particulates in San Francisco Bay (Conomos and Peterson 1974, 1977; Peterson et al. 1975a; Arthur and Ball 1979). The location along the longitudinal axis of the estuary where net bottom currents are nullified by seaward-flowing river water, the "null zone" is generally characterized by high accumulations of suspended particulates (see, for example, Meade 1972), including planktonic algae, because net advective displacement is relatively slow. Peterson et al. (1975b) demonstrated that the location of this null zone varies seasonally in response to seasonal variations in Delta outflow, and that it is often located near San Pablo Bay during spring, when Delta outflow is high. As marine diatoms enter San Pablo Bay in the bottom density current, they encounter the opposing current from the Sacramento-San Joaquin rivers (Fig. 9), and are either entrained or advected into the surface layer. There, they are either quickly retransported seaward or they disperse laterally into the shallows of San Pablo Bay.

Since waters of the northern reach are generally very turbid, growth rates of planktonic algae are small in the deep (≥ 10 m) channel where average irradiance is low. However, the shallow

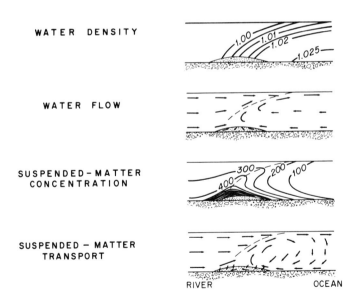

Fig. 9. Schematic representation of the null zone of partially mixed estuaries, showing spatial variations in water density and circulation, concentration of suspended particulates, and the transport of suspended particulates (including dense algal cells). Reproduced with permission from Meade (1972).

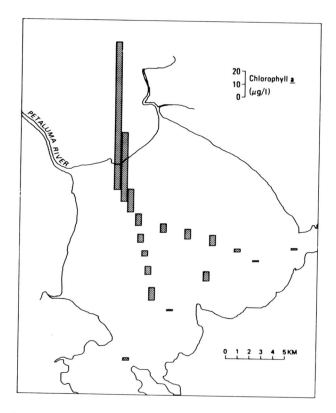

Fig. 10. Distribution of chlorophyll *a* in surface waters of San Pablo Bay, 15 March 1978.

(≤ 2 m) depth over San Pablo Bay tidal flats (Fig. 1) provides for increased exposure of the suspended algae to light and thus for rapid growth rates (Cloern 1978). Rapid growth of seed populations that enter San Pablo Bay shallows can maintain large population densities, particularly in northern San Pablo Bay where the exchange rate between the shallows and channel is presumably slow and water residence time is highest. During March 1978, the spatial distribution of phytoplankton biomass (chlorophyll *a*) in San Pablo Bay (Fig. 10) was consistent with the hypothesis that the shallows are sites of rapid population growth. The extremely high chlorophyll *a* concentrations in northern San Pablo Bay represented a large phytoplankton population that was almost exclusively *Skeletonema costatum*.

The proposed mechanism that causes increased phytoplankton abundance in San Pablo Bay requires: (1) an allochthonous source (in this case, coastal waters) of viable planktonic algae that (2) rapidly disperse over the tidal flats because of location of the null zone, and (3) divide rapidly because of shallow depth (i.e., increased availability of light). The decline of phytoplankton populations during summer (or sometimes as late as fall - Fig. 4) probably results from decreased inputs of marine diatoms in the bottom density current as upwelling frequency declines offshore, and movement of the null zone landward in response to decreased Delta outflow.

Suisun Bay

The most dramatic feature of phytoplankton dynamics in San Francisco Bay is the large standing stock typically seen in the vicinity of Suisun Bay during summer (Figs. 2, 4; see also Ball and Arthur 1979). The rapid population growth seen during most summers is coupled to the timing of decreased freshwater discharge through the Sacramento-San Joaquin Delta (Fig. 11). The null zone is found near Suisun Bay during the summer low flow period (Peterson et al. 1975b; Arthur and Ball 1979) where, as in San Pablo Bay during spring, the presence of an estuarine circulation cell adjacent to a shallow region allows for population growth.

The summer growth of phytoplankton populations in Suisun Bay is accompanied by a dramatic shift in species composition that is reflected in the partitioning of chlorophyll *a* into different size fractions. For example, between September 1977 and July 1978, phytoplankton populations in Suisun Bay had a diverse composition and were not dominated by forms within one size class (e.g. Fig. 8b). However, as the summer maximum developed in 1978, netplankton constituted a growing proportion of total phytoplankton biomass. Between July and September 1978, over 80% of the particles comprising the chlorophyll *a* maximum of Suisun Bay (Fig. 12a) were larger than 22 μ (Fig. 12b). Both upstream and downstream of this chlorophyll maximum, the relative importance of netplankton diminished in more diverse phytoplankton assemblages. Microscopic examination revealed that these large netplankton populations were dominated by several species of neritic diatoms (mainly *Thalassiosira* [*Coscinodiscus*] *excentricus*, *Skeletonema costatum* and *Chaetoceros* spp.), and that frustules of these diatoms were coated with a dense layer of clay-size lithogenous material. Adsorption of inorganic particulates onto the silica surface of diatoms is probably governed by the same physicochemical factors that cause aggregation of suspended particulates in estuaries (e.g. Meade 1972; Edzwald et al. 1974; Krone 1978; Zabawa 1978), and clearly is a dominant force in the creation of the phytoplankton maximum in Suisun Bay (Arthur and Ball 1979). Just as dense inorganic particulates accumulate in an estuarine circulation cell (Meade 1972), rapidly-settling diatoms sink into the landward-flowing bottom density current before they can be transported away from Suisun Bay in the surface layer (Fig. 9).

A numerical model of phytoplankton population dynamics in the northern reach (J. Cloern

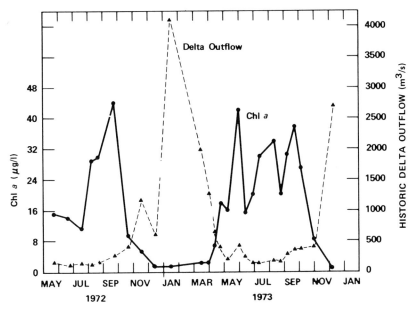

Fig. 11. Historic Delta outflow (from USBR, Sacramento, Calif.) and chlorophyll *a* concentration in surface waters of Suisun Bay, May 1972 to December 1973 (USBR et al. 1977).

SAN FRANCISCO BAY

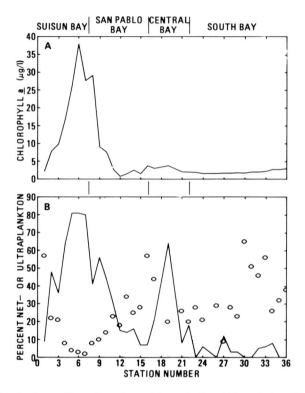

Fig. 12. Longitudinal profiles of surface chlorophyll *a* (A) and size distribution of phytoplankton (B) in the San Francisco Bay system, 19-20 September 1978. See Fig. 8 for details.

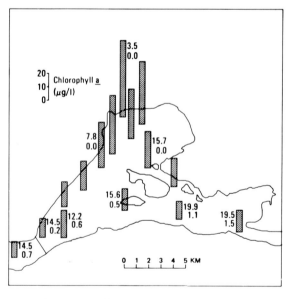

Fig. 13. Distribution of chlorophyll *a* in surface waters of Grizzly Bay, 13 July 1978. For some stations, concentration (μg-atoms·liter^{-1}) of dissolved nitrate + nitrite (top number) and ammonium (bottom number) are also given.

and R. Cheng unpublished data) suggests that the physical accumulation of diatoms by gravitational circulation is not sufficient in itself to create the very large phytoplankton densities observed in the channel of Suisun Bay. Rather, it is the presence of an estuarine circulation cell in the vicinity of shallow areas where algal growth is rapid that accounts for the high chlorophyll a concentration seen in Suisun Bay. This hypothesis is supported by the observations that phytoplankton densities are consistently higher in the shallow Grizzly Bay than in the Suisun Bay channel (e.g. Fig. 13), and that during summers of unusually high or low Delta outflow (when the null zone is moved downstream or upstream from the Grizzly-Honker Bay shallows), phytoplankton standing stocks are relatively low (Arthur and Ball 1979; Ball and Arthur 1979).

Although phytoplankton standing stocks are typically large in Suisun Bay, net rates of autotrophic carbon assimilation are apparently lower in the Suisun Bay channel than in other parts of the Bay system. During March and August 1976, mean chlorophyll a concentrations were four to five times higher in the Suisun Bay channel than in Central Bay, but areal productivity (i.e. rate of photosynthetic carbon fixation averaged over the day and depth of the water column) was up to 10 times higher in Central Bay than in the channel of Suisun Bay (Table 1). The low net productivity in Suisun Bay seems attributable to the fact that its extreme turbidity restricts the photic zone to a shallow depth, especially since primary productivity is inversely related to extinction coefficient (Table 1). These results suggest, again, that large phytoplankton densities in the channel of Suisun Bay are not a consequence of rapid population growth there.

TABLE 1. MEAN PRIMARY PRODUCTIVITY P ($gC \cdot m^{-2} \cdot d^{-1}$), CHLOROPHYLL A CONCENTRATION ($\mu g \cdot liter^{-1}$) AND EXTINCTION COEFFICIENT ϵ (m^{-1}) MEASURED AT THREE LOCATIONS IN SAN FRANCISCO BAY (DATA FROM COLE AND HERNDON 1979)

Location	March 1976				August 1976			
	P	Chl a	ϵ	n*	P	Chl a	ϵ	n*
Central Bay	0.52	4.6	1.0	4	0.93	2.7	0.7	2
Suisun Bay	.05	22.5	9.0	3	.26	10.8	3.4	2
South Bay					.38	2.4	1.3	2

* number of observations

PERSPECTIVES FOR FUTURE RESEARCH

The capability of San Francisco Bay waters to support or promote the growth of planktonic populations is continually altered by effluents from power generating plants, sewage treatment facilities, chemical, petroleum and other industries, surface runoff from agricultural and urban areas, and oil spills. Further, physical characteristics of San Francisco Bay that affect plankton dynamics (temperature, salinity, concentration of suspended particulates, bathymetry and circulation patterns) change in response to man's activities. Clearly, a detailed understanding of man's impact on plankton dynamics (and, hence, dynamics of other phenomena in the water column) will not be attained until the relationships between important physicochemical factors and phytoplankton growth are better defined. The capability to forecast even gross changes in the density and composition of phytoplankton populations will only be attained if basic studies are implemented that have as major objectives the description and interpretation of natural changes in the phytoplankton community. Our ability to predict the impacts of major perturbations, and our understanding of the variable success of fisheries, will certainly be restricted until such studies are initiated.

Future research programs should give a high priority to basic ecological investigations that attempt to define geographic areas of significant algal carbon assimilation and determine which algal species (or functional groups) are the important primary producers in San Francisco Bay. Most work in the Bay system has been confined to study of surface waters of the main channel. But the limited field work done outside the channel, and results of modeling studies (Cloern 1978), have demonstrated that shallow waters of San Pablo, Grizzly and Honker bays may be the sites of significant autotrophic productivity and population growth by planktonic algae; the same may be true for the shallow areas of South Bay. An intensive effort is needed to measure rates of carbon assimilation both in the waters of the channel and in the shallows, over an annual cycle, to determine the importance of productivity over shoals to the total autotrophic assimilation of carbon within the Bay system. Related studies are needed to define the relative contribution of different communities of planktonic algae to total productivity. Results of previous studies have shown considerable spatial and temporal variation in the relative biomass of nanoplankton and netplankton. But no attempt has yet been made to measure the contribution of different size fractions to total community productivity, and no attempt has been made to determine the relative significance of fresh-water, brackish, and coastal plankton, all of which exist within the Bay system. The expansive tidal flats around the perimeter of San Francisco Bay apparently support a considerable biomass of epibenthic diatoms (F. Nichols pers. comm.). Autotrophic production by this assemblage may be an important source of reduced carbon for benthic invertebrates and ultimately vertebrate predators, but the relative importance of this contribution is yet unknown.

A second major thrust required from the research community is an effort to trace the fate of organic carbon produced in the estuary and to quantify the relative importance of phytoplankton productivity for other trophic levels, especially those having economic value or ecological significance. Central to this effort is the need to integrate results of studies that measure rates of specific processes into a budget of organic carbon transfers in San Francisco Bay. Particularly important processes that require measurement include: (1) the rate of *in situ* autotrophic production by phytoplankton and benthic algae; (2) rates of exportation to the sea and importation of allochthonous plankton and detritus from both coastal waters and tributaries; (3) rates of accrual to the sediments; (4) the rate of bacterial decomposition (mineralization of organic carbon to CO_2) in the water column and sediments; (5) the rate of transformation of particulate carbon to dissolved organic carbon; (6) the rate of organic loading from waste discharges; and (7) the rates of energy and material flow along important pathways between producers and consumers.

Some of these processes have been quantified in San Francisco Bay (see for example Peterson 1979 and Spiker and Schemel 1979), but plankton ecologists have made little progress in studies of trophic dynamics. Particularly important unanswered questions relate to the interactions between phytoplankton and their consumers (presumably zooplankton and benthic invertebrates), and the link between herbivores and the ultimate consumers that are harvested. Storrs et al. (1964), Painter (1966), Heuback (1969), USBR (1976), Siegfried and Knight (1976) and Orsi and Knutson (1979), in their studies of the northern reach, demonstrated that zooplankton are ubiquitous, that composition of the zooplankton community varies temporally and spatially, and that large standing stocks ($> 5 \times 10^4$ crustaceans·m^{-3}) are common. A. Hutchinson (pers. comm.) has measured densities of the copepod *Acartia clausi* (all life stages) as high as 15×10^4 individuals·m^{-3} in South Bay. We know, then, that zooplankton are abundant in the San Francisco Bay system, but we know nothing about the importance of zooplankton in controlling phytoplankton population growth and species composition. Laboratory studies of Richman et al. (1977) suggest that selective grazing of netphytoplankton by *A. clausi* alone may account for the small biomass of diatoms observed in South Bay during 1978. Hutchinson (pers. comm.) also observed that, at times,

South Bay waters contain large numbers of microzooplankton (tintinnid ciliates and rotifers) which, presumably, graze selectively on the microflagellates. A sustained effort is needed (1) to define the impacts of the zooplankton community on regulating the size and composition of phytoplankton populations in the Bay system and, in turn, (2) to define the relationships between phytoplankton abundance and composition and the fluctuating population dynamics of the zooplankton. The impact of benthic filter-feeders has not yet been considered, although their high densities on mudflats of San Pablo and South bays (Nichols 1979) suggest that they may be very important regulators of phytoplankton standing stocks over the shoals.

Of particular interest in the northern reach and Delta is the relationship between algal primary productivity and the success of fisheries, especially the valuable striped bass (Stevens 1979). We know that planktonic invertebrates, primarily the mysid *Neomysis mercedis* (Orsi and Knutson 1979) and copepods, are an important food source for larval and juvenile fishes, and that phytoplankton are an important component in the diet of these zooplankters (e.g. Kost and Knight 1975). The correlation between fluctuations in the abundance of striped bass and Delta outflow (Stevens 1979) may be an indirect consequence of the relationship between Delta outflow and standing stock of phytoplankton in Suisun Bay. So, a link between phytoplankton and fisheries dynamics is probable but not quantified. We do not, for example, know what rates of autotrophic production are required to optimize the yield of fisheries. Although we do know that netplanktonic diatoms are important in the diet of *Neomysis mercedis* and herbivorous copepods, we do not know if the large standing stock of netphytoplankton in Suisun Bay during summer is a prerequisite for the survival of juvenile striped bass. Therefore, we cannot yet forecast the impacts of potential shifts in the species composition of phytoplankton on fisheries.

Similar questions can be asked of the importance of phytoplankton in Central Bay and San Pablo Bay to the success of the dungeness crab. Crab larval stages are common in coastal waters off San Francisco, and first-year crabs utilize San Pablo Bay as a nursery ground (Tasto 1979). Identification of those planktonic algae that are required by crab zooeae, and determination of environmental factors that regulate primary productivity of these phytoplankters may offer important clues to the solution of the problem of declining dungeness crab catches along central California over the past 15 years (Tasto 1979).

Finally, basic research is needed to determine which physicochemical factors are most important in regulating the composition and productivity of phytoplankton populations in San Francisco Bay. Growth rates of algal cells are governed by light intensity, temperature, water chemistry (including salinity levels and nutrient and toxin concentrations), while population changes result from the net balance between rates of growth, transport and losses. Results of past studies allow us to make gross generalizations about interactions between physicochemical factors and phytoplankton population growth, but we require a higher level of understanding before a predictive capability can be attained. Results of past surveys indicate that light availability (including surface irradiance, water transparency and ratio of photic depth to mixed depth) is a primary factor that limits the growth of planktonic algae in the Bay system. Concentrations of dissolved phosphate and silicate consistently exceed levels that limit algal growth (Peterson et al. 1975a; Conomos et al. 1979), but inorganic nitrogen (nitrate plus ammonium) fell to relatively low levels (< 1 μg-atoms·liter^{-1}) in Suisun Bay during the summer phytoplankton maximum of 1972 (Conomos and Peterson 1975), and dissolved ammonium disappeared in Grizzly Bay during periods of summer, when inputs of nitrogen are relatively small and algal densities are large, low nitrogen concentrations may override light availability as a limiting factor for algal growth. We know very little about the existence of toxins and their impacts on phytoplankton population growth. And our understanding of the interactions between physical forces (including algal settling

SAN FRANCISCO BAY

rates, estuarine circulation, longitudinal advective-dispersive processes, and rates of exchange between water masses of the main channel and lateral shoals) and phytoplankton population dynamics is primitive. Answers to these difficult questions are required by decision makers to permit wise choices concerning proposed major perturbations (the Peripheral Canal, for example, will alter the flow regime, salinity distribution, and turbidity of waters in the northern reach; the San Luis Drain will increase nitrogen concentrations near Suisun Bay), and they will come only from long-term interdisciplinary studies based upon cooperative efforts of hydrodynamicists, biologists, chemists and geologists.

ACKNOWLEDGMENTS

I gratefully acknowledge the efforts of A. E. Alpine, B. E. Cole, Anne Hutchinson and R. L. J. Wong who gathered 1977-78 data reported here. Thoughtful manuscript reviews by T. J. Conomos, D. S. McCulloch, F. H. Nichols and D. H. Peterson were very helpful, as were the insights generated by discussions with J. F. Arthur and M. D. Ball.

LITERATURE CITED

Arthur, J. F., and M. D. Ball. 1979. Factors influencing the entrapment of suspended material in the San Francisco Bay-Delta estuary. Pages 143-174 *in* T. J. Conomos, ed. San Francisco Bay: The Urbanized Estuary. Pacific Division, Amer. Assoc. Advance. Sci., San Francisco, Calif.

Ball, M. D., and J. F. Arthur. 1979. Planktonic chlorophyll dynamics in the northern San Francisco Bay and Delta. Pages 265-285 *in* T. J. Conomos, ed. San Francisco Bay: The Urbanized Estuary. Pacific Division, Amer. Assoc. Advance. Sci., San Francisco, Calif.

Bolin, R. L., and D. P. Abbott. 1963. Studies on the marine climate and phytoplankton of the central coastal area of California, 1954-60. Calif. Coop. Ocean. Fish. Invest. Rep. 9:23-45.

Bowden, K. F. 1967. Circulation and diffusion. Pages 15-36 *in* G. H. Lauff, ed. Estuaries. Amer. Assoc. Advance. Sci. Publ. No. 83. Washington, D.C.

Cloern, J. E. 1978. Empirical model of *Skeletonema costatum* photosynthetic rate, with applications in the San Francisco Bay estuary. Advance. Water Res. 1:267-274.

Cole, B. E., and R. E. Herndon. 1979. Hydrographic properties and primary productivity of San Francisco Bay waters, March 1976-July 1977. U. S. Geol. Surv. Open-File Rep. 79-983.

Conomos, T. J., D. S. McCulloch, D. H. Peterson, and P. R. Carlson. 1971. Drift of surface and near-bottom waters of the San Francisco Bay system: March 1970 through April 1971. U. S. Geol. Surv. Open-File Map.

Conomos, T. J., and D. H. Peterson. 1974. Biological and chemical aspects of the San Francisco Bay turbidity maximum. Mem. Inst. Geol. Bassin d'Aquitaine 7:45-52.

Conomos, T. J., and D. H. Peterson. 1975. Longitudinal distribution of selected micronutrients in northern San Francisco Bay during 1972. Pages 103-126 *in* R. L. Brown, ed. Proceedings of a Workshop on Algae Nutrient Relationships in the San Francisco Bay and Delta (8-10 November 1973, Clear Lake, Calif.). San Francisco Bay and Estuarine Assoc., San Francisco, Calif.

Conomos, T. J., and D. H. Peterson. 1977. Suspended-particle transport and circulation in San Francisco Bay: an overview. Pages 82-97 *in* M. Wiley, ed. Estuarine Processes. Vol. 2. Circulation, sediments, and transfer of material in the estuary. Academic Press, New York.

Conomos, T. J., F. H. Nichols, R. T. Cheng, and D. H. Peterson. 1978. Field and modeling studies of San Francisco Bay. Pages 1917-1927 *in* Coastal Zone '78; Proc. Symp. Technical, Environmental, Socioeconomic and Regulatory Aspects of Coastal Zone Management, ASCE, San Francisco, Calif., March 14-16, 1978.

Conomos, T. J., R. E. Smith, D. H. Peterson, S. W. Hager, and L. E. Schemel. 1979. Processes affecting seasonal distributions of water properties in the San Francisco Bay estuarine system. Pages 115-142 *in* T. J. Conomos, ed. San Francisco Bay: The Urbanized Estuary. Pacific Division, Amer. Assoc. Advance. Sci., San Francisco, Calif.

Conomos, T. J. 1979. Properties and circulation of San Francisco Bay waters. Pages 47-84 *in* T. J. Conomos, ed. San Francisco Bay: The Urbanized Estuary. Pacific Division, Amer. Assoc. Advance. Sci., San Francisco, Calif.

Dyer, K. R. 1973. Estuaries: a physical introduction. Wiley-Interscience, New York. 140 pp.

Edzwald, J, K., J. B. Upchurch, and C. R. O'Melia. 1974. Coagulation in estuaries. Environ. Sci. Technol. 8:58-63.

Garrison, D. L. 1976. Contribution of the net plankton and nanoplankton to the standing stocks and primary productivity in Monterey Bay, California during the upwelling season. Fish. Bull. 74:183-194.

Heuback, W. 1969. *Neomysis awatschensis* in the Sacramento-San Joaquin River estuary. Limnol. Oceanogr. 14:533-546.

Imberger, J., W. B. Kirkland, Jr., and H. B. Fischer. 1977. The effect of Delta outflow on the density stratification in San Francisco Bay. ABAG (Assoc. of Bay Area Governments) Rep. HBF-77/02. 109 pp.

Kost, A. L. B. and A. W. Knight. 1975. The food of *Neomysis mercedis* Holmes in the Sacramento-San Joaquin Delta estuary system. Calif. Fish Game 61:34-41.

Krone, R. B. 1978. Aggregation of suspended particles in estuaries. Pages 177-190 *in* B. Kjerfve, ed. Estuarine Transport Processes. Belle W. Baruch Library in Marine Science No. 7. Univ. South Carolina Press, Columbia, S. C.

McCulloch, D. S., D. H. Peterson, P. R. Carlson, and T. J. Conomos. 1970. A preliminary study of the effects of water circulation in the San Francisco Bay estuary--some effects of freshwater inflow on the flushing of South San Francisco Bay. U. S. Geol. Surv. Circ. 637-A. 27 pp.

Malone, T. C. 1971. The relative importance of nanoplankton and netplankton as primary producers in the California current system. Fish. Bull. 69:799-820.

Meade, R. H. 1972. Transport and deposition of sediments in estuaries. Pages 91-120 *in* B. W. Nelson, ed. Environmental Framework of Coastal Plain Estuaries. Geol. Soc. Amer. Mem. 133.

Nichols, F. H. 1979. Natural and anthropogenic influences on benthic community structure in San Francisco Bay. Pages 409-426 *in* T. J. Conomos, ed. San Francisco Bay: The Urbanized Estuary. Pacific Division, Amer. Assoc. Advance. Sci., San Francisco, Calif.

Orsi, J. J., and A. C. Knutson, Jr. 1979. The role of mysid shrimp in the Sacramento-San Joaquin estuary and factors affecting their abundance and distribution. Pages 401-408 *in* T. J. Conomos, ed. San Francisco Bay: The Urbanized Estuary. Pacific Division, Amer. Assoc. Advance. Sci., San Francisco, Calif.

Painter, R. E. 1966. Zooplankton of San Pablo and Suisun Bays. Pages 18-39 *in* D. W. Kelley, ed. Ecological studies of the Sacramento-San Joaquin estuary. Part I. Zooplankton, zoobenthos, and fishes of San Pablo and Suisun Bays, zooplankton and zoobenthos of the Delta. Calif. Fish Game, Fish. Bull. 133.

Peterson, D. H., T. J. Conomos, W. W. Broenkow, and E. P. Scrivani. 1975a. Processes controlling the dissolved silica distribution in San Francisco Bay. Pages 153-187 *in* L. E. Cronin, ed. Estuarine Research. Vol. 1. Chemistry, Biology and the Estuarine System. Academic Press, New York.

Peterson, D. H., T. J. Conomos, W. W. Broenkow, and P. C. Doherty. 1975b. Location of the non-tidal current null zone in northern San Francisco Bay. Estuarine Coastal Mar. Sci. 3:1-11.

Peterson, D. H. 1979. Sources and sinks of biologically reactive oxygen, carbon, nitrogen, and silica in northern San Francisco Bay. Pages 175-193 *in* T. J. Conomos, ed. San Francisco Bay: The Urbanized Estuary. Pacific Division, Amer. Assoc. Advance. Sci., San Francisco, Calif.

Richman, S., D. R. Heinle, and R. Huff. 1977. Grazing by adult estuarine calanoid copepods of the Chesapeake Bay. Mar. Biol. 42:69-84.

Seckler, D., ed. 1971. California water: A study in resource management. University of California Press, Berkeley, Calif. 348 pp.

Siegfried, C. A., and A. W. Knight. 1976. A baseline ecological evaluation of the western Sacramento-San Joaquin Delta. University of California, Davis, Water Science and Engr. Papers, No. 4504.

Spiker, E., and L. E. Schemel. 1979. Distribution and stable-isotope composition of carbon in San Francisco Bay. Pages 195-212 *in* T. J. Conomos, ed. San Francisco Bay: The Urbanized Estuary. Pacific Division, Amer. Assoc. Advance. Sci., San Francisco, Calif.

Stevens, D. E. 1979. Environmental factors affecting striped bass (*Morone saxatilis*) in the Sacramento-San Joaquin estuary. Pages 469-478 *in* T. J. Conomos, ed. San Francisco Bay: The Urbanized Estuary. Pacific Division, Amer. Assoc. Advance. Sci., San Francisco, Calif.

Storrs, P. N., R. E. Selleck, and E. A. Pearson. 1963. A comprehensive study of San Francisco Bay 1961-62. Univ. Calif. Sanit. Engr. Res. Lab. Rep. No. 63-3.

Storrs, P. N., R. E. Selleck, and E. A. Pearson. 1964. A comprehensive study of San Francisco Bay 1962-63. Univ. Calif. Sanit. Engr. Res. Lab. Rep. No. 64-3.

Storrs, P. N., E. A. Pearson, and R. E. Selleck. 1966. A comprehensive study of San Francisco Bay, a final report. Vol. V. Summary of physical, chemical and biological water and sediment data. Univ. Calif. Sanit. Engr. Res. Lab. Rep. No. 67-2.

Tasto, R. N. 1979. San Francisco Bay: Critical to the dungeness crab? Pages 479-490 *in* T. J. Conomos, ed. San Francisco Bay: The Urbanized Estuary. Pacific Division, Amer. Assoc. Advance. Sci., San Francisco, Calif.

U. S. Bureau of Reclamation. 1976. Delta-Suisun Bay ecological studies. A report of water quality in the Sacramento-San Joaquin estuary during the low flow year, 1976. USBR, Water Quality Branch, Sacramento, Calif.

U. S. Bureau of Reclamation, California Department of Water Resources, and California Department of Fish and Game. 1977. Delta-Suisun Bay ecological studies. Biological methods and data for 1968-74. USBR, Water Quality Branch, Sacramento, Calif.

Whedon, W. F. 1939. A three-year survey of the phytoplankton in the region of San Francisco. Int. Rev. Hydrobiol. 38:459-476.

Zabawa, C. F. 1978. Microstructure of agglomerated sediments in northern Chesapeake Bay estuary. Science 202:49-51.

PLANKTONIC CHLOROPHYLL DYNAMICS IN THE NORTHERN SAN FRANCISCO BAY AND DELTA

MELVIN D. BALL AND JAMES F. ARTHUR
U. S. Bureau of Reclamation, 2800 Cottage Way, Sacramento, California 95825

Diatoms were the dominant phytoplankters throughout the Sacramento-San Joaquin Delta into San Pablo Bay during 1969 through 1977. Green algae seldom exceeded 20% of the total. Chlorophyll a concentrations seldom exceeded 6 μg·liter^{-1} at the most upstream station in the Sacramento River, the major water source to the Delta, except during the 1977 drought when 40 μg·liter^{-1} was measured. Conversely, peak summer chlorophyll concentrations entering the Delta from the San Joaquin River were the highest (100-350 μg·liter^{-1} and were inversely related to riverflow. During spring through fall, export pumping from the southern Delta caused a net flow reversal in the lower San Joaquin River, drawing Sacramento River water across the central Delta to the export pumps. The relatively deep channels and short water residence time apparently resulted in the chlorophyll concentrations remaining low from the northern Delta and in the cross-Delta flow to the pumps. Chlorophyll concentrations in the shallower eastern Delta sloughs and channels were often quite high and variable. Western Delta spring blooms reached concentrations of 25-50 μg·liter^{-1}. Spring blooms of 30-40 and summer blooms of 40-100 μg·liter^{-1} were observed in Suisun Bay. The entrapment zone location adjacent to the shallows of Suisun and Honker bays appears to increase the Suisun Bay phytoplankton standing crop. Chlorophyll a concentrations in Suisun Marsh generally peaked to 40-100 μg·liter^{-1} in the late spring except during 1977. Chlorophyll levels in central San Pablo Bay seldom exceeded 6 μg·liter^{-1}, although blooms as high as 40 μg·liter^{-1} were observed in the northern shallow portion of the Bay.

Percent chlorophyll a (of the total chlorophyll a plus the pheo-pigments) in near-surface water generally varied from 50-80% during the spring-fall months throughout most of the study area. Upstream of the entrapment zone, percent chlorophyll a near the bottom averaged about 5% lower than that near the surface. Downstream of the entrapment zone, percent chlorophyll a was as much as 40% lower near the bottom.

Phytoplankton are important primary producers in the estuarine environment since they form the base of the water-column food web. However, in many aquatic environments, high phytoplankton concentrations can cause oxygen concentration reductions to a point detrimental to higher aquatic organisms. High phytoplankton levels can also adversely affect water supplies, or create aesthetically undesirable environments. In the study area, phytoplankton problems have been negligible, with the exception of the southern and eastern Delta.

In this chapter we summarize results of a cooperative multiagency study of phytoplankton biomass, measured as chlorophyll, in the Sacramento-San Joaquin Delta, Suisun Bay, Suisun Marsh, and San Pablo Bay during 1969 through 1977. We discuss, in particular, factors influencing phytoplankton biomass distributions, occurrence and distribution of dominant phytoplankton classes, and the chlorophyll - pheo-pigment relationships.

SAN FRANCISCO BAY

This coordinated monitoring program is one of two major studies involving routine chlorophyll measurements in the Sacramento-San Joaquin Delta Suisun Bay area (Arthur and Ball 1979). The other study was conducted during 1966 and 1967 by California Departments of Water Resources (DWR) and Fish and Game (DFG) at 10 stations throughout the Delta and Suisun Bay. Only total chlorophyll (chlorophyll $a+b+c$ not corrected for pheo-pigments) measurements were reported (California DWR and DFG 1972).

Previous water quality studies in the system were prepared by the U. S. Bureau of Reclamation (USBR 1972, 1974) for 1968-70 data. The 1969-74 chlorophyll data were evaluated by Ball (1975, 1977).

Phytoplankton chlorophyll analysis, a relatively rapid and quantitatively accurate method of determining the total phytoplankton standing crop, can be directly related to total phytoplankton carbon with conversion factors determined in any given study (Strickland and Parsons 1968).

THE STUDY AREA

The Delta, with over 1,100 km of waterways, is formed by the confluence of the Sacramento and San Joaquin rivers (Fig. 1). The Sacramento River system contributes approximately 80% of the freshwater entering the Delta, and the smaller San Joaquin River system contributes about 15%. The remaining flow (5%) is primarily from the smaller streams entering the eastern Delta.

During spring through fall, southern Delta pumping by the USBR and DWR for export in the Delta-Mendota Canal and California State Aqueduct generally exceed the San Joaquin River flow, drawing most of this water to the pumps. This combination of low flow and export pumping

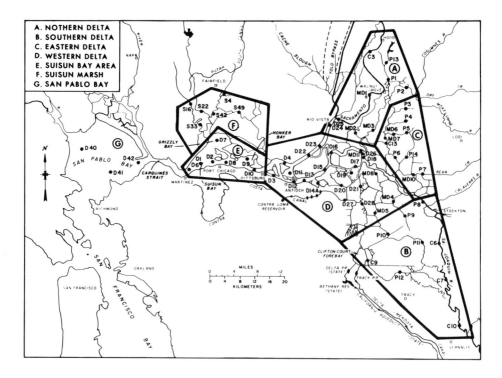

Fig. 1. Study area, subareas, and sample site locations.

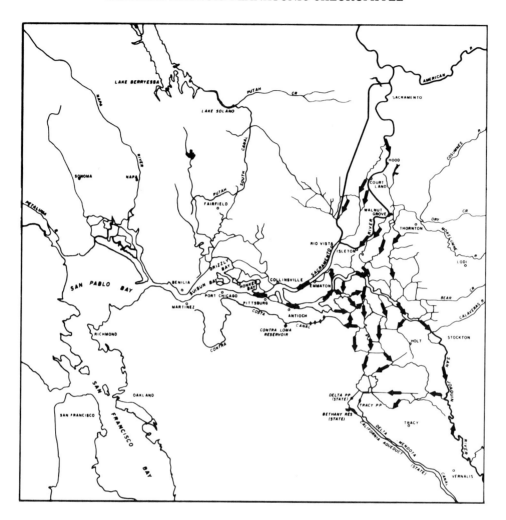

Fig. 2. Typical Delta summer net flow patterns. Note the reverse flow in the San Joaquin River system toward the export pumps. (From Ball 1977).

causes a net flow reversal in the lower San Joaquin River and draws Sacramento River water across the central Delta to the export pumps (Fig. 2). The water residence time from the northern Delta to the export pumps during periods of high summer export is only a few days.

Because of unmeasured inflow in several minor streams, consumptive water use within the Delta, and tidal current action, the net inflow to San Francisco Bay has not accurately been measured (Conomos 1979). Instead, an estimate termed the Delta Outflow Index (Sacramento and San Joaquin River flow less export and consumptive use [see Arthur and Ball 1979] is computed daily for operations of the state and federal water projects. This error during typical summer outflow conditions may be as great as ±30 to 60 $m^3 \cdot sec^{-1}$.

The quantity of river inflow to the Delta is important to phytoplankton growth in that it regulates nutrient concentrations, affects water transparency, determines water and phytoplankton residence times, and regulates the extent of salinity intrusion and the location of the entrapment zone. These and other factors all interact to determine the quantity and quality of the phytoplankton.

Environmental Factors

Water and/or phytoplankton residence time, nutrients, temperature, and light are the main factors interacting and influencing the algal standing crop. Other factors such as toxic materials, zooplankton, zoobenthos and algal parasites may also have inhibiting effects on the algal growth rate but are not discussed.

The residence time of water varies greatly throughout the study area. During winter, high river inflow rates greatly shorten residence times as compared to the summer, when export effects are more significant. Geographic effects also modulate water residence time. For example, eastern Delta dead-end sloughs have very long water-residence times, whereas major tributaries have short residence times. Currents generated by tidal and wind action also influence residence times.

Water residence time and degree of vertical and horizontal mixing are very influential on the algal growth rate, standing crop, and the bloom development. High algal growth rates occur where there is sufficient light penetration relative to the water depth. Assuming other growth factors equal, the longer the phytoplankton remain in areas of high relative light penetration, the greater the standing crop that will develop.

Certain species of phytoplankton tend to accumulate in the entrapment zone as a result of two-layered flow (see also Arthur and Ball 1979). This phenomenon increases their residence time over that of the water.

There are great seasonal and spatial variations in the nutrient concentrations (USBR 1972, 1974). During summer, highest concentrations of inorganic nitrogen typically occur in dead-end sloughs and near agricultural drains and sewage effluents. Lower concentrations tend to occur in the western Delta and San Pablo Bay. During winter, river inflow largely controls concentrations of inorganic nitrogen, phosphorus, silica, and other nutrients required for algal growth (see also Peterson 1979; Conomos et al. 1979).

Results of past studies have demonstrated that nitrogen typically was the most limiting nutrient and was often depleted during algal blooms. Phosphorus and the micronutrients were always present at sufficient levels to cause higher algal growth concentrations when additional nitrogen was supplied to the water in laboratory test (USBR 1972).

Seasonal increases in incident solar radiation (insolation) and water transparency greatly increase algal growth rates. Incident sunlight in the Delta is approximately five times greater in the summer than in winter (Fig. 3A). As the water transparency is generally greater during the summer low-flow period than during the winter high-flow period (see Arthur and Ball 1979), the covariance of these parameters increase in the summer algal growth rate severalfold.

The temperature increase from winter to summer (Fig. 3B) also has a profound effect on algal growth rates. The growth rate of most planktonic algal species can increase 2 to 4 times with each 10°C temperature increase, until the optimum temperature (about 20 to 25° C; Fogg 1965) is reached. Additional increases above optimum cause rapid growth rate decline.

METHODS

The sampling sites in our study area are grouped into various subareas (northern Delta, southern Delta, eastern Delta, western Delta, Suisun Bay, Suisun Marsh, and San Pablo Bay) to simplify discussion (Fig. 1). The boundaries are based on geographical location and on similarity in seasonal phytoplankton growth patterns. Sampling frequency typically varied from twice monthly and data for nearly 70 sites studied through 1974.

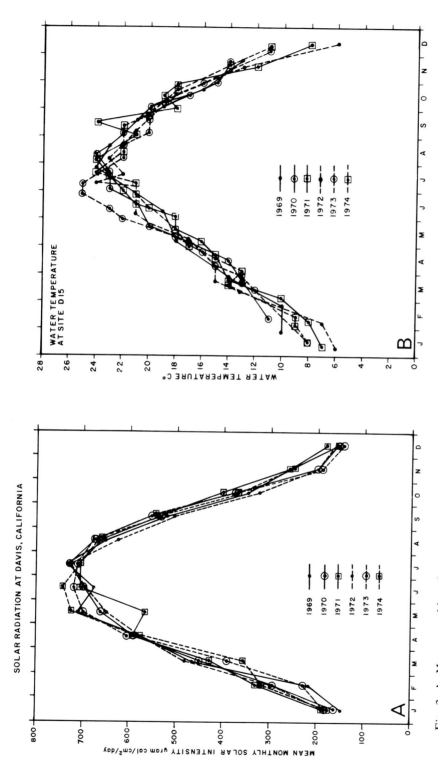

Fig. 3. A. Mean monthly solar intensity (insolation) measured at Davis, California for the years 1969-1974. B. Water temperature at Jersey Point (site D14) for the years 1969-1974. Data from the routine monitoring program. (From Ball 1977).

SAN FRANCISCO BAY

All samples, unless otherwise stated, were collected using submersible pumps or Van Dorn-type water bottles within approximately 1 hour of the time of high slack tides (except in the eastern and southern Delta) and approximately 1 m below the water surface. Chlorophyll a and pheopigment sample volumes ranged from 0.1 to 1 liter. Sample handling and filtrations were according to the methods of Strickland and Parsons (1968) with modifications described by Ball (1977).

The accuracy of the chlorophyll a measurements following May 1970 is probably within ±10% while the accuracy of the measurements made prior to that date is slightly lower.

Phytoplankton samples for identification and counting were stored in glass bottles containing 1% Lugol's solution, and at times, stored up to two years before analysis. The samples were identified and counted by several persons using inverted microscopes and various types of settling chambers, including 3.2-mm deep specially designed chambers and the standard Utermohl chamber. The organisms were identified and enumerated to genus except for a few easily recognized species.

DOMINANT PHYTOPLANKTON GROUPS

The dominant genera varied with time of year and location in the estuary. The diatoms (class Bacillarophyceae) dominated (cell number) the spring through fall algal community (Table 1; the green algae (class Chlorophyceae) were the second most numerous group, and generally constituted less than 20% of the total count. In the dead-end sloughs of the eastern Delta, however, the green algae percentage was often higher. Cryptomonads (class Cryptophyceae) were occasionally dominant at some sites during winter. Cryptomonads, dinoflagellates, and other flagellates also were in high concentrations in low water velocity areas such as marinas and the Stockon Ship Channel turning basin.

TABLE 1. DOMINANT GENERA OF PHYTOPLANKTON AND AREAS OF OCCURRENCE DURING SPRING THROUGH FALL, 1969-1974[1]

Area	Dominant genera
Northern Delta	*Coscinodiscus, Cyclotella, Melosira*
Western Delta	*Coscinodiscus, Cyclotella, Melosira, Skeletonema, Stephanodiscus, Microsiphona*
Suisun Bay	*Chaetoceros, Coscinodiscus, Cyclotella, Melosira, Skeletonema, Stephanodiscus*
San Pablo Bay	*Chaetoceros, Coscinodiscus, Cyclotella, Skeletonema*
Suisun Marsh	*Coscinodiscus, Cyclotella, Stephanodiscus*
Eastern Delta	*Coscinodiscus, Cyclotella, Stephanodiscus, Melosira*
Southern Delta	*Coscinodiscus, Cyclotella, Stephanodiscus, Melosira*

[1] From Ball 1977.

GEOGRAPHIC AND SEASONAL CHLOROPHYLL DISTRIBUTIONS

The phytoplankton standing crop, defined by averaged chlorophyll *a* concentration, indicates that the major rivers contribute vastly differing amounts of chlorophyll *a* to the Delta (Fig. 4). Very low chlorophyll concentrations occurred in the Sacramento River water entering the Delta whereas very high concentrations occurred in the shallower and smaller San Joaquin River. We present these separately as each area has differing and complicating characteristics.

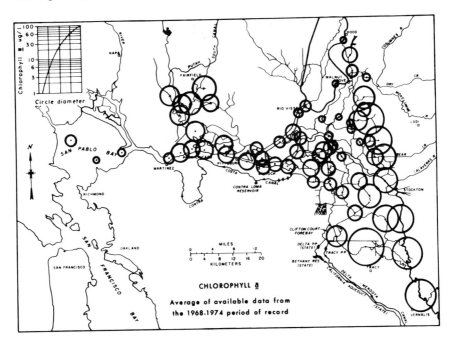

Fig. 4. Average chlorophyll *a* concentration from September 1968-1974. The area of each circle is proportional to the concentration.

Northern Delta

Phytoplankton primarily enter the northern Delta from the Sacramento River. The Yolo Bypass, containing mostly Sacramento River overflow, may contribute the greatest water discharge to the Delta during high flow and flood periods (see also Conomos 1979).

Chlorophyll concentrations at Hood (C3), representing Sacramento River inputs, were uniformly low and seldom exceeded 6 μg·liter^{-1} during the spring through fall months (Fig. 5). An important exception was the occurrence of a high phytoplankton standing crop during the 1977 drought year (spring and summer flow was less than one-half normal), which we think is the result of longer water residence time.

At Rio Vista (D24) the chlorophyll *a* concentrations seldom exceed 12 μg·liter^{-1} (Fig. 5). During typical summer low flow conditions the chlorophyll levels increased nearly twofold between sites C3 and D24.

Chlorophyll *a* concentrations below the confluence of the Mokelumne and Cosumnes rivers (P2) were usually less than 3 μg·liter^{-1} excepting years of very low or no flow (1972, 1976, 1977).

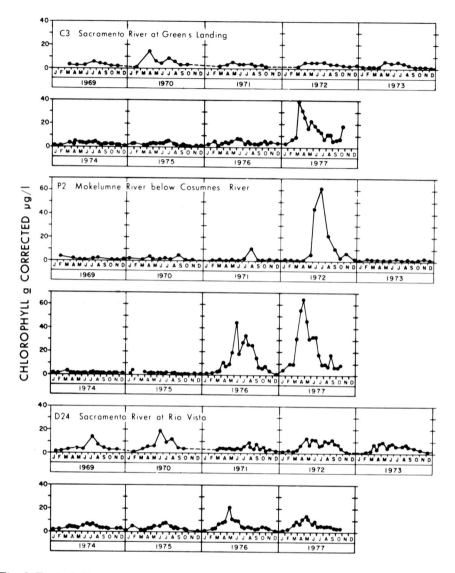

Fig. 5. Typical chlorophyll *a* concentrations in the northern Delta.

Southern Delta

In contrast to the Sacramento-derived low chlorophyll concentrations, the levels in the San Joaquin River (consisting largely of agricultural and municipal waste water) entering the Delta from the south were extremely high (Figs. 6A, B). The winter chlorophyll *a* concentrations were typically between 10 and 20 μg·liter^{-1}, whereas during summer, the concentrations were an order of magnitude higher. The June and July (1969-77) chlorophyll *a* concentrations at Vernalis in the San Joaquin River (C10) were inversely related to riverflow (r = -0.75; Fig. 7).

The standing crop of phytoplankton at site P8 (Fig. 6A) typically declined following spring blooms. Since average water depth at this site is the greatest of the southern Delta sites, the higher concentrations during the spring (100 μg·liter^{-1}) prior to flow reversal in the San Joaquin River

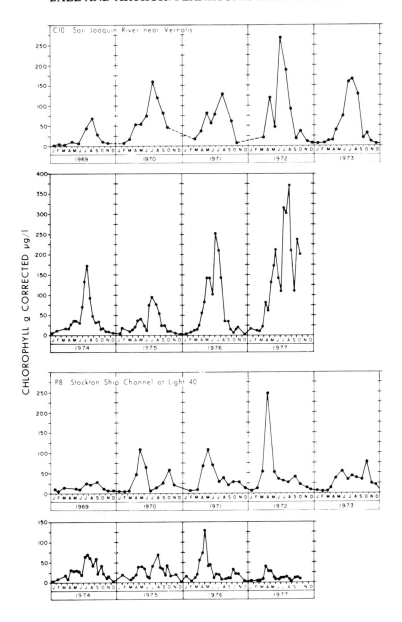

Fig. 6A. Typical chlorophyll *a* concentration in the southern Delta.

suggest that phytoplankton produced in the shallow upstream portion of the San Joaquin River may be transported to the site. Phytoplankton contained in the Stockton wastewater treatment plant effluent and other discharges may also influence the distribution.

Site P12 on the shallow and nutrient-laden Old River had consistently higher chlorophyll concentrations than any other Delta site (Fig. 6B). High phytoplankton levels (chlorophyll $a\,\mu$ 200 μg·liter^{-1} occurred in spring, reached peak levels during late spring or summer, and did not subside until fall. Concentrations reached 350 μg·liter^{-1} in the spring of dry years (1976-77). Most of this highly productive, algae-laden San Joaquin River water is pumped into the Delta-Mendota Canal

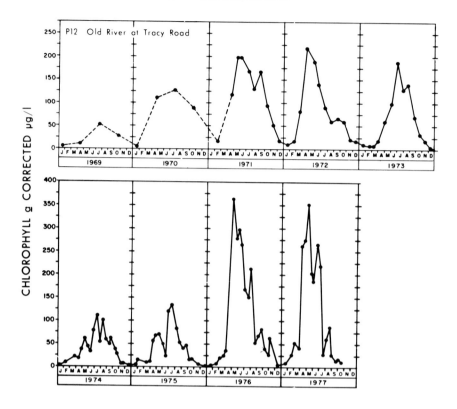

Fig. 6B. Typical chlorophyll *a* concentrations in the southern Delta (continued).

during low tides and is returned to the San Joaquin Valley instead of flowing into the Delta.

Eastern Delta

Initially the eastern Delta sites were chosen because of their close alignment with the proposed Peripheral Canal (a canal to convey Sacramento River water from the northern Delta to the export pumps in the southern Delta for transport southward). Most of the sites are located on dead-end sloughs and have high chlorophyll concentrations.

The timing of maximum chlorophyll concentration was variable at these and other sites in the eastern Delta and did not tend to follow seasonal patterns (Fig. 8). Of the dead-end sloughs studied, Sycamore Slough (P5) had the highest summer chlorophyll *a* concentrations (50-150 $\mu g \cdot liter^{-1}$) and Hog Slough (P4), the lowest (35 $\mu g \cdot liter^{-1}$).

Western Delta

Most of the western Delta sites are on the lower San Joaquin River system (Fig. 1) and had similar chlorophyll concentration patterns. Typically spring blooms peaking at 20-50 $\mu g \cdot liter^{-1}$ were followed by summer declines (Fig. 9). In some years a second peak occurred during the fall. Spring blooms were not measured in 1969 and 1977 (years with extreme variations in river flow; see Fig. 16A, B).

Vertical chlorophyll *a* measurements made in 1973 were quite uniform between surface (1-m depth) and bottom (0.5 m above the bottom) because of tidal mixing. The surface sample concentrations averaged 1% higher.

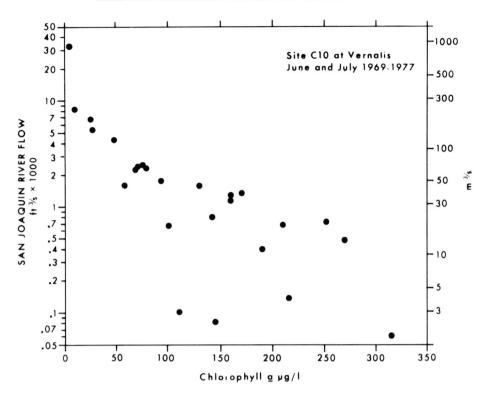

Fig. 7. Relationship of chlorophyll *a* concentration at site C10 to the San Joaquin River flow.

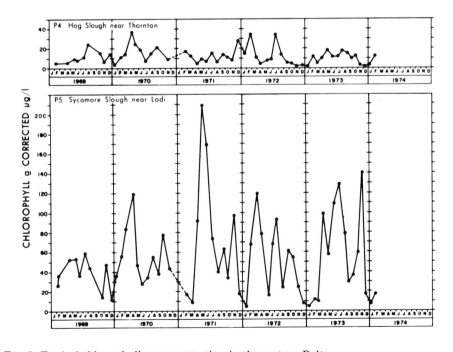

Fig. 8. Typical chlorophyll *a* concentration in the eastern Delta.

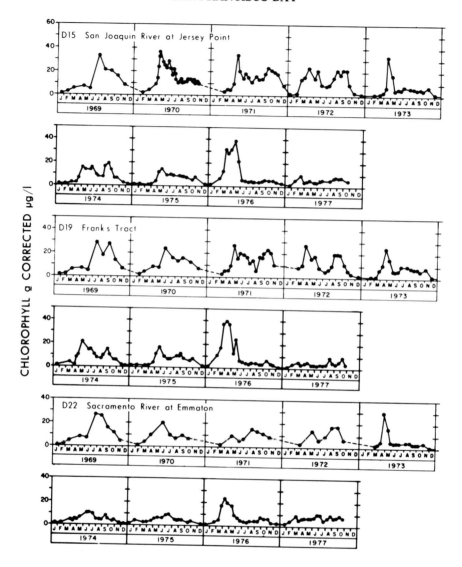

Fig. 9. Typical chlorophyll *a* concentrations in the western Delta.

Suisun Bay

The timing of phytoplankton blooms and the level of standing crops in Suisun, Grizzly and Honker bays were highly variable (Fig. 10). The blooms were generally associated with different dominant phytoplankton species than those found in the western Delta, and usually peaked in the two areas at different times.

All sites in Suisun Bay had similar concentration patterns, with maximum values (50 to 100 $\mu g \cdot liter^{-1}$) generally measured in August. During an extremely wet year (1969) only one summer peak was observed, whereas during 1976, a dry year, a late winter and a spring peak appeared but no summer bloom occurred. No bloom was detected during the 1977 drought.

During phytoplankton blooms, the highest chlorophyll concentrations in the river channel

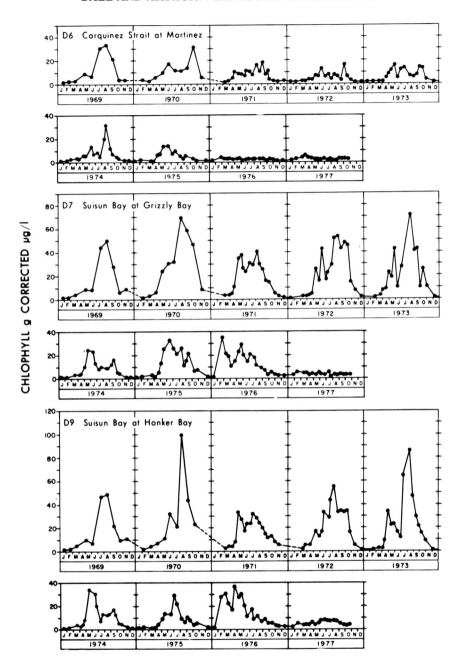

Fig. 10. Typical chlorophyll *a* concentrations in Suisun Bay.

typically occurred in waters having specific conductances of approximately 2-10 millimho/cm (1 to 6 ‰ salinity) or slightly higher (see also Arthur and Ball 1979). Intensified sampling starting in September 1973 demonstrated that there were often severalfold differences between the chlorophyll *a* concentration at surface and bottom (Fig. 11). Highest surface concentrations frequently

SAN FRANCISCO BAY

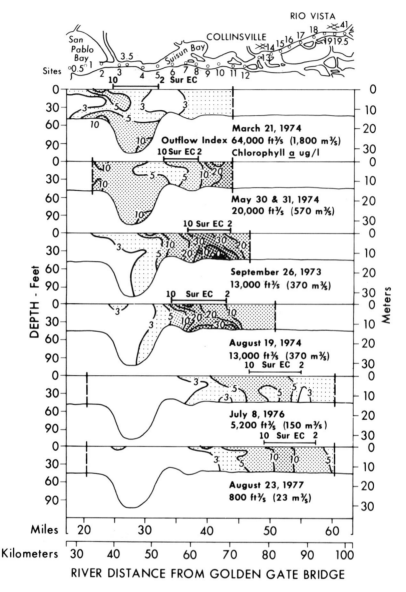

Fig. 11. Chlorophyll *a* distribution in the Sacramento River Channel relative to surface salinity (2-10 millimho/cm specific conductance is approximately 1 - 6 °/oo salinity) on high slack tides at various Delta outflow indices. (From Arthur and Ball 1978).

occurred downstream of the maximum in bottom waters.

Suisun Marsh

Maximum chlorophyll *a* concentrations generally occurred during spring and summer, seldom exceeding 60-80 µg·liter^{-1}. The seasonal concentration patterns were similar throughout the marsh. As for Suisun Bay, chlorophyll concentrations in Suisun Slough were unusually low during 1977 (Fig. 12).

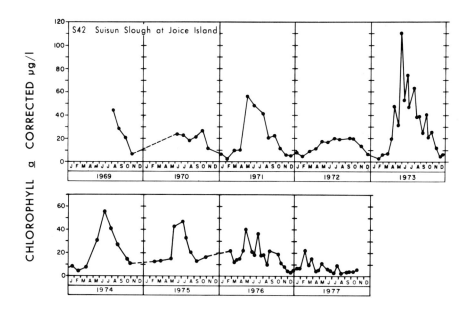

Fig. 12. Typical chlorophyll *a* concentrations in Suisun Marsh.

San Pablo Bay

The three sites in San Pablo Bay demonstrated significant yearly variations in chlorophyll concentrations; however, both the temporal and spatial sampling frequency has been less than for the other areas (Fig. 13). The greatest variation in chlorophyll levels among years sampled occurred at shallow-water site D40. The highest concentration (38 µg·liter^{-1}) was measured during spring 1973 and occurred two months earlier than a similar concentration in Suisun Bay. Central San Pablo Bay (D41) consistently had the lowest chlorophyll concentrations (seldom exceeding 7 µg·liter^{-1}) for the entire study area.

CHLOROPHYLL - PHEO-PIGMENT RELATIONSHIPS

The percent chlorophyll *a* (of the total chlorophyll *a* plus pheo-pigments) was generally in the range reported by Yentsch (1965a, b) and Spence and Steven (1974) for other productive major water bodies. Typical levels were 50 to 80% (20 to 50% pheo-pigments) between spring and fall for most areas except in some of the dead-end sloughs, where at times the percentages were quite low. Winter levels in much of the study area often dropped below 50%.

Furthermore, the percent chlorophyll *a* appeared to be highest during the algal bloom periods and usually dropped 10 to 20% during the weeks immediately following the bloom peaks.

Surface and near-bottom sampling in the channels indicated the percent chlorophyll *a* levels were approximately 5% lower near the bottom upstream of the entrapment zone (Fig. 14). In the channels of Suisun and San Pablo bays, where salinity stratification occurred, the percent chlorophyll *a* levels were generally 10 to 40% lower near the bottom than the surface.

DISCUSSION

The phytoplankton standing crop in the northern and southern Delta is greatly influenced by chlorophyll concentrations in river inflow (Fig. 4). Summer insolation and water temperatures maxima largely determine the timing of the blooms (Figs. 5, 6A,B). The low chlorophyll levels in the Sacramento River entering the Delta combined with the relatively short residence time result in low levels in the northern Delta. In contrast, the very high concentrations in the southern Delta are promoted by the high concentrations entering the Delta in the San Joaquin River, the shallowness of this portion of the Delta, and the high levels of agricultural and municipal wastewater rich in inorganic nitrogen and phosphorus.

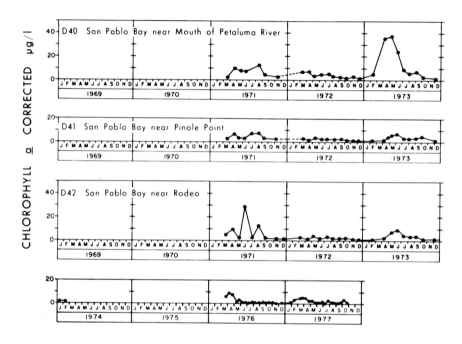

Fig. 13. Typical chlorophyll *a* concentrations in San Pablo Bay.

In the central portion of the Delta, the relatively deep channels and short residence time created by reverse flow in the San Joaquin River (Fig. 2) apparently are primarily responsible for the low chlorophyll concentrations. The central and western Delta would probably be much more productive during the summer if the San Joaquin River water were not drawn to the export pumps but allowed to flow seaward.

The high concentrations in the eastern Delta reflect the long water residence time due to the sluggish circulation patterns, the high agricultural and municipal waste inflows, and the shallow water depths.

In the western Delta and Suisun Bay a more complicated series of interactions occur. These two areas typically were dominated by different algal species and experienced peak chlorophyll concentrations at different times of the year (Figs. 9, 10; Ball 1977). Also, the growth patterns for each area varied considerably from year to year. Summer insolation and temperature maxima did not correspond to peak bloom periods in all years, such as in the southern Delta. However, blooms did not occur in midwinter. We believe the following factors primarily influence the differences in phytoplankton abundance between years in these two distinct areas (Ball 1975, 1977; Arthur and Ball 1979): (1) average length of time phytoplankton spend in the photic zone as influenced by the water transparency, (2) entrapment zone location (including several possible

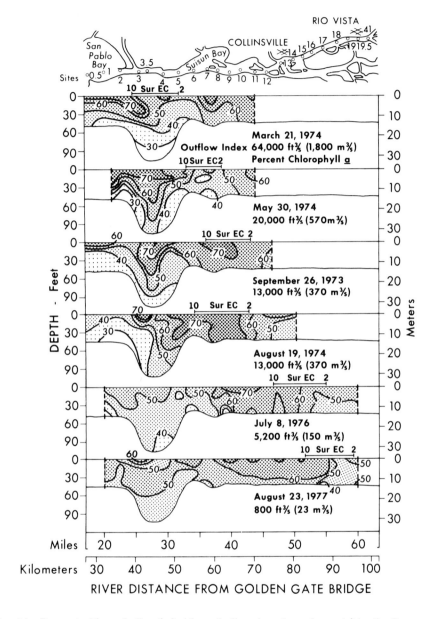

Fig. 14. Percent chlorophyll *a* (of chlorophyll *a* plus pheo-pigments) in the Sacramento River Channel relative to surface salinity (2-10 millimho/cm specific conductance is approximately 1-6 °/oo salinity) on high slack tides at various Delta outflow indices. (From Arthur and Ball 1978).

effects), (3) nutrient concentrations, and (4) transport of phytoplankton into the area.

In a well-mixed system such as the western Delta, the proportion of time phytoplankton are in the photic zone depends on the water depth and transparency. In addition, these factors control their growth rate which influences the standing crop. The water transparency varies greatly from year to year and appears to have directly influenced the phytoplankton growth rate and standing crop.

Several environmental factors influence the water transparency: suspended particle concentrations in the water entering the Delta, particle flocculation and settling, waste discharges within the Delta, location of the entrapment zone (see also Arthur and Ball 1979), tidal current- and wind-induced sediment resuspension, and possibly the resuspension and upstream transport of a portion of the winter sediment load deposited in the bays downstream. We do not understand fully the interaction of these factors.

The water transparency in the western Delta varies with river flow. During the winter and spring flood-prone periods the transparency tends to be lowest and inversely related to the amount of Delta outflow. During the spring months the outflow decreases and the water transparency increases. The variation in water transparency during spring from 1969 to 1977 appeared to be influential on the development of the spring phytoplankton blooms. Earlier spring bloom development and relatively higher chlorophyll a concentrations occurred during years of greater water transparency (Fig. 15; Ball 1977). During summer months, the transparency in the western Delta tends to be directly related to the amount of outflow as the entrapment zone moves upstream into

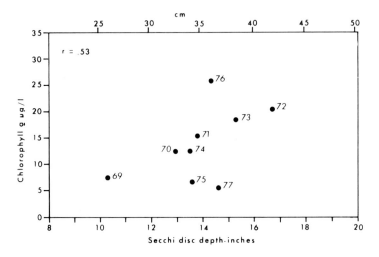

Fig. 15. Relationship of the average April and May chlorophyll a concentrations to water transparency in the western Delta. Averages are for sites D11, D12, D14A, D15, and D19.

this area and wind velocities are at their peak. The phytoplankton populations also decline during summer. Although water transparency values may be representative for the time of measurement, the great reductions in transparency due to tidal current- and wind-induced sediment resuspension may make these averages questionable.

Water source and algal residence time may also influence phytoplankton standing crops. The water source during late spring through early fall is primarily the Sacramento River which typically has low chlorophyll concentrations. In contrast, San Joaquin River water with several times the chlorophyll concentration is drawn to the export pumps in the southern Delta. During high flows of the San Joaquin River in 1969 the flow reversal did not occur until midsummer, and the highest midsummer chlorophyll concentrations ever measured in the western Delta may have partially originated in the southern Delta.

Inorganic nitrogen normally is the nutrient that at times limits algal growth in the study area (USBR 1972). Measurements in the western Delta and Suisun Bay indicated prior to the spring blooms sufficient nitrogen was available to support higher algal concentrations than were observed. In most years during spring blooms, nitrogen was depleted to limiting ($\leqslant 0.02$ mg·liter^{-1}) or

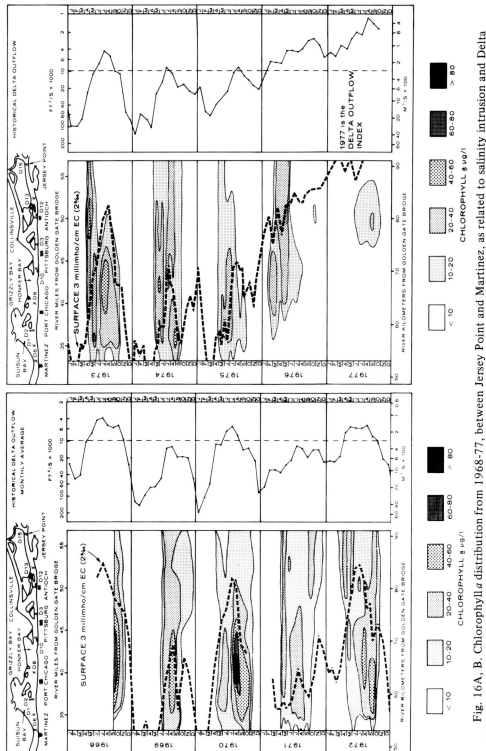

Fig. 16A, B. Chlorophyll *a* distribution from 1968-77, between Jersey Point and Martinez, as related to salinity intrusion and Delta outflow. The 3 millimho/cm specific conductance contour represents the approximate location of the upstream edge of the entrapment zone at high slack tides. The tidally averaged center location of the entrapment zone would be nearly 10 km downstream.

near-limiting levels at many sites. The greatest nitrogen limitation was observed in 1971, a year with high water transparency. In 1971 the spring blooms in these two areas generally depleted the nitrogen to limiting levels and subsequent phytoplankton growth depleted nitrogen from the inflowing water until fall. Chlorophyll concentrations were at lowest recorded levels during 1977, except near Antioch (D12) and Emmaton (D22), the approximate center of the entrapment zone during that time. Also, inorganic nitrogen concentrations remained near winter high levels.

It is difficult to determine which are the dominant factors that develop and maintain phytoplankton blooms in the Suisun Bay area (Ball 1977). The highest chlorophyll peaks between 1969 and 1977 were measured during the period of maximum water temperature (23 to 25°C) and slightly after the period of maximum insolation. These data suggest that high water temperatures and solar radiation levels stimulate the phytoplankton growth rate to produce high level standing crops. However, low standing crops occurred during July-September of 1974, 1975, 1976, and throughout 1977. These differences indicate that high standing crops will occur only if factors in addition to insolation, water temperature, and available nutrients are favorable for growth.

Prior to 1976 we believed that the Suisun Bay phytoplankton standing crop was directly related to the water transparency and indirectly related to Delta outflow (Ball 1975). But, while the 1976 and 1977 summer water transparencies were nearly double those of the previous seven years and the outflows were at record lows, only a late winter and a spring bloom developed in 1976 and no bloom occurred in 1977. This pronounced inconsistency with preceding trends prompted special studies in 1977 which evaluated all possible factors influencing phytoplankton standing crop (see also Arthur and Ball 1979).

It became apparent that chlorophyll a concentrations above 20 μg·liter^{-1} (Figs. 16A,B) developed only when the Delta outflow was between about 700 (25,000) and 110 m^3·s^{-1} (4,000 ft^3·s^{-1}). Below about 110 m^3·s^{-1} (4,000 ft^3·s^{-1}), the standing crop either declined or remained low.

The tidally-averaged location of the entrapment zone is adjacent to the shallows of the Suisun-Honker Bay area when the outflow varies between 110 and 700 m^3·s^{-1}. Such a location appears to provide the potential for a high phytoplankton standing crop if other environmental factors are favorable (hypotheses are discussed by Arthur and Ball 1979).

The shallowness of Suisun Marsh may promote relatively high chlorophyll concentrations and minimize the growth-limiting effects of high water turbidity. Because the salinity distribution is similar to that of Suisun Bay, many of the factors influencing growth there are probably present in Suisun Marsh. The few available nutrient data indicate that inorganic nitrogen is depleted to limiting conditions during periods of high phytoplankton standing crops.

We do not know which factors most control phytoplankton growth in San Pablo Bay. As in Suisun Bay, estuarine circulation and nutrient limitation undoubtedly influence growth (see also Cloern 1979).

The 1977 drought year has provided us with interesting but puzzling findings. Areas of the southern and northern Delta experienced two- to six-fold increases in summer chlorophyll concentrations, respectively, whereas the western Delta, Suisun Bay and Marsh experienced two- to four-fold reductions. The increases are apparently due to increased water residence time and nutrient buildup during that period. The decreases in the western Delta and Suisun Bay are apparently caused by the landward migration of the entrapment zone (Arthur and Ball 1979). The cause of the chlorophyll a decrease in Suisun Marsh is unknown, but may also relate to salinity intrusion.

ACKNOWLEDGMENTS

These data were obtained from a coordinated monitoring program involving USBR and DWR and are stored in the Environmental Protection Agency's (EPA) national data storage and retrieval system (STORET).

LITERATURE CITED

Arthur, J. F. and M. D. Ball. 1978. Entrapment of suspended materials in the San Francisco Bay-Delta Estuary. U. S. Bureau of Reclamation. Sacramento, Calif. 106 pp.

Arthur, J. F. and M. D. Ball. 1979. Factors influencing the entrapment of suspended material in the San Francisco Bay-Delta estuary. Pages 143-174 *in* T. J. Conomos, ed. San Francisco Bay: The Urbanized Estuary. Pacific Division, Amer. Assoc. Advance. Sci., San Francisco, Calif.

Ball, M. D. 1975. Chlorophyll levels in the Sacramento-San Joaquin Delta to San Pablo Bay. *In* R. L. Brown, ed. Proceedings of a Workshop on Algal Nutrient Relationships in the San Francisco Bay and Delta, November 1973. California Dep. Water Resources, Sacramento, Calif.

Ball, M. D. 1977. Phytoplankton growth and chlorophyll levels in the Sacramento-San Joaquin Delta through San Pablo Bay. U. S. Bureau of Reclamation. Sacramento, Calif. 96 pp.

California State Department of Water Resources and Department of Fish and Game. 1972. Delta fish and wildlife protection study, dissolved oxygen dynamics, Sacramento-San Joaquin Delta and Suisun Bay. 129 pp.

Cloern, J. E. 1979. Phytoplankton ecology of the San Francisco Bay system: The status of our current understanding. Pages 247-264 *in* T. J. Conomos, ed. San Francisco Bay: The Urbanized Estuary. Pacific Division, Amer. Assoc. Advance. Sci., San Francisco, Calif.

Conomos, T. J. 1979. Properties and circulation of San Francisco Bay waters. Pages 47-84 *in* T. J. Conomos, ed. San Francisco Bay: The Urbanized Estuary. Pacific Division, Amer. Assoc. Advance. Sci., San Francisco, Calif.

Conomos, T. J., R. E. Smith, D. H. Peterson, S. W. Hager, and L. E. Schemel. 1979. Processes affecting seasonal distributions of water properties in the San Francisco Bay estuarine system. Pages 115-142 *in* T. J. Conomos, ed. San Francisco Bay: The Urbanized Estuary. Pacific Division, Amer. Assoc. Advance. Sci., San Francisco, Calif.

Fogg, G. E. 1965. Algal cultures and phytoplankton ecology. University of Wisconsin Press, Madison, Wis. 126 pp.

Peterson, D. H. 1979. Sources and sinks of biologically reactive oxygen, carbon, nitrogen, and silica in northern San Francisco Bay. Pages 175-193 *in* T. J. Conomos, ed. San Francisco Bay: The Urbanized Estuary. Pacific Division, Amer. Assoc. Advance. Sci., San Francisco, Calif.

Spence, C. and D. M. Steven. 1974. Seasonal variation of the chlorophyll a:pheo-pigment ratio in the Gulf of St. Lawrence. J. Fish. Res. Bd. Can. 31:1263-1268.

Strickland, J. D. H. and T. R. Parsons. 1968. A practical handbook of seawater analysis. Bull. Fish. Res. Bd. Can. 167. 311 pp.

U. S. Bureau of Reclamation. 1972. Delta-Suisun Bay surveillance program. A progress report on the Delta San Luis Drain surveillance portion of the program. Sacramento, California. 95 pp.

U. S. Bureau of Reclamation. 1974. Delta-Suisun Bay surveillance program. A water quality progress report on the Peripheral Canal study program. Sacramento, Calif. 123 pp.

Yentsch, C. S. 1965a. The relationship between chlorophyll and photosynthetic carbon production with reference to the measurement of decomposition products of chloroplastic pigments. Primary productivity in aquatic environments. Mem. Inst. Ital. Idrobiol. 18 (suppl.), *also* University of California, Berkeley, Calif. 1966.

Yentsch, C. S. 1965b. Distribution of chlorophyll and phaeophytin in the open ocean. Deep-Sea Res. 12:653-666.

THE BENTHIC ALGAL FLORA
OF CENTRAL SAN FRANCISCO BAY

PAUL C. SILVA
Department of Botany, University of California, Berkeley, CA 94720

The benthic algae of San Francisco Bay received only sporadic attention prior to 1968. The present account, based on a sustained program begun in 1968, is the first publication devoted entirely to the Bay flora.

The Bay offers a diversity of habitats ranging from moderately exposed to fully protected. The absence of fully exposed sites precludes those species that are restricted to such situations on the outer coast.

The benthic algae of the Bay constitute a large biomass, the bulk of which at any particular site is composed of an abundance of luxuriant plants belonging to only a few species. Those species found throughout the Bay flourish at least to the extent that they do on the outer coast. *Cryptopleura violacea* and *Polyneura latissima* are more abundant and luxuriant in the Bay than on the outer coast. On the other hand, many species that barely penetrate the Bay are less abundant and well developed than in physiographically similar sites on the outer coast.

The benthic algal flora of the Bay comprises about 170 specific and infraspecific taxa, a number that ranks the Bay, when considered a single locality, as fairly rich. The presence of at least 30 protected-water forms compensates for the absence of those outer-coast species that are unable to penetrate the Bay. The richest localities are Fort Point, Lime Point-Point Cavallo, and Point Blunt (Angel Island). Almost all major orders and families of the outer-coast flora of central California are represented in the Bay, although many common outer-coast species are absent. The major floristic component (at least 53%) is constituted by outer-coast cool-temperatate species endemic to Pacific North America. The second largest component (about 26%) is constituted by species with wide distributions, especially in the Northern Hemisphere. The warm-water element is negligible. There are at least two recently introduced weeds (*Sargassum muticum* and *Codium fragile* subsp. *tomentosoides*).

Protected-water species are limited to mudflats, marshes, yacht harbors, and commercial port areas without regard to distance from the Golden Gate. Outer-coast species, on the other hand, generally have ranges extending inward from the Golden Gate to varying distances. Groups of species characteristic of the upper, middle, and lower intertidal zones, respectively, on the outer coast all exhibit the same spectrum of ability to penetrate the Bay. The association of species characteristic of the sanded-in habitat on the outer coast exhibits this same spectrum.

The varying degree of penetration of the Bay by outer-coast species is in large part related to the salinity gradient extending from the Golden Gate to Point San Pablo, but precisely which aspects of salinity are limiting for individual species remains to be demonstrated. Wave action has a similarly oriented gradient and may restrict certain species to the extreme western part of the Bay, although most species occur in less exposed situations within the Bay than on the outer coast. Variation in temperature in the central Bay can be assumed to have a negligible effect on the distribution of outer-coast species since the great majority have ranges extending over many degrees of latitude. The unusually low and high temperatures that occasionally obtain in marshes and mudflats, especially in the

Copyright ©1979, Pacific Division, AAAS.

SAN FRANCISCO BAY

northern and southern reaches of the Bay, preclude all but eurythermal forms in those habitats. The Bay offers a wide variety of substrates so that this factor is not limiting except in the case of the sanded-in habitat. Nutrients are maintained at a sufficiently high level to eliminate them as a factor affecting distribution of outer-coast species within the central Bay. Turbidity is probably not a factor in excluding outer-coast species from the central Bay, although heavy silting of the intertidal zone may have local adverse effects. As far as benthic algae are concerned, the central Bay, with the exception of the port area of San Francisco from Aquatic Park to the Bay Bridge and the Berkeley-Richmond shore, is relatively free from deleterious effects of urbanization.

HISTORY OF PHYCOLOGICAL RESEARCH IN SAN FRANCISCO BAY

San Francisco Bay! The mere mention of the name evokes a deep emotional response from countless persons. The Bay area, with its equable over-all climate and cluster of intriguing microclimates, exceptional scenic beauty and diversity, nearness to an abundance of a large variety of natural resources, strategic position in relation to the vast agricultural lands of the Central Valley, and its romantic history, is internationally known and loved. Within California, San Francisco Bay far exceeds in size and economic importance all other estuarine and lagoonal areas combined. Thus, while one may calmly accept the fact that little is known about the biology of estuaries in general compared to that of the outer coast, it is startling to discover that no comprehensive account of the marine flora of San Francisco Bay has ever been published. This situation seems even more amazing when it is realized that two major universities, the University of California (Berkeley) and Stanford University, both renowned for their contributions to our knowledge of marine biology, are located close to the shores of the Bay. A brief historical review will clarify, if not justify, the shortcoming.

Just as San Francisco Bay lay concealed from the eyes of European explorers for more than two centuries after the outer coast of California had been tentatively charted, so did its flora lie neglected long after the seaweeds of Pacific shores were reasonably well known. True, ships of various 19th century voyages of exploration anchored off the Presidio or in Yerba Buena Cove, in the lee of Loma Alta (Telegraph Hill), but only one instance is known in which the naturalists on board sampled the seaweeds along the shore. H.M.S. *Blossom* under the command of Captain Frederick William Beechey was in the Bay from 7 November to 28 December 1826, and again for a short period beginning 17 November 1827. The naturalist (George Tradescant Lay) and the surgeon (Alexander Collie) are said to have collected algae at San Francisco (Papenfuss 1976), but the specimens themselves (originally deposited at the Royal Botanic Gardens, Kew, but later transferred to the British Museum [Natural History], with some duplicates at Trinity College, Dublin) do not support this statement. In view of my present knowledge of the flora of the Bay, I conclude that few of the Lay & Collie specimens represent species or portions of morphological spectra of species likely to be found in a cursory sampling along the north shore of San Francisco. On the other hand, all of their specimens can be matched by plants growing either near the Presidio at Monterey or near the Mission San Carlos Borromeo at Carmel, sites also visited by Lay & Collie.

H.M.S. *Sulphur* under the command of Captain Edward Belcher visited San Francisco Bay from 19 October to 30 November 1837, and again in September 1839. In this instance there is unequivocal evidence that one of the surgeons (Andrew Sinclair) collected algae there, presumably between Telegraph Hill and Fort Point, but possibly farther west. Sinclair's collections, together with a few collections made on the Monterey Peninsula by David Douglas, a Scottish botanical explorer employed by the Horticultural Society of London, formed the basis for a short list of California seaweeds published by the eminent Irish phycologist William Henry Harvey (1840). This list includes 14 species from San Francisco, 8 from Monterey, and 3 from both localities. Douglas visited San Francisco in July 1831 and in November 1833. He is said to have collected at least one alga there (Papenfuss 1976), but the specimens in question (originally sent to Kew) are labeled merely "California".

The first American to collect California seaweeds for at least partly scientific purposes seems to have been a New Yorker by the name of A. D. Frye, who boarded at Jones's Hotel in San Francisco shortly after the Gold Rush. His collections, which were exhibited in New York in 1851 and sent to Harvey in 1852, served as the basis for several new species (Harvey 1853, 1858). The provenance was given as "Golden Gate", but the specimens provide presumptive evidence that Frye collected on both sides of Fort Point and thus not entirely within San Francisco Bay as defined for purposes of the present account.

A Swedish botanist, Sven Berggren, visited San Francisco in 1875 in the course of a personal excursion that included New Zealand, Australia, and the Hawaiian Islands. His collections of seaweeds, labeled simply "Golden Gate," were entrusted to Jacob Georg Agardh of the University of Lund, a renowned phycologist. Several served as the basis for new species (J. Agardh 1876, 1898). Again, the specimens provide presumptive evidence that both sides of Fort Point were visited.

Domination of Western American phycology by Europeans came to a sudden end in 1895. In that year William Albert Setchell, a protégé of the illustrious cryptogamic botanist William Gilson Farlow of Harvard University, was brought to the University of California at Berkeley to develop the newly established Department of Botany (Silva 1977). For the first time there was now a professional phycologist residing in California. No longer would specimens have to be sent to European workers or even to Farlow. Setchell's first collections in California were made along the Berkeley shore of the Bay, but except for repeated visits to Fort Point, his subsequent field experience was largely limited to the outer coast. In view of his keen interest in the classification and distribution of marine plants, one can only surmise that even in that relatively pristine period the mud and quiet waters of the Bay could not compete with the clean sand and pounding surf of the open coast as an attractive area of study.

Setchell's horizons eventually expanded beyond California, and he was caught up in a quest for answers to certain questions that took him to distant shores. Moreover, a proliferation of interests and administrative responsibilities left him little time for local problems. Fortunately for phycology in California, one of Setchell's students, Nathaniel Lyon Gardner, was invited in 1909 to join the faculty at Berkeley, where he became *de facto* an assistant to his former professor. They formed an efficient, productive team. Unlike Setchell, Gardner thought "small" and proceeded to probe, scrape, and section his way into the poorly known world of microscopic marine algae, especially those forms that grow on and in other organisms. Although Gardner turned to the Bay—just 5 km west of his laboratory—for much of his material, he did not systematically survey its flora.

The decades passed and the Bay underwent profound physical and biological alterations as its shores became more highly populated. Much of this change was undetected, or if detected then unrecorded, or if recorded then unassessed, since baseline data were not available. In fairness to a few farsighted persons—Charles Atwood Kofoid in particular—it should be mentioned that in 1911 a biological survey of the Bay was instigated by the University of California in cooperation with Stanford University, the California State Fish and Game Commission, and the U. S. Bureau of Fisheries (see also Hedgpeth 1979). The survey, undertaken by the U. S. Fisheries steamer *Albatross* in 1912-1913, was well intended but suffered from lack of support in the critical post-survey stages. None of the numerous samples was fully analyzed. The principal published results are the reports on physical factors by Sumner et al. (1914), mollusks by Packard (1918a,b) and crustaceans by Schmitt (1921) and Tattersall (1932).

SAN FRANCISCO BAY

Substantial interest in the seaweeds of San Francisco Bay first arose during the past decade, concomitant with the development of a public awareness of our environment and its importance to our welfare. Impacts on the Bay—both anticipated, such as the laying of a sewage outfall into the Gulf of the Farallones proposed by the City of San Francisco in 1969, and actual, as the release of 3,200 m^3 (840,000 gallons) of Bunker-C fuel oil from one of two tankers that collided at the Golden Gate on 18 January 1971 (Conomos 1975)—brought into sharp focus the paucity of information on the Bay, especially its biota. In 1968, faced with an increased number and urgency of requests for data on the algae of the Bay, I converted a passing interest into a sustained program of observation and collection. The results constitute the basis for this chapter, in which I have attempted to characterize the benthic algal flora of the Bay with respect to abundance, luxuriance, richness, diversity, habitats, geographic distribution, and environmental factors. While much work remains to be done, the main lines of the picture seem clear.

Of the meager material available from sources other than my own field work, mention should be made of collections by students at San Francisco State University, especially those of James D. Ripley, who undertook a floristic and ecological study of Angel Island (1969). Ripley's record of 55 species of marine algae constitutes the first and only such list for the Bay.

PHYSICAL SETTING

The physiographic and hydrographic distinctness of San Francisco Bay as compared with the outer coast is obvious, yet it is necessary not only to mention it but to elaborate on it in order to understand the striking differences between the benthic floras of the two areas. San Francisco Bay is vast (ca. 1,240 km^2 at mean high water) in relation to the narrowness of its opening to the Pacific Ocean (the Golden Gate, only 1.7 km wide). It has an extensive shoreline (ca. 445 km) and is exceedingly complex hydrographically. The northern parts are estuarine, receiving large quantities of fresh water sporadically from the Sacramento-San Joaquin river system which drains the great Central Valley of California (Conomos 1979). In former times, there were two distinct peaks of discharge: one during the height of the rainy season (December-March), the other in conjunction with snow melt (May-June). Today, however, the river system is programmed by engineers, so that not only is the discharge less regular, but the volume is markedly reduced by diversion of water to south central and southern California by way of aqueducts. The southern part of San Francisco Bay is like a tidally oscillating lagoon, with large areas of marshland. Here the inflow of fresh water is slight, being attributable to creeks and to waste-water effluents. Salinity and temperature are relatively high, especially in summer. Both northern and southern reaches of the Bay are shallow and thus prone to fluctuation in environmental factors. In addition, both areas receive large amounts of industrial wastes, resulting in further environmental stress. In these quiet muddy waters the benthic algal flora is reduced to a small number of species with broad physiological tolerances, species that for the most part are represented in bays and harbors throughout the North Temperate Zone.

The central part of the Bay (that area bounded by the Golden Gate Bridge, the San Francisco-Oakland Bay Bridge, and the Richmond-San Rafael Bridge) is relatively deep, the maximum depth being 110 m at the Golden Gate, and is scoured each day by four tidal flows of high velocity (see also Rubin and McCulloch 1979). This flushing action has a homeostatic effect on the waters of the central Bay, moderating the extremes of salinity and temperature (Conomos 1979) and lessening the adverse effects of pollution. Indeed, as one stands on the southwest shore of Angel Island and looks to the left at the manhattanized skyline of San Francisco and to the right at the Golden Gate, listening to the waves break against a boulder beach interspersed with clean sand, one experiences the exhilaration of seeing a vigorous intertidal biota within the collective shadow of almost five million persons. Here is where the distinctive character of the algal flora of the central Bay is underscored: why, at an ecologically diverse locality bathed by clean ocean water pouring through the Golden Gate and splashing against the shore with considerable force—a locality that even smells like the ocean rather than a bay—are so many outer-coast species missing? At first glance, the intertidal flora looks rich, but closer examination reveals that the number of species is not commensurate with abundance and luxuriance. It is this central part of the Bay that I have emphasized in my research, primarily because my point of reference was the outer-coast flora, which with very few exceptions does not penetrate the Bay beyond Alameda to the south and Richmond to the north. Moreover, my studies have been restricted to the intertidal and uppermost subtidal zones. The turbidity of the water, the muddiness of the bottom in shallow areas, and the swift currents in hard-bottom areas make any diving program unattractive. If and when a suitable combination of weather, sea, and personnel develops, however, diving will be attempted at such potentially rewarding places as Alcatraz Island. Plants (often fragments) cast ashore at Bonita Cove, Lime Point, and Fort Point provide evidence that at least a few unusual algae grow in the deep waters adjacent to the Golden Gate. Some information on subtidal algae in the Bay has come from dredging and trawling, but most of the stations sampled are in the northern and southern reaches rather than in the central part.

INTERTIDAL HABITATS FOR BENTHIC ALGAE

The rocky shore from Lime Point, at the north pier of the Golden Gate Bridge, to Point Cavallo supports a rich flora, including many species whose distribution in the Bay is limited to this area or to this area and the south portal (Fort Point). The richness is partly attributable to the relatively high degree of wave action along this shore. The complete protection offered by the breakwater at Fort Baker provides local habitat diversity. Far fewer species are to be found on the rocky shore between Point Cavallo and Sausalito, which incidentally is accessible only with difficulty. The riprap, pilings, and floats of Sausalito provide substrates within a protected situation comparable to Berkeley and the shore of San Francisco from the Yacht Harbor to Gas House Cove. Richardson Bay is marshy, similar to much of the East Bay. The rocky points of the Tiburon Peninsula facing Raccoon Strait along with the west and south sides of Angel Island and Alcatraz Island benefit from the flushing action through the Golden Gate, allowing the development of a relatively rich flora. Wave action at Point Blunt on Angel Island and on the southwest side of Alcatraz Island is sufficient to cause the intertidal zonation to be skewed upward about a meter compared with that of less exposed sites in the Bay. At Point Blunt, the presence of a fairly well-developed sanded-in habitat with a characteristic assemblage of species accounts for an especially rich flora. Pilings and floats at Tiburon provide local habitat diversity. The rocky east side of the Tiburon Peninsula is protected and thus supports a decreased number of species, although the presence of a small amount of sand at Paradise Beach causes that site to be relatively rich. The area from California Point to San Quentin Point is marshy.

Returning to the Golden Gate, the shore between Fort Point and Crissy Field—with boulders, cobbles, sand, concrete, and wood—is one of the richest localities in the central Bay. The shore from the San Francisco Yacht Harbor to Gas House Cove offers a variety of protected substrates, including pilings, floats, riprap, and cobbles. Fort Mason (Black Point) is surprisingly rich for a very small area. As at Point Blunt and Alcatraz Island, direct exposure to surf coming through the Golden Gate causes the intertidal zonation to be skewed upward about a meter. Aquatic Park offers a sea wall for algal growth, but it is better known as a place where large quantities of drift may occasionally be found, especially *Gracilaria* and *Stenogramma*. The port of San Francisco between Fisherman's Wharf and Rincon Point (the west end of the San Francisco-Oakland Bay

SAN FRANCISCO BAY

Bridge) remains to be studied. Yerba Buena Island and its man-made adjunct Treasure Island are rewarding places to visit, supporting many species not found along the nearby East Bay shore.

The physiography of the East Bay has been changed markedly by urban development, especially in the past two decades. Much of the original marshland has been eliminated by dredge-and-fill operations. On the other hand, riprap installed to protect the filled land from erosion has increased the amount of substrate available to seaweeds. The riprap at the entrance to the Berkeley Yacht Harbor supports a surprisingly rich and luxuriant flora. The marshes and mudflats from Berkeley to the Richmond-San Rafael Bridge are in themselves interesting, since a large number of species are specially adapted to such a habitat, but in addition they provide some unusual and rewarding substrates, including sheets of polyethylene plastic and the fabric of decomposing automobile tires. A discarded tire in its usual upright position in a mudflat is particularly offensive, but when caught in a horizontal position it may become transformed into a beautiful tide pool! Several naturally rocky points (Golden Gate Fields, Point Isabel, and Point Richmond) interrupt the muddy shores of Albany and Richmond, providing for local increases in richness. Leaves of eel grass (*Zostera*) growing in shallow-water beds in various parts of the Bay often serve as substrate for a fairly large number of species of algae (see also Atwater et al. 1979).

CHARACTERIZATION AND ANALYSIS OF THE BENTHIC ALGAL FLORA

Abundance and Luxuriance

The abundance and luxuriance of the benthic algal flora of San Francisco Bay may surprise someone whose experience has been limited to bays that enclose uniformly protected waters. Sites supporting a large biomass are not limited to those localities directly exposed to the flushing action through the Golden Gate, however, but include such protected places as Paradise Beach on the east side of the Tiburon Peninsula and the riprap at the entrance to the Berkeley Yacht Harbor.

At any site within the central Bay the bulk of the algal biomass is composed of an abundance of luxuriant plants belonging to only a few species. Those species that are found throughout the Bay appear to be flourishing at least to the degree that they do on the outer coast. *Cryptopleura violacea* and especially *Polyneura latissima* are notable in being more abundant and luxuriant in the Bay than on the outer coast. On the other hand, many species that barely penetrate the Bay (not extending beyond Fort Point and Point Cavallo) and some that reach only as far as Alcatraz Island and Fort Mason are less abundant and less well developed than in physiographically similar sites on the outer coast. *Agardhiella gaudichaudii* and *Rhodymenia californica*, common on the outer coast, are known from the Bay only as occasional isolated plants. *Analipus japonicus*, which on the outer coast produces clusters of erect shoots each spring from a perennial crust, grows at four sites in the western part of the Bay, where even in summer only a few depauperate shoots can be found. By contrast, certain species that are restricted to the western part of the Bay, such as *Alaria marginata, Laminaria sinclairii,* and *Gymnogongrus linearis,* thrive.

Richness and Diversity

Richness as used in this chapter is a subjective assessment based on a comparison of the number of specific and infraspecific taxa in the flora of a given locality with that of another. The degrees of richness indicated in the preceding section on habitats are based on comparisons between the Bay localities and numerous central California outer-coast sites that I have studied in detail. It would be highly desirable, and in some respects more pertinent, to compare San Francisco Bay with similar estuarine-lagoonal areas in other parts of the world. Hopefully, data that would make such comparisons possible will eventually become available.

SILVA: BENTHIC ALGAL FLORA

Diversity as used in this chapter is a subjective assessment of the extent to which the total taxonomic spectrum of the flora of a region is represented by the flora of a given locality. One measure is the number of genera, families, and orders represented in the flora, but since there is disagreement among workers as to the number of higher taxa (especially orders) that should be recognized, a better way to assess the situation is simply to turn the pages of a regional manual (in this instance Abbott and Hollenberg 1976), noting how well or how poorly the various parts of the taxonomic spectrum are represented. This concept of taxonomic diversity is unusual, if not unique, but hopefully the ensuing discussion will show that it is meaningful.

The number of taxa in each of the four major groups (classes) of benthic algae in the San Francisco Bay flora is given in Table 1. In addition to the 156 specific and infraspecific taxa listed in the systematic catalog, there are about 14 species remaining to be determined in the following genera, which are already represented by at least one species in the flora: *Callithamnion, Ceramium, Grateloupia, Peyssonnelia,* and *Polysiphonia* (Rhodophyceae); *Ectocarpus* and *Laminaria* (Phaeophyceae); and *Bryopsis* and *Cladophora* (Chlorophyceae). In addition, there is an undetermined species of *Lithophyllum* (Rhodophyceae: Corallinaceae).

TABLE 1. NUMBER OF TAXA IN THE BENTHIC ALGAL
FLORA OF CENTRAL SAN FRANCISCO BAY.

Classes	Orders	Families	Genera	Specific and Infra-specific taxa
Rhodophyceae	8	24	61	94
Phaeophyceae	9	12	22	27
Xanthophyceae	1	1	1	1
Chlorophyceae	5	9	17	34
	23	46	101	156[a]

[a] There are about 14 additional species at hand that have not yet been identified.

To be assessed for richness, the Bay flora, with its approximately 170 specific and infraspecific taxa, must be compared with floras of areas of equal habitat diversity, although not necessarily identical habitats. Duxbury Reef (Marin County) and Moss Beach (San Mateo County) both offer such diversity, although the habitats that they provide differ sharply from those in the Bay in the lack of fully protected waters and the presence of fully exposed situations. If the Bay is considered a single locality, the richness of its flora ranks close behind that of those two sites. The presence in the Bay of at least 31 protected-water forms compensates for the absence of many outer-coast species. The Bay flora, as one would expect, is far richer than that of sites outside the Golden Gate that offer essentially only one habitat (e.g. Lands End in San Francisco, Bonita Cove and Rocky Point in Marin County).

With regard to taxonomic diversity, the Bay flora includes representatives of almost all major orders and families present in the outer-coast flora of central California. A notable deficiency is in the Rhodymeniales, with a single species barely entering the Bay and not thriving. The absence of certain common outer-coast species, however, is striking. The absentees include such well-known algae as *Calliarthron tuberculosum, Prionitis andersoniana, Plocamium oregonum, Gigartina harveyana, Rhodoglossum affine, R. californicum, Halosaccion glandiforme, Palmaria palmata* var. *mollis, Gastroclonium coulteri, Botryoglossum farlowianum, Laurencia spectabilis, Leathesia difformis, Haplogloia andersonii, Soranthera ulvoidea* (although both of its hosts, *Odonthalia* and

SAN FRANCISCO BAY

Rhodomela, occur in the Bay), *Colpomenia peregrina, Macrocystis integrifolia,* and *Cystoseira osmundacea* (with its epiphytes *Pterochondria woodii* and *Coilodesme californica*).

The Bay flora includes at least 25 species, representing 17 genera, that are microscopic and thus not readily detectable in the field.

Analysis of Components by Geographic Range

The 156 specific and infraspecific taxa listed in the systematic catalog can be categorized on the basis of their known geographic ranges. The results, summarized in Table 2, clearly show that the San Francisco Bay flora is composed predominantly of outer-coast cool-temperate species endemic to Pacific North America. This component would total 53% even if most of the 14 undetermined species prove to belong to other categories (as I expect). Species with wide distributions, especially in the Northern Hemisphere, constitute the second largest component, totaling 26%. The 16 species (10%) with bipolar distributions (temperate Pacific North and South America) are of special phytogeographic interest. The warm-water element is negligible.

TABLE 2. ANALYSIS OF THE BENTHIC ALGAL FLORA OF CENTRAL SAN FRANCISCO BAY IN TERMS OF GEOGRAPHIC RANGES

Protected-water species	
North Pacific species	4
Widespread temperate species	21
Weeds	2
Endemic species (in need of reappraisal)	4
Total protected-water species	31
Outer-coast species	
North Pacific species	
Ranging to San Francisco	5
Ranging to Monterey Co.	4
Ranging to San Luis Obispo Co.	5
Ranging to Santa Barbara Co.	19
Ranging to S. Calif.	10
Ranging to Baja Calif. and/or Gulf of Calif.	45
Ranging from San Francisco to S. Calif.	2
	90
Bipolar species (temperate Pacific North and South America)	16
Widespread boreal species	
Ranging to San Francisco or San Mateo Co.	2
Ranging to S. Calif.	2
Ranging to Baja Calif. or beyond	13
	17
Widespread warm-water species	2
Total outer-coast species	125
Total species	156

Analysis of Components by Distribution within San Francisco Bay

When the distribution of each species within San Francisco Bay is plotted, a pattern is readily perceived. The protected-water species, as would be expected, are limited to mudflats, marshes, yacht harbors, and commercial port areas at various distances from the Golden Gate and hence apparently independent of its flushing action. The outer-coast species, on the other hand, generally have ranges that extend inward from the Golden Gate to varying distances. Continuity of the ranges may be broken by the absence of specifically required habitats, but it is clear that these species penetrate the Bay progressively until a limiting factor comes into play. As will be discussed below, the overriding environmental factor appears to be some aspect of salinity, with wave action playing a secondary role.

Coincidence of distributional patterns of groups of species suggests the recognition of zones of penetration into the Bay. Purposes of analysis seem best served by recognizing five such zones (Fig. 1). In Tables 3-7 the outer-coast species are listed according to the five zones. (Of the 125 outer-coast species listed in the systematic catalog, 14—mostly crusts and microscopic forms—for which adequate distributional data are not yet available have been omitted.) Twenty-seven species (Table 3) appear to be limited to Zone I—the area just east of the portals (Fort Baker and Fort Point). Another 17 species (Table 4) penetrate the Bay only as far as Alcatraz Island and Fort Mason (Zone II). Five of this group have not yet been found in Zone I, but may be expected to occur there. Twenty-six species (Table 5) penetrate the Bay only as far as the southwest side of Angel Island and Tiburon (Zone III), of which three are known only from Angel Island. Eighteen species (Table 6) penetrate the Bay only as far as Sausalito, Paradise Beach (east side of the

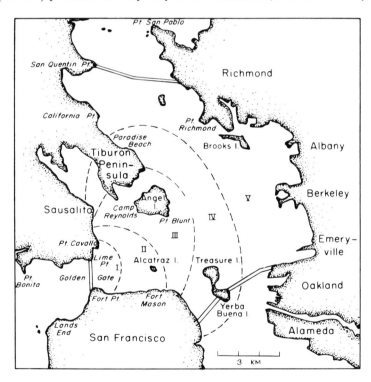

Fig. 1. Map of the central part of San Francisco Bay showing zones of penetration by outer-coast species of benthic algae.

SAN FRANCISCO BAY

TABLE 3. OUTER-COAST SPECIES THAT PENETRATE SAN FRANCISCO BAY ONLY AS FAR AS ZONE 1.

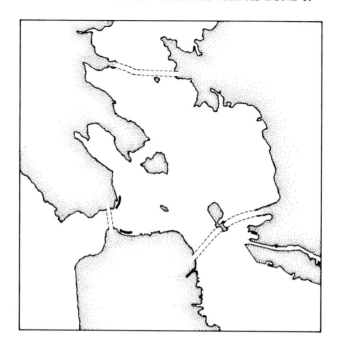

Species (27 total)	Lime Pt.- Pt. Cavallo	Fort Point
Acrochaetium coccineum		x
A. porphyrae		x
Agardhiella gaudichaudii	x	
Alaria marginata	x	
Asterocolax gardneri	x	x
Bossiella dichotoma		x
Chaetomorpha aerea		x
Cryptonemia obovata	x	x
Cumagloia andersonii		x
Ectocarpus dimorphus		x
Farlowia mollis		x
Gigartina volans	x	x
Hymenena flabelligera	x	x
Neoptilota californica		x
Nienburgia andersoniana	x	x
Phycodrys setchellii		x
Plocamium cartilagineum subsp. *pacificum*	x	x
Porphyra kanakaensis	x	x
Prionitis filiformis		x
Pseudodictyon geniculatum		x
Ralfsia fungiformis	x	
Rhodophysema elegans var. *polystromaticum*		x
Rhodymenia californica	x	
Scytosiphon dotyi		x
Spongomorpha mertensii		x
Streblonema pacificum		x
Urospora doliifera		x

TABLE 4. OUTER-COAST SPECIES THAT PENETRATE SAN FRANCISCO BAY ONLY AS FAR AS ZONE II.

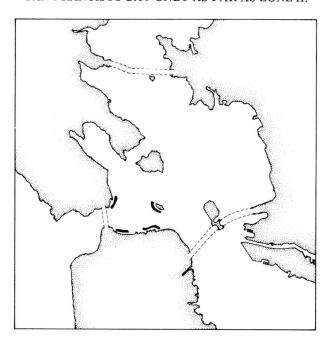

Species (17 total)	Lime Pt.-Pt. Cavallo	Fort Point	Alcatraz Island	Fort Mason
Acrochaetium amphiroae				x
A. arcuatum	x	x	x	x
A. pacificum		x	x	
Ahnfeltia plicata	x	x		x
Analipus japonicus	x	x	x	x
Bossiella plumosa	x	x	x	x
Corallina vancouveriensis	x	x	x	x
Cylindrocarpus rugosus				x
Dilsea californica		x	x	
Erythrotrichia porphyroides		x	x	
Gigartina ornithorhynchos				x
Lithothamnium pacificum	x	x	x	
Melanosiphon intestinalis f. tenuis		x		x
Peyssonnelia meridionalis			x	x
Pseudolithophyllum neofarlowii	x	x	x	
Ralfsia pacifica		x	x	x
Schizymenia pacifica			x	

Tiburon Peninsula), the northeast side of Angel Island, Yerba Buena Island, and Treasure Island (Zone IV). Of this group, one species has been found only at Paradise Beach. Finally, 23 species (Table 7) are found at various, if not most, localities throughout the central Bay.

A similar pattern of distributions within San Francisco Bay has been noted for barnacles by Newman (1967).

SAN FRANCISCO BAY

Analysis of Components with Respect to their Outer-Coast Habitat

According to Day (1951), "It is usually assumed that the marine animals [of estuaries] are necessarily intertidal forms which, being already adapted to the changes of salinity in tidal pools, find no great difficulty in extending their range into estuaries." Day observed that while this assumption appears to be generally true for European shores and estuaries where wave action is slight, it must be accepted with reserve in South Africa where the majority of the species found on the wave-battered coasts are different from those found in the muddy estuaries. Moore (1958: 402), presumably referring only to animals in the Tamar estuary (Devonshire, England), concluded that "only those forms which, under marine conditions, live at low levels are able to penetrate the estuary". Along the same lines, Filice (1959) found that the invertebrate fauna of the estuarine portion of San Francisco Bay was derived largely from the shallow subtidal regions of the adjacent coast. With regard to seaweeds, however, Dixon and Irvine (1977:46) stated that intertidal species have greater tolerance to increased or reduced salinities than those of the subtidal. In order to test these partly conflicting ideas, I plotted the distribution within the Bay of three groups of species: those characteristic of the upper, middle, and lower intertidal zones, respectively. (Almost all species in the third group also occur in the upper subtidal zone, while a few are more commonly found subtidally than intertidally.) As shown in Table 8, all three groups encompass the complete spectrum of patterns progressing inward from Zone I to Zone V. No support is thus given to the idea that benthic marine algae of a particular tide level are especially capable of penetrating estuaries.

The sanded-in habitat—a lower intertidal area in which sand seasonally covers (and scours) underlying rocks to a depth of up to 2 m—is the almost exclusive home of about 20 species of algae in central California. In order to see how this association behaves in the Bay, I plotted the distribution of its components (Table 9). Again, the entire spectrum of patterns progressing inward from Zone I to Zone V is revealed. This varying penetration of the Bay, coupled with the absence of six species of this association that are found just outside the Bay (at Rockaway Beach, San Mateo County), emphasizes the fact that associations are held together by the coincidence of tolerance ranges for several factors, a change in any one of which may be sufficient to dismember the association. Within San Francisco Bay, it is obviously some aspect of salinity that has sharply decreased the membership of the sanded-in association.

ENVIRONMENTAL FACTORS AND THEIR EFFECTS ON THE DISTRIBUTION OF BENTHIC ALGAE

Salinity

Salinity depends primarily upon the interaction of three factors: inward flow of saline oceanic water; dilution by rainfall, runoff, and sewage effluent; and evaporation (Conomos 1979). Short-term changes in salinity are produced by tidal currents, sudden variations in the hydraulic head of incoming streams, and wind. Prior to the construction of major dams on the Sacramento-San Joaquin river system, there was a marked rhythm in the over-all salinity of the central Bay, with a maximum in summer and a minimum in winter. Today, with the discharge from that system regulated, seasonal changes in salinity are subject to significant modification, but the general pattern still prevails. The amplitude of these seasonal changes varies greatly from year to year depending upon the frequency, intensity, and distribution of storms. The lowest value for surface salinity at Fort Point, for example, varied from 13.6 °/oo in 1965 to 26 °/oo in 1968 (National Ocean Survey 1970). During the dry season of most years, surface salinity at Fort Point reaches approximately that of the outer coast, which averages 32.9-33.8 °/oo (Conomos 1979; Conomos

TABLE 5. OUTER-COAST SPECIES THAT PENETRATE SAN FRANCISCO BAY ONLY AS FAR AS ZONE III.

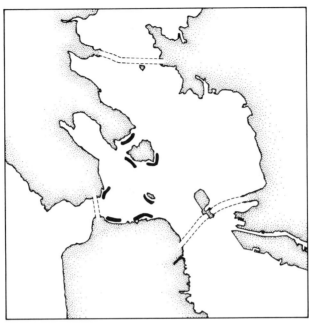

Species (26 total)	Lime Pt.-Pt. Cavallo	Fort Point	Alcatraz Island	Fort Mason	Angel I.-Tiburon
Ahnfeltiopsis pacifica					x
Callophyllis crenulata	x	x	x	x	x
C. pinnata	x	x	x	x	x
Cladophora columbiana	x	x		x	x
Desmarestia herbacea	x	x		x	x
Ectocarpus acutus	x			x	x
Endocladia muricata	x	x	x		x
Gigartina agardhii	x	x	x	x	x
G. canaliculata	x	x		x	x
Gymnogongrus linearis		x			x
Iridaea heterocarpa	x	x			x
Laminaria sinclairii	x	x	x	x	x
Melobesia marginata		x			x
M. mediocris		x			x
Odonthalia floccosa	x				x
Pelvetia compressa					x
Pelvetiopsis limitata	x			x	x
Petalonia fascia					x
Phaeostrophion irregulare		x		x	x
Plocamium violaceum	x	x	x	x	x
Porphyra lanceolata	x	x			x
Ptilota filicina		x			x
Rhodomela larix	x	x			x
Smithora naiadum		x			x
Spongomorpha coalita	x	x	x		x
Stenogramma interrupta		x	x	x	x

TABLE 6. OUTER-COAST SPECIES THAT PENETRATE SAN FRANCISCO BAY ONLY AS FAR AS ZONE IV

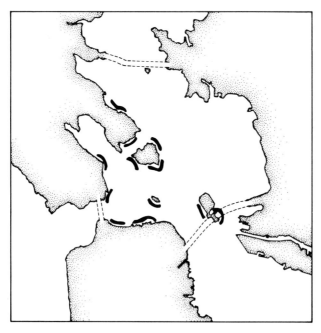

Species (18 total)	Lime Pt.-Pt. Cavallo	Fort Point	Alcatraz Island	Fort Mason	Angel I.-Tiburon	Yerba Buena	Paradise Beach
Antithamnion kylinii					x	x	
Blidingia minima var. *vexata*	x		x		x	x	
Callithamnion pikeanum	x	x		x	x	x	
Choreocolax polysiphoniae	x				x	x	
Cryptosiphonia woodii	x	x	x	x	x	x	x
Egregia menziesii	x	x	x	x	x	x	
Gymnogongrus chiton	x	x	x	x	x	x	x
G. leptophyllus	x	x			x		x
Halymenia schizymenioides	x	x	x	x	x	x	
Microcladia borealis	x	x	x	x	x	x	
M. coulteri	x	x	x	x	x	x	
Pikea californica	x	x	x	x	x	x	x
Prionitis lanceolata	x		x	x	x		x
P. lyallii	x	x	x	x	x	x	
Psammophyllum californicum							x
Pterosiphonia bipinnata	x			x	x		x
Tiffaniella snyderae	x			x	x	x	
Urospora penicilliformis	x	x	x		x		x

et al. 1979). In the intertidal and shallow subtidal zones of the central Bay, mixing effected by wind is sufficient to render negligible any vertical salinity gradient. The horizontal salinity gradient of the central Bay, on the other hand, is roughly linear, with a maximum at the Golden Gate and a minimum at Point San Pablo. The northern reaches of the Bay are brackish throughout the year, with the more saline bottom water penetrating farthest during the dry season. The situation in the southern part of the Bay is more complicated. During the rainy season salinity is lowered by river runoff transported from the northern part of the Bay, while during the dry season the local inflow of fresh water is low relative to evaporation, causing the salinity to approach or even exceed the value for the outer coast (see also Conomos 1979, Fig. 16) and Conomos et al. 1979, Fig. 5). Change of salinity during a tidal cycle (from HHW to LLW) amounts to about 2 °/oo at the Golden Gate and in the central Bay, increasing to about 4 °/oo in the extreme north and south (Imberger et al. 1977).

TABLE 7. OUTER-COAST SPECIES FOUND
THROUGHOUT CENTRAL SAN FRANCISCO BAY

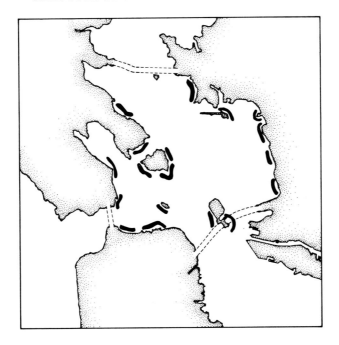

Species (23 total)

Acrochaetium subimmersum
Antithamnionella glandulifera
Bangia fusco-purpurea
Blidingia minima var. *minima*
Ceramium eatonianum
Cryptopleura violacea
Endophyton ramosum
Fucus gardneri
Gigartina exasperata
G. papillata
Gracilaria sjoestedtii

Gracilariophila oryzoides
Grateloupia doryphora
Iridaea splendens
Platythamnion villosum
Polyneura latissima
Polysiphonia paniculata
Porphyra perforata
Pterosiphonia dendroidea
Rhodochorton purpureum
Ulva angusta
U. lobata

SAN FRANCISCO BAY

TABLE 8. DISTRIBUTION OF SPECIES IN CENTRAL SAN FRANCISCO BAY ARRANGED BY OUTER-COAST ZONATION

Zones	1	2	3	4	5
Upper intertidal species					
Cumagloia andersonii	x				
Scytosiphon dotyi	x				
Analipus japonicus	x	x			
Cladophora columbiana	x	x	x		
Endocladia muricata	x	x	x		
Gigartina agardhii	x	x	x		
Pelvetia compressa			x		
Pelvetiopsis limitata	x	x	x		
Porphyra lanceolata	x		x		
Blidingia minima var. *vexata*	x	x	x	x	
Urospora penicilliformis	x	x	x	x	
Bangia fusco-purpurea	x	x		x	x
Blidingia minima var. *minima*	x	x	x	x	x
Fucus gardneri	x	x	x	x	x
Gigartina papillata	x	x	x	x	x
Porphyra perforata	x	x	x	x	x
Middle intertidal species					
Bossiella plumosa	x	x			
Corallina vancouveriensis	x	x			
Dilsea californica	x	x			
Gigartina canaliculata	x	x	x		
Iridaea heterocarpa	x		x		
Odonthalia floccosa	x		x		
Plocamium violaceum	x	x	x		
Rhodomela larix	x		x		
Callithamnion pikeanum	x	x	x	x	
Cryptosiphonia woodii	x	x	x	x	
Microcladia borealis	x	x	x	x	
Gelidium coulteri	x	x	x	x	x
Lower intertidal (and upper subtidal) species					
Agardhiella gaudichaudii	x				
Alaria marginata	x				
Bossiella dichotoma	x				
Cryptonemia obovata	x				
Gigartina volans	x				
Hymenena flabelligera	x				
Neoptilota californica	x				
Nienburgia andersoniana	x				
Phycodrys setchellii	x				
Plocamium cartilagineum subsp. *pacificum*	x				
Prionitis filiformis	x				
Rhodymenia californica	x				
Ahnfeltia plicata	x	x			
Lithothamnium pacificum	x	x			
Peyssonnelia meridionalis		x			
Schizymenia pacifica		x			
Ahnfeltiopsis pacifica			x		
Callophyllis crenulata	x	x	x		
C. pinnata	x	x	x		
Desmarestia herbacea	x	x	x		

TABLE 8 (Continued)

Zones	1	2	3	4	5
Lower intertidal (and upper subtidal) species (continued)					
Gymnogongrus linearis	x		x		
Laminaria sinclairii	x	x	x		
Phaeostrophion irregulare	x	x	x		
Ptilota filicina	x		x		
Spongomorpha coalita	x	x	x		
Stenogramma interrupta	x	x	x		
Egregia menziesii	x	x	x	x	
Gymnogongrus chiton	x	x	x	x	
G. leptophyllus	x		x	x	
Halymenia schizymenioides	x	x	x	x	
Pikea californica	x	x	x	x	
Psammophyllum californicum				x	
Tiffaniella snyderae	x	x	x	x	
Cryptopleura violacea	x	x	x	x	x
Gigartina exasperata	x	x	x	x	x
Gracilaria sjoestedtii	x	x	x	x	x
Antithamnionella glandulifera	x	x	x	x	x
Platythamnion villosum	x	x	x	x	x
Polyneura latissima	x	x	x	x	x
Polysiphonia paniculata	x	x	x	x	x
Pterosiphonia dendroidea	x	x	x	x	x

A single salinity value is of little biological significance. Extreme values may in themselves be limiting for certain organisms, but in most cases they operate in conjunction with the temporal factors of rate of change and duration. One would expect an organism to have greater ability to adapt to gradual change than to abrupt change, but no pertinent experimental data seem to be available. Similarly, one would expect greater ability to adapt to a change of short duration than to one of long duration, and numerous studies bear this out. The distribution of an organism can be related to the salinity regime only after physiological tolerances and preferences for both vegetative growth and reproduction have been ascertained from experimental cultures and after its life history has been delineated. The research necessary to understand fully the relationship between the distribution of organisms and their environment is a herculean task that could keep generations of workers occupied. Meanwhile, much can be inferred from simple observations, such as presence, absence, abundance, luxuriance, and reproductive state, when even rudimentary measurements of environmental factors are taken into account. A review of the role of salinity in the ecology of marine plants is given by Gessner and Schramm (1971).

The discovery of *Sargassum muticum* on the riprap at the entrance to the Berkeley Yacht Harbor suggested the desirability of obtaining a general picture of the salinity (and temperature) for this site to help predict the spread of this weed. The period of surveillance encompassed an average rainy season (1974/75), two years of drought (1975/1977), and a season of heavy rainfall (1977/78). Measurements, begun on a monthly schedule, were made weekly during January and February 1978 in order to keep pace with the drastic changes brought about by the rapid succession of storms at that time (Fig. 2). Two extreme situations were embraced by this series—one in which there was very little change of a high salinity over a long period (February 1976 to 12 January 1978), the other in which there were marked and rapid fluctuations, dropping as low as one-third open-ocean salinity (12 January to 1 March 1978). The response of the intertidal flora was striking. During the first period, luxuriant summer growth persisted through the following winter; during the second period, the annual portions of the algae were badly damaged or even eliminated, as they were also along stretches of outer coast bathed by heavy river runoff. Coincidence of heavy rainfall with exposure by extreme low tide was another factor in the deterioration of the intertidal flora early in 1978, both inside and outside the Bay.

While there is little experimental information, from observation of simple estuaries (if any estuary can be called simple), in which there are essentially linear horizontal and vertical salinity gradients, it can be inferred that most marine algae cannot tolerate a sustained reduction of salinity below two-thirds normal sea water. A classical illustration is provided by the Baltic Sea, whose brackish waters support a benthic flora that is drastically reduced in richness (and modified morphologically) compared to that of the North Sea. Since the salinity of the East Bay, even in an average rainy season, can be two-thirds that of normal sea water (or less) for three to four months, it is not surprising that only 23 of the 125 outer-coast species that penetrate San Francisco Bay extend into Zone V. It is obvious that the varying degree of penetration into San Francisco Bay by outer-coast species is in large part related to the over-all linear salinity gradient extending from the Golden Gate to Point San Pablo, but precisely which aspects of salinity are limiting for individual species remains to be demonstrated. It seems likely that rate, frequency, magnitude, and duration of change will all prove more important than mean annual value. A similar conclusion was reached by von Wachenfeldt (1975) in a study of the effect of salinity on the distribution of benthic algae in the Öresund, part of the transitional passage between the Baltic Sea and the North Sea.

Temperature

There are even fewer temperature data for the shores of San Francisco Bay than salinity measurements. Monthly readings at the Berkeley Yacht Harbor for the period October 1974 to March 1978 (Fig. 2) show a rhythmical change from a high of 20.0-20.8°C in late summer to a low of 8.7-12.2°C in mid-winter (see also Conomos 1979, Fig. 16, and Conomos et al. 1979, Fig. 5). Sporadic measurements that I have made in the central Bay over the past eight years indicate that from mid-spring through autumn there is a temperature gradient of about 3°C in surface water along a line from the Golden Gate to Berkeley, with Berkeley being warmest. In winter, depending upon the frequency and intensity of cold weather fronts, the trend may be reversed, with the temperature at Berkeley as much as 3°C lower than at the Golden Gate. Insolation, evaporation, and air temperature obviously affect the temperature of the shallow waters of the East Bay more than that of the deeper portion of the central Bay. Temperature is the overriding factor determining the geographic distribution of individual species of marine algae in the various oceans. However, since the great majority of those outer-coast species that penetrate the Bay have ranges extending over

TABLE 9. DISTRIBUTION OF SANDED-IN ASSOCIATION
IN CENTRAL SAN FRANCISCO BAY

Zones	1	2	3	4	5
Gigartina volans	x				
Phycodrys setchellii	x				
Ahnfeltia plicata	x	x			
Ahnfeltiopsis pacifica			x		
Gymnogongrus linearis	x		x		
Phaestrophion irregulare	x	x	x		
Stenogramma interrupta	x	x	x		
Gymnogongrus leptophyllus	x		x	x	
Psammophyllum californicum				x	
Tiffaniella snyderae	x	x	x	x	
Gracilaria sjoestedtii	x	x	x	x	x

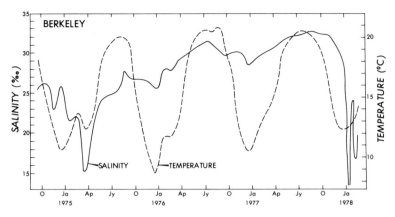

Fig. 2. Salinity and temperature of surface water at entrance to Berkeley Yacht Harbor.

many degrees of latitude, their distribution within the central Bay can be assumed to be unaffected by temperature. Protected sites—those not exposed to the flushing homeostatic effect of the Golden Gate—are subject to greater temperature ranges, with depth of water, present and recent weather, time of day, and time of low tide becoming important factors. The temperature in marshland pools may drop to 6°C, especially in the northern reaches of the Bay, and rise to 28.5°C, especially in the southern portion. Sustained temperatures of 28°C in summer and 8°C in winter would preclude all outer-coast species and many protected-water species, limiting the flora to such eurythermal forms as *Enteromorpha*. A review of the role of temperature in the ecology of marine plants is given by Gessner (1970).

Wave Action

Many seaweeds, for reasons not completely elucidated, require at least some turbulence. Some species occur only at sites pounded by extremely heavy surf, while a larger number are generally restricted to moderately exposed habitats. No quantitative data on the geographic variations of wave action in San Francisco Bay other than those in Conomos 1979 (Fig. 8) are available to me, but from observations in various seasons and under various conditions of weather and sea a meaningful picture has emerged. Within Zones I-III, Point Blunt at the southeast tip of Angel Island, Alcatraz Island, and Fort Mason (Black Point) are exposed to a surf line which sweeps through the Golden Gate. In response, the intertidal zonation is skewed upward relative to that of such less-exposed sites as Lime Point, Point Cavallo, and the east side of Fort Point. The shores of the East Bay may be surprisingly turbulent from waves produced by strong westerly winds, especially on summer afternoons (Conomos 1979, Fig. 7). Even sites that do not face the Golden Gate (such as Paradise Beach) are at times subjected to some wave action produced by the wind and the wakes of passing boats. Nowhere in the Bay, however, is wave action sufficiently strong to permit the development of those seaweeds that are restricted to fully exposed sites. Nonetheless, many species that are usually found in moderately to fully exposed habitats on the outer coast are able to survive (and even thrive) in more protected situations in the Bay. In fact, many of the species found throughout the central Bay, even in fully protected situations, usually grow in at least moderately exposed sites on the outer coast. It is the absence at moderately exposed sites (such as Point Blunt) of certain species that commonly inhabit similar sites on the outer coast that suggests salinity as the overriding factor affecting distribution within the Bay.

SAN FRANCISCO BAY

Substrate

While some algae (usually protected-water forms) can continue to grow and reproduce after being detached from their substrate, most must remain attached to some object—rocks, concrete, plastic, metal, wood, glass, shells, or living organisms. Parasites and obligate epiphytes are obviously dependent upon living hosts, often with a high degree of specificity. When the substrate serves solely as a place of attachment, contributing neither nutrients nor growth substances, its chemical composition is largely immaterial. Its physical properties, on the other hand, are of great importance. Differences in texture and hardness combine to allow sandstone, for example, to support a flora distinctive from that of granite or basalt, although many species thrive on both substrates. The indigenous rocks of San Francisco Bay comprise a heterogeneous assemblage of sediments (sandstone, shale, and limestone), contemporaneous volcanics, and intrusives (see, for example, Schlocker 1974), thus providing a range of texture and hardness sufficient to render negligible this aspect of the substrate.

Stability of the substrate is of overriding importance in determining the ability of marine algae to colonize an area. As the substrate becomes increasingly dispersed—from solid rock to colloidal-sized clay particles—the number of species that can grow on it is correspondingly reduced. In quiet waters, pebbles and small shells may be sufficiently stable, but on surf-swept beaches even cobbles and boulders may be scoured clean by constantly moving sand and gravel. Certain species, however, if firmly attached to stable rocks, can tolerate being covered periodically by as much as 2 m of sand. The upper portion of the plant may be abraded, but a new frond will be produced from the crustose base after the sand has been washed away by winter storms. Within San Francisco Bay, this sanded-in habitat is best represented at Point Blunt. The behavior of the sanded-in association of algal species in the Bay was discussed earlier in this paper (cf. Table 9).

Some of the more unusual substrates in the Bay include the fabric of automobile tires, sheets of polyethylene plastic embedded in the mud, and the fur of a sick Harbor Seal that spent much time lying in the uppermost intertidal zone at Richardson Bay.

Nutrients

Along the coast of central California, upwelling (Conomos 1979) insures an abundance of nitrate and phosphate for intertidal seaweeds. Some of this nutrient-rich water enters the Bay. Most of the nutrient supply, however, is provided by sewage effluents and river inflow (Conomos et al. 1979; Peterson 1979). Although measurements of either nitrate or phosphate at specific collecting sites are lacking, from the deep color of seaweeds throughout the central Bay one may judge that these nutrients do not limit growth through depletion (see also Peterson 1979; Cloern 1979).

Turbidity

Seaweeds, being primarily photosynthetic organisms, depend upon the receipt and absorption of a certain minimum quantity of light. In general, they have very low light requirements, so that in clear waters they can grow at depths of at least 100 m. Colloidal- to silt-sized particles greatly reduce the penetration of light, so that seaweeds in turbid waters may be restricted to depths not exceeding 2 m or even less (see also Peterson 1979). While there is no information on turbidity as related to benthic subtidal algae in San Francisco Bay, to judge from the amount and the composition of material cast ashore it seems likely that the bottom of the Bay supports a dense growth of algae wherever a suitable substrate is present (e.g. shells) and a scouring current is absent, notably the area between Alameda and Coyote Point. Turbidity may have an effect in the intertidal zone in that certain algae do not tolerate being covered with sediment, although many forms with a delicateness suggesting a preference for clear water thrive in muddy situations as long as there is a suitable substrate.

Urbanization

Domestic sewage, because of its high biological oxidation demand (BOD) and high concentration of reduced nitrogenous fractions, has a marked deleterious effect in the immediate vicinity of an outfall, but progressively farther away, as dilution is accompanied by microbiological oxidation (mineralization), the resulting nutrients have an equally marked stimulating effect on the flora. This picture is changed drastically, however, by treatment of the effluent with chlorine, which is toxic to most seaweeds at the level usually used (Stone et al. 1973). The effect of industrial wastes on the marine biota, while generally deleterious, depends upon the concentration, solubility, persistence, toxicity, and synergistic relationships of the chemical components.

Urbanization of much of the shore of the central Bay has been limited to the construction of military bases. This fact, coupled with the strong flushing action of the tidal currents through the Golden Gate, accounts for the relatively pristine condition of Fort Point, Lime Point, Point Cavallo, Angel Island, Alcatraz Island, Fort Mason, and Yerba Buena Island. Those parts of the Bay that are heavily industrialized and that receive the bulk of the domestic sewage lie largely outside the central area (see Peterson 1979, Table 4) and hence beyond the scope of this chapter with the exception of the port area of San Francisco from Aquatic Park to the Bay Bridge and the East Bay. Difficulty of access has precluded a satisfactory assessment of the San Francisco port area, but it can be assumed that the replacement of the original shore with a solid line of piers standing in heavily polluted water has resulted in profound changes in the flora, which now comprises mostly widespread protected-water species. While the East Bay is under stress (and part of Richmond has undergone the same transformation as the port area of San Francisco), it seems likely that this naturally muddy shore, subject to drastic salinity changes, at no time supported a benthic algal flora significantly different from the one at present. In fact the addition of man-made substrates, especially riprap, has tended to increase the biomass. No part of the central Bay shows the obvious impoverishment of the flora exhibited, for example, at Lands End to the west of the Golden Gate or at Palos Verdes near Los Angeles, where chlorine is the chief culprit.

EPILOGUE

Benthic algae are alive and well in San Francisco Bay, despite being ignored by botanists (Howell et al. 1958) as well as by zoologists (Lee 1977). In fact, they constitute the most conspicuous part of the intertidal biota. Contrary to the belief mistakenly held even by many phycologists, an estuarine-lagoonal system such as San Francisco Bay is not a wasteland of uninteresting, common, cosmopolitan, pollution-tolerant forms, but rather the home of a flora of moderate richness that poses a wealth of important taxonomic and ecological problems. Aside from their critical position in the trophic structure of the Bay due to the fact that they are photosynthetic organisms, benthic algae constitute a resource of great scientific, economic, educational, recreational, and esthetic value.

ACKNOWLEDGMENTS

Impetus to my research during the period 1970-1972 was provided by grants from Environmental Quality Analysts, a division of Brown and Caldwell, Consulting Engineers, San Francisco. At all times I have received full support from the Herbarium of the University of California at Berkeley. I am grateful to those students and visiting phycologists who have served as able companions in the field: Thomas C. DeCew, Thomas Glimme, H. William Johansen, Terrie L. Klinger, Phillip A.

SAN FRANCISCO BAY

Lebednik, Richard L. Moe, and Alan R. Polanshek. I am especially indebted to Richard Moe, whose keen eye has resulted in many new records, especially of crustose and microscopic forms. Phillip Lebednik has kindly provided his expert opinion on the crustose Corallinaceae as has Mitchell D. Hoyle on *Gracilaria* and W. F. Prud'homme van Reine on *Sphacelaria*.

Dennis E. Shevlin has generously shared unpublished results of research on *Bryopsis* and John A. West on *Gigartina*. John West also made helpful comments on the manuscript. I thank Ida Geary of San Francisco Community College for her hospitality at the Fort Point Promenade Classroom. I am pleased to acknowledge the cooperation of the staff of the Golden Gate National Recreation Area (GGNRA) in arranging access to Alcatraz Island. Finally, I should like to express my deep appreciation to all those, in and out of government, who have worked hard and achieved success in placing the priceless shores of the central Bay in a public trust (the GGNRA). All of us who love the Bay should be grateful to the military for preserving these lands from commercial development for more than a century.

LITERATURE CITED

Abbott, I. A. 1972. On the species of *Iridaea* (Rhodophyta) from the Pacific coast of North America. Syesis 4:51-72.

Abbott, I. A., and G. J. Hollenberg. 1976. Marine algae of California. Stanford University Press, Stanford, Calif. 827 pp.

Agardh, J. G. 1849. Algologiska bidrag. Öfvers. Förh. Kongl. Svenska Vetensk.-Akad. 6:79-89.

Agardh, J. G. 1876. Species genera et ordines algarum. Vol. 3, pars 1. Weigel, Lipsiae. 724 pp.

Agardh, J. G. 1885. Till algernes systematik. Nya bidrag. (Fjerde afdelningen.) Lunds Univ. Årsskr. 21, Afd. Math. Naturvetensk. 8. 117 pp.

Agardh, J. G. 1898. Species genera et ordines algarum. Vol. 3, pars 3. Gleerup, Lundae. 239 pp.

Agardh, J. G. 1899. Analecta algologica. Continuatio V. Lunds Univ. Årsskr. 35, Afd. Kongl. Fysiogr. Sällsk. Handl. 4. 160 pp.

Atwater, B. F., S. G. Conard, J. N. Dowden, C. W. Hedel, R. L. MacDonald, and W. Savage. 1979. History, landforms, and vegetation of the estuary's tidal marshes. Pages 347-380 *in* T. J. Conomos, ed. San Francisco Bay: The Urbanized Estuary. Pacific Division, Amer. Assoc. Advance. Sci., San Francisco, Calif.

Cloern, James E. 1979. Phytoplankton ecology of the San Francisco Bay system: The status of our current understanding. Pages 247-264 *in* T. J. Conomos, ed. San Francisco Bay: The Urbanized Estuary. Pacific Division, Amer. Assoc. Adv. Sci., San Francisco, Calif.

Collins, F. S. 1911. *Spongomorpha arcta* (Dillw.) Kuetz. forma *limitanea* Collins, n.f. *In* F. S. Collins, I. Holden, and W. A. Setchell, Phycotheca boreali-americana. Malden, Massachusetts. Fasc. XXXV, no. 1736.

Collins, F. S. 1918. The green algae of North America, second supplementary paper. Tufts Coll. Stud. (Sci.) 4(7). 106 pp.

Conomos, T. J. 1975. Movement of spilled oil as predicted by estuarine nontidal drift. Limnol. Oceanogr. 20:159-173.

Conomos, T. J. 1979. Properties and circulation of San Francisco Bay waters. Pages 47-84 *in* T. J. Conomos, ed. San Francisco Bay: The Urbanized Estuary. Pacific Division, Amer. Assoc. Advance. Sci., San Francisco, Calif.

Conomos, T. J., R. E. Smith, D. H. Peterson, S. W. Hager, and L. E. Schemel. 1979. Processes affecting seasonal distributions of water properties in the San Francisco Bay estuarine system. Pages 115-142 *in* T. J. Conomos, ed. San Francisco Bay: The Urbanized Estuary. Pacific Division, Amer. Assoc. Advance. Sci., San Francisco, Calif.

Dawson, E. Y. 1961. Marine red algae of Pacific Mexico. Part 4. Gigartinales. Pac. Nat. 2:191-343.

Dawson, E. Y. 1963. Marine red algae of Pacific Mexico. Part 8. Ceramiales: Dasyaceae, Rhodomelaceae. Nova Hedwigia 6:401-481.

Day, J. H. 1951. The ecology of South African estuaries. Part 1. A review of estuarine conditions in general. Trans. R. Soc. S. Afr. 33:53-91.

DeCew, T. C., and P. C. Silva. 1979. A guide to the seaweeds of British Columbia, Washington, Oregon, and northern California. Poseidon Press, Berkeley, Calif. (in press).

Dixon, P. S., and L. M. Irvine. 1977. Seaweeds of the British Isles. Vol. 1. Rhodophyta. Part 1. Introduction, Nemaliales, Gigartinales. British Museum (Natural History), London. 252 pp.

Doty, M. S. 1947. The marine algae of Oregon. Part 1. Chlorophyta and Phaeophyta. Farlowia 3:1-65.

Drew, K. M. 1928. A revision of the genera *Chantransia, Rhodochorton,* and *Acrochaetium,* with descriptions of the marine species of *Rhodochorton* (Naeg.) gen. emend. on the Pacific coast of North America. Univ. Calif. Publ. Bot. 14:139-224.

Filice, F. P. 1959. Invertebrates from the estuarine portion of San Francisco Bay and some factors influencing their distributions. Wasmann J. Biol. 16:159-211.

Gardner, N. L. 1910. Variations in nuclear extrusion among the Fucaceae. Univ. Calif. Publ. Bot. 4:121-136.

Gardner, N. L. 1917. New Pacific coast marine algae. I. Univ. Calif. Publ. Bot. 6:377-416.

Gardner, N. L. 1922. The genus *Fucus* on the Pacific coast of North America. Univ. Calif. Publ. Bot. 10:1-180.

Gardner, N. L. 1927a. New species of *Gelidium* on the Pacific coast of North America. Univ. Calif. Publ. Bot. 13:273-318.

Gardner, N. L. 1927b. New Rhodophyceae from the Pacific coast of North America. VI. Univ. Calif. Publ. Bot. 14:99-138.

Gessner, F. 1970. Temperature. Plants. Pages 363-406 *in* O. Kinne, ed. Marine Ecology. Vol. 1, part 1. Wiley-Interscience, London.

Gessner, F. and W. Schramm. 1971. Salinity. Plants. Pages 705-820 *in* O. Kinne, ed. Marine Ecology. Vol. 1, part 2. Wiley-Interscience, London.

Hamel, G. 1931. Chlorophycées des côtes françaises (fin). Rev. Algol. 6:9-73.

Harvey, W. H. 1840. Algae. Pages 406-409 *in* W. J. Hooker and G. A. W. Arnott, The botany of Captain Beechey's voyage. Bohn, London.

Harvey, W. H. 1853. Nereis boreali-americana. Part II. Rhodospermeae. Smithson. Contrib. Knowl. 5(5). 258 pp.

Harvey, W. H. 1858. Nereis boreali-americana. Part III. Chlorospermeae. Smithson. Contrib. Knowl. 10(2). 140 pp.

Hedgpeth, J. W. 1979. San Francisco Bay—the unsuspected estuary. A history of researches. Pages 9-29 *in* T. J. Conomos, ed. San Francisco Bay: The Urbanized Estuary. Pacific Division, Amer. Assoc. Advance. Sci., San Francisco, Calif.

Hollenberg, G. J. 1944. An account of the species of *Polysiphonia* on the Pacific coast of North America. II. *Polysiphonia.* Amer. J. Bot. 31:474-483.

Howe, M. A. 1914. The marine algae of Peru. Mem. Torrey Bot. Club 15:1-185.

Howell, J. T., P. H. Raven, and P. Rubtzoff. 1958. A flora of San Francisco, California. Wasmann J. Biol. 16:1-157.

Imberger, J., W. B. Kirkland, Jr., and H. B. Fischer. 1977. The effect of delta outflow on density stratification in San Francisco Bay: Rept. to Assoc. of Bay Area Governments (Rept. HBF-77/02), Berkeley, Calif. 109 pp.

Kohlmeyer, J., and E. Kohlmeyer. 1972. A new genus of marine Ascomycetes on *Ulva vexata* Setchell et Gard. Bot. Jahrb. Syst. 92:429-432.

Kylin, H. 1925. The marine red algae in the vicinity of the Biological Station at Friday Harbor, Wash. Lunds Univ. Årsskr., N. F., Avd. 2, 21(9). 87 pp.

Lee, W. L. 1977. The San Francisco Bay Project: a new approach to using systematics. ASC [Association of Systematics Collections] Newsletter 5:15-17.

Leister, G. L. 1977. Taxonomy and reproductive morphology of *Iridaea cordata* (Turner) Bory and *Iridaea crispata* Bory (Gigartinaceae, Rhodophyta) from southern South America. Ph.D. Thesis. Duke University, Durham, N.C. 186 pp.

Montagne, C. 1842. Troisième centurie de plantes cellulaires exotiques nouvelles. Décades V, VI, VII et VIII. Ann. Sci. Nat. Bot., Ser. 2, 18:241-282.

Moore, H. B. 1958. Marine ecology. J. Wiley & Sons, New York. 493 pp.

Mumford, T. F. 1973. A new species of *Porphyra* from the west coast of North America. Syesis 6:239-242.

National Ocean Survey. 1970. Surface water temperature and density. Pacific coast. North and South American and Pacific Ocean islands, 3rd ed. NOS Publ. 31-3. 88 pp.

Newman, W. A. 1967. On physiology and behaviour of estuarine barnacles. Mar. Biol. Assoc. India, Symp. Ser. 2:1038-1066.

Norton, T. A. 1976. Why is *Sargassum muticum* so invasive? Brit. Phycol. J. 11:197-198. (Abstr.)

Packard, E. L. 1918a. A quantitative analysis of the molluscan fauna of San Francisco Bay. Univ. Calif. Publ. Zool. 18:299-336.

Packard, E. L. 1918b. Molluscan fauna from San Francisco Bay. Univ. Calif. Publ. Zool. 14:199-452.

Papenfuss, G. F. 1976. Landmarks in Pacific North American marine phycology. Pages 21-46 *in* I. A. Abbott and G. J. Hollenberg, Marine Algae of California. Stanford University Press, Stanford, Calif.

Peterson, D. H. 1979. Sources and sinks of biologically reactive oxygen, carbon, nitrogen, and silica in northern San Francisco Bay. Pages 175-193 *in* T. J. Conomos, ed. San Francisco Bay: The Urbanized Estuary. Pacific Division, Amer. Assoc. Advance. Sci., San Francisco, Calif.

Polanshek, A. R., and J. A. West. 1975. Culture and hybridization studies on *Petrocelis* (Rhodophyta) from Alaska and California. J. Phycol. 11:434-439.

Polanshek, A. R., and J. A. West. 1977. Culture and hybridization studies on *Gigartina papillata* Rhodophyta). J. Phycol. 13:141-149.

Postels, A., and F. Ruprecht. 1840. Illustrationes algarum. Petropoli. 22 pp., 40 pls.

Ripley, J. D. 1969. A floristic and ecological study of Angel Island State Park, Marin County, California. M. A. Thesis. San Francisco State College, San Francisco, Calif. 113 pp.

Rubin, D. M., and D. S. McCulloch. 1979. The movement and equilibrium of bedforms in central San Francisco Bay. Pages 97-113 *in* T. J. Conomos, ed. San Francisco Bay: The Urbanized Estuary. Pacific Division, Amer. Assoc. Advance. Sci., San Francisco, Calif.

Schlocker, J. 1974. Geology of the San Francisco North Quadrangle, California. U. S. Geol. Surv. Prof. Paper 782. 109 pp.

Schmitt, W. L. 1921. The marine decapod Crustacea of California with special reference to the decapod Crustacea collected by the United States Bureau of Fisheries steamer "Albatross" in connection with the Biological Survey of San Francisco Bay during the years 1912-1913. Univ. Calif. Publ. Zool. 23:1-470.

Setchell, W. A. 1899. *Pikea pinnata* Setchell mss. *In* F. S. Collins, I. Holden, and W. A. Setchell, Phycotheca boreali-americana. Malden, Massachusetts. Fasc. XIII, no. 648.

Setchell, W. A., and N. L. Gardner. 1920a. Phycological contributions. I. Univ. Calif. Publ. Bot. 7:279-324.

Setchell, W. A., and N. L. Gardner. 1920b. The marine algae of the Pacific coast of North America. Part II. Chlorophyceae. Univ. Calif. Publ. Bot. 8:139-374.

Setchell, W. A., and N. L. Gardner. 1925. The marine algae of the Pacific coast of North America. Part III. Melanophyceae. Univ. Calif. Publ. Bot. 8:383-898.

Setchell, W. A. and N. L. Gardner. 1937. *Iridophycus* in the Northern Hemisphere. Proc. Nat'l. Acad. Sci. 23:169-174.

Silva, P. C. 1977. Phycological resources of the Herbarium of the University of California at Berkeley. Poseidon Press, Berkeley, Calif. 16 pp.

Stone, R. W., W. J. Kaufman, and A. J. Horne. 1973. Long-term effects of toxicants and biostimulants on the waters of central San Francisco Bay. University of California Sanitary Engineering Research Laboratory Rep. 73-1. 112 pp.

Sumner, F. B., G. D. Louderback, W. L. Schmitt, and E. C. Johnston. 1914. A report upon the physical conditions in San Francisco Bay, based upon the operations of the United States Fisheries steamer "Albatross" during the years 1912 and 1913. Univ. Calif. Publ. Zool. 14:1-198.

Tattersall, W. M. 1932. Contributions to a knowledge of the Mysidacea of California. II. The Mysidacea collected during the survey of San Francisco Bay by the U.S.S. "Albatross" in 1914. Univ. Calif. Publ. Zool. 37:315-347.

Wachenfeldt, T. von. 1975. Marine bentic [sic] algae and the environment in the Öresund. 328 pp. (Privately published.)

West, J. A. 1966. The life histories of several marine Bangiophycidae and Florideophycidae (Rhodophycophyta, Rhodophyceae) in laboratory culture. Ph.D. Thesis. University of Washington, Seattle, Wash. 209 pp.

Wynne, M. J. 1969. Life history and systematic studies of some Pacific North American Phaeophyceae (brown algae). Univ. Calif. Publ. Bot. 50. 62 pp.

APPENDIX A

SYSTEMATIC CATALOGUE
of the
BENTHIC ALGAL FLORA
of
SAN FRANCISCO BAY

SILVA: BENTHIC ALGAL FLORA

SYSTEMATIC CATALOGUE

Class RHODOPHYCEAE (Red Algae)

Subclass Bangiophycidae

Goniotrichales

Goniotrichaceae

Goniotrichum alsidii (Zanardini) Howe

Minute irregularly branched filaments to 5 mm long, cells rose-red, loosely arranged in gelatinous sheath, mostly uniseriate; on various algae and animals in lower intertidal zone and subtidal to 12 m; widely distributed (type locality: Italy), British Columbia to Chile.

San Francisco Bay: Camp Reynolds, Angel I. (on *Polysiphonia pacifica*); Yerba Buena I. (on *Pterosiphonia dendroidea*); Berkeley (on *Sargassum muticum*); to be expected throughout the Bay.

Bangiales

Erythropeltidaceae

Erythrotrichia carnea (Dillwyn) J. Agardh

Minute bright-pink unbranched uniseriate filaments to 8 mm long, attached by lobed basal cell or by short rhizoidal outgrowths from basal cell; on various algae and animals in intertidal zone and subtidal to 12 m; widely distributed (type locality: Wales), Alaska to Costa Rica.

San Francisco Bay: Fort Mason and Alcatraz I. (on *Laminaria sinclairii*); Yerba Buena I. (on *Sargassum muticum*); Alameda (on *Cladophora*); to be expected throughout the Bay.

Erythrotrichia porphyroides Gardner

Minute red ribbons of irregular width, to 2 mm long, on blades of *Laminaria sinclairii* in San Francisco Bay (type locality: Fort Point) and on *Halidrys dioica* at Corona del Mar (Orange Co.), Calif.

San Francisco Bay: Fort Point and Alcatraz I.

Smithora naiadum (Anderson) Hollenberg

Clusters of membranous purple blades 1-5 cm long arising from pulvinate perennial bases, obligately epiphytic on surf grass (*Phyllospadix*) and eel grass (*Zostera*); Alaska to Baja Calif. (type locality: Farallon Is.).

San Francisco Bay: Fort Point and Pt. Blunt (Angel I.), on *Phyllospadix*.

Bangiaceae

Bangia fusco-purpurea (Dillwyn) Lyngbye

Dense stands of lubricous brownish-purple filaments, reaching maximum length (6-10 cm) in winter; on cobbles, boulders, cliffs, and wood in upper intertidal zone; widely distributed in both hemispheres (type locality: Wales), Alaska to Costa Rica, Chile.

San Francisco Bay: Various localities throughout the central Bay.

A collection made by A. D. Frye at "Golden Gate," probably at Fort Point or at the Presidio, was used by Harvey (1858) as the basis of a new species, *Bangia vermicularis*, but current taxonomic opinion considers the local populations to be conspecific with *B. fusco-purpurea*.

Porphyra kanakaensis Mumford

Membranous umbilicate fronds to 50 cm diameter, vegetative fronds steel-gray to olive-green in center, purple at edges, reproductive fronds with reddish-brown female areas interrupted by cream-colored male areas; on moderately exposed cobbles in middle and lower intertidal zones; British Columbia to San Francisco Bay (type locality: Kanaka Bay, San Juan I., Wash.).

San Francisco Bay: Fort Point, Lime Point.

This species has recently been segregated from *P. perforata* primarily on the basis of anatomical differences (two chloroplasts per cell; carpospores arranged in two tiers). The two species have similar external appearances and may grow side-by-side (Mumford 1973).

Porphyra lanceolata (Setchell & Hus) G. M. Smith

Dense stands of membranous linear or lanceolate blades to 40 cm (-1 m) long; male blades narrowly linear, spirally twisted, with attenuated apex and continuous cream-colored margin; female blades lanceolate, deeply ruffled at margins, with brownish-red fertile patches; on boulders in upper middle intertidal zone; Ore. to Gaviota (Santa Barbara Co.), Calif. (type locality: Carmel Bay, Calif.).

San Francisco Bay: Fort Point, Pt. Cavallo, Camp Reynolds (Angel I.).

Porphyra perforata J. Agardh

Membranous deeply ruffled orbiculate or broadly lanceolate fronds, 20-30 cm broad, 30-50 cm long, steel-gray to brownish-purple, pale male areas and reddish female areas in patches along margin; on moderately exposed rocks in upper and middle intertidal zones; Alaska to Baja Calif. (type locality: "Golden Gate," probably Fort Point).

San Francisco Bay: Widely distributed throughout the central Bay, but less abundant and luxuriant in the eastern part.

Edible species of *Porphyra* bear various common names, including "laver" and "asakusa-nori."

Subclass Florideophycidae
Nemaliales
Acrochaetiaceae

Acrochaetium amphiroae (Drew) Papenfuss

Tufts of minute creeping and erect filaments growing in the joints of various articulated coralline algae; British Columbia to Baja Calif. (type locality: San Pedro, Calif.), Chile.

San Francisco Bay: Fort Mason (on *Bossiella plumosa*).

Acrochaetium arcuatum (Drew) Tseng

Dense tufts of minute filaments (less than 1 mm tall) on various intertidal algae; Wash. to Monterey Peninsula, Calif. (type locality: Moss Beach, San Mateo Co., Calif.), Gulf of Calif.

San Francisco Bay: Fort Point (on *Spongomorpha mertensii*), Fort Mason (on *Ahnfeltia plicata* and *Laminaria sinclairii*), Alcatraz I. (on *Laminaria sinclairii*), Pt. Cavallo (on *Alaria marginata*).

The Fort Point collection served as the basis of *Rhodochorton densum* Drew (1928:168), a species that has been merged with *A. arcuatum* by Abbott & Hollenberg (1976:310), downgrading the importance of the alleged difference in spore-germination pattern. Culture studies by West (1966) provided evidence supporting this merger.

Acrochaetium coccineum (Drew) Papenfuss

Systems of minute filaments in and on blades and stipes of *Desmarestia*, *Egregia*, and *Laminaria*, the epiphytic portions forming bright-red velvety stripes; Duxbury Reef (Marin Co.) to Little Sur (Monterey Co.), Calif. (type locality: Fort Point, San Francisco).

San Francisco Bay: Fort Point (in and on blades of *Laminaria sinclairii*).

Rhodochorton obscurum Drew, also growing in and on *Laminaria sinclairii*, was recorded from Fort Point by Drew (1928:193). This species has been merged with *A. coccineum* by Abbott & Hollenberg (1976:313). The alleged differences between the two species are indeed minor.

Acrochaetium daviesii (Dillwyn) Nägeli

Bright-pink tufts of filaments 4-6 (-8) mm tall on various algae and animals, lower intertidal and subtidal zones; widely distributed in temperate waters (type locality: Wales), Wash. to Baja Calif.

San Francisco Bay: Fort Point (on *Phyllospadix scouleri*), Alcatraz I. (on *Laminaria sinclairii*).

Acrochaetium pacificum Kylin

Tufts of minute filaments 1-3 mm tall on various algae (especially Laminariales) and animals, lower intertidal and subtidal zones; Alaska to Sonora, Mexico (type locality: Brown I., Wash.)

San Francisco Bay: Fort Point (on air bladder of *Egregia menziesii*), Alcatraz I. (on *Laminaria sinclairii*).

The Fort Point collection served as the basis of *Rhodochorton plumosum* Drew (1928:173), a species that has been merged with *A. pacificum* by Abbott & Hollenberg (1976:315).

Acrochaetium porphyrae (Drew) G. M. Smith

Systems of minute filaments forming bright-red patches 1-5 cm broad in basal portion of blades of *Porphyra*, at times causing distortion; British Columbia to Baja Calif. (type locality: Fort Point, San Francisco).

San Francisco Bay: Fort Point (in *Porphyra perforata*).

Acrochaetium subimmersum (Setchell & Gardner) Papenfuss

Systems of minute filaments forming deep-red patches and at times causing distortion in blades of various foliose red algae, especially *Grateloupia doryphora, Halymenia schizymenioides,* and *Schizymenia pacifica*; Japan, British Columbia to Pt. Arguello (Santa Barbara Co.) and Channel Is., Calif. (type locality: Whidbey I., Wash.).

San Francisco Bay: Various localities in the central Bay, limited by the distribution of the hosts.

Rhodochorton purpureum (Lightfoot) Rosenvinge

Extensive deep-red velvety coating on shaded rocks (especially sandstone and shale) and wood in uppermost intertidal zone; widely distributed in N. Hemisphere (type locality: Scotland), Bering Sea to Baja Calif.

San Francisco Bay: Various localities throughout the central Bay and at Coyote Point.

Helminthocladiaceae

Cumagloia andersonii (Farlow) Setchell & Gardner

Erect axes annual (Feb.-Nov.), arising from discoid base, mostly 15-30 cm long, olive-brown to purplish-red, gelatinous but tough, becoming flattened and even saccate, clothed with short branchlets; on rocks in upper and upper middle intertidal zones; British Columbia to Baja Calif. (type locality: Santa Cruz, Calif.).

San Francisco Bay: Near ranger's office at Fort Point.

This alga is widely but spottily distributed on the outer coast, being restricted to certain rocks where it occurs abundantly year after year.

Gelidiales

Gelidiaceae

Gelidium coulteri Harvey

Dense mats or clumps of very narrow compressed axes mostly 4-8 cm tall bearing numerous distichous branchlets, olive-purple, on rocks or shells of living invertebrates (barnacles, mussels, limpets); British Columbia to Baja Calif. (type locality: Monterey Peninsula, Calif.).

San Francisco Bay: Abundant on most shores of the central Bay and at Coyote Point; an important component of the flora.

Gelidium sinicola Gardner (1927a:278)

This species was described on the basis of tetrasporangial plants collected in a rock pool near high-tide line at Pt. Cavallo. As described this species is not distinctive; the axes were said to be subcylindrical and provided with subdistichous branchlets. A special effort to collect similar material at the type locality has been unsuccessful. In the absence of cystocarpic plants, it is impossible even to ascertain the proper generic placement.

Cryptonemiales

Dumontiaceae

Cryptosiphonia woodii (J. Agardh) J. Agardh

Clumps of blackish-purple cylindrical fronds mostly 5-10 cm tall, the lower portion unbranched, the upper portion bearing numerous short radially arranged branchlets; on rocks in middle intertidal zone and in shallow pools or crevices at higher levels; Alaska to Government Point (Santa Barbara Co.), Calif. (type locality: Vancouver I., Canada).

San Francisco Bay: Abundant on all rocky shores within Zones I-IV (absent from the East Bay).

Dilsea californica (J. Agardh) Kuntze

Clusters or dense stands of brownish-red blades usually deeply dissected into sickle-shaped divisions ca. 10 cm tall; on rocks in lower middle intertidal zone and subtidal to 20 m; Alaska to San Luis Obispo Co., Calif. (type locality: Ore.).

San Francisco Bay: Fort Point and Alcatraz I.; to be expected elsewhere in Zones I-III.

Farlowia mollis (Harvey & Bailey) Farlow & Setchell

Isolated small clusters of blackish-red fronds 10-15 cm tall, alternately branched in one plane, the main branches compressed, to 5 mm wide, slippery, bearing hair-like branchlets which eventually erode away; on rocks in lower middle intertidal zone and subtidal to 20 m, at times associ-

ated with sand; Alaska to San Diego, Calif. (type locality: Puget Sound).
San Francisco Bay: Fort Point.

Pikea californica Harvey

Isolated bushy clusters of dark-red fronds mostly 6-12 cm tall with profuse pinnate branching; on rocks in lower intertidal zone and subtidal to 20 m; British Columbia to Baja Calif. (type locality: "Golden Gate," probably Fort Point).

San Francisco Bay: Fairly abundant on all rocky shores within Zones I-IV (but less common on Angel I. than elsewhere); a characteristic component of the flora.

A form from Fort Point with long, delicate, and regularly pinnate ultimate branchlets has been described as a segregate species, *Pikea pinnata* Setchell (1899).

Peyssonneliaceae

Peyssonnelia meridionalis Hollenberg & Abbott

Thin circular reddish-brown or purple crusts, mostly less than 2 cm diam., on rocks and shells in lower intertidal zone; Ore. to San Luis Obispo Co., Calif. (type locality: Pacific Grove, Monterey Co., Calif.).

San Francisco Bay: Fort Mason and Alcatraz I.; to be expected elsewhere in Zones I-III.

A thick, extensively calcified form of *Peyssonnelia* from Fort Point awaits further study.

Rhodophysema elegans (Crouan frat.) Dixon var. *polystromaticum* (Batters) Dixon

Thin subcircular rose-red crusts on rocks or large algae in lower intertidal zone; widespread in N. Atlantic (type locality: England), Wash. to Corona del Mar (Orange Co.), Calif.

San Francisco Bay: Fort Point (on *Gymnogongrus linearis*).

Hildenbrandiaceae

Hildenbrandia occidentalis Setchell

Dark purplish-red crusts, indefinitely expanded, 1-2 mm thick, with tetrasporangia produced in flask-shaped or deeply cylindrical cavities; on rocks in upper intertidal zone; Alaska to Baja Calif. (type locality: Lands End, San Francisco), Galapagos Is.

San Francisco Bay: Various localities in the western part of the central Bay (Zones I-III).

Hildenbrandia prototypus Nardo

Pale rose-red to bright-red crusts, indefinitely expanded, mostly 0.2-0.4 mm thick, with tetrasporangia produced in shallow depressions; on rocks from middle intertidal zone down to upper subtidal zone; nearly cosmopolitan (type locality: Italy), Alaska to Panama, Galapagos Is.

San Francisco Bay: Golden Gate Fields (Albany), Pt. Isabel (Richmond).

Whether continued sampling will uphold the presently perceived distributional pattern of these two species within the Bay remains to be seen.

Corallinaceae

This family is remarkable in that all of its members are at least partially calcified. They have several widely divergent habits. Some are crusts covering large areas of rock. These crusts may be smooth, knobby, or with simple or branched papillae. Others are crusts on marine plants, forming discrete or confluent patches or discs. Still others are articulated—that is, they have erect branches in which uncalcified joints alternate with calcified segments. Articulated coralline algae grow on rocks, animals, or other algae. The Corallinaceae often form a conspicuous pink or lavender zone beginning at the lowest tide level and extending subtidally to varying depths. Many are restricted to fully exposed situations. Within San Francisco Bay, they are confined to Zones I-III.

Crustose species on rocks

This group has not yet been studied in detail, but the aid of a specialist (Dr. Phillip A. Lebednik) has been enlisted. Preliminary work has established the presence of at least three genera: *Lithothamnium*, *Lithophyllum*, and *Pseudolithophyllum*. Two species have been identified with certainty.

Lithothamnium pacificum (Foslie) Foslie

Rose-red crust with small rough papillae; on pebbles, cobbles, boulders or shells in lower intertidal zone; British Columbia to S. Calif. (type locality: Pebble Beach, Monterey Co., Calif.).

San Francisco Bay: Fort Point, Pt. Cavallo, Alcatraz I.

Pseudolithophyllum neofarlowii (Setchell & Mason) Adey

Crust with surface roughened by subspherical outgrowths to 2 mm diam., these often anastomosing, varying from white in upper intertidal zone to purple in lower intertidal zone; on sides of rocks and floors of tide pools; Alaska to Baja Calif. (type locality: Monterey, Calif.).

San Francisco Bay: Fort Point, Pt. Cavallo, Alcatraz I.

Crustose species on marine plants

Melobesia marginata Setchell & Foslie

Thin purple subcircular patches 5-8 mm diam., becoming confluent, on various red algae in lower intertidal zone; British Columbia to Baja Calif., Costa Rica, Peru (type locality: Bodega Bay, Sonoma Co., Calif.).

San Francisco Bay: Fort Point and Pt. Blunt, Angel I. (on *Gymnogongrus linearis*).

Melobesia mediocris (Foslie) Setchell & Mason

Thin pink to rose-red subcircular patches to 2 mm diam., becoming densely crowded and angular, on leaves of surf grass (*Phyllospadix*) and eel grass (*Zostera*); British Columbia to Baja Calif. (type locality: Santa Cruz, Calif.).

San Francisco Bay: Fort Point and Pt. Blunt, Angel I. (on *Phyllospadix scouleri*).

Articulated species

Bossiella dichotoma (Manza) Silva

Isolated clusters of reddish-purple dichotomously branched fronds to 8 cm tall; on rocks in lower intertidal and upper subtidal zones, often associated with sand; Alaska to Baja Calif. (type locality: Moss Beach, San Mateo Co., Calif.).

San Francisco Bay: Fort Point Promenade (on seaplane ramp and nearby cobbles).

Bossiella dichotoma is treated by Johansen (in Abbott & Hollenberg 1976:414) as a subspecies of *B. orbignyana* (Decaisne) Silva, but in my opinion the two species should be kept separate.

Bossiella plumosa (Manza) Silva

Dense stands of reddish-purple pinnately branched fronds mostly 3-4 cm tall, on rocks and mussels from lower middle intertidal zone to shallow subtidal depths; Alaska to Pt. Conception, Calif. (type locality: Moss Beach, San Mateo Co., Calif.).

San Francisco Bay: Present but not abundant at Fort Point, Pt. Cavallo, Fort Mason, and Alcatraz I.

Corallina vancouveriensis Yendo

Dense stands of violet to purple densely pinnately branched fronds mostly 4-8 cm tall; around edges of tide pools and on rocks and mussels from lower middle intertidal zone to shallow subtidal depths; Alaska to Baja Calif. (type locality: Port Renfrew, Vancouver I.), Galapagos Is.

San Francisco Bay: Present but not abundant at Fort Point, Pt. Cavallo, Fort Mason, and Alcatraz I.

This species and the preceding one are characteristic components of the mussel-bed association on the outer coast.

Endocladiaceae

Endocladia muricata (Endlicher) J. Agardh

Dense stands of bushy clusters of blackish-brown harsh-textured fronds, 4-6 cm tall; on rocks in upper intertidal zone; Alaska (type locality) to Baja Calif.

San Francisco Bay: Present but not abundant at Fort Point, Pt. Cavallo, Alcatraz I., and on the west side of Angel I.

It is of interest to find that this ubiquitous high-growing outer-coast alga, its subjection to long periods of desiccation suggesting hardiness, does not thrive in San Francisco Bay.

Cryptonemiaceae

Cryptonemia obovata J. Agardh

Isolated reddish-brown foliose blades mostly 20-30 cm tall, tending to be obovate and lobed; on rocks, usually at depths of 10-50 m; Alaska to Baja Calif. (type locality: "Golden Gate," probably

SAN FRANCISCO BAY

Fort Point, San Francisco).

San Francisco Bay: Fort Point, Lime Point, Pt. Cavallo (all at extreme low-tide mark).

Grateloupia doryphora (Montagne) Howe (Fig. 4)

Lanceolate blades to 1 (-2) m long, extremely slippery, highly variable in color (yellowish-brown to purple), frequently with marginal proliferations; on rocks and animals in lower intertidal zone; Wash. to Peru (type locality).

San Francisco Bay: Luxuriant plants have been found in abundance at Pt. Cavallo, Paradise Beach, Yerba Buena I., Ayala Cove (Angel I.), and Berkeley, thus covering all five zones, but spottily.

Plants apparently indistinguishable morphologically from those in the Bay grow on the outer coast. In view of the fact that this species has recently been recognized as having been introduced into the native flora of southern England, there is a possibility that the populations within San Francisco Bay have resulted from an introduction from distant waters rather than from a migration from the adjacent outer coast. Historical collections of this species from the Bay, which would rule out the assumption of a recent introduction, do not exist. It would be interesting to determine whether the outer-coast and Bay populations represent distinct physiological races.

Other plants of *Grateloupia* that do not seem to be variants of this polymorphic species have been found at Coyote Point and await further study.

Halymenia schizymenioides Hollenberg & Abbott

Small clusters of lanceolate blades mostly 20-30 cm long, slippery, wine-red to rose-red, often deeply cleft; on rocks in lower intertidal and upper subtidal zones; Wash. to Santa Barbara Co., Calif. (type locality: Mission Point, Monterey Peninsula, Calif.).

San Francisco Bay: Common at most localities within Zones I-IV.

Prionitis filiformis Kylin

Dense clusters of reddish-brown wiry fronds to 30 cm long, dichotomously branched, the segments flattened but only ca. 1 mm wide, bearing very short marginal proliferations; on rocks covered with sand in lower intertidal zone; British Columbia, Ore. to Gaviota (Santa Barbara Co.), Calif. (type locality: "San Francisco," probably Fort Point).

San Francisco Bay: Fort Point (known *in situ* only from collections made in 1895 and 1915, but often cast ashore).

Prionitis lanceolata (Harvey) Harvey

Clusters of tough cartilaginous reddish-brown fronds mostly 20-30 cm long with a basic dichotomous branching pattern often masked by an abundance of well-developed lateral proliferations; on rocks in high tide pools and from middle intertidal zone to depths of 30 m in varying degrees of exposure to surf; Alaska to Baja Calif. (type locality: Monterey Peninsula, Calif.).

San Francisco Bay: This polymorphic species is fairly common in the western part of the central Bay (Zones I-III) and at Paradise Beach.

Prionitis lyallii Harvey

Clusters of reddish-brown fronds mostly 20-30 cm long, the axes tough, the lateral proliferations soft and often slippery; on rocks from middle intertidal zone to upper subtidal zone; British Columbia to Pt. Conception, Calif. (type locality: Esquimalt, Vancouver I.).

San Francisco Bay: Several forms of a *Prionitis* from the western part of the central Bay have been referred tentatively to this poorly defined species.

No available classification of the California representation of *Prionitis* is satisfactory, in my opinion. The polymorphism within the genus and its gradation into its sister genus *Grateloupia* are causes for extreme frustration.

Kallymeniaceae

Callophyllis crenulata Setchell

Deep-red foliose blades mostly to 15 cm tall, crisp, flabellately divided, with undulate or crenulate margins; on rocks or algae in lower intertidal zone to depths of 30 m; British Columbia to Diablo Cove (San Luis Obispo Co.), Calif. (type locality: Whidbey I., Wash.).

San Francisco Bay: Abundant throughout Zones I-III.

Callophyllis pinnata Setchell & Swezy

Deep-red foliose blades mostly less than 35 cm tall, plane, slippery, flabellately divided, lower segments usually with well-developed marginal proliferations; on rocks, animals, and algae (especially *Laminaria sinclairii*) in lower intertidal and upper subtidal zones; Alaska to Baja Calif. (type locality: Duxbury Reef, Marin Co., Calif.).

San Francisco Bay: Abundant throughout Zones I-III.

The two species of *Callophyllis* usually occur together, but are easily distingusihable in the field.

Choreocolacaceae

Choreocolax polysiphoniae Reinsch

Colorless globular mass ca. 1 mm diam., parasitic on *Polysiphonia, Pterosiphonia,* or *Pterochondria*, intertidal to subtidal (-10 m); N. Atlantic (type locality), Alaska disjunctly to Baja Calif.

San Francisco Bay: Pt. Cavallo, Pt. Blunt (Angel I.), Yerba Buena I. (all on *Polysiphonia paniculata*).

Gigartinales

Nemastomataceae

Schizymenia pacifica (Kylin) Kylin

Brownish-red foliose blades mostly to 30 cm tall, ovate to broadly lanceolate, often deeply split, slippery, surface finely granular when partially dry; on rocks exposed to at least moderate surf in lower intertidal zone and subtidal to 18 m; Japan, Alaska to Baja Calif. (type locality: Friday Harbor, Wash.).

San Francisco Bay: Alcatraz I.; to be expected elsewhere in Zones I-III.

Solieriaceae

Agardhiella gaudichaudii (Montagne) Silva et Papenfuss, comb. nov. (*Gigartina gaudichaudii* Montagne 1842:255)

Bright-pink to deep-red cylindrical axes mostly to 30 cm tall, 2-3 mm diam., radially or subdistichously branched; on rocks in lower intertidal zone and subtidal to 30 m; often in sandy pools, Alaska to Baja Calif., Peru (type locality).

San Francisco Bay: Pt. Cavallo (an isolated plant).

This species is listed in Abbott & Hollenberg (1976:483) as *Neoagardhiella baileyi* (Kützing) Wynne & Taylor. Reasons for the change of name are given in DeCew & Silva (in press).

Plocamiaceae

Plocamium cartilagineum (Linnaeus) Dixon subsp. *pacificum* (Kylin) Silva, comb. nov. (*P. pacificum* Kylin 1925:42)

Flat fern-like rose-red fronds mostly 10-20 cm tall arising from stoloniferous base; on rocks in lower intertidal zone, often associated with *Phyllospadix* and with sand, subtidal to 40 m; British Columbia to Mexico (type locality: San Juan I., Wash.), Galapagos Is.

San Francisco Bay: Fort Point, Pt. Cavallo.

I agree with Dawson (1961) in giving taxonomic recognition at the infraspecific level to the Pacific coast populations of this species because of their general coarseness, despite the occurrence (especially in warmer waters) of finely branched forms apparently indistinguishable from European populations.

Plocamium violaceum Farlow

Dense clumps of reddish-violet flat fronds mostly 4-5 cm tall, with congested branching, arising from stoloniferous base; on rocks and mussels in middle intertidal zone exposed to heavy surf; British Columbia to Baja Calif. (type locality: Santa Cruz, Calif.).

San Francisco Bay: This alga is a member of the mussel-bed association on the outer coast. Within the Bay it is restricted to those localities in the western part that receive some surf: Fort Point, Pt. Cavallo, Fort Mason, Alcatraz I., and the west side of Angel I.

Gracilariaceae

Gracilaria sjoestedtii Kylin (Fig. 8)

Stringy plants to 1 (-2) m long, yellowish-brown to reddish-purple, irregularly branched mostly in lower portion, branches cylindrical, 0.5-1.5 mm diam. (to 3.5 mm in male plants); on rocks covered with sand in lower intertidal zone and subtidal to 15 m on outer coast, on rocks and

shells along muddy shores of bays; British Columbia to Costa Rica (type locality: Pacific Grove, Calif.).
San Francisco Bay: Throughout the central Bay.

Gracilaria verrucosa (Hudson) Papenfuss
Stringy plants to 60 cm long, greenish-purple to dark-purple, profusely branched, branches cylindrical, 0.5-1.5 mm diam. (to 3.5 mm in male plants), often with abundant spine-like proliferations; on rocks and shells in lower intertidal and upper subtidal zones in sheltered waters; worldwide distribution (type locality: England), Alaska to Baja Calif.
San Francisco Bay: Uncertain, but probably throughout the central Bay.
Initially I assumed that the stringy *Gracilaria* common throughout the central Bay was *G. verrucosa*, a species that has been repeatedly recorded for harbors in many parts of the world, in distinction to the outer-coast *G. sjoestedtii*, which seemed to be restricted to the sanded-in habitat and with fronds reportedly less branched and more robust. The intermixing within the Bay of stringy fronds of varying diameter suggested that my assumption was incorrect. Within the past year the problem was approached by a specialist in the genus, Dr. Mitchell D. Hoyle. Preliminary investigations of presumably diagnostic anatomical and reproductive characters have revealed the presence of *G. sjoestedtii* throughout the central Bay, the muddy habitat notwithstanding. While *G. verrucosa* has been shown to occur with *G. sjoestedtii* in at least two localities, its distribution within the Bay is as yet only vaguely outlined.

Gracilariophila oryzoides Setchell & Wilson
Colorless globular tubercles 1-2 mm diam. parasitic on lower portions of fronds of *Gracilaria sjoestedtii* and *G. verrucosa*; Wash. to Baja Calif. (type locality: Fort Point, San Francisco).
San Francisco Bay: Following the hosts through most of the central Bay.

Phyllophoraceae

Members of this family have tough, often rigid fronds. The four genera represented in San Francisco Bay are strongly associated with the sanded-in habitat.

Ahnfeltia plicata (Hudson) Fries
Bushy clumps of profusely and mostly dichotomously branched wiry fronds 5-10 cm tall, the branches reddish-purple to purplish-black, 0.25-0.5 mm diam., intertangled; on rocks partially covered with sand in lower intertidal and upper subtidal zones; colder waters of N. Hemisphere (type locality: England), Bering Sea to Baja Calif.
San Francisco Bay: Fort Point, Pt. Cavallo, and Fort Mason; to be expected on the southwest side of Angel I.

Ahnfeltiopsis pacifica Silva & DeCew
Clumps of wiry fronds mostly 10-20 cm tall, deep-red to purplish-black, 10-15 times dichotomous, the branches to 1.5 mm diam.; on rocks partially covered with sand in lower intertidal and upper subtidal zones; British Columbia to Baja Calif. (type locality: Dillon Beach, Marin Co., Calif.).
San Francisco Bay: Pt. Blunt (Angel I.); to be expected at Fort Point and Pt. Cavallo.
This species is listed in Abbott & Hollenberg (1976:503) as *Ahnfeltia gigartinoides* J. Agardh (type locality: Oaxaca, Mexico). Reasons for establishing the genus *Ahnfeltiopsis* and for recognizing *A. pacifica* as a species distinct from *Ahnfeltia gigartinoides* are given in DeCew & Silva (in press).

Gymnogongrus chiton (Howe) Silva et DeCew, comb. nov. (*Actinococcus chiton* Howe 1914:115 adnot.) (Fig. 3)
Small clusters of rigid dull-red fronds mostly 8-15 cm tall, dichotomously branched, the branches divaricate, thin, 3-6 mm broad, pendant when exposed at low tide; on rock faces in lower intertidal zone and subtidal to 18 m; British Columbia to Baja Calif. (type locality: Duxbury Reef, Marin Co., Calif.).
San Francisco Bay: Abundant on all rocky shores within Zones I-IV.
This species is listed in Abbott & Hollenberg (1976:508) as *G. platyphyllus* Gardner. Reasons for the change of name are given in DeCew & Silva (in press). It is surprising that this species, which is found only occasionally on the outer coast, is common within the Bay. Unlike *G. leptophyllus* and *G. linearis*, it is not associated with sand, and the lack of this special requirement

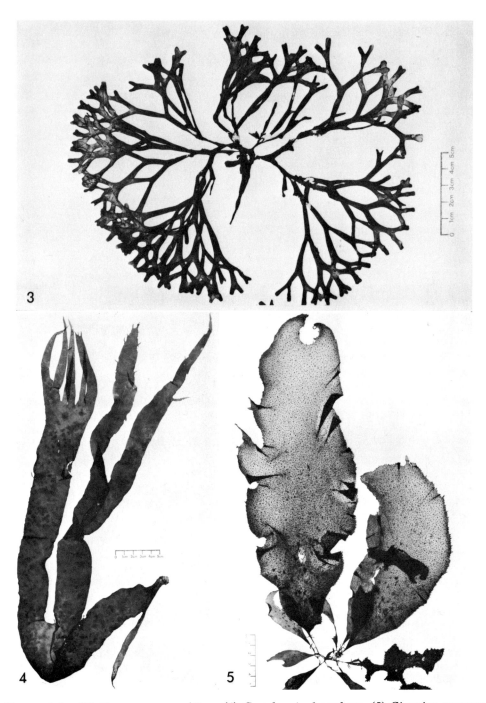

Figures 3-5. (3) *Gymnogongrus chiton.* (4) *Grateloupia doryphora.* (5) *Gigartina exasperata.*

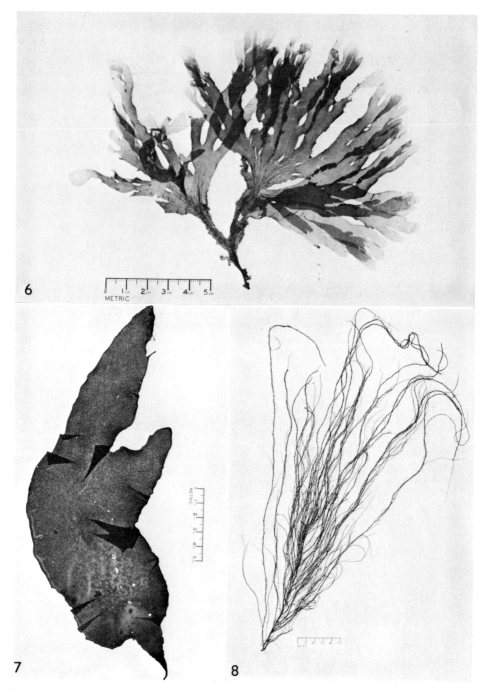

Figures 6-8. (6) *Cryptopleura violacea.* (7) *Iridaea splendens.* (8) *Gracilaria sjoestedtii.*

explains its broader distribution within the Bay.

Gymnogongrus leptophyllus J. Agardh

Dense hemispherical clumps of fronds ca. 6 cm tall, the branching mainly dichotomous but lateral proliferations common, the branches dull-red, firm but not rigid, flattened, varying in width in different populations from 0.75 to 2.5 mm; on sand-covered rocks in lower intertidal zone and subtidal to 30 m; Alaska to Baja Calif. (type locality: Santa Cruz, Calif.).

San Francisco Bay: Penetrating as far as Zone IV, but restricted to sandy areas (Fort Point, Pt. Cavallo, Camp Reynolds and Pt. Blunt (Angel I.), Paradise Beach).

The narrow form prevails in the Bay, offering an opportunity for confusion with *Ahnfeltia plicata*, which, however, has cylindrical wiry fronds.

Gymnogongrus linearis (C. Agardh) J. Agardh

Dense clumps of rigid brownish-purple fronds mostly 15-25 cm tall, dichotomously branched, the branches flattened, 5-8 mm broad, remaining erect when exposed at low tide; on rocks in lower littoral, near or partially covered by sand; British Columbia to Pt. Conception, Calif. (type locality: Trinidad, Humboldt Co., Calif.).

San Francisco Bay: Fort Point and southwest side of Angel I. (frequently bearing the lavender patches of the crustose coralline *Melobesia marginata*).

This common outer-coast alga is a conspicuous and reliable indicator of the sanded-in habitat.

Stenogramma interrupta (C. Agardh) Montagne

Clusters of papery deep-red fronds mostly 10-20 cm tall, dichotomously branched, the branches 5-10 mm broad, often with small lateral proliferations; female plants with fertile area resembling an interrupted midrib, tetrasporangial plants with a mottled surface; on rocks partially covered by sand in lower intertidal zone and subtidal to 30 m; Atlantic Europe and N. Africa (type locality: Cádiz, Spain), Japan, New Zealand, Alaska to Baja Calif., Galapagos Is.

San Francisco Bay: Fort Point to Aquatic Park (San Francisco), Alcatraz I., Pt. Blunt (Angel I.).

The genus *Stenogramma* was based on *S. californica* Harvey (1840), a species described from a collection made by Sinclair at "San Francisco" (probably Fort Point). Although the consensus of present-day workers is to treat that species as a later synonym of *S. interrupta*, no critical comparative study has been made to prove the point.

An abundance of luxuriant plants is often cast ashore in San Francisco, indicating the existence of extensive subtidal stands. The Alcatraz I. plants were exceptional in not being associated with sand.

Gigartinaceae

Gigartina agardhii Setchell & Gardner

Dense clumps of rigid reddish-brown fronds ca. 10 cm tall, 3-5 times dichotomously branched, the branches 2-5 mm wide, furrowed, contorted, papillate; on rocks in upper and middle intertidal zones; British Columbia to Pt. Conception, Calif. (type locality: Pyramid Point, Monterey Peninsula, Calif.).

San Francisco Bay: Restricted to the western part (Zones I-III).

This species has the appearance of a highly dissected *G. papillata*, to which it is closely related. Unpublished work by Dr. John A. West of the University of California at Berkeley shows that it has both sexually reproductive and apomictic populations (as does *G. papillata*). The two species are not interfertile. *Gigartina agardhii*, unlike *G. papillata*, has not been shown definitely to include a Petrocelis stage in its life history.

Gigartina canaliculata Harvey

Extensive stands of soft olive-purple fronds mostly 5-10 cm tall with entangled bases, branching ultimately distichous, the branches cylindrical or somewhat compressed, 2-3 mm wide; on rocks in lower middle intertidal zone; Ore. to Baja Calif. (type locality: "San Francisco," but possibly Monterey Peninsula.).

San Francisco Bay: Widespread but not abundant in Zones I-III.

The type locality was stated to be San Francisco and the collector David Douglas (who visited there in 1830), but the type specimen is labeled merely "California." Only one *in situ* collection from Fort Point exists, made in 1906, and the species has not been detected either at Lands End or at Pt. Lobos, San Francisco, suggesting that the type came from the Monterey Peninsula, the usual provenance of Douglas's California algae.

SAN FRANCISCO BAY

Gigartina exasperata Harvey & Bailey (Fig. 5)
Small clumps of soft deep-red foliose fronds mostly 20-40 cm (-1 m) tall, broadly lanceolate, often deeply incised, usually with an abundance of lateral proliferations, conspicuously papillate; on rocks and shells in lower intertidal zone and subtidal to 18 m in protected waters; Alaska to Elkhorn Slough (Monterey Co.), Calif. (type locality: opposite Fort Nisqually, Wash.).
San Francisco Bay: Extremely common everywhere in the central Bay; forming extensive stands on shells in the area between Alameda and San Mateo Co., the shells (each with one plant) being transported onto the shore at Alameda.
Of all the conspicuous components of the San Francisco Bay flora, this alone (among the presumed native species) stands apart from anything on the outer coast. Gardner described it as a distinct species in an unpublished manuscript. My personal experience with similar populations has been limited to Tomales Bay, so that it is with much uncertainty that I refer the Bay material to *G. exasperata*. My concept of the species excludes the thick reddish-brown fronds of the outer-coast *G. californica* J. Agardh, which was merged by Abbott & Hollenberg (1976:521) into *G. exasperata*.

Gigartina ornithorhynchos J. Agardh
Small clumps of heavy coarse greenish-purple fronds mostly 20-30 cm tall, cylindrical below, flattened above, once or twice dichotomously branched, the branches bearing marginal proliferous bladelets which in turn may be proliferous, conspicuously papillate; on rocks exposed to heavy surf in lower intertidal zone; San Francisco Bay, Monterey to Baja Calif. (type locality: "California," probably Monterey Peninsula, Calif.).
San Francisco Bay: Fort Mason.
It is with some uneasiness that I report a single collection of this species from the Bay in view of the fact that it has not been reported on the outer coast north of Monterey. In the field it was distinctive in both color and form from the ubiquitous *G. exasperata*.
Gigartina ornithorhynchos J. Agardh (1849) is the oldest available name for the complex previously known as *G. spinosa* (Kützing) Harvey. The latter name, dating from 1853, is untenable because of the prior existence of *G. spinosa* (Linnaeus) Greville, a binomial dating from 1830 in the synonymy of *Eucheuma spinosa* (Linnaeus) J. Agardh. The type specimen of *G. ornithorhynchos* was collected by David Douglas in 1833, presumably on the Monterey Peninsula.

Gigartina papillata (C. Agardh) J. Agardh
Clumps of reddish-brown or blackish-brown fronds mostly 10-15 cm tall, dichotomously branched, the branches narrow or foliose, papillate (male plants yellowish and smooth); on rocks, forming a nearly continuous belt in upper and middle intertidal zones; Bering Sea to Gaviota (Santa Barbara Co.), Calif., Baja Calif. (type locality: uncertain).
San Francisco Bay: Abundant throughout the central Bay.
This extremely polymorphic taxon is a complex of forms, some sufficiently distinct to have been recognized as species. Polanshek & West (1977) have shown that certain populations comprise sexual plants that alternate with a tetrasporangial crust previously considered an independent alga, *Petrocelis*. Other populations have an apomictic life history in which the erect fronds are recycled, eliminating the Petrocelis stage. Recent unpublished studies show that individual plants may possess the option of reproducing either sexually or apomictically, but the extent of this option has not been established. Recognizing the existence of these asexually reproducing populations helps explain the polymorphism of this complex. Various forms are abundantly represented throughout the central Bay. The Petrocelis stage grows only in the western part (Zones I-III), so that all East Bay populations of *G. papillata* would appear to be apomictic.
The Petrocelis stage is a reddish-black gummy crust that expands indefinitely over rocks in the upper and middle intertidal zones, in proximity to the erect fronds. The central California crust was described by Setchell & Gardner (in Gardner 1917) as *Petrocelis franciscana* on the basis of a specimen from Fort Point noted as being 2.25 mm thick. While outer-coast plants are usually no thicker than 2.5 mm, some patches at Fort Point are up to 5 mm thick. After studying populations of the *Gigartina papillata* complex from Alaska, Polanshek & West (1975) concluded that the Petrocelis stage of that area, *P. middendorfii* (Ruprecht) Kjellman, described in 1850 from the Okhotsk Sea, showed no significant differences from *P. franciscana*.
On the basis of two specimens of *G. papillata* collected at "Golden Gate" (almost certainly Fort Point) by Sven Berggren in 1875, J. Agardh (1899) described a new species, *G. obovata*. On the type sheet, however, Agardh wrote *G. papillata* var. *obovata*. This form, which does not deserve

taxonomic recognition, is common on the outer coast as well as in San Francisco Bay.

Gigartina volans (C. Agardh) J. Agardh
Loose clumps of purplish-brown fronds mostly 15-25 cm tall, once or twice dichotomously branched, the branches foliose above a flattened stipe; tetrasporangial blades plane, male blades with superficial bladelets, female blades with marginal papillae; on rocks covered by sand in lower intertidal zone and subtidal to 10 m; Ore. to Baja Calif. (type locality: uncertain).
San Francisco Bay: Fort Point and Lime Point.

Iridaea heterocarpa Postels & Ruprecht
Isolated small clumps of thin reddish-brown fronds mostly 10-15 cm tall, often broader than high with divaricate dichotomous lobes, cystocarps unusually large (-4 mm diam.); on rocks in middle intertidal zone; Siberia, Alaska to Ventura, Calif. (type locality: N. Pacific).
San Francisco Bay: Fort Point, Lime Point, Pt. Cavallo, Tiburon, Pt. Blunt (Angel I.).

Iridaea splendens (Setchell & Gardner) Papenfuss (Fig. 7)
Small clumps of rubbery purple or brownish-purple iridescent blades mostly 20-40 (-120) cm long, usually entire but at times longitudinally incised; on rocks from lower middle intertidal zone to depths of 7 m, often forming belt; British Columbia to Baja Calif. (type locality: Carmel, Calif.).
San Francisco Bay: Abundant at nearly all localities within Zones I-IV and at Pt. Richmond in Zone V.
Certain plants from the western part of the Bay can be assigned to *I. heterocarpa* with a reasonable degree of certainty. The remaining plants in the Bay, variable in form but usually lanceolate and drying uniform purple, possibly represent more than one species, but most seem to belong to the same taxon. On the basis of such plants, but with a very brief description and without making comparisons with other species, Setchell & Gardner (1937) described *I. sinicola* (as *Iridophycus sinicola*; type locality: Pt. Cavallo). Abbott (1972) included *I. sinicola* in the morphological spectrum of *I. splendens*, which she treated as a variety of *I. cordata* (Turner) Bory. It has been shown, however, that the type of *I. cordata* was probably collected at Isla de los Estados, Argentina, rather than at Banks Island, British Columbia, as traditionally thought (Leister 1977). Morphological studies of South American material of *I. cordata* indicate that *I. splendens* can no longer be associated with that species. *Iridaea lilacina* Postels & Ruprecht (1840) is probably referable to the species from the Northwest previously (but incorrectly) known as *I. cordata*, but re-examination of the type specimen will be necessary to establish the relationship between that taxon and the San Francisco Bay forms.

Rhodymeniales

Rhodymeniaceae

Rhodymenia californica Kylin
Small clusters of papery but rigid bright-red fronds, mostly 2.5-6 cm tall, arising from stoloniferous base, blades 2-5 times dichotomously divided, segments linear, 1.5-3 mm broad; on shaded rocks in lower intertidal zone and subtidal to 8 m, often forming extensive stands; British Columbia to Nayarit, Mexico (type locality: Pacific Grove, Calif.), Galapagos Is.
San Francisco Bay: Pt. Cavallo (isolated plant).

Ceramiales

Ceramiaceae

Members of this family are mostly finely branched filamentous plants. The branches are composed of a single series of cells which may, however, become partially or completely covered by an ensheathing layer of cells called a cortex.

Antithamnion kylinii Gardner
Small red tufts of oppositely branched filaments to 2 cm tall; on shells, wood, animals, or algae in lower intertidal zone and subtidal to 20 m; Alaska to Baja Calif. (type locality: Victoria, Vancouver I.).
San Francisco Bay: Various localities within Zones I-IV.

Antithamnionella glandulifera (Kylin) Wollaston
Small silky rose-red tufts of mostly oppositely branched filaments to 5 cm tall, the branches

simple; on stones, shells, debris, or algae in lower intertidal and upper subtidal zones; British Columbia to Baja Calif. (type locality: Friday Harbor, Wash.).

San Francisco Bay: Throughout the central Bay; abundant at certain places.

The incidence of gland cells seems to be significantly lower in Bay populations than in those on the outer coast.

Callithamnion pikeanum Harvey

Brownish-purple woolly rope-like fronds mostly 10-20 cm long, on sides of rocks exposed to at least moderate surf in upper and middle intertidal zones; Alaska to Pt. Dume (Los Angeles Co.), Calif. (type locality: "Golden Gate," probably Fort Point).

San Francisco Bay: On almost all rocky shores in Zones I-IV.

This species is a member of the mussel-bed association on the outer coast.

Ceramium eatonianum (Farlow) De Toni

Loose tufts of purplish-brown or blackish dichotomously branched filaments, the branches spreading, completely corticated but appearing banded; on rocks and algae, especially in high tide pools; British Columbia to Baja Calif. (type locality: Oregon).

San Francisco Bay: Throughout Zones I-III; also at Berkeley (Zone V).

There is a form of *Ceramium* with cortication confined to the nodes that occurs in protected waters throughout the Bay on rocks, floats, and algae. Although it is similar to *C. gardneri* Kylin, a species of exposed coasts originally described from the Monterey Peninsula, it probably represents a widespread species not yet recorded from California.

Microcladia borealis Ruprecht

Clusters of grayish-red densely branched fronds mostly 6-14 cm tall arising from a rhizomatous base, the branches unilateral and in one plane; on rocks, animals, or algae in upper middle intertidal zone exposed to at least moderate surf; Alaska to Pt. Conception, Calif. (type locality: Unalaska I.).

San Francisco Bay: Most rocky shores in Zones I-IV.

Microcladia coulteri Harvey (Fig. 10)

Rose-red densely branched lacy fronds mostly 10-30 cm tall, the branches alternate in one plane; on various foliose algae (especially *Gigartina, Iridaea,* and *Prionitis*) in middle and lower intertidal zones and subtidal to 10 m; British Columbia to Baja Calif. (type locality: "California," probably Monterey Peninsula).

San Francisco Bay: Most localities in Zones I-IV.

Neoptilota californica (Ruprecht ex Harvey) Kylin

Clusters of dark-red densely branched feathery fronds mostly 10-25 cm tall, the branches opposite and in one plane, the ultimate branchlets with smooth or minutely serrate margins; on algae (especially *Laminaria sinclairii*) in lowest intertidal and upper subtidal zones; British Columbia to Pt. Conception, Calif. (type locality: Sonoma, Calif.).

San Francisco Bay: Fort Point.

Platythamnion villosum Kylin

Small feathery rose-red tufts of verticillately branched filaments to 6 cm tall; on rocks, shells, or wood in lower intertidal zone and subtidal to 10 m; Alaska to Baja Calif. (type locality: Friday Harbor, Wash.).

San Francisco Bay: Common throughout the central Bay.

Ptilota filicina J. Agardh

Clusters of dark-red densely branched feathery fronds mostly 10-20 cm tall, the branches opposite and in one plane, the ultimate branchlets with sharply serrate margins; on rocks in the lower intertidal and upper subtidal zones; Bering Sea to Baja Calif. (type locality: Vancouver I.).

San Francisco Bay: Fort Point; west side of Angel I.

Tiffaniella snyderae (Farlow) Abbott

Turf of grayish-red sparsely branched stiff filaments 2-5 cm tall; on rocks partially covered by sand, lower intertidal zone and subtidal to 20 m; British Columbia to Baja Calif. (type locality: Pacific Beach, San Diego Co., Calif.), Gulf of Calif., Chile.

San Francisco Bay: Various localities in Zones I-IV.

Delesseriaceae

This family merits comment because of its size, widespread distribution, beauty, and importance to the San Francisco Bay flora. It comprises about 75 genera, with an exceptionally large number represented in the N. Pacific. Twenty genera are known from central California, many infrequently encountered, of which seven grow in San Francisco Bay. Four of these seven genera (*Asterocolax, Hymenena, Nienburgia,* and *Phycodrys*) are restricted to Zone I, not occurring beyond Pt. Cavallo or Fort Point. One genus (*Psammophyllum*) penetrates as far as Paradise Beach on the east side of the Tiburon Peninsula while two (*Cryptopleura* and *Polyneura*) are abundant on all shores of the central Bay. Delesseriaceae are usually easily recognized by their membranous leafy fronds, often with midribs or veins (or both). Some are large and strikingly beautiful. Most are confined to the subtidal or the lowest intertidal zone.

Asterocolax gardneri (Setchell) J. & G. Feldmann

Clusters of needle-like or flattened but narrow colorless leaflets to 3 mm long, parasitic on lower portions of fronds of other Delesseriaceae (*Nienburgia andersoniana, Phycodrys setchellii, Polyneura latissima,* and *Psammophyllum californicum*) in lowest intertidal zone and subtidal to 50 m; Puget Sound, Duxbury Reef (Marin Co.), Calif. to Baja Calif. (type locality: Pt. Cavallo, San Francisco Bay).

This alga is an adelphoparasite—an organism of which the vegetative part is greatly reduced but the reproductive details closely resemble those of the host. Adelphoparasites are believed to have evolved from the ancestral stock of their hosts or possibly from the hosts as we know them at present. The four known hosts of *Asterocolax gardneri* are closely related, so that a single derivation from an ancestral stock seems more probable than multiple origins. Although it has not previously been recorded on *Psammophyllum*, I have collected specimens on that host at Santa Cruz, Carpinteria, and in the Channel Islands.

Cryptopleura violacea (J. Agardh) Kylin (Fig. 6)

Membranous fronds mostly 12-20 cm tall, violet but often with a greenish-brown cast, dissected flabellately or subdichotomously into ribbon-like segments with undulate margins frequently bearing rows of proliferous bladelets; on rocks, mussels, or algae in lower intertidal and upper subtidal zones; British Columbia to Baja Calif. (type locality: "Golden Gate," probably Fort Point).

San Francisco Bay: All shores of the central Bay, even areas of industrial pollution; a constant companion of *Polyneura latissima* and *Gigartina exasperata*.

Although this species is common in central California, nowhere along the open coast does it thrive to the extent that it does in San Francisco Bay.

Hymenena flabelligera (J. Agardh) Kylin

Fronds mostly 10-25 cm tall, membranous but fairly rigid, salmon-pink to dull-red, flabellately or subdichotomously divided into linear-lanceolate segments; on rocks (rarely on algae) in lowest intertidal and upper subtidal zones on fully exposed coasts; British Columbia to Pt. Conception, Calif. (type locality: "Golden Gate," probably Fort Point).

San Francisco Bay: Abundant at Lime Point and Pt. Cavallo, but known from Fort Point only from a collection made in 1920.

While many red algae that supposedly have isomorphic reproductive stages show varying degrees of subtle heteromorphism, *Hymenena flabelligera* has three distinct forms. Tetrasporangial plants have conspicuous linear sori in closely spaced parallel or forked longitudinal rows in the upper portion of the blade. Cystocarpic plants are smaller and of a lighter color, with narrower segments, the conspicuous cystocarps scattered near the margins. Spermatangial plants are pale and have broad segments with ruffled margins, the sori lying along the margins.

Another species of *Hymenena* that was described on the basis of a collection labeled simply "Golden Gate" is *H. multiloba* (J. Agardh) Kylin. This seaweed is a member of the *Mytilus californianus* association, forming extensive stands in mussel beds exposed to heavy surf, and is apparently absent from the Bay. The type specimen, if collected *in situ*, seems likely to have come from east of the postal (e.g. Lands End).

The third species of *Hymenena* described on the basis of material from "Golden Gate" is *H. fryeana* (Harvey) Kylin. The two specimens constituting this collection are both sterile, making positive identification difficult if not impossible, but it seems probable that they are referable to *H. multiloba*.

SAN FRANCISCO BAY

Nienburgia andersoniana (J. Agardh) Kylin

Clumps of fronds to 30 cm tall arising from a rhizomatous base, membranous but fairly rigid, rose-red to dull-red, alternately and flabellately branched, the branches in one plane, highly variable in width (1-16 mm), with coarsely dentate margins and a flattened percurrent midrib; on rocks exposed to at least moderate surf in lower intertidal zone and subtidal to 20 m; Alaska to Wash., Tomales Bay (Marin Co.), Calif. to Baja Calif. (type locality: Santa Cruz, Calif.).

San Francisco Bay: Fort Point (abundant), Pt. Cavallo (occasional).

Phycodrys setchellii Skottsberg

Membranous leaf-like fronds arising from a stoloniferous base, mostly 6-13 cm tall, dark-pink to brownish-red, with elliptical to obovate lobes, conspicuous percurrent midribs, and paired veins; on rocks in lower intertidal zone, often at edge of sandy pools and channels, more commonly subtidal to 100 m; Alaska, Ore., Bodega Head (Sonoma Co.), Calif. to Baja Calif. (type locality: Moss Beach, San Mateo Co., Calif.), Gulf of Calif.

San Francisco Bay: Fort Point (abundant).

Polyneura latissima (Harvey) Kylin (Fig. 13)

Membranous pinkish-red fronds mostly 10-20 cm tall, irregularly incised, the blades with a conspicuous network of coarse veins; on rocks in lower intertidal zone and subtidal to 68 m along outer coast, on various substrates from mean low tide level to undetermined depths in San Francisco Bay; Alaska to Baja Calif. (type locality: Esquimalt, Vancouver I.).

San Francico Bay: Abundant throughout the central Bay; on rocks, concrete, wood, sessile animals, or algae.

When well-developed and free of epiphytic diatoms, this alga is among the most beautiful inhabitants of the Bay. While large specimens (30 cm tall and larger) are known from the outer coast (particularly from Puget Sound and the Monterey Peninsula), nowhere does this species thrive to the extent that it does in the Bay. Its broad salinity tolerance is evidenced by its ability to penetrate San Pablo Bay. Along the outer coast I have observed unusually large plants growing near sewage outfalls. Even the industrial pollution of the Richmond area does not seem to affect it adversely. Especially luxuriant plants are often found in the drift at Fort Point. One such plant, collected by Sven Berggren in 1875, was described as a distinct species (*Nitophyllum macroglossum* J. Agardh 1898).

Psammophyllum californicum (J. Agardh) Silva et Moe (*Delesseria californica* J. Agardh 1885:69; *Erythroglossum californicum* (J. Agardh) J. Agardh 1898:176)

Rose-pink linear dentate blades 1.5-3 (-5) cm tall arising from a rhizomatous base; on rocks covered by sand for at least part of the year, especially at the edges of pools and channels or entangled with rhizomes of surf grass (*Phyllospadix*); Dillon Beach (Marin Co.), Calif. to Baja Calif. (type locality: Santa Barbara, Calif.), Chile.

San Francisco Bay: Paradise Beach; also to be expected at Fort Point and at Pt. Blunt (Angel I.).

This species is listed in Abbott & Hollenberg (1976:653) as *Anisocladella pacifica* Kylin. Reasons for establishing the genus *Psammophyllum* and for considering *Delesseria californica* an earlier synonym of *Anisocladella pacifica* are given in DeCew & Silva (in press).

Rhodomelaceae

Odonthalia floccosa (Esper) Falkenberg

Extensive stands of blackish-brown profusely branched fronds mostly 10-30 cm tall, major branches ca. 1 mm diam., alternately distichous with clusters of lateral branchlets; on rocks in lower middle intertidal zone, often with *Rhodomela larix*, and in lower intertidal zone, especially in surge channels; Bering Sea to Pt. Conception, Calif. (type locality: Trinidad, Humboldt Co., Calif.).

San Francisco Bay: Lime Point, Pt. Cavallo, and southwest side of Angel I.

Genus *Polysiphonia*

This genus of filamentous red algae encompasses a bewildering array of forms exhibiting minute differences in anatomy. Hundreds of species have been described, many of which have been accredited (rightly or wrongly) with widespread distributions. The representation in San Francisco Bay has not been worked out completely. One inherent difficulty is the possibility that certain species may have been introduced within modern times by ships from distant ports.

Polysiphonia brodiaei (Dillwyn) Sprengel
Soft reddish-brown tufts mostly 10-20 cm tall, segments with 6-7 pericentral cells, older axes fully corticated; on rocks, concrete, or wood in lowest intertidal and upper subtidal zones of harbors; N. Europe (type locality: Scotland), Puget Sound to Santa Monica, Calif.
San Francisco Bay: San Francisco Yacht Harbor, Sausalito.

Polysiphonia flaccidissima Hollenberg var. *smithii* Hollenberg
Soft reddish-brown tufts 2-5 cm tall, segments with 4 pericentral cells, all branches uncorticated; on rocks, concrete, or wood, in sheltered water, lower intertidal zone and subtidal to 12 m; San Francisco to La Jolla, Calif. (type locality: Newport Harbor, Orange Co., Calif.).
San Francisco Bay: San Francisco Yacht Harbor.

Polysiphonia pacifica Hollenberg var. *pacifica*
Soft deep-red tufts mostly 10-20 cm tall, the erect axes profusely branched in upper portions, mostly naked below, ultimate branchlets indeterminate, segments with 4 pericentral cells, uncorticated; on rocks, concrete, or wood in lower intertidal zone and subtidal to 30 m; Alaska to Baja Calif. (type locality: Santa Cruz, Calif.), Peru.
San Francisco Bay: Fort Point.

Polysiphonia pacifica Hollenberg var. *disticha* Hollenberg
Reddish-brown tufts 3-7 cm tall, ultimate branchlets more or less distichous, determinate, short, relatively rigid; on rocks in upper intertidal zone; British Columbia to San Francisco (type locality: near Cape Flattery, Wash.).
San Francisco Bay: Sea wall at Fort Point.

Polysiphonia paniculata Montagne
Soft brownish-red tufts or carpets 10-20 (-30) cm tall arising from densely matted prostrate branches, erect axes sparsely branched below, densely branched above, segments with 10-12 pericentral cells, uncorticated; on rocks in lower intertidal zone, especially in sandy areas; British Columbia to Baja Calif., Gulf of Calif., Peru (type locality).
San Francisco Bay: Common throughout the central Bay.
Polysiphonia californica Harvey (1853), which is considered to be conspecific with *P. paniculata* by Hollenberg (1944:480), was described on the basis of a collection from "Golden Gate" (probably Fort Point).

Pterosiphonia bipinnata (Postels & Ruprecht) Falkenberg
Reddish-brown tufts 6-25 (-50) cm tall, erect axes cylindrical, uncorticated, bipinnately branched, recurved branchlets common but sometimes absent, segments with 11-18 pericentral cells; on rocks, lower middle intertidal zone to upper subtidal zone; Japan, Siberia (type locality), Alaska to San Pedro, Calif.
San Francisco Bay: Common throughout Zones I-IV.
As conceived by Abbott & Hollenberg (1976:706), this species embraces a broad spectrum of morphological variability. Gardner (1927b:102) established a segregate species *P. robusta* (type locality: Moss Beach, San Mateo Co., Calif.) to accommodate those plants that are exceptionally large and have recurved branchlets near the base of each order of branching. On the shore between Sausalito and Pt. Cavallo he encountered a population of plants which were even more luxuriant than those typical of *P. robusta*. These plants lacked recurved branchlets and served, along with a population from Neah Bay (Wash.), as the basis for a new variety, *P. robusta* var. *inermis* Gardner (loc. cit.) (type locality: Neah Bay).

Pterosiphonia dendroidea (Montagne) Falkenberg
Brownish-red fern-like fronds mostly 2-10 cm tall arising from creeping branches, compressed, pinnately branched; forming extensive stands on rocks in lower intertidal zone and subtidal to 10 m, especially in sandy areas; Alaska to Baja Calif., Pacific S. America (type locality: Peru).
San Francisco Bay: Common throughout the central Bay (Zones I-V).

Pterosiphonia pennata (C. Agardh) Falkenberg
Several reddish laxly erect slightly compressed axes to 2 cm tall arising from a rhizomatous base, once-pinnate or partly bipinnate, the branches cylindrical, slightly incurved; on rocks in lower intertidal zone; Europe (type locality: Mediterranean), Japan, Oakland (Calif.) to Baja Calif., Gulf of Calif.
San Francisco Bay: Abbott & Hollenberg (1976:711) list Oakland as the northern limit for this

species in California, but this record is not borne out by collections available to me. Previously, the northern limit in California had been cited as Ventura (Dawson 1963:428).

Rhodomela larix (Turner) C. Agardh

Clusters of firm brownish-black axes mostly 10-20 cm tall closely beset with spirally arranged simple or compound needle-like branchlets 5-10 mm long; forming dense stands on rocks and mussels in middle intertidal zone; Bering Sea to Goleta (Santa Barbara Co.), Calif. (type locality: Nootka Sound, Vancouver I.).

San Francisco Bay: Fort Point, Pt. Cavallo, and the southwest side of Angel I.

Class PHAEOPHYCEAE (Brown Algae)

Ectocarpales

Ectocarpaceae

Ectocarpus acutus Setchell & Gardner

Feathery tufts to 7 cm long on *Desmarestia*; British Columbia to San Pedro, Calif. (type locality: Carmel, Calif.).

San Francisco Bay: Zones I-III (Pt. Cavallo, Fort Mason, southwest side of Angel I.).

Ectocarpus dimorphus Silva

Tufts of sparingly branched filaments 5-20 mm long arising from prostrate base epiphytic on Laminariales; British Columbia to Baja Calif. (type locality: Pacific Grove, Calif.), Gulf of Calif.

San Francisco Bay: Fort Point (on *Egregia menziesii* and *Laminaria sinclairii*); undoubtedly at other localities in Zones I-III.

This species is listed in Abbott & Hollenberg (1976:126) as a synonym of *E. parvus* (Saunders) Hollenberg (type locality: San Pedro, Calif.), but the two entities should probably be kept separate since the latter was originally described as growing on rocks.

Ectocarpus siliculosus (Dillwyn) Lyngbye

Feathery tufts to 30 cm long on various aquatic plants or wood, or floating in quiet waters; widely distributed in N. Atlantic (type locality: England), Coos Bay, Ore. and San Francisco Bay.

San Francisco Bay: Sausalito, Richmond (published as f. *subulatus* by Setchell & Gardner 1925:411), Alameda.

A collection from logs at Fort Point, made by Gardner in April, 1916, was reported with doubt as *E. confervoides* (Roth) Le Jolis f. *typicus* by Setchell & Gardner (1925:414). Similar material has recently been found elsewhere in the Bay and awaits further study.

Giffordia sandriana (Zanardini) Hamel

Silky tufts of profusely branched filaments to 20 cm long on aquatic plants, shells, stones, or glass, especially in quiet waters; widely distributed in N. Atlantic (type locality: Yugoslavia), Puget Sound to San Diego, Calif.

San Francisco Bay: Lying on shore exposed at low tide at Candlestick Point (San Francisco) and Coyote Point (San Mateo Co.).

Pilayella littoralis (Linnaeus) Kjellman

Tufts of profusely branched filaments 2-5 (-60) cm long on wood, stones, or other algae; widely distributed (type locality: N. Europe), Bering Sea to San Pedro, Calif., Chile.

San Francisco Bay: Fort Mason (on *Melanosiphon intestinalis* f. *tenuis*).

Streblonema pacificum Saunders

Circular patches 2-4 mm diam. on sporophylls of *Alaria*; Yakutat Bay, Alaska (type locality), Patrick's Point (Humboldt Co.), and San Francisco Bay.

San Francisco Bay: Fort Point (on *Alaria marginata*).

Ralfsiales

Ralfsiaceae

Analipus japonicus (Harvey) Wynne

Extensive thick perennial crusts with annual erect shoots to 35 cm long on rocks exposed to at least moderate surf in middle intertidal zone; Japan (type locality), Bering Sea to Pt. Conception, Calif.

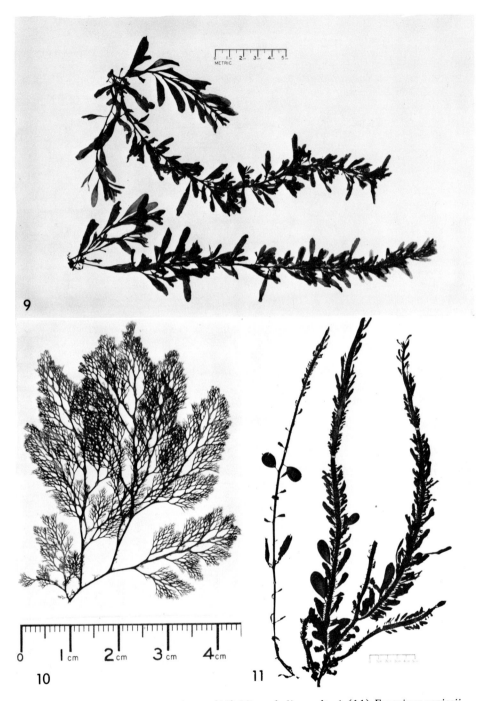

Figures 9-11. (9) *Sargassum muticum.* (10) *Microcladia coulteri.* (11) *Egregia menziesii.*

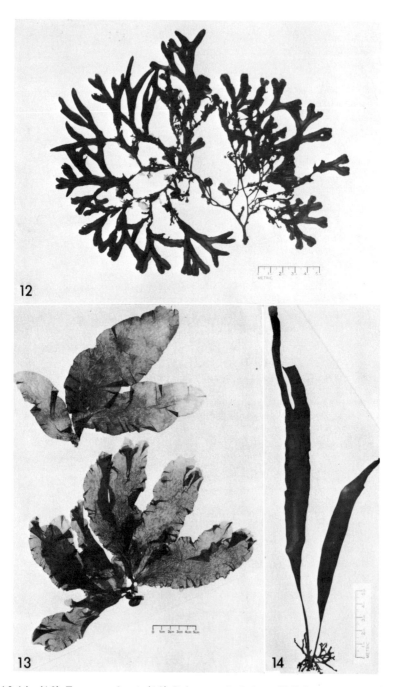

Figures 12-14. (12) *Fucus gardneri.* (13) *Polyneura latissima.* (14) *Laminaria sinclairii.*

San Francisco Bay: The crustose base may be detected throughout the year at Pt. Cavallo, Fort Point, Fort Mason, and Alcatraz I., but even in summer only a few depauperate shoots have been found.

Petroderma maculiforme (Wollny) Kuckuck

Thin brown crusts on stones in lower intertidal zone and subtidal to ca. 20 m; widely distributed in N. Atlantic (type locality: Helgoland), Puget Sound, Humboldt Bay, Duxbury Reef (Marin Co.) to Monterey, Calif.

San Francisco Bay: Golden Gate Fields (Albany) and Pt. Isabel (Richmond), on sheet of polyethylene plastic embedded in mud.

This species has only recently been detected on the Pacific Coast. It has been found to be the algal symbiont of a marine lichen (*Verrucaria*) growing intertidally at Moss Beach (San Mateo Co.) (Wynne 1969:9). *Verrucaria* grows at several localities within San Francisco Bay (e.g. Fort Mason, Tiburon), but the algal symbiont of these populations has not yet been studied.

Ralfsia fungiformis (Gunnerus) Setchell & Gardner

Leathery dark-brown crusts 2-6 cm broad, loosely attached, with free overlapping lobes, on rocks in upper and middle intertidal zones; widely distributed in N. Atlantic (type locality: Norway), Japan, Bering Sea to San Francisco Bay.

San Francisco Bay: Pt. Cavallo (extending the known range southward from Trinidad, Humboldt Co., Calif.).

Ralfsia pacifica Hollenberg

Firmly attached thin brown crusts, common on intertidal rocks; Alaska to Baja Calif. (type locality: Corona del Mar, Orange Co., Calif.), Gulf of Calif., Chile.

San Francisco Bay: Zones I-III (Fort Point, Fort Mason, Alcatraz I.).

Chordariales

Myrionemataceae

Myrionema corunnae Sauvageau f. *uniforme* Setchell & Gardner

Minute circular patches (ca. 1 mm diam.) on blades of various Laminariales; San Francisco Bay (type locality: Fort Point) and Moss Beach (San Mateo Co.), Calif.

San Francisco Bay: Fort Point (on *Alaria marginata*), Pt. Cavallo (on tip of blade of juvenile *Egregia menziesii*).

Other forms of this widespread N. European species have been reported from Humboldt Co., Monterey Co., and Pt. Lobos, San Francisco.

Corynophlaeaceae

Cylindrocarpus rugosus Okamura

Dark-brown, spongy, convoluted roundish crusts on rocks in upper middle intertidal zone; Japan (type locality), MacKerricher Beach State Park (Mendocino Co., Calif.) to Baja Calif.

San Francisco Bay: Fort Mason.

Dictyosiphonales

Punctariaceae

Melanosiphon intestinalis (Saunders) Wynne f. *tenuis* (Setchell & Gardner) Wynne

Tufts of dark reddish-brown unbranched cylindrical shoots to 2.5. cm tall on rocks in uppermost intertidal zone; Coos Bay, Ore., Duxbury Reef (Marin Co., Calif), San Francisco (type locality: Fort Point).

San Francisco Bay: Fort Point, Fort Mason.

This alga closely resembles *Scytosiphon dotyi* and even occurs in the same habitat. It can be distinguished with certainty only by anatomical examination. The nominal form of this species (f. *intestinalis*), in which the shoots are much larger and usually twisted, is widely distributed in northern waters (type locality: Popof I., Alaska).

Phaeostrophion irregulare Setchell & Gardner

Dense stands of blackish-brown papery blades, wedge-shaped, irregularly notched, mostly 6-12 cm tall, arising from an indefinitely expanded coalescence of crusts on rocks in lower middle and

lower intertidal zones, usually associated with sand; Yakutat Bay, Alaska to Pt. Conception, Calif. (type locality: Bolinas, Marin Co., Calif.).
San Francisco Bay: Fort Point, Fort Mason, Pt. Blunt (Angel I.).

Scytosiphonales

Scytosiphonaceae

Petalonia fascia (O. F. Müller) Kuntze
Isolated clusters of greenish-brown linear-lanceolate blades to 35 cm tall, arising from discoid base, on rocks in middle intertidal zone; widely distributed in N. Atlantic (type locality: Norway), Japan, Alaska to Baja Calif., Chile.
San Francisco Bay: Camp Reynolds (west side of Angel I.), Tiburon.

Scytosiphon dotyi Wynne
Tufts of yellowish-brown unbranched cylindrical shoots to 12 cm tall on vertical faces of boulders and cliffs in uppermost intertidal zone; Coos Bay, Ore. to Pt. Conception, Calif. (type locality: Pillar Point, San Mateo Co., Calif.), Channel Is., Islas Coronados (Baja Calif.).
San Francisco Bay: Fort Point.
Another, much larger species of *Scytosiphon, S. lomentaria* (Lyngbye) Link, which is found throughout the world in high tide pools on exposed coasts and on stones and wood at lower levels in harbors, seems to be absent from San Francisco Bay. A form of *Scytosiphon lomentaria* (f. *complanatus minor*) was described by Setchell & Gardner (1925:534) on the basis of material from Fort Point which Wynne (1969:34) referred to his new species *S. dotyi.*

Sphacelariales

Sphacelariaceae

Sphacelaria furcigera Kützing
Tufts of stiff, sparsely branched filaments to 1 cm tall arising from a base of entangled creeping filaments, with slender bifurcate propagula; on rocks or algae in lower intertidal and upper subtidal zones; widely distributed in temperate and tropical waters (type locality: Persian Gulf), sporadic from Alaska to Baja Calif., Costa Rica, Galapagos Is., Chile.
San Francisco Bay: On rocks at Paradise Beach. (This determination was kindly made by Dr. W. F. Prud'homme van Reine of the Rijksherbarium, Leiden, Netherlands.)

Desmarestiales

Desmarestiaceae

Desmarestia herbacea Lamouroux (Fig. 16)
Erect fronds golden-brown, to 4 m long, arising from discoid holdfast, oppositely branched in one plane, the axes compressed, the blades foliose; on rocks or wood in lower intertidal zone and subtidal to 15 m; Alaska to Baja Calif. (type locality: "North-west coast of America").
San Francisco Bay: Fort Point, Lime Point, Pt. Cavallo, Fort Mason, west side of Angel I.
A broad form, possibly representing a distinct species (*D. munda* Setchell & Gardner), has been found at Pt. Blunt (Angel I.) and on floats at the San Francisco Yacht Harbor. Both *D. herbacea* and *D. munda* are listed in Abbott & Hollenberg (1976:222) as synonyms of *D. ligulata* (Lightfoot) Lamouroux (type locality: Scotland), but the argument offered in support of this treatment is not convincing.
Desmarestia herbacea, like certain other species in the genus, has a high concentration of sulfuric acid in its cell sap. When the alga is collected, the acid is released, producing an acrid odor and decomposing any plant material that it touches.

Laminariales

Members of this order are known as "kelps." One of many distinctive features is the presence in each frond of a localized meristematic tissue—that is, a group of cells responsible for producing new cells. The general plan of organization includes an attachment organ (holdfast), a stem (stipe), and a blade, with the meristem located at the juncture of blade and stem. In some genera the stipe is branched repeatedly, each branch having its own meristem. The ultimate result of this elaboration is *Macrocystis*, the giant kelp, individual plants of which may be 45 m long and with hundreds of stipes and blades. Kelps are especially diverse and abundant in the colder waters of the N. Pacific. Since most kelps grow in situations exposed to heavy surf, it is not surprising that only three genera of the eleven known from central California occur in San Francisco Bay.

Laminariaceae

Laminaria sinclairii (Hooker f. & Harvey) Farlow, Anderson & Eaton (Fig. 14)
Dense stands of branched rhizomes bearing numerous stiff stipes to 30 (-50) cm tall, 3-5 mm diam., each terminating in a linear blade to 4 (-11) cm broad, 55 cm long, on rocks in lower intertidal zone, often partially covered by sand; British Columbia to Ventura, Calif. (type locality: "San Francisco," probably Fort Point).

San Francisco Bay: Abundant on most shores of the western part of the central Bay (Zones I-III), even in localities (such as Alcatraz I.) without sand; an important component of the flora.

While the stipes persist for more than a year, the blades may be regenerated, especially in early spring. Irregular growth forms, such as branched stipes, are often encountered. Since growth of the blade is initiated by the meristem at its base, the tip is the oldest portion and is thus vulnerable to colonization by minute algae, especially filamentous brown and red algae. Small black dots, however, usually turn out to be infestations of a blue-green alga, *Entophysalis conferta* (Kützing) Drouet & Daily.

Alariaceae

Alaria marginata Postels & Ruprecht
Single unbranched stipe 2-7 cm long attached by mass of branched haptera, terminating in single blade 2-4 (-6) m long, 10-20 (-30) cm broad, with conspicuous percurrent midrib; 10-20 elliptical sporophylls borne pinnately at juncture of blade and stipe; on rocks in lower intertidal zone to shallow subtidal depths, exposed to at least moderate surf; Alaska to Pt. Buchon (San Luis Obispo Co.), Calif. (type locality: Unalaska I.).

San Francisco Bay: Although this species barely gets beyond the portals (to Fort Point and Pt. Cavallo), it thrives at Lime Point, plants reaching more than 2 m in length with blades 20 cm broad.

Egregia menziesii (Turner) J. E. Areschoug (Fig. 11)
Stipe arising from conical holdfast, bearing numerous strap-like branches to 15 m long, densely fringed with filiform to spatulate blades to 6 cm long, some of which develop spindle-shaped to subspherical air bladders; straps and blades tuberculate; forming belt on rocks in lower middle intertidal zone to shallow subtidal depths; British Columbia to Pt. Conception, Calif. (type locality: Nootka Sound, Vancouver I.).

San Francisco Bay: Abundant in the western half of the central Bay, extending into Zone IV (Yerba Buena I.). In protected areas there is a tendency for the blades to become dissected and for the incidence of tubercles to decrease.

Fucales

Members of this order are known popularly as "rockweeds." Almost all cool north temperate shores support a high or mid-level intertidal band of Fucaceae and a lower intertidal and shallow subtidal band of Cystoseiraceae. The colder waters of the southern hemisphere abound in a great variety of Cystoseiraceae and other Fucales, while tropical waters everywhere are noted for their abundance of Sargassaceae.

Fucaceae

Fucus gardneri Silva (Fig. 12)
Flat dichotomously branched dark-brown fronds mostly 12-30 cm tall with flattened and sometimes swollen reproductive terminal segments, arising from spongy base, forming belt on rocks in upper and middle intertidal zones; Alaska (type locality: "Unalaschka," possibly in error) to Pt. Conception, Calif.

San Francisco Bay: Abundant on all shores of the central Bay; an important component of the flora.

Members of this genus thrive in harbors throughout the cool part of the North Temperate Zone. Fronds are often epiphytized by the green algae *Enteromorpha* and *Ulva* and by various minute filamentous brown algae.

SAN FRANCISCO BAY

Gardner (1922) recognized the San Francisco Bay populations of *Fucus* from Pt. Cavallo inward as representing a distinct species, which he named *F. nitens*. He considered a series of populations between Pt. Cavallo and "some distance west of Lime Point" to be intermediate morphologically between *F. nitens* and the outer-coast species *F. gardneri*. Future studies of the Bay flora should include a reappraisal of *F. nitens*.

Fucus gardneri is included by Abbott & Hollenberg (1976:261) in their circumscription of *F. distichus* subsp. *edentatus* (de la Pylaie) Powell (type locality: Newfoundland), but my experience with the latter in Nova Scotia leads me to believe that this treatment is incorrect.

Pelvetia compressa (J. Agardh) De Toni

Subcylindrical dichotomously branched yellowish-brown or greenish-brown fronds mostly 20-40 cm tall with swollen reproductive terminal segments, on rocks in upper intertidal zone; Shelter Cove (Humboldt Co.), Calif. to Baja Calif. (type locality: Monterey Peninsula).

San Francisco Bay: This common outer-coast rockweed is scarce inside the Bay, having been found only on the southwest side of Angel I.

This species is listed in Abbott & Hollenberg (1976:261) as *P. fastigiata* (J. Agardh) De Toni. Reasons for the change of name are given in DeCew & Silva (in press).

Pelvetiopsis limitata (Setchell) Gardner

Flat dichotomously branched yellowish-brown fronds mostly 4-8 cm tall, arcuate, with swollen reproductive terminal segments, on rocks in upper intertidal zone; British Columbia to Pt. Buchon (San Luis Obispo Co.), Calif. (type locality: Lands End, San Francisco).

San Francisco Bay: Most rocky localities within Zones I-III.

Populations with broader, longer (-15 cm), and straight branches and larger reproductive segments have been described as f. *lata* by Gardner (1910). This form is found at Pt. Cavallo (type locality), Tiburon, and on the southwest side of Angel I.

[*Ascophyllum nodosum* (Linnaeus) Le Jolis]

A common N. Atlantic rockweed, used in packing bait worms, lobsters, and other live animals for shipment to the Pacific coast. Pieces of *Ascophyllum* are frequently encountered in San Francisco Bay, following the trail of fishermen and restaurant workers. Since *Ascophyllum* thrives in estuarine situations and has a form that lies loose on mud flats, the probability that it will eventually become established in San Francisco Bay seems high.

Sargassaceae

Sargassum muticum (Yendo) Fensholt (Fig. 9)

Bushy plant with main axes arising from spongy holdfast or as branches from lower portion of primary axes, to 2 (-10) m long, profusely branched, ultimately fringed with branchlets composed of leaves and axillary air bladders and reproductive structures; on rocks in lower intertidal zone and subtidal to 5 m, forming dense stands in quiet waters; Japan (type locality), England, France, sporadic from Alaska to Punta Baja, Baja Calif.

San Francisco Bay: San Francisco Yacht Harbor; inside breakwater at Fort Baker; east side of Yerba Buena I.; East Bay (Alameda to Albany).

Most species of *Sargassum* inhabit warm, fully saline water, and the two that are native to California follow this pattern, growing on the outer coast south of Pt. Dume (Los Angeles Co.). The exceptional distribution of *S. muticum*, encompassing a relatively wide range of temperature and salinity, is directly related to its remarkably broad physiological tolerances, rapid growth rate, and high reproductive capacity (Norton 1976). It is a weed in the sociological sense of a plant growing outside its original range and in such a manner as to be unwanted. It was introduced to northwestern North America, presumably in shipments of oyster spat from Japan, just before World War II. It was first collected in British Columbia in 1944 and near Coos Bay, Ore. in 1947. Subsequently, it has spread sporadically both northward and southward into protected waters. It was discovered in northern California in 1963 (Crescent City), southern California in 1970 (Santa Catalina I.), and in San Francisco Bay in 1973 (Berkeley). It was detected in southern England in 1973, possibly having been introduced in shipments of Japanese oysters from British Columbia. The English established a task force to devise means of eradicating the weed, fearing that it would displace eel grass (*Zostera marina*) and other native plants and that the long dense fronds would foul small boats in harbors. The eradication program was largely unsuccessful. In San Francisco Bay, *S. muticum* has spread rapidly and is especially abundant on the east side of Yerba Buena I.

Undoubtedly it will soon become established at Sausalito.

There is no evidence that *S. muticum* is displacing the native biota of San Francisco Bay. It is a popular host for various epiphytes, and it will be of interest to determine whether the epiphytes in San Francisco Bay are native or introduced. Unfortunately, most genera of these epiphytes include many poorly defined species that have been assigned wide geographic distributions and hence offer great taxonomic difficulties.

Class XANTHOPHYCEAE (Yellow-Green Algae)

Vaucheriales

Vaucheriaceae

Vaucheria longicaulis Hoppaugh

Dense blackish-green mats on mudflats; Bodega Bay (Sonoma Co.) to Elkhorn Slough (Monterey Co.) (type locality), Calif.

San Francisco Bay: East Bay mudflats.

Class CHLOROPHYCEAE (Green Algae)

Ulotrichales

Ulotrichaceae

Genus *Ulothrix*

Members of this genus occupy fresh-water, brackish-water, and marine habitats. Several species have been accredited with almost cosmopolitan distribution in harbors, where they form soft green coatings on various substrates in the upper intertidal zone, often intermixed with one another and with *Urospora* and *Bangia*. In the Herbarium of the University of California at Berkeley are collections from San Francisco Bay determined by Setchell & Gardner as belonging to three species:

Ulothrix flacca (Dillwyn) Thuret. Type locality: Wales. San Francisco Bay: Berkeley (on wood).

Ulothrix implexa (Kützing) Kützing. Type locality: Netherlands. San Francisco Bay: San Francisco Marina (on wood).

Ulothrix pseudoflacca Wille. Type locality: Norway. San Francisco Bay: San Francisco Marina (on boulders).

A comparison of the treatment of these cosmopolitan species in various well-known floristic accounts reveals so many significant conceptual differences as to destroy all confidence in the ability to determine the San Francisco Bay representatives with any degree of meaningfulness. Certain recent collections agree with the descriptions provided by Setchell & Gardner (1920b), but others have combinations of characters that would seem to preclude the application of any existing name. Culture studies are essential to elucidating the diagnostic value of such traditional characters as width of the filament, shape of the cells, number of pyrenoids, and whether the chloroplast is a complete or incomplete ring.

Ulvales

Monostromataceae

Monostroma oxyspermum (Kützing) Doty

Attached monostromatic sac soon expanding into an irregularly lobed and folded free-floating membrane to 30 cm broad, 40 cm tall, soft, delicate, pale-green, in quiet brackish waters; widespread (type locality: Baltic Sea), British Columbia to S. Calif.

San Francisco Bay: Salt marshes.

In the Herbarium of the University of California at Berkeley are collections from San Francisco Bay determined by Setchell & Gardner as belonging to three widespread species: *M. latissimum* Wittrock, *M. orbiculatum* Thuret, and *M. quaternarium* (Kützing) Crouan frat. Doty (1947), following Hamel (1931), considered all three species to be conspecific with *M. oxyspermum*.

SAN FRANCISCO BAY

Ulvaceae

Blidingia minima (Kützing) Kylin var. *minima*
　　Clusters of light-green tubes mostly 5-15 cm tall arising from cushion-like base, tubes cylindrical or compressed, unbranched or occasionally with basal proliferations, forming nearly continuous belt on rocks or wood in upper intertidal zone; widespread (type locality: Helgoland, Germany), Alaska to Mexico, Chile.
　　San Francisco Bay: Abundant throughout the Bay.

Blidingia minima (Kützing) Kylin var. *subsalsa* (Kjellman) Scagel
　　Tubes light-green, highly proliferous, often entangled and distorted, on rocks, shells, wood, or other aquatic vegetation, or free-floating on mudflats, common in protected brackish waters; widespread (type locality: Arctic Ocean), Alaska to Ventura Co., Calif.
　　San Francisco Bay: Richardson Bay; Alameda.

Blidingia minima (Kützing) Kylin var. *vexata* (Setchell & Gardner) J. Norris
　　Tubes deep-green to nearly black, flattened, linear to spatulate, usually unbranched, distorted by black or brown wart-like fruiting bodies of an infesting fungus, on rocks in upper intertidal and splash zones, British Columbia to Monterey Co., Calif. (type locality: Fort Point, San Francisco).
　　San Francisco Bay: Throughout Zones I-IV (not yet detected in East Bay).
　　This type of algal-fungal association, in which the habit of the alga predominates, is termed a mycophycobiosis in distinction to a true lichen. The fungus is an ascomycete, *Turgidosculum ulvae* (Reed) J. & E. Kohlmeyer (1972).

Enteromorpha clathrata (Roth) Greville var. *clathrata*
　　Light-green profusely branched plants to 40 cm tall, the branches in the form of a monostromatic tube, filiform, cylindrical to compressed, with multiseriate endings, the cells in longitudinal rows; on rocks, wood, or other algae or free-floating on mudflats, upper and middle intertidal zones in protected brackish waters; widespread (type locality: Baltic Sea, Germany), Alaska to Mexico.
　　San Francisco Bay: Quiet brackish habitats, as at Sausalito and Alameda.

Enteromorpha clathrata (Roth) Greville var. *crinita* (Roth) Hauck
　　Branch-endings uniseriate; on rocks, wood, or other aquatic vegetation or free-floating on mudflats, upper and middle intertidal zones in protected waters; widespread (type locality: North Sea, Germany), Alaska to Mexico.
　　San Francisco Bay: Protected habitats, such as Belvedere, Alameda, San Francisco Yacht Harbor, San Mateo Co.

Enteromorpha compressa (Linnaeus) Greville
　　Light-green profusely branched plants to 40 cm tall, the branches in the form of a monostromatic flattened tube, rarely rebranched, the cells irregularly arranged; on rocks, shells, or rarely on other algae on semiprotected outer coast or in estuaries, throughout the intertidal zone; widespread (type locality: Sweden), Bering Sea to Costa Rica.
　　San Francico Bay: Sausalito; to be expected in other protected habitats.

Enteromorpha flexuosa (Roth) J. Agardh
　　Dark-green usually unbranched plants mostly 4-8 cm tall, cylindrical in basal half, at times slightly compressed in upper half, cells in longitudinal rows in narrow portions; on rocks, wood, or other algae or free-floating on mudflats in estuaries, throughout the intertidal zone; widespread (type locality: Adriatic Sea), British Columbia to Central America.
　　San Francisco Bay: Lake Merritt; to be expected in other protected habitats.

Enteromorpha intestinalis (Linnaeus) Link
　　Yellowish-green or grass-green unbranched tubes mostly less than 20 cm tall, cylindrical throughout or often compressed, crisped and contorted distally, the cells irregularly arranged; on rocks, wood, or other algae or free-floating in protected bays and estuaries, upper and middle intertidal zones; widespread (type locality: N. Europe), Alaska to Mexico.
　　San Francisco Bay: Common in many parts of the Bay.

Enteromorpha linza (Linnaeus) J. Agardh
　　Grass-green unbranched plants mostly 15-40 cm tall, the basal portion a cylindrical tubular stipe, the upper portion a flattened distromatic blade with hollow margins; on rocks, wood, or

other algae in semiprotected habitats of outer coast and bays, middle and lower intertidal zones; widespread (type locality: N. Europe), Alaska to Mexico.
San Francisco Bay: Common throughout the Bay.

Enteromorpha prolifera (O. F. Müller) J. Agardh
Grass-green profusely branched plants to 60 cm (-2 m) tall, main axis distinct, tubular, cylindrical to compressed, with few to numerous tubular proliferations, the cells of younger portions in longitudinal rows; on rocks or wood or free-floating in bays and estuaries, upper and middle intertidal zones; widespread (type locality: Denmark), Alaska to Mexico.
San Francisco Bay: Tiburon; to be expected in other protected habitats.

Percursaria percursa (C. Agardh) Rosenvinge
Entangled masses of pale-green unbranched filaments composed of 2 cell rows, floating in high pools in salt marshes; widespread (type locality: Denmark), Alaska to Washington, Tomales and San Francisco bays in Calif.
San Francisco Bay: Sausalito, Richmond.

Ulva angusta Setchell & Gardner
Membranous blades mostly to 35 cm tall, pale grass-green, distromatic, undivided, linear to oblanceolate, spirally twisted with ruffled margins, on rocks or other algae, intertidal, Ore. to Baja Calif. (type locality: Moss Beach, San Mateo Co., Calif.).
San Francisco Bay: Throughout the central Bay.

Ulva expansa (Setchell) Setchell & Gardner
Membranous blades mostly to 1 m long, pale-green to medium-green, distromatic, orbicular to irregularly expanded, undivided but with deeply ruffled margins; weakly attached to rocks, shells, or other algae, usually becoming free-floating, in protected waters, lower intertidal and upper subtidal zones, British Columbia to Baja Calif. (type locality: Monterey, Calif.).
San Francisco Bay: Salt marshes.

Ulva lobata (Kützing) Setchell & Gardner
Membranous blades mostly 10-30 cm tall, to 15 cm broad, rich grass-green, distromatic, broadly obovate, deeply lobed with ruffled margins; on rocks or occasionally on other algae, middle intertidal zone to upper subtidal zone, Ore. to Mexico, Pacific S. America (type locality: Chile).
San Francico Bay: Throughout the central Bay.

The various species of *Ulva* are commonly called "sea lettuce."

Chaetophorales

Chaetophoraceae

Endophyton ramosum Gardner
Green filamentous endophyte, forming irregular dark-colored areas in lower part of host, in blades of various red and brown algae (especially *Iridaea*), lower intertidal zone, Wash. to Redondo Beach (Los Angeles Co.), Calif. (type locality: Fort Point, San Francisco).
San Francisco Bay: In *Iridaea splendens* at various localities in Zones I-IV; in *Gigartina exasperata* at Alameda and undoubtedly elsewhere.

Pseudodictyon geniculatum Gardner
Green filamentous endophyte, forming greenish areas on host, in blades of various red and brown algae (especially Laminariales), lower intertidal zone, Wash. to Redondo Beach (Los Angeles Co.), Calif. (type locality: Fort Point, San Francisco).
San Francisco Bay: In *Laminaria sinclairii* at Fort Point; to be expected throughout range of host.

Pseudulvella consociata Setchell & Gardner
Thin dark-green crusts composed of closely adjoined radiating prostrate filaments, on shells of various mollusks in intertidal zone, South Bay, Ore. and San Francisco Bay (type locality: Alameda).
San Francisco Bay: Alameda (on *Ilyanassa obsoleta*, the dog whelk).

Cladophorales

Codiolaceae

SAN FRANCISCO BAY

Urospora doliifera (Setchell & Gardner) Doty
Dark-green filaments 3-9 cm long, uniseriate, unbranched, forming a slippery coating on rocks in upper intertidal zone, Ore. and San Francisco (type locality: San Francisco Marina).
San Francisco Bay: San Francisco Marina.
This apparently rare species is also known from Lands End (San Francisco).

Urospora penicilliformis (Roth) J. Areschoug
Dark-green filaments 3-7 cm long, uniseriate, unbranched, forming a slippery coating on rounded boulders in upper intertidal and upper middle intertidal zones, often intermixed with *Bangia* and *Ulothrix*; widespread (type locality: Germany), Alaska to S. Calif.
San Francisco Bay: Most localities in Zones I-IV.

These two species of *Urospora* are distinguishable only by such microscopic characters as the size and shape of the cells and the number of pyrenoids in each cell.

Acrosiphoniaceae

Spongomorpha coalita (Ruprecht) Collins
Profusely branched uniseriate filaments entangled in strands with the appearance of a frayed rope, to 25 cm tall, grass-green to dark-green; on rocks in lower intertidal zone; Alaska to Pt. Conception, Calif. (type locality: Sonoma Co., Calif.).
San Francisco Bay: Fort Point, Lime Point, Alcatraz I., west side of Angel I.

Spongomorpha mertensii (Ruprecht) Setchell & Gardner
Branched uniseriate filaments entangled basally, fairly rigid and free distally, to 10 cm tall, bright-green; on intertidal rocks or algae; Alaska (type locality: Sitka) to Carmel, Calif.
San Francisco Bay: Fort Point.
The Fort Point population was described as a new form (f. *limitanea*) of *Spongomorpha arcta* (Dillwyn) Kützing by Collins (1911). This entity was later referred to *S. mertensii* by Setchell & Gardner (1920a:280). Most species in *Spongomorpha* are poorly defined, however, so that the present placement must be considered highly tentative. It is important to note that the Fort Point plants were "growing on a stone wall, just above high water mark, with fresh water running over it," whereas *S. arcta* and *S. mertensii* are said to be found in the middle and lower intertidal zones.

Cladophoraceae

Chaetomorpha aerea (Dillwyn) Kützing
Dense stands of coarse dark-green unbranched uniseriate filaments usually 5-15 (-25) cm long, the cells 0.12-0.3 mm diam., 0.5-2 times as long; on rocks or concrete, often in high tide pools but also at lower intertidal levels; widespread (type locality: England), Ore. to Baja Calif.
San Francisco Bay: Fort Point (in depressions in the upper middle intertidal zone and on cobbles and concrete in the lower middle intertidal zone).

Chaetomorpha californica Collins
Dense stands of fine green unbranched uniseriate filaments to 20 cm long, the cells 0.03-0.04 mm diam., about 4 times as long; on sand in shallow pools in upper intertidal zone, Calif. (Shell Beach, Sonoma Co. to San Francisco Bay; Laguna Beach; type locality: La Jolla) and on limpets at depth of 10 m, Strait of Juan de Fuca, Wash.
San Francisco Bay: Pt. Blunt (Angel I.).
Except for its small stature (6-7 mm), the Pt. Blunt material fits the original description of this species perfectly.

Cladophora amphibia Collins
Basal layer of densely branched prostrate uniseriate filaments giving rise to short erect filaments, which in turn may produce slender descending rhizoids; forming a turf on the ground among pickle weed (*Salicornia*) along high-tide line, Alameda, San Francisco Bay (type locality).
San Francisco Bay: As above.
This species is known only from the type collection and thus needs to be reinvestigated. It may possibly be referrable to *Spongomorpha*.

Cladophora columbiana Collins
Bright-green matted tufts, hemispherical to laterally expanded, mostly 3-5 cm thick, accumulating sand, the filaments uniseriate, branching di- or trichotomously in lower portion, unilaterally

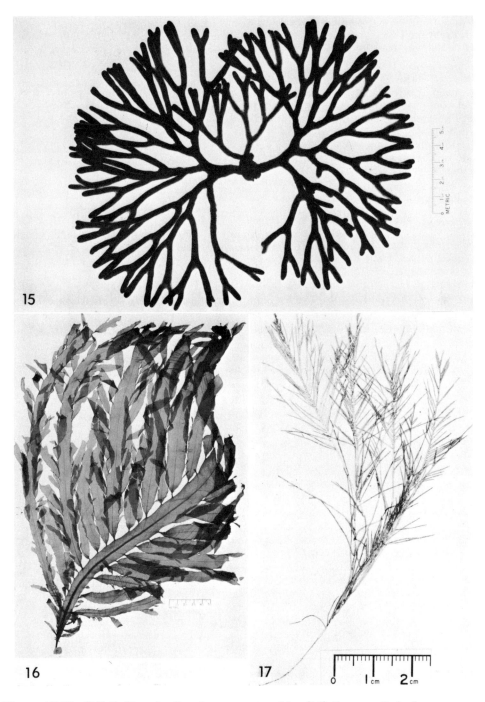

Figures 15-17. (15) *Codium fragile* subsp. *tomentosoides*. (16) *Desmarestia herbacea*. (17) *Bryopsis* sp.

above, on rocks in upper intertidal zone, especially in cracks and depressions; British Columbia (type locality: Port Renfrew, Vancouver I.) to Baja Calif.

San Francisco Bay: Present, but neither abundant nor luxuriant, at Fort Point, Pt. Cavallo, Fort Mason, and Stuart Point (west side of Angel I.).

Cladophora sericea (Hudson) Kützing

Yellowish-green to light-green plants to 30 (-50) cm tall, composed of slender uniseriate filaments, main axes irregularly dichotomously branched, the branches unilateral or bilateral, floating in quiet pools; widespread (type locality: England), Alaska to San Diego, Calif.

San Francisco Bay: Oakland.

A collection of sparsely branched *Cladophora* from a warm-water pool near the old Key System Power House in Oakland was described as a new form (f. *eramosa*) of *C. rudolphiana* (C. Agardh) Kützing by Gardner (in Collins 1918:81). This entity has been referred to *C. sericea* by Abbott & Hollenberg (1976:108).

Two large and beautiful species of *Cladophora* growing on shells at Alameda have not yet been satisfactorily identified.

Lola lubrica (Setchell & Gardner) A. & G. Hamel

Pale-green or yellowish plants 30-50 cm long, composed of straight unbranched uniseriate filaments, forming dense lubricous mats, entangled with other algae or free-floating on mudflats; N. Atlantic, Alaska, Wash. to Costa Rica (type locality: Oakland, Calif.).

San Francisco Bay: Oakland.

The generic placement of this species is questionable inasmuch as the concept of the genus *Lola* was based on material from Atlantic France identified with *Rhizoclonium lubricum* Setchell & Gardner rather than on material from the type locality (Oakland).

Rhizoclonium implexum (Dillwyn) Kützing

Yellowish-green mats of unbranched uniseriate filaments, floating, on mud, or entangled with other algae, on outer coast but especially frequent in protected areas such as sloughs and marshes; widespread (type locality: Ireland), Alaska to Mexico, Galapagos Is.

San Francisco Bay: Camp Reynolds, west side of Angel I. (entangled with *Gigartina canaliculata* in lower middle intertidal zone); Fort Mason (with *Rhodochorton purpureum* on sides of trench and on stick in upper intertidal zone).

Codiales

Bryopsidaceae

Bryopsis corticulans Setchell

Clusters of blackish-green coenocytic fern-like fronds 4-10 (-16) cm tall, main axes to 1 mm diam., percurrent, naked below, pinnately branched 2 or 3 times above, branches tending to be progressively longer toward base, ultimate pinnules mostly 0.12-0.25 mm diam., male and female gametes produced on separate plants; usually on vertical sides of rocks in lower intertidal zone; British Columbia to Baja Calif. (type locality: Carmel Bay, Calif.).

San Francico Bay: Southwest side of Angel I.

The delineation of species within this genus is notoriously difficult. There are few morphological characters and strong suspicion that at least some of these are markedly influenced by temperature, salinity, and the amount of wave action. In exposed situations on the outer coast of central and northern California the only species of *Bryopsis* is *B. corticulans*, which, while widespread, is seldom abundant. *Bryopsis hypnoides* Lamouroux (type locality: Mediterranean France) has been reported occasionally from protected waters (e.g. on sand-covered rocks at Pebble Beach, Carmel Bay, and on docks in Humboldt Bay). Within San Francisco Bay, *Bryopsis* has not yet been found in Zones I or II. It grows sparingly on docks and piling at Tiburon, Ayala Cove (Angel I.), and Yerba Buena I. and on rocks at Pt. Blunt (Angel I.). In the East Bay and at Coyote Point (San Mateo Co.), by contrast, it is seasonally abundant.

The doctoral research of Dennis E. Shevlin at the University of California at Berkeley—a study of the reproductive biology of *Bryopsis*—clearly shows that more than one species is represented in the Bay. Dr. Shevlin has kindly allowed me to report those results that are pertinent to the taxonomy of the genus. He can distinguish in the field (specifically, on the docks at the Berkeley Yacht Harbor) three entities growing in close proximity.

One form (Fig. 17) is a robust, densely branched plant that does not reproduce sexually, either in nature or in culture. This entity is an important member of the flora of the East Bay from Alameda to Richmond and at Coyote Point. We have observed it especially at Golden Gate Fields (Albany), where it appears in early spring on various substrates in the lower intertidal zone and develops explosively until it becomes one of the more conspicuous components of the flora in late June and early July. This same entity has been cast ashore at Alameda in large quantities every summer since at least 1951 (when I was introduced to the phenomenon), constituting a significant, though minor part of an algal mass that becomes a nuisance as it decomposes. Such behavior is clearly that of a weed, but the original home and identity of this plant is undeterminable in the absence of sexual reproduction. Branching is inconsistently distichous, with the ultimate pinnules of irregular length but frequently long, tapered at both ends, mostly 0.07-0.22 mm diam., and often discontinuous along the axis. Each primary branch develops a cluster of irregularly branched rhizoids at its base. Rapid propagation is undoubtedly facilitated by abscission of these plantlets.

A second entity is monoecious, the fertile pinnules (gametangia) producing both male and female gametes. In some plants the male gametes are formed in the proximal portion of the gametangium, in others in the distal portion, the pattern being maintained in culture as a strain-specific character. This entity forms clusters with fewer and less robust fronds than the weed. Branching is mainly radial, but subdistichous at the tips, with the ultimate pinnules cylindrical or gradually tapering distally, long, slender (mostly 0.035-0.07 mm diam.), and closely investing the axis.

A third entity is dioecious, with separate male and female plants, as in *B. corticulans*. Branches arise radially from the main axis, but the ultimate pinnules are distichous and more or less in one plane, fusiform, mostly 0.07-0.14 (-0.165) mm diam., and of approximately the same length along the axis, giving the frond a fern-like appearance.

Comparative studies of such cytological features as the plastid, which would undoubtedly help in the identification of the various forms of *Bryopsis* in San Francisco Bay, have not yet been undertaken. The present assignment of certain collections from Angel I. to *B. corticulans* is tentative in that the reproduction of those populations has not been investigated.

Codiaceae

Codium fragile (Suringar) Hariot subsp. *tomentosoides* (van Goor) Silva (Fig. 15)

Small clumps of blackish-green plants mostly 10-25 cm tall, dichotomously branched, spongy, formed by a complex interweaving of coenocytic filaments; on rocks or shells in lower intertidal and upper subtidal zones; N. Atlantic (type locality: Netherlands), Mediterranean, New Zealand, San Francisco Bay.

San Francisco Bay: Alameda; Coyote Point (San Mateo Co.).

This alga, which is a weed in the sociological sense of a plant that grows rankly in unwanted places, has been introduced into San Francisco Bay within the last few years by some yet undetermined means. At the present time it occurs in abundance only at Coyote Point, but it may be expected to spread into the central Bay. Although this subspecies was first described from the Netherlands (as an introduced plant), its original home probably is Japan (type locality for *C. fragile*), where it forms an undetectable part of the morphological and physiological spectrum encompassed by the highly variable parental species. In this respect it is similar to *Sargassum muticum*, the other well-publicized weed in San Francisco Bay. Its weediness is due to its broad physiological tolerance and unusually great reproductive capacity. Wherever this weed has been studied, the populations have been shown to be solely female, with the female gametes capable of parthenogenetic development. It thus comprises a single clone or at most a few clones.

Several geographic non-weedy subspecies of *Codium fragile* have been recognized. The subspecies native to California (ranging from Alaska to Baja Calif.) has a sporadic distribution on the outer coast, being locally abundant but confined to specific rocks. It has not been found within San Francisco Bay.

HISTORY, LANDFORMS, AND VEGETATION OF THE ESTUARY'S TIDAL MARSHES

BRIAN F. ATWATER
U. S. Geological Survey, 345 Middlefield Road, Menlo Park, CA 94025
SUSAN G. CONARD
Botany Department, University of California, Davis, CA 95616
JAMES N. DOWDEN
California State Lands Commission, 1807 13th Street, Sacramento, CA 95814*
CHARLES W. HEDEL
U. S. Geological Survey, 345 Middlefield Road, Menlo Park, CA 94025
RODERICK L. MACDONALD
Botany Department, University of California, Davis, CA 95616
WAYNE SAVAGE
Biology Department, San Jose State University, San Jose, CA 95192

Around 8,000 to 10,000 years ago, sharply rising sea levels nursed a newborn San Francisco Bay estuary whose tidal marshes probably covered less area than open water. Thereafter the rate of submergence decreased about 10-fold, and by 6,000 years ago sediment began to maintain marshes that later spread across marginal parts of San Francisco Bay. By thus counteracting or overtaking submergence, sedimentation created marshes that, as of 1850, covered about 2200 km^2, nearly twice as much area as the bays. People have leveed or filled all but approximately 85 km^2 of these marshes during the past 125 years. Concurrently, human activities have caused the delivery of enormous quantities of sediment to the bays and the slackening of tidal currents in sloughs, thereby contributing to the creation of nearly 75 km^2 of marsh, about half of which remains pristine. Plains situated near high-tide levels are the most extensive landforms of both historic and modern marshes. Tides rather than upland tributaries created most sloughs around the bays, but riverine floods erected natural levees that confined tidal water in the Delta. Tidal marshes around San Francisco Bay typically contain 13 or 14 species of vascular plants characteristic of salt marshes and are dominated by common pickleweed (*Salicornia pacifica*) and California cordgrass (*Spartina foliosa*). In the Delta, tidal marshes support about 40 species characteristic of fresh-water marshes and are dominated by tules and bulrushes (*Scirpus* spp.), cat-tails (*Typha* spp.), and common reed (*Phragmites communis*). These contrasting communities overlap around San Pablo Bay, Carquinez Strait, and Suisun Bay. Damage to tules and bulrushes during the drought of 1976-1977 confirms that intolerance of salt causes these plants to disappear toward San Francisco Bay. The disappearance of California cordgrass and common pickleweed toward the Delta, alternatively, may result from unsuccessful competition against tules, bulrushes, and other species. If export equals one quarter of net above-ground productivity, then vascular plants of the tidal marshes collectively contribute about 10 billion grams of carbon per year to other parts of the estuary.

---

* The views expressed herein are not necessarily those of the California State Lands Commission (CSLC) or of other bureaus of the State.

SAN FRANCISCO BAY

Though initially regarded as wastelands, tidal marshes of the San Francisco Bay estuary have gained considerable human significance during the past 125 years. The monetary value of these marshes derives chiefly from their historical conversion to farmlands, salt ponds, and sites for commerce, industry, recreation, and residence. Areas of former marshland in the Sacramento-San Joaquin Delta, for example, currently yield about $300 million in crops, including one quarter of America's domestic asparagus (Delta Advisory Planning Commission 1976:77, 80). From impounded marshes around San Francisco and San Pablo bays, the Leslie Salt Company annually harvests approximately 400,000 metric tons of crude salt worth $7 million (M. Armstrong pers. comm.). Other leveed areas of former tidal marsh attract hunters, particularly north of Suisun Bay, where the annual duck kill equals about 10% of California's total (Jones and Stokes Associates Inc., and EDAW, Inc. 1975:46).

Few people deliberately maintained tidal marshes in their pristine condition until the 1960's, when concern about human encroachment on the bays led to the creation of the San Francisco Bay Conservation and Development Commission (BCDC) (Gilliam 1969). Since 1969, this regulatory agency has mandated the preservation of most remaining tidal marshes around San Francisco, San Pablo, and Suisun bays. According to the Commission's findings, tidal marshes warrant such protection because, directly or indirectly, they nourish and shelter many estuarine animals (BCDC 1969: 11). Some people also value pristine tidal marshes as sties for outdoor education or recreation. In response to this interest, local governments have established parks that preserve tidal marshes for students, bird-watchers, and strollers.

This chapter reflects the current concern for pristine tidal marshes by providing an overview of their history, landforms, and vegetation. Drawn partly from unpublished observations by the authors, the overview also depends on information from published sources, particularly topographic surveys by the U. S. Coast and Geodetic Survey (USC&GS); geologic investigations by Gilbert (1917), Pestrong (1963, 1972), and Atwater et al. (1977); and botanical studies by Cooper (1926), Marshall (1948), Hinde (1954), Mason (1957), Cameron (1972), Mahall and Park (1976a, b, c) and Conard et al. (1977). Much additional information remains to be gathered and assimilated, so we expect that others will improve many of the ideas put forth in this synthesis.

HISTORY

The discovery of gold at Sutter's mill in 1848 initiated human activities that have worked vast changes in tidal marshes of the San Francisco Bay estuary. Before the Gold Rush, people interfered with few of the natural processes that create, maintain, or destroy tidal marshes. Since the Gold Rush, however, people have leveed or filled most pre-existing marshes, accidentally promoted the erosion of others, and created some new marshes by both accident and design.

Events Before the California Gold Rush

Rates of submergence (rise in sea level relative to the land) and sedimentation largely controlled the areal extent of tidal marshes in the San Francisco Bay estuary between the inception of the estuary and the arrival of the Forty-Niners about 10,000 years later. Known tidal-marsh deposits older than 8,000 years form lenses no more than a few meters thick and underlie sediments that accumulated in open-water bay environments. This distribution implies that 8,000 to 10,000 years ago a discontinuous fringe of tidal marsh retreated from a rising, spreading bay, presumably because sediments accumulated in tidal marshes less rapidly than the level of the Bay climbed.

By about 6,000 years ago, the rate of submergence had slowed by nearly 10-fold to its subsequent average of 1-2 m per millenium (Atwater 1979, Fig. 5), thereby allowing sedimentation to counterbalance submergence in some parts of the estuary. In the western Delta, peat as thick as 20 m indicates that vertical accretion in marshes has kept pace with submergence during the past 4,000-6,000 years (Weir 1950; Schlemon and Begg 1973). A balance between sedimentation and submergence likewise accounts for thick accumulations of tidal-marsh deposits in Massachusetts (Mudge 1858; Davis 1910; Redfield 1972).

The establishment of extensive tidal marshes around southern San Francisco Bay appears to have occurred later than in the Delta, probably close to 4,000 years after the rate of submergence reached 1-2 m per millenium (Atwater et al. 1977). This delay is evidenced by deposits of tidal-flat mud that typically underlie peaty tidal-marsh sediments at elevations close to modern mean tide level (MTL) (Table 1). The boundary between tidal-flat and tidal-marsh sediment records the colonization of mudflats by marsh plants, so it marks the inception of a marsh (Shaler 1886: 364-365). The date at which MTL equalled the elevation of this boundary approximates the minimal ages of the marsh because California cordgrass (*Spartina foliosa*), the pioneer vascular plant of San Francisco Bay's mudflats, colonizes surfaces near MTL (Pestrong 1972; Hinde 1954). Approximately equating MTL with mean sea level, estimating former mean sea levels from radiocarbon-dated marsh deposits elsewhere in southern San Francisco Bay (Atwater 1979, Fig. 5), and correcting elevations for local subsidence due to withdrawal of groundwater (Poland 1971), we infer that marshes such as Palo Alto Baylands originated within the past 2,000 years. The 4,000-yr lag between the inception of a slow rate of submergence and the creation of such marshes probably represents the time required for sedimentation to make up for the effects of earlier, more rapid submergence.

TABLE 1. DATUMS FOR TIDE LEVELS AND HEIGHTS.[a]

Datum	Abbreviation	Definition
Mean higher high water	MHHW	Average height of the higher of the daily high tides
Mean high water	MHW	Average height of all high tides
Mean tide level	MTL	Plane halfway between mean high water and mean low water, also called half-tide level
Mean low water	MLW	Average height of all low tides
Mean lower low water	MLLW	Average height of the lower of the daily low tides. Adopted as plane of reference for hydrographic surveys and nautical charts of the west coast of the United States
Mean sea level	MSL	Average height of the water surface for all stages of the tide, determined from hourly readings
National Geodetic Vertical Datum of 1929	NGVD	The standard datum for heights across the nation. Formerly called the "U.S. Coast and Geodetic Survey sea-level datum of 1929," and originally determined from mean sea levels at 26 tide stations in the United States and Canada. Generally differs from local mean sea level (Fig. 1), so it is best regarded as an arbitrary datum that happens to be close to mean sea level.

[a] Tidal datums are ideally determined from 19 years of measurement, but shorter series of observations may be compared with a long-term record to determine mean values.

SAN FRANCISCO BAY

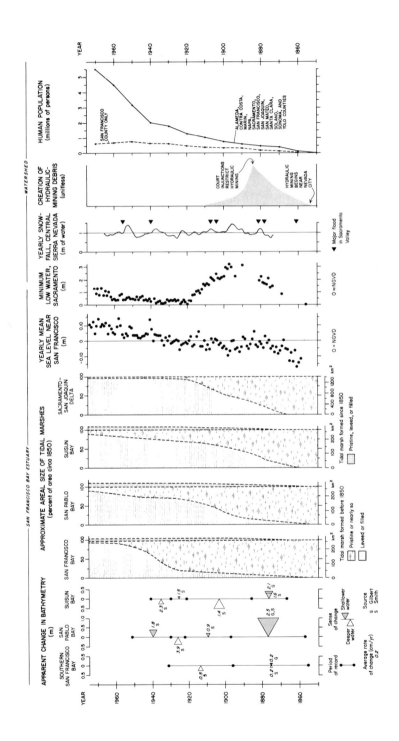

Fig. 1. Some historic changes in the San Francisco Bay estuary and its watershed.

BATHYMETRY. Apparent changes in water depth are adapted from comparisons of archival hydrographic maps as performed by Gilbert (1917:32-37) and Smith (1965). Few of the average apparent changes in bathymetry can be attributed with confidence to net erosion or deposition on the Bay floor because of possible errors introduced by methods of bathymetric comparison. The average decrease in bathymetry for all of San Pablo Bay during the late 19th century, however, seems too large to ascribe to such errors, particularly because some areas of this bay appear to have shoaled by 1.5-2.5 m (Gilbert 1917:35).

TIDAL MARSHES. Initial size of tidal marshes follows Gilbert (1917:78) and Nichols and Wright (1971). Approximate areas at later dates are compiled and modified from textual and cartographic information of Gilbert (1917:86-88); Matthew and Blackie (1931:96-98); Matthew et al. (1931:158); Cosby (1941: 16); Van Royen and Siegel (1959); Atwater and Hedel (1976); Kingsley et al. (1977:3); archival topographic maps of the USC&GS examined during preparation of Figs. 3, 6; and planning maps of the BCDC (1969). Precision is greatest for changes at the Delta and San Francisco Bay, least for Suisun Bay.

MEAN SEA LEVEL. Dots show yearly mean of the surface elevation of San Francisco Bay at Fort Point (1857-1877), Sausalito (1877-1896), and the Presidio (1897-1974). The compilation was prepared by the NOS for the CSLC. NGVD denotes National Geodetic Vertical Datum (Table 1).

LOW WATER. This statistic denotes the lowest reported level of the Sacramento River at Sacramento as compiled by Gilbert (1917: 29-30) and the U. S. Weather Bureau (1914-1971). Lowest levels occur during the low-flow summer or autumn months. The broad peak centered between 1890 and 1900 reflects changes in the level of the river bed. These changes presumably indicate deposition and subsequent erosion of hydraulic-mining debris, which progressed downriver like a wave (Gilbert 1917:29-31). Some smaller changes may reflect variation in the discharge of water, and others may be artifacts of infrequent (once daily) measurement.

SNOWFALL. The curve, from Curry (1969:28) shows smoothed yearly records of snowfall at Donner Pass. Dates of major floods on the Sacramento River (triangles) follow Thompson (1958:446-467).

MINING DEBRIS. This graph, from Gilbert (1917:36), sketches the relative output of sediment from hydraulic gold mines of the western Sierra Nevada and from Sierran streams that carried debris from these mines.

POPULATION. Graph shows decennial changes in counties surrounding the San Francisco Bay estuary, as reported by the U.S. Census Office (1901: 11) and by the County Supervisors Association of California (1975:22).

SAN FRANCISCO BAY

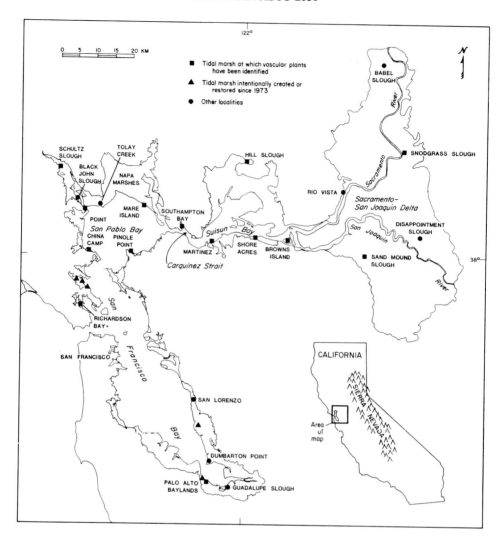

Fig. 2. San Francisco Bay estuary and the Sacramento-San Joaquin River Delta. Locations of newly created marshes follow Kingsley and Boerger (1976), Knutson et al. (1976), and Harvey et al. (1977:85). Thin lines approximate shorelines or margins of tidal marsh ca. 1860 as shown on compilations by Gilbert (1917:76) and Nichols and Wright (1971).

Events since the California Gold Rush

Early surveys by Ringgold (1852) and the USC&GS portray the San Francisco Bay estuary as it appeared during the California Gold Rush. Collectively, the tidal marshes and open-water bays covered about 3400 km², an area slightly larger than Rhode Island. The area of tidal marsh was nearly double the area of the bays, with the Delta marshes making up about 1400 km² and the marshes bordering San Francisco, San Pablo, and Suisun bays accounting for another 800 km² (Gilbert 1917:78).

Approximately 95% of the estuary's tidal marshes have been leveed or filled since the Gold Rush (Figs. 1, 3). The typical age of levees varies with location and appears to depend on the

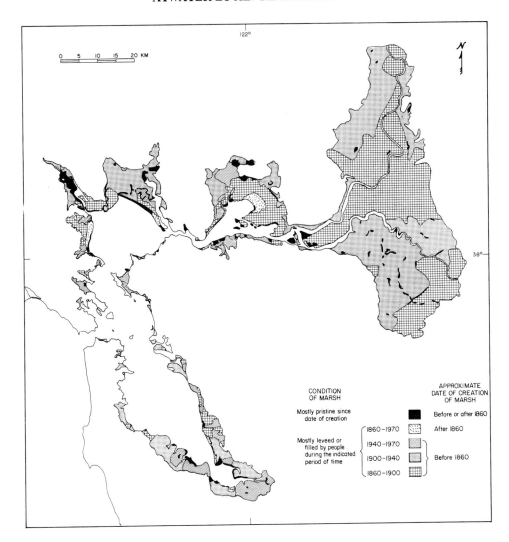

Fig. 3. Generalized and approximate historic changes in aerial distribution of tidal marshes. The map scale requires that patterned areas exceed about 0.5 km in width and about 0.5 km² in area. Dates for levees and fill follow Van Royen and Siegel (1959) for San Francisco Bay. Around San Pablo and Suisun Bays, dates are estimated by inspection of archival maps of the USC&GS (scale generally 1:10,000) and the USGS (scale 1:31,680; Suisun Bay only); Van Royen and Siegel incompletely discriminate for these bays between marshes enclosed by levees and marshes subject to natural inundation by tides. Changes in the Delta follow Matthew et al. (1931, pl. 34). Areas of modern tidal marsh are compiled from topographic maps of all areas by the USGS (1968 and 1973 editions, scale 1:24,000), from a landuse map of San Francisco, San Pablo, and Suisun bays by the BCDC (1969; approximate scale 1:250,000), and from maps by the California Department of Fish and Game (CDF&G) showing dominant vascular plants north of Suisun Bay as of 1973 (H. A. George, unpublished map). Some modern marshes have undoubtedly been omitted or misrepresented.

suitability of marshland for agriculture. Fresh-water marshes characterized the pristine Delta, and most of these were leveed for farming before 1920. Marshes nearer the Golden Gate, on the other

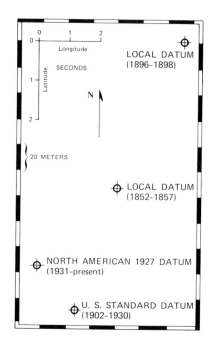

Fig. 4. Horizontal datums for maps of southern San Francisco Bay. The circled crosses locate intersecting lines of the geographic projection for latitude and longitude. For example, the intersection of a latitude and longitude according to the local datum of 1896-1898 is located 176 m N and 117 m E of the intersection of the same latitude and longitude according to the North American 1927 Datum. Such relationships provide precise geodetic means for comparing features shown on archival and modern maps (Fig. 5). The local datums are reconciled to each other and to the nationwide datums by comparing the published positions, as referenced to the various horizontal datums, of persistent triangulation stations. These positions appear in: (1) annual reports of the superintendent of the USC&GS, 1851-1868, 1904, and 1910; (2) "Old Registers" maintained by the Early Drawing Division, U. S. Coast Survey, 1850-1927; and (3) a special publication on triangulation in California (Mitchell 1936). Early geodesists in California, led by George Davidson, established local datums astronomically and projected them according to a mathematical model of the shape of the earth called the Bessel Spheroid of 1841. These datums yielded to others as a result of several advances in geodesy: (1) a more accurate determination of longitude in the 1870's; (2) the application of the Clarke Spheroid of 1866; (3) the completion of the transcontinental triangulation network in the late 1880's, which led to the adoption of the U. S. Standard Datum in 1902; and (4) readjustment of the network during the early 1920's, which resulted in the North American 1927 Datum (Shalowitz 1964:112-114, 141-158).

hand, contained more salt in their soils and sloughs. This salt presumably limited the agricultural value of the marshland and therefore delayed the construction of levees until other human uses were found. For example, whereas agricultural levees in 1900 enclosed about half of the historic marshes of the Delta and Suisun Bay, less than one fifth of the Bay's marshes had been leveed by that time, and these mostly for production of salt. Tidal marshes around the Bay remained largely pristine until later in the 20th century when, coincident with a rapid increase in the population of suburban areas, most of the marshes were converted to salt-evaporation ponds or to sites for residential and industrial structures, transportation facilities, and garbage dumps.

Changes in mapped shorelines (Figs. 3, 5, 6) indicate that approximately 75 km² of new tidal marsh have appeared around San Francisco, San Pablo, and Suisun bays since the Gold Rush. It seems likely that humans accidentally created much of this marshland by supplying sediment to the bays and by building levees and jetties that promoted deposition.

The widespread and rapid expansion of marshland during the late 19th century (Fig. 6) probably resulted in large measure from contemporaneous hydraulic mining in the Sierra Nevada. Between 1853 and 1884, gold miners washed prodigious quantities of sediment into Sierran streams. Further downstream, this debris caused damage to farmlands and waterways and thereby led to court injunctions that effectively halted hydraulic mining (Gilbert 1917:11; Briscoe 1979). Much debris travelled even further and entered the San Francisco Bay estuary, as evidenced by the wave of river-bottom sediment, presumably sand and gravel, that crested near the northern end of the Delta between 1890 and 1900 (Fig. 1, "minimum low water"). Clay and fine silt from the hydraulic mines should have reached the estuary sooner because they move in suspension, and deposition of this fine sediment apparently caused both shoaling of subtidal areas in San Pablo and Suisun bays and rapid horizontal expansion of marshlands into mudflats of northern Suisun Bay, western San Pablo Bay, and southern San Francisco Bay during the late 19th century (Gilbert 1917:36, 86-88; Figs. 1, 6). The delivery of mining debris to the margins of southern San Francisco Bay may be doubted because few subtidal areas of this bay shoaled greatly during the late 19th century (Fig. 6; Krone 1979) and because Ferdinand Westdahl, the topographer who mapped much of the expanded marshland, designated commercial oyster shells and Oakland's growing port as the causes of tidal-marsh accretion (Westdahl 1897). Nevertheless, large quantities of silt- and clay-size mining debris certainly reached San Pablo Bay; some of this sediment undoubtedly entered the layer of low-salinity water that, according to McCulloch et al. (1970) and Carlson and McCulloch (1974), spreads across southern San Francisco Bay during periods of high discharge from the Sacramento River; and, once delivered to southern San Francisco Bay, mining debris might have preferentially accumulated on marginal tidal flats as clay and silt appear to be doing today (Conomos and Peterson 1977).

Tidal marshes have probably received additional sediment from farmlands (Gilbert 1917:36), urbanized uplands (Knott 1973), and dredged channels and harbors. Moreover, many future marshes may rest entirely on dredged material if, as currently planned, public agencies mitigate the disposal of dredged material by intentionally creating tidal marshes on spoils. Beginning with the experiments of H. T. Harvey during the 1960's, both independent parties and members of the U.S. Army Corps of Engineers have demonstrated that people can establish such marshes by planting seeds, seedlings, and cuttings of California cordgrass in previously barren areas (Knutson et al. 1976; Kingsley and Boerger 1976; Fig. 2).

Levees and jetties have also contributed to the creation of marshlands, particularly during the 20th century. Construction of levees around tidal marshes almost invariably preceded and probably caused the historic appearance of new marshes along the banks of sloughs that formerly served the marshes (K. Dedrick pers. comm.). Presumably, sediment accumulated because the levees prevented exchange of water with the former tidal marshes and thereby slackened currents in the sloughs (Gilbert 1917:102-103). Levees may have also promoted expansion of marshland into the bays by reducing the area in which sediment could accumulate. According to Robert Nadey (pers. comm.), this effect may account for part of the spectacular growth of marshes into western San Pablo Bay during the period of hydraulic mining. Extension of a jetty south of Mare Island during the 20th century coincides with shoaling of nearby tidal flats (Smith 1965) and expansion of nearby marshland. The jetty probably caused these changes because, during the late 19th century, the marsh eroded rather than advanced, even though other parts of San Pablo Bay were trapping large quantities of hydraulic-mining debris (Figs. 6, 7).

Many changes in mapped shorelines indicate erosion rather than deposition at the bayward

SAN FRANCISCO BAY

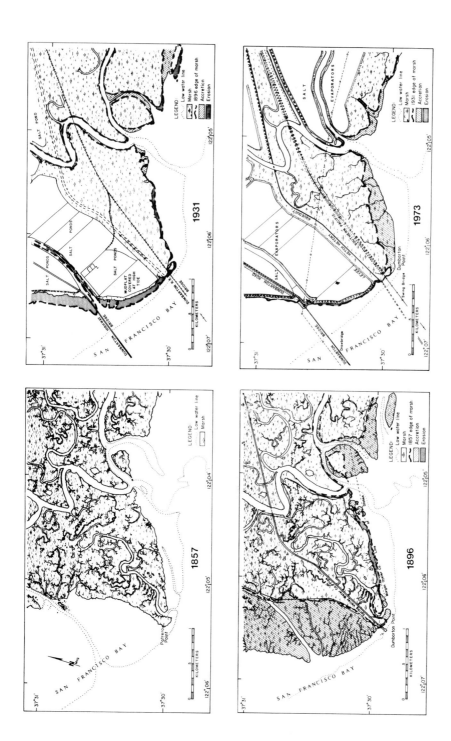

margins of tidal marshes (Fig. 6). As inferred by Gilbert (1917:21-22), such retreat may result in part from a rise in sea level relative to the land. Other contributing factors include burrowing by an introduced isopod (Carlton 1979).

Available inventories of the areal size of tidal marshes at a given date employ different means of distinguishing "tidal" from "leveed" and rarely discriminate between marshes formed before and after 1850. These problems, together with the infrequency of such inventories, currently preclude a detailed summary of historic changes in tidal-marsh areas. Even the generalized graphs and maps (Figs. 1, 3) imply unwarranted precision, as the reader can determine by consulting the cited sources. Despite such deficiencies, our estimates justify several conclusions about the effects of levees and fill on the San Francisco Bay estuary: (1) The present area of tidal marsh within the entire estuary is about 125 km^2, one third of which has originated since 1850. Both Van Royen and Siegel (1959) and Nichols and Wright (1971) offer a much higher figure for the total area of marshland because their tallies include many non-tidal leveed marshes. (2) Even excluding the Delta, leveed or filled tidal marshes cover far more area than the 140 km^2 of open-water baylands that have been leveed, filled, or converted to tidal marsh since the Gold Rush (Nichols and Wright 1971). Thus, tidal marshes rather than open-water bays have provided most of the leveed and filled areas of the San Francisco Bay estuary.

LANDFORMS

Natural topographic features of the tidal marshes of the San Francisco Bay estuary include broad, nearly flat surfaces; narrower surfaces that descend into tidal flats, some precipitously; beach ridges and related berms; tidal sloughs; riverine channels and their natural levees; shallow ponds and pans; and islands of pre-existing bedrock and sand dunes. The following discussion emphasizes the principal kinds of marshlands and waterways.

Plains near High-Tide Levels

Around the turn of the century, the "typical tidal marsh" of the San Francisco Bay estuary

---

Fig. 5. Historic changes in tidal marshes near Dumbarton Point, southern San Francisco Bay. The archival maps copy lines and names from 1:10,000-scale plane-table sheets of the USC&GS: T-634, surveyed in 1857 by David Kerr; T-2258, surveyed in 1896 by Fremont Morse and Ferdinand Westdahl (low-water line from contemporary hydrographic surveys H-2304 and H-2413); and T-4626, surveyed in 1931 by H. G. Conerly (low-water line from contemporaneous hydrographic survey H-5135). The modern map is traced from a 1:24,000-scale photogrammetric map (Newark 7.5-min quadrangle). Additional symbols show old shorelines and inferred areas of erosion and deposition. Registration of archival maps to the North American 1927 Datum (Fig. 4) controls the comparisons of charted shorelines. Though subject to uncertainties related to method and season of surveying, shorelines approximate the bayward limit of vegetation (Gilbert 1917:86; Shalowitz 1964:177). The density and distribution of sloughs reflects methods of mapping and changes related to human structures. Morse and Westdahl apparently traced Kerr's lines for most of the mid-19th century sloughs (R. Nadey pers. comm.); according to Westdahl's (1896) description of a nearby survey "on the salt marshes only the sloughs used for navigation, the shore-line, the area between the old and new bayshore, and improvements, such as dykes, houses, and saltworks, have been surveyed." Conerly apparently neglected all but the largest sloughs. Differences between modern sloughs and those mapped by Kerr imply that the railroad and salt ponds have disrupted the original pattern of drainage.

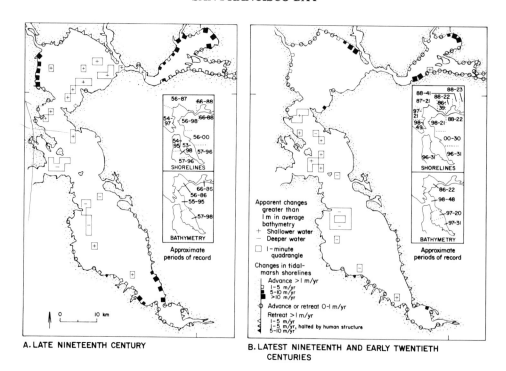

Fig. 6. Historic changes in tidal-marsh shorelines and subtidal bathymetry (A) during the late 19th century and (B) during the latest 19th century and early 20th centuries.

SHORELINES. Changes in shoreline area measured by comparing archival topographic maps that were prepared by the USC&GS. Comparisons make use of 1:24,000-scale photographic reductions of the CSLC. Most maps are registered to one another by matching persistent features such as hills, rocky shorelines, intricately meandering sloughs, railroad tracks, and occasional triangulation stations. Precise geodetic registration (Figs. 4, 5) is limited to the southern and eastern shores of San Francisco Bay. Elsewhere the uncertainties in registration, together with errors in surveying, possible distortion of original map paper, and possible differences in notation for tidal-marsh shorelines (Gilbert 1917:86), prevent resolution of changes that average less than 1 m·yr^{-1} over 20- to 40-yr intervals.

BATHYMETRY. Changes in bathymetry for subtidal areas excluding sloughs are adapted from Smith's (1965) comparison of archival hydrographic maps. Smith compared average depths within 1/8-min quadrangles and reported the sum of changes for 1-min quadrangles. The distribution of areas showing large changes in bathymetry must be interpreted with reference to index maps because the magnitude of change depends partly on the length of the period of record. Dates on index maps omit numerals for century and millennium.

resembled "a plain traversed by a branching system of sloughs" (Gilbert 1917:75). Excluding sloughs, the relief on such "plains" must have been slight because topographers such as Ferdinand Westdahl (1897) used "the level of the salt-marsh in its natural state" as a datum plane for upland elevations.

Nearly flat surfaces appear to remain the most extensive landforms of the tidal marshes of the San Francisco Bay estuary (Fig. 7; Bodnar et al. 1975: Figs. 32, 33; Hinde 1954). These plains characterize marshland formed both before and after 1850, and they cross historic shorelines without appreciable change in level or relief (China Camp and Mare Island marshes, Fig. 7).

Within uncertainties of measurement[1], most of the broad surfaces (Fig. 7) are probably situated within a few decimeters of MHHW. This coincidence implies a widespread tendency of tidal-marsh surfaces to approach high-tide levels. Presumably, such heights equilibrate deposition, erosion, and subsidence (Pestrong 1972).

Differences in elevation between some tidal-marsh plains, however, exceed probable errors in measurement. The flat surface of the marsh at Richardson Bay appears to be situated about 0.2 m below MHHW (Fig. 7), and broad parts of several tidal marshes also occupy elevations below MHHW along the western shore of southern San Francisco Bay northwest of Palo Alto Baylands (K. Dedrick, pers. comm.). Typical elevations near Point Pinole, on the other hand, approximately equal MHHW according to third-order leveling (Bodnar et al. 1975: Figs 32, 33), and less precise measurements at the nearby China Camp marsh (Fig. 7) suggest similar elevations. Furthermore, marshland near Mare Island appears to rise 0.2-0.5 m above MHHW (Fig. 7). Both here and at Palo Alto Baylands, however, probable but unmeasured subsidence of the bench mark at the origin of the transect (Table 2, footnotes 3, 5) may erroneously heighten the measured elevations.

These geographic variations in the elevation of marshlands with respect to tidal datums imply that local conditions influence topography. The entrapment of suspended sediment in San Pablo and Suisun bays (Conomos and Peterson 1977), for example, may partly explain why post-1850 marshes at Richardson Bay have reached lower levels than contemporaneous marshes at China Camp. Anomalously high elevations near Mare Island may reflect not only such entrapment but also southerly high winds, which potentially pile water above normal high-tide levels, and the nearby jetty, which has promoted intertidal deposition (see above discussion of levees and jetties). Finally, subsidence due to withdrawal of groundwater (Poland 1971) dropped the tidal marsh at Palo Alto Baylands to a lower level between 1954 and 1965, as evidenced by the spread of cordgrass into areas whose former elevation may have excluded this plant (Harvey 1966). Most of this marsh is nevertheless situated at or near MHHW (Fig. 7), so it seems likely that sedimentation has, on the average, largely maintained the level of the marsh against subsidence, which amounts to nearly 1 m since 1931 (Table 2, footnote 3).

Uncertainties and geographic variation in the elevations of modern tidal-marsh plains compound the problems of defining, relative to tidal datums, the "level of salt-marsh in its natural state" as of the 19th century. One possible solution presumes similarity between the elevations of modern and historic marshes and therefore must allow for differences between localities. Additional complications arise if marshlands have reached unnaturally high levels because of human activities such as hydraulic gold-mining, disposal of dredge spoils, and construction of levees and jetties. A remark by Gilbert (1917:77) indirectly supports this hypothesis by equating areas vegetated by tules (*Scirpus* spp.) and California cordgrass with the "broader parts" of marshes. Currently, these plants mostly grow along narrow surfaces that descend into mudflats or sloughs (Fig. 7), so it seems possible that the few remaining pristine marshlands have risen to extraordinary heights during this century, perhaps by trapping hydraulic-mining debris and dredge spoils. Alternatively, Gilbert erred, perhaps by attributing to all marshlands the characteristics of those that spread across mudflats during the late 19th century (Figs. 5, 6) and initially supported California cordgrass and tules rather than common

---

[1] With respect to tide levels, the elevations of modern surfaces reported in Fig. 7 may err by 0.1 m or more because of undetermined changes in the published elevations of bench marks and tidal datum planes (Table 2), extrapolation or interpolation of datums from distant tide stations, and imprecise methods of leveling. The leveling generally fails to meet several of the National Ocean Survey's (NOS) standards for third-order work (NOS 1974): (1) maximum length of sights—some sights exceed the 90-m standard by 10-30 m; (2) difference in length of forward and backward sights—the 10-m standard is met for turning points, but most elevations along transects represent unbalanced forward and backward sights between turning points; and (3) minimum error in closure—complete closure was not attempted at marshes shown in Fig. 7, and partial closure at China Camp and Palo Alto Baylands indicate cumulative errors of about 5 cm, roughly three times as large as the standard.

SAN FRANCISCO BAY

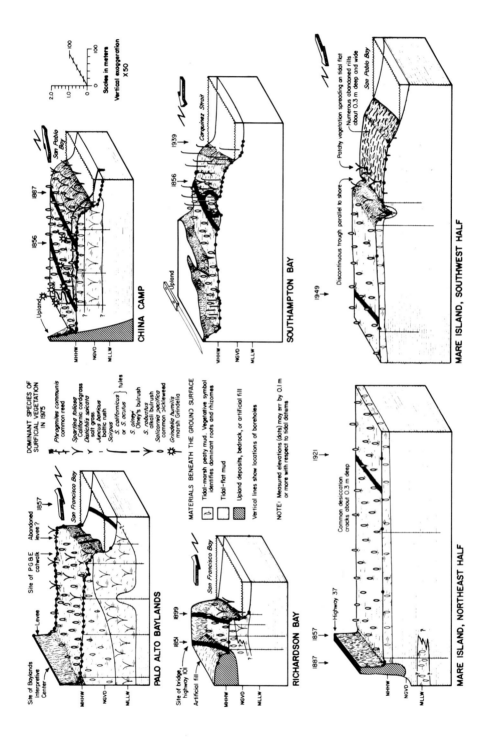

TABLE 2. REFERENCE STATIONS FOR ELEVATIONS AND TIDAL DATUMS SHOWN IN FIG. 7

MARSH	BENCH MARK[a]			TIDAL DATUMS				
				Location of gauge		Elevations at gauge (m)[b]		
	Designation	Elevation (m)	Year of leveling	Place name	Distance from marsh (km)	MLLW	MTL	MHHW
Palo Alto Baylands	Tidal No. 1	1.09[c]	1965	Palo Alto Yacht Harbor	1	-1.3	0.2	1.4
Richardson Bay	R481	2.60[d]	1955	San Francisco (Presidio)	10	-0.8	0.1	0.9
China Camp	D552	12.56[d]	1956	Pinole Point	11	-0.9	0.2	1.0
Mare Island	N466	1.21[e]	1956	Hercules	11	-0.8	0.2	1.0
Southampton Bay	C467	4.94[d]	1951	Crockett	4	-0.8	0.2	1.0

[a] From "Vertical Control Data" compiled by the National Geodetic Survey. Datum is NGVD.

[b] Compiled by the National Ocean Survey (1977a, 1977b) and referenced to leveling completed in 1956 or, for Palo Alto, 1967. Datum is NGVD. See Table 1 for definition of reference planes.

[c] Repeated leveling by the U. S. Coast and Geodetic Survey indicates that the elevation of Tidal No. 1 decreased 0.76 m between 1931 and 1965, chiefly because of regional subsidence accompanying ground-water withdrawal (Poland: 1971). Hinde (1954: 217) apparently used the initial elevation of the benchmark, and his measured elevations may therefore be too high by 0.5 m, the change in elevation of Tidal No. 1 between 1931 and 1955. Changes in elevation since the 1965 leveling are ignored here because they are probably 0.2 m or less; artificial recharge of ground water halted subsidence near Palo Alto by 1971 (Poland: 1971).

[d] Leveled only once by the U. S. Coast and Geodetic Survey. Changes in elevation since year of leveling presumably do not exceed 0.05 m because monument rests on bedrock (D552, C467) or on a concrete pier supporting a 10-lane bridge (R481).

[e] According to repeated leveling by the U. S. Coast and Geodetic Survey, bench mark N466 subsided 0.04 m between 1951 and 1956. No correction for unmeasured, subsequent movement is attempted here, but at least 0.10 additional subsidence seems likely because the road embankment beneath the monument overlies compressible estuarine sediments and because fill has been added to the embankment or adjacent road since 1956.

Fig. 7. Generalized landforms, vegetation, and subsurface sediments of some salt- and brackish-water tidal marshes. (Vertical exaggeration 50X). Dots on profiles denote places where elevation was surveyed in 1975 or 1976. Elevations were transferred with a rod and tripod-mounted level from the nearest geodetic or tidal bench mark (Table 2). No correction is made for probable but unmeasured subsidence of bench marks near Palo Alto Baylands and Mare Island. Methods of surveying meet only some of the standards for third-order leveling (see footnote 1 in text). Tidal datums are extrapolated from the nearest long-term tide gauge for which the relationship between MLLW and NGVD has been determined (Table 2). Collectively, these procedures may cause measured elevations to err by 0.1 m or more with respect to tidal datums. Elevations away from dots are estimated by extrapolation, chiefly with reference to vegetation. Small channels are generally omitted. The water surface is at MTL. Gray bands approximate the horizontal position of the bayward limit of vascular plants at the indicated date, as interpreted from archival maps (Figs. 5, 6). The distribution of vascular plants on the surface shows approximate conditions in 1975 (Fig. 10). Appendix A lists native species at all localities except Mare Island. Fossil rhizomes (below-ground stems) and roots are tentatively identified by macroscopic examination of core samples. Most of the tidal-flat mud lacks roots or rhizomes in growth position. The Mare Island diagrams join at center.

pickleweed (*Salicornia pacifica*) and salt grass (*Distichlis spicata*) (see description by Westdahl 1896).

Sloping Surfaces Bordering Mudflats

Marshes unquestionably vary in the slope of surfaces that descend into tidal mudflats. Near China Camp and at Southampton Bay, marshland dips gently into adjacent mudflats (Fig. 7). At Palo Alto Baylands and Richardson Bay, on the other hand, most of the bayward edge of the marsh drops precipitously.

Gently sloping margins imply net deposition and precipitous margins imply net erosion. Several lines of evidence support these inferences: (1) gently sloping margins correspond with shorelines that typically migrated toward the Bay during the late 19th and early 20th centuries, and precipitous margins characterize shorelines that generally retreated during this period of time (Fig. 6); and (2) precipitous slopes locally correlate with ongoing erosion, as indicated around parts of southern San Francisco Bay by blocks of tidal-marsh mud that slump from vertical or overhanging scarps onto the adjacent mudflat.

Low beach ridges historically bordered some tidal marshes of the San Francisco Bay estuary (Gilbert 1917:86), and a few ridges remain today. Beach ridges apparently impounded marshlands near San Lorenzo and thereby created natural salt ponds (Fig. 8). By analogy with sandy barriers that fringe marshes of Delaware Bay (Kraft et al. 1976:98-104), these ridges may have contained sand that had been derived from eroding headlands, particularly the ancient dune sands near Oakland (Atwater et al. 1977). Other beach ridges of the San Francisco Bay estuary are made of shell (Westdahl 1897). A possibly related feature of unknown origin is the broad berm of mud that currently appears to rise above MHHW at the bayward margin of a marsh near Mare Island (Fig. 7). A similar berm probably forced sloughs near this locality to drain away from San Pablo Bay in 1856 (Fig. 8).

Waterways

Patterns of tidal-marsh drainage around San Francisco, San Pablo, and Suisun bays depend partly on the age of surrounding marshland. Whereas prominent meanders characterize the sloughs of marshes created before 1850, the sloughs of younger marshes follow relatively straight paths that trend nearly perpendicular to the bayward edge of the marsh. Such direct paths cross modern marshlands (Fig. 7, China Camp) as well as their 19th-century ancestors (Fig. 5; Westdahl 1897) and therefore appear to have gained little sinuousity since formation. The contrast between straight and meandering sloughs may reflect differences in the rate of formation of marshland if, as seems likely from enormous changes in historic shorelines (Fig. 6), marshes drained by straight channels initially spread and rose at an extraordinarily rapid pace during the late 19th century.

Tidal water rather than the discharge of upland creeks controls the dimensions of most sloughs around the bays. Pestrong (1965:32-33) and Gilbert (1917:102-103) implicitly advocated such control, Pestrong by adopting Chapman's (1960:30) conclusion that sloughs grow because of the flow of tidal water to and from an upward-building marsh, and Gilbert by proposing that reduction of this flow, owing to impoundment of marshland behind levees, caused shoaling of a slough near Mare Island. Moreover, although the widths of waterways commonly increase with discharge (Myrick and Leopold 1963), the widths of historic tidal sloughs north of San Pablo Bay greatly tapered toward upland creeks (Nichols and Wright 1971), so it seems likely that the widths of these sloughs depended mainly on the areas of their tidal-marsh drainage basins. Such drainage basins must also account for the considerable widths of sloughs near Guadalupe Slough and Mare Island (Fig. 8) that drained no major upland creeks.

Riverine floods, on the other hand, probably restricted the reach of tides in the northern

Delta by creating natural levees along the channels of rivers and distributaries. Near Babel Slough, the Sacramento River built natural levees about 1 km wide and up to 5 m high (Fig. 9). Such levees diminished in height toward Suisun Bay but extended as far downstream as the confluence of the Sacramento and San Joaquin Rivers (Thompson 1958:26; Ringgold 1852). At autumnal low stages of the rivers, high tides probably could not surmount many of the levees in the northern Delta, so perhaps only riverine floods inundated low-lying marshes that were enclosed by naturally leveed channels. Thus, some areas designated as historical tidal marsh in Figs. 1-3 may have actually been isolated from autumnal tides.

Natural levees in the southern Delta generally reached much lower elevations, as evidenced by archival records (Thompson 1958:37), by tidal sloughs that transect levees of the San Joaquin River (Fig. 9), and by peaty soils along the San Joaquin River that contrast with the bands of inorganic soil bordering waterways of the northern Delta (Cosby 1941). Consequently, it seems probable that the southern part of the pristine Delta was flooded and drained more nearly like tidal marshes of the bays than like the naturally leveed marshes near the Sacramento River.

VEGETATION

Vascular plants[2] visually dominate the vegetation of tidal marshes and distinguish the marshes from mudflats. Our discussion of these plants considers their distribution with respect to geographic location, elevation, and other environmental variables. In addition, we attempt to estimate the quantity of organic material that vascular plants of tidal marshes export to the rest of the estuary.

Distribution of Species

Geographic and vertical trends. About 125 species of vascular plants have been reported from tidal marshes of the San Francisco Bay estuary. Most of these species are native to California (Appendix A), but some have been introduced from other parts of the world (Table 3).

Diversity generally increases from San Francisco Bay to the Delta. Whereas individual marshes around San Francisco Bay typically contain 13 or 14 species of native plants, specific sites in the Delta contain about 40 species. Composite regional lists imply even greater differences in diversity: only 15 native species reportedly live in tidal marshes around San Francisco Bay, but about 30 reportedly live around San Pablo Bay and Carquinez Strait, 40 around Suisun Bay, and 80 in the Delta.

San Francisco Bay and the Delta differ in kinds as well as numbers of tidal-marsh plants. Inhabitants of San Francisco Bay's marshes belong to the group of plants that characterize California salt marshes (Macdonald 1977). Few species from San Francisco Bay, however, have also been reported from tidal marshes of the Delta. Rather, the Delta's marshes are dominated by other plants that typically inhabit low-altitude fresh-water marshes in California (Mason 1957).

Common pickleweed (*Salicornia pacifica*) and California cordgrass (*Spartina foliosa*) dominate the tidal-marsh vegetation around San Francisco Bay. Common pickleweed generally monopolizes tidal-marsh plains at elevations near and above MHHW (Hinde 1954:218). Excepting salt-marsh dodder (*Cuscuta salina,* a parasite on common pickleweed), additional species on tidal-marsh plains typically grow in scattered patches next to sloughs, natural uplands, and man-made levees. These plants include salt grass (*Distichlis spicata*), marsh Grindelia (*Grindelia humilis*), halberd-leaved saltbush (*Atriplex patula* ssp. *hastata*), alkali heath (*Frankenia grandifolia*), and

---

[2] Vascular plants (Phylum *Tracheophyta*) contain veinlike channels that convey metabolic materials between roots, stems, and leaves. Other kinds of tidal-marsh plants, such as diatoms, are not described in this chapter.

SAN FRANCISCO BAY

Fig. 8. Marshes of San Francisco and San Pablo bays as mapped before significant human disturbance. Locations on Fig. 2.

SOURCES OF INFORMATION. Channels and ponds are traced from unpublished 1:62,500-scale compilations, by D. R. Nichols and N. A. Wright, of 1:10,000-scale topographic maps prepared shortly after the California Gold Rush by A. F. Rodgers and David Kerr of the U. S. Coast Survey. Topographic contours, shown near Palo Alto Baylands only, are generalized from a modern 1:24,000-scale topographic map.

INTERPRETATIONS. Sloughs near Palo Alto Baylands, surveyed in 1857, show relation of tidal-marsh channels to active and abandoned mouths of an ephemeral fresh-water stream, San Francisquito Creek. The active mouth of this stream joins an average-size slough. The abandoned mouth lacks a comparable connection with San Francisco Bay, and a finger of marsh occupies a vestige of the old stream channel. Natural levees of both the active and abandoned courses of San Francisquito Creek, built by the stream when it overtopped its banks (Westdahl 1897; Gerow and Force 1968:24-27), cause the topographic contours to point downstream, as on a ridge, rather than upstream, as in a valley.

Guadalupe Slough followed a shortcut to San Francisco Bay when Rodgers and Kerr surveyed its course in 1857. At some earlier time, marshland presumably intervened between the starred meander and the bay (K. Lajoie pers. comm.). Erosion along the edge of the Bay probably removed this marsh. Similar erosion took place in this area during the late 19th century despite the predominance of deposition along most other shorelines (Fig. 6).

Ridges at the bayward margins of marshland may have caused water to collect in large ponds near San Lorenzo and to drain away from San Pablo Bay near Mare Island. Ponds near San Lorenzo appear on maps as old as F. W. Beechey's chart of San Francisco Bay, surveyed in 1827-1828 (Harlow 1850:64). When Kerr mapped them in detail 30 years later, he labelled the largest, "crystal salt pond." Predictably, commercial production of salt from San Francisco Bay began in this area (Ver Planck 1958:107). The berm along the bayward edge of a modern marsh near Mare Island (see Fig. 7) probably resembles the landform that caused the sloughs to drain northward when A. F. Rodgers surveyed them in 1856. A possible ancestor of the discontinuous trough at the southern edge of the modern marsh supplied Rodgers with a name, "Long Pond", for the triangulation station at left.

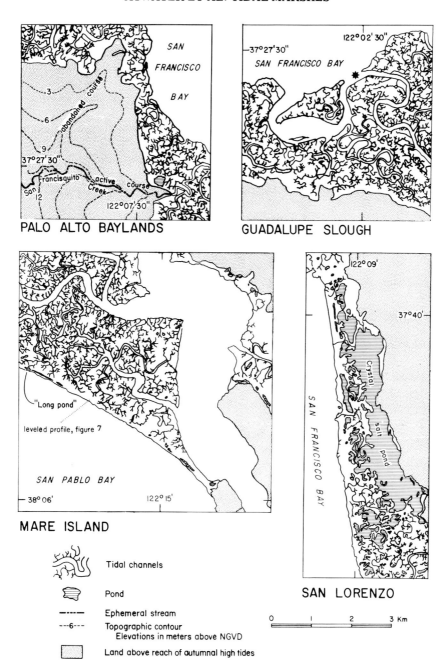

Fig. 9. Marshes of the Sacramento-San Joaquin Delta as mapped before significant human disturbance. Locations on Fig. 2.

SOURCES OF INFORMATION. The USC&GS prepared no detailed maps of pristine marshes in the Delta until 1930-1940. Most channels and topographic contours on these diagrams are based on 1:31,680-scale plane-table sheets surveyed in 1906-1908 by the USGS. Some marshes had been leveed (Fig. 3) and some channels modified before these maps were made. Allowing for errors in map-making, the courses of river channels match the meanders shown on a 1:250,000-scale map by Ringgold (1852). The approximate courses of tributaries to Disappointment Slough are sketched from Cosby's (1941) 1:63,360-scale base map and, where highlighted by tonal differences between soils, from modern aerial photographs. Additional waterways probably existed before construction of dikes, but tall, dense stands of tules (*Scirpus* spp.) and other plants undoubtedly prohibited detailed mapping by plane-table methods. Elevations of natural levees along the Sacramento and San Joaquin rivers are consistent with verbal descriptions assembled by Thompson (1958: 36-37).

INTERPRETATIONS. The Sacramento River created most of the landforms near Babel Slough. The complex lobes of high ground, the largest of which enclosed Babel Slough, were built by sediment-laden flood waters that surged over or through the broad natural levees that flank the Sacramento River. In the bird-foot delta of the Mississippi River, such lobes are called crevasse or overbank splays (Coleman and Gagliano 1964). Paired fingers at the distal ends of the lobes represent the narrow levees of distributaries. Floods converted the Yolo Basin into a lake or river (Gilbert 1917:14-15) that accommodated so much more water than its parent that, on occasion, the discharge from the Yolo Basin transected and hydraulically dammed the Sacramento River near Rio Vista (Thompson 1958:448, 453). The 1.5-m contour locates the approximate northern limit of tidal water in the historic Yolo Basin during times of low Sacramento River discharge. During such low river stages, tides in Yolo Basin probably communicated with the rest of the estuary via the basin's outlet near Rio Vista. The top edge of the map approximates the northern boundary of tidal marsh as mapped in 1906-1908 and as generalized in Figs. 2, 3. Additional marshes covered higher parts of the Yolo Basin according to the USGS plane-table sheets.

Disappointment Slough and its tributaries more nearly resemble the typical drainages of tidal marshes bordering the bays. Lacking a river at its head, Disappointment Slough was probably created and maintained by tidal water that flowed in and out of nearby marshes. Low levees apparently forced some adjoining marshes to drain away from the San Joaquin River but, unlike the high borders of the Sacramento River near Babel Slough, these levees allowed tidal water to traverse the banks of the river in such channels as Disappointment Slough and Twenty-one-mile Slough.

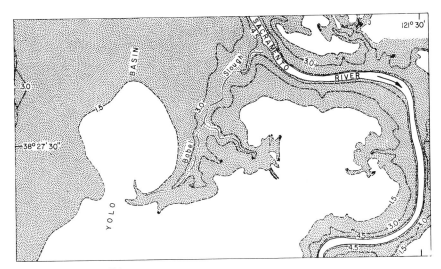

BABEL SLOUGH

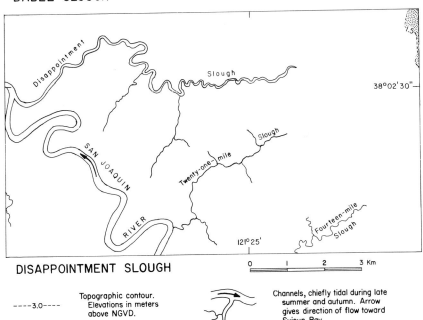

DISAPPOINTMENT SLOUGH

- - - 3.0 - - - Topographic contour. Elevations in meters above NGVD.

Land above reach of most autumnal high tides

Channels, chiefly tidal during late summer and autumn. Arrow gives direction of flow toward Suisun Bay.

— — — — Ephemeral distributary

TABLE 3. COMMON INTRODUCTIONS IN TIDAL MARSHES
OF THE SAN FRANCISCO BAY ESTUARY.[a]

FAMILY	SPECIES	
	Linnean name	Common name
(Monocotyledons)		
GRAMINEAE	*Bromus diandrus* Roth var. *gussonei* (Parl.) Coss & Durieu	Gussone's ripgut grass
Grass family	*B. mollis* L.	Soft chess
	Cortaderia selloana (Schult.) Asch. & Graebn.	Pampas grass
	Festuca elatior L.	Meadow fescue
	Hordeum leporinum Link.	Hare barley
	Polypogon monspeliensis Buckl.	Rabbit's-foot grass
	Spartina patens (L.) Greene	Salt hay
(Dicotyledons)		
CAROPHYLLACEAE	*Spergularia media* (L.) Presl.	Sand-spurrey
Chickweed family		
CHENOPODIACEAE	*Atriplex semibaccata* R. Br.	Australian saltbush
Goosefoot family	*Chenopodium album* L.	Lamb's quarters
COMPOSITAE	*Cirsium vulgare* (Savi) Ten.	Common thistle
Sunflower family	*Cotula australis* (Sieber) Hook.	Australian Cotula
	C. coronopifolia L.	Brass buttons
CRUCIFERAE	*Lepidium latifolium* L.	Broad-leaved pepper-grass
Mustard family		
DIPSACACEAE	*Dipsacus fullonum* L.	Fuller's teasel
Teasel family		
LABIATAE	*Mentha piperita* L.	Peppermint
Mint family		
LEGUMINOSEA	*Melilotus albus* Desr.	White sweet clover
Pea family		
PLANTAGINACEAE	*Plantago major* L.	Common plantain
Plantain family		
POLYGONACEAE	*Rumex crispus* L.	Curly dock
Buckwheat family		
PONTEDERIACEAE	*Eichhornia crassipes* (Mart.) Solms.	Water hyacinth
Pickerel-weed family		
SOLANACEAE	*Solanum dulcamara* L.	Climbing nightshade
Nightshade family	*S. nodifolium* Jacq.	Small-flowered nightshade
UMBELLIFERAE	*Apium graveolens* L.	Celery
Carrot family	*Conium maculatum* L.	Poison-hemlock
	Foeniculum vulgare Mill.	Sweet fennel
VERBENACEAE	*Lippia nodiflora* Michx. var. *rosea* (D. Don) Munz	Garden Lippia
Vervain family		

[a] The list draws from the same sources as Appendix A. In addition, it includes R. E. Mall's report of salt hay at Southampton Bay (Munz 1968:195), a find which we have not duplicated either at Southampton Bay or anywhere else in the estuary. Among grasses other than salt hay, all commonly inhabit the landward fringes of tidal marshes around San Pablo and Suisun Bays except for pampas grass, which grows mainly in the Delta. The principal species among dicotyledons include Australian saltbush (all bays), curly dock (San Pablo and Suisun bays in 1975 but not, with a few exceptions, in 1977), brass buttons (wet places near high-tide levels around Suisun Bay), and garden Lippia (the Delta).

fleshy Jaumea (*Jaumea carnosa*). California cordgrass fringes tidal-marsh plains where they descend into mudflats. Near MTL it forms pure stands, but midway between MTL and MHHW it intermingles with red pickleweed (*Salicornia rubra*), and at higher elevations it yields to common pickleweed. Subsidence due to ground-water withdrawal probably accounts for the anomalous presence of California cordgrass on the tidal-marsh plain at Palo Alto Baylands (Fig. 7; Harvey 1966).

Common tule (*Scirpus acutus*), Olney's bulrush (*Scirpus olneyi*), cat-tails (*Typha* spp.), common reed (*Phragmites communis*) and arroyo willow (*Salix lasiolepis*) dominate islands of pristine marsh in the Delta. Typical associates of these plants include swamp knotweed (*Polygonum coccineum*), broadfruited bur-reed (*Sparganium eurycarpum*) and Pacific silverweed (*Potentilla egedei*) (*Scirpus-Phragmites-Typha* association, Table 4). Another associated species is marsh bindweed (*Calystegia sepium*), a morning glory that twines around tules and reeds. Below MTL these plants yield to monotonous stands of tules (*Scirpus acutus* and *Scirpus californicus*) and, in areas of quiet water, to floating aquatic species (*Ludwigia* association, Table 4).

Tidal-marsh plants of San Pablo Bay, Carquinez Strait, and Suisun Bay provide an intricate, mutable transition between salt marshes of San Francisco Bay and freshwater marshes of the Delta (Table 4, Appendix A; Figs. 7, 10). Details of this transition include: (1) Species from opposite ends of the spectrum overlap to varying degrees in the middle. Most salt-marsh plants of San Francisco Bay live around San Pablo Bay and Carquinez Strait (*Spartina* and *Salicornia pacifica* associations, Table 4) and also around Suisun Bay (Appendix A). Salt grass and marsh Grindelia even grow in the western Delta. Neither California cordgrass nor red pickleweed, however, appear to grow east of Carquinez Strait. Cosmopolitan species of the Delta include tules and bulrushes (*Scirpus acutus, S. californicus, S. olneyi*), cat-tails, and common reed. All of these plants range as far west as the large sloughs north of San Pablo Bay. East of San Pablo Bay they generally supplant California cordgrass (Fig. 10; *Scirpus californicus* association, Table 4). (2) Some common plants of San Pablo and Suisun bays are scarce or absent in tidal marshes of the Bay and Delta. These species include alkali bulrush (*Scirpus robustus*), sea milkwort (*Glaux maritima*), and soft bird's beak (*Cordylanthus mollis*). (3) The vertical range and relative abundance of many species vary with geographic location. Common pickleweed, for instance, shortens its vertical range and reduces its ubiquity and abundance from west to east (Fig. 10). (4) Plant communities change not only from the Pacific Ocean to the Sacramento and San Joaquin rivers but also from mouths to heads of sloughs that drain major upland creeks north of San Pablo and Suisun bays. (5) The vertical and geographic ranges of some species, most conspicuously the tules and bulrushes, can change significantly within one or two years (Figs. 10, 11).

Reasons for trends. Environmental variables that may influence the distribution of vascular plants in tidal marshes include the reproductive methods of the plants, the frequency and duration of tidal flooding, and characteristics of the soil such as particle size, salinity, aeration, moisture, and nutrients (Chapman 1960). Competition between species may also restrict the ranges of some plants. Available evidence from the San Francisco Bay estuary used to test several of these possibilities implies that soil salinity, tidal inundation, and interspecific competition largely control the distribution of local species.

High soil salinity related to saline tidal water causes many plants to disappear toward San Francisco Bay. Too much salt inhibits growth, as evidenced in the case of bulrushes and tules west of the Delta by the decrease in their size and abundance during the drought of 1976-1977 (Figs. 10, 11). The damage or demise of these plants mostly reflects the increased salinity of tidal water rather than the decreased local rainfall because daily high tides inundate the soils of most bulrushes and tules. Excessive salt likewise appears to discourage the growth of bulrushes, cattails, and rushes (*Juncus* spp.) in leveed marshes north of Suisun Bay (Mall 1969:36). Consistent with its reduced seed production in tidal marshes during 1976 and 1977, alkali bulrush produces few seeds in these leveed marshes if vernal soils contain more than 24 °/oo salt (Mall 1969:38).

The salinity of soils may also contribute to the vertical zonation of vascular plants if, as reported from north of San Pablo Bay, salinity during the growing season increases with elevation (Fig. 12). According to field and greenhouse studies by Mahall and Park (1976b), salt rather than aeration or nutrients probably favors pickleweed over cordgrass at high elevations near Black John Slough and Mare Island. Similar considerations may account for the scarcity of tules and bulrushes above high-tide levels around San Pablo Bay, Carquinez Strait, and Suisun Bay.

SAN FRANCISCO BAY

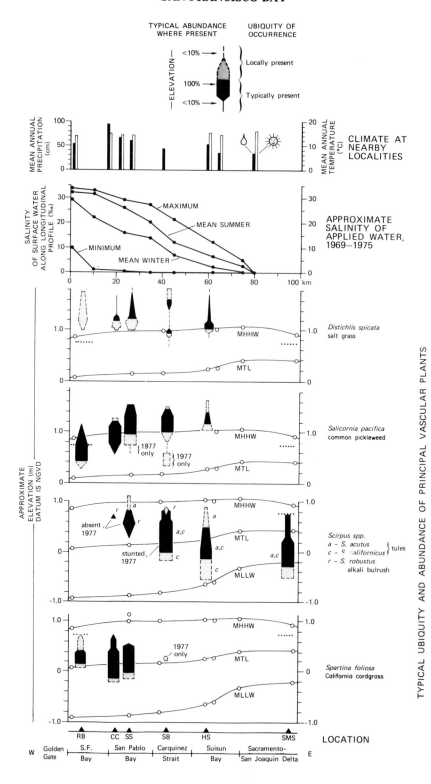

The frequency and duration of tidal flooding commonly correlate with vertical ranges of vascular plants in tidal marshes (Johnson and York 1915; Purer 1942; Hinde 1954). For the San Francisco Bay estuary, this correlation implies causation according to two lines of evidence: (1) the scarcity of cordgrass, tules, and bulrushes above MHHW (Fig. 10) may indirectly result from tidal inundation, if, as seems likely, the vertical increase in salinity reflects more prolonged desiccation at higher elevations; and (2) tidal water may prevent pickleweed from growing at low elevations by dislocating, suffocating, or leaching seeds and seedlings (Chapman 1960:45-49; Mahall and Park 1976c).

Though the disappearance of species toward the Golden Gate reflects the physiological hardships of saline water, the disappearance of species toward the Delta may represent a sociological consequence of fresh water. According to greenhouse experiments, the principal vascular plants of San Francisco Bay's tidal marshes grow better in fresh water than in saline water (Barbour and Davis 1970; Barbour 1970; Phleger 1971). The paradoxical disappearance of these species toward the Delta therefore implies either that saline soils uniquely contain vital nutrients or that other species competitively exclude salt-marsh plants from brackish- and fresh-water areas. The vertical ranges of coexisting, potential competitors (Fig. 10) suggest a role for competition. California cordgrass, for instance, seems to yield to alkali bulrush at elevations greater than 0.5 m at Schultz Slough, but below 0.5 m the abundance of California cordgrass remains the same as at more saline marshes such as China Camp. Similarly, bulrushes and tules appear to eliminate

---

Fig. 10. Regional and vertical distribution of the principal vascular plants in six tidal marshes of the northern San Francisco Bay estuary. SYMBOLS FOR UBIQUITY AND ABUNDANCE OF PLANTS (top). Solid lines and black shading indicate widespread occurrence at or near a given elevation; dashed lines and stippled shading show relatively sparse occurrence. The width of each figure represents abundance and ranges from 1-10% (one line-width) to 100% (broadest part of figure). Abundance approximates the area, relative to other vascular plants, covered by the projected canopy of the live individuals of a given species within a 3-m^2 circle centered at a point of measured elevation. Symbols depict conditions as of autumn 1975, and principal changes observed in autumn 1977. DISTRIBUTION OF PLANTS WITH RESPECT TO APPROXIMATE TIDE LEVELS, BAY-WATER SALINITY, AND CLIMATE (main figure). All localities are projected to the nearest point along a longitudinal profile of the estuary. This procedure generalizes the comparison of vegetation with longitudinal trends in environmental variables; for example, the water serving the marsh near Schultz Slough can contain less salt (Matthew et al. 1931:340-364) and rise to slightly higher levels (see MHHW for Lakeville, identified elsewhere in this caption) than water at the nearest point along the longitudinal profile in southeastern San Pablo Bay. Vertical ranges of plants were measured along or near leveled transects (Fig. 7; Atwater and Hedel 1976). With respect to tidal datums these ranges may err by 0.1 m or more (see text). Horizontal rows of dots show the highest elevation of pristine tidal marsh near transects at Richardson Bay and Sand Mound Slough; plants above this level are rooted in artificial levees. Marshes are abbreviated as follows: RB, Richardson Bay; CC, China Camp; SS, Schultz Slough; SB, Southampton Bay; HS, Hill Slough, SMS, Sand Mound Slough (see Fig. 2 for locations). Open circles along lines for tidal datums represent gauges for which differences between various planes of reference have been determined by the NOS (1977a, 1977b; Table 1). Locations of tide gauges are, from west to east: Presidio (San Francisco); Pinole Point and, for the higher MHHW, on the bottom graph, Lakeville (3 km SE of SS); Crockett (4 km W of SB); entrance of Suisun Slough (about 13 km SW of HS); Port Chicago (between Martinez and Shore Acres), Pittsburg (3 km W of SMS); and Old River at Orwood (10 km SE of SMS). Surface-water salinities follow Conomos and Peterson (1977). Climatic data (U. S. Department of Commerce 1964) refer to the following localities, listed from west to east: downtown San Francisco; San Rafael (between RB and CC); Hamilton Air Force Base (2 km N of CC); Petaluma (3 km NW of SS); Crockett; Port Chicago; Fairfield (3 km W of HS); and Antioch (15 km W of SMS).

TABLE 4. Groups of principal vascular plants in nine tidal marshes of the northern San Francisco Bay estuary during 1976. Symbols in the body of the matrix denote the approximate percentage of the ground surface covered by the projected canopy of species within areas (stands) of 10-20 m^2 having a roughly homogeneous distribution of species: R, single individual; +, 1%; 1, 1-5%; 2, 6-25%; 3, 26-50%; 4, 51-75%; 5, 76-100%. Stands were subjectively located and described in 1976 by the "Braun-Blanquet" or "relevé" method (Mueller-Dombois and Ellenberg 1974:45-66). Upper case headings indicate localities (Fig. 2): BI, Browns Island; BP, Black Point; CC, China Camp; MI, Mare Island (stands located approximately 5 km WNW of square on figure); SB, Southampton Bay; SGS, Snodgrass Slough; SMS, Sand Mound Slough; SS, Schultz Slough. Lower-case headings label individual stands. Species and stands are grouped by similarity in occurrence and species composition, respectively, with the assistance of a computer program developed by Ceska and Roemer (1971) and adapted by David Randall and Dean Taylor (University of California, Davis). Associational groups at bottom designate the five tabular groups at left by their diagnostic species, and subgroups suggest several divisions of the *Salicornia* group. Most species inhabit at least 40% of the stands in their indicated associational group and less than 25% of the stands in other groups. Each stand in the Delta contains more than 50% of the species in the corresponding associational group, but more flexible criteria arrange stands from Carquinez Strait and San Pablo Bay because most contain a small number of species or a large percentage of cosmopolitan species.

GENERAL LOCATION									SAN PABLO BAY									CARQUINEZ STRAIT						SACRAMENTO-SAN JOAQUIN DELTA			
MARSH	MI	PP	MI	CC	CC	BP	MI		CC	CC	BP		SB			CC	BP	PP	SS	SB	SMS	SB	BI	SMS			SGS
STAND	f	i	e	e	d	h	d	f	c	a	h	g	c	c	a	d	e	i	e	c	b	k	j	h	f	d	g	d	b	f	f	g	e	l	g	c	b	j	m	b	a	m	l	i	h	e	g	d	c	n	k	b

Species

Ludwigia peploides — columns show: + (h), 1 (d), + (g), + (d/c area); SGS: 5 2, 1 2, 2 -, 2 2
Myriophyllum spicatum
Azolla filiculoides
Potamogeton pectinatus
Elodea canadensis

Scirpus acutus
Scirpus olneyi
Phragmites communis
Typha domingensis
Polygonum ssp.
Potentilla egedei
Sparganium eurycarpum

Scirpus californicus
Typha spp.

Salicornia pacifica
Distichlis spicata
Atriplex patula
Scirpus robustus
Frankenia grandifolia
Lythrum hyssopifolia
Juncus balticus
Cuscuta salina
Achillea borealis
Grindelia humilis
Limonium californicum
Jaumea carnosa

Spartina foliosa
Salicornia rubra

Associational group	Spar- tina	Salicornia	Salicornia pacifica		Salicornia- Grindelia	Scirpus californicus	Scirpus- Typha- Phragmites	Lud- wigia
			Salicornia- Distichlis					

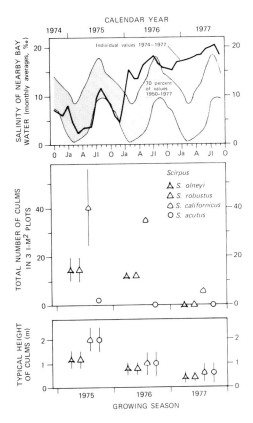

Fig. 11. Decrease in size and abundance of *Scirpus* spp. (bulrushes and tules) bordering Carquinez Strait during the drought of 1976-1977. The plots are located near the leveled transect at Southampton Bay (Fig. 7) at approximate elevations of 0.9 m, 0.4 m, and -0.5 m. Conditions in 1975 are estimated by comparing (qualitatively) living culms (above-ground stems) in plots along the transect in October 1975, with dead culms in September 1976. Measurements and counts of dead culms attempt to exclude those that grew before 1975, but similarities among dead culms of differing vintage result in large uncertainties, particularly for *S. californicus*. Conditions in 1976 and 1977 were determined from measurements and counts of living plants within the plots, excepting heights for 1977, which had to be scaled elsewhere because of the scarcity and absence of *Scirpus* within the plots. Vertical bars approximate the range of observed or estimated values. The top graph shows monthly averages of salinity of near-surface water at the eastern end of Carquinez Strait (USBR station D-6). The shaded area spans 1 SD (approximately 70%) of the monthly averages from October 1974 to September 1977. Data show that Carquinez Strait contained unusually saline water during the winter and spring of 1976 and 1977.

common pickleweed from the lower part of its salt-water range, as confirmed by the reciprocal spread of common pickleweed into low-lying areas denuded of tall tules and bulrushes during the drought of 1976-1977 (Fig. 10).

Productivity

According to classic investigations in the southeastern United States, the vascular plants of extensive tidal marshes supply most of the organic material on which local estuarine animals depend (Teal 1962; Day et al. 1973). Recent studies in Georgia and Holland, however, point to estuarine algae and riverine or marine debris as principal sources of estuarine food (Haines 1976, 1977; Wolff 1976). Given current controversy about these studies and shortcomings of related information about the San Francisco Bay estuary, we can hardly guess what percentage of food in this estuary originates in the vascular plants of its tidal marshes. In the following discussion we merely assemble information about the production of food in tidal marshes, estimate how much of this food enters other parts of the estuary, and offer a tentative comparison with the production of food by floating algae.

Conventional methods equate export by tidal-marsh plants with a calculated or arbitrary percentage of their net productivity. Net productivity refers to the quantity of organic matter that living plants store in excess of what they respire (Odum 1971:43). Bacteria, insects, and other organisms may consume some of this organic matter in the marsh, and high tides may move another fraction toward upland areas; hence, only a fraction of net productivity in a marsh can reach other parts of the estuary. The simplest measure of net productivity is the seasonal peak in the weight of live, above-ground, annual tissues (peak standing crop). Peak standing crop underestimates net productivity, however, because living tissue disappears during the growing season (Hardisky and Reimold 1977; Reimold and Linthurst 1977:87; Kirby and Gosselink 1976).

Local measurements of standing crop (reported herein as grams of dry plant material per

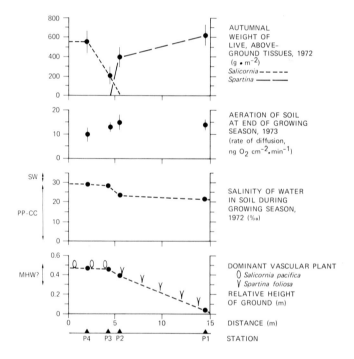

Fig. 12. Autumnal weight (dry) of above-ground tissues of *Spartina foliosa* and *Salicornia pacifica* and environmental variables of a profile near Black John Slough. Aeration of soil (upper 30 cm) at end of 1973 growing season and salinity of water in soil (upper 35 cm) in 1972 are data of Mahall 1974 and Mahall and Park 1976a, c. These data are compared with normal ocean salinities (SW) and the range of salinity of Bay water between Pinole Point (PP) and China Camp (CC) during 1969-75 (station 14 of Conomos and Peterson 1977). Topographic profile is referenced to estimated MHW datum. Vertical bars represent 1 SD of the measurements at each station; the diameter of dots on lower two graphs exceeds the length of bars.

square meter, $g \cdot m^{-2}$) suggest that vascular plants of the San Francisco Bay estuary produce at least as much organic material as their counterparts in the eastern United States (Fig. 13). Peak standing crops of California cordgrass (*Spartina foliosa*) range from 300 to 1700 $g \cdot m^{-2}$, comparable with its eastern relative, smooth cordgrass (*Spartina alterniflora*). Common pickleweed (*Salicornia pacifica*) creates standing crops of 500-1200 $g \cdot m^{-2}$, likewise similar to the salt hay (*Spartina patens*), salt grass (*Distichlis spicata*), and short variety of smooth cordgrass which commonly inhabit high parts of Atlantic-coast marshes. The largest reported above-ground standing crop in North American tidal marshes may belong to tules in the Sacramento-San Joaquin Delta which, at low elevations along sloughs, grow 3-4 m tall and weigh about 2500 $g \cdot m^{-2}$.

Adjusted for slight loss during the growing season and extrapolated to other species and localities, the standing crops of plants from a variety of marshes (Fig. 13) imply that net above-

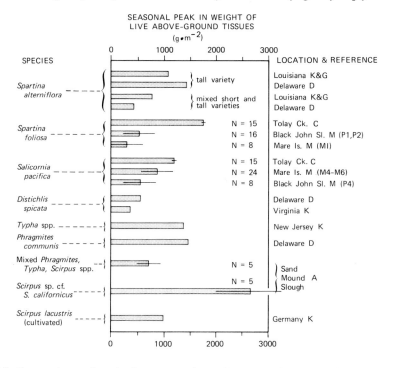

Fig. 13. Comparison of peak above-ground standing crops (dry weight) for some tidal-marsh plants. Weights refer to annual tissues that were harvested from multiple plots near the end of the growing season. Error bar shows 1 SD and N denotes the number of plots. Abbreviations for references: A—Brian Atwater unpublished data. Harvests were made in October 1977. Plots, 0.50 m² for mixed vegetation and 0.12 m² for *Scirpus*, are located on a remnant of pristine marsh near the transect of Atwater and Hedel (1976, pl. 8). Elevations relative to NGVD are 0.6±0.2 m for mixed plots and -0.4±0.2 m for *Scirpus*. Brown leaves attached to green *Phragmites* are included with live standing crop. Samples were oven-dried to constant weight at 100° C. Infertile flowers prevented definite identification of *Scirpus* sp. C—Cameron (1972:61, 64, 66; pers. comm.). Graph shows harvests of July 1969. Peak standing crops at Tolay Creek in autumn 1968 were 1400 $g \cdot m^{-2}$ (dry weight) for *Spartina* and 1050 $g \cdot m^{-2}$ for *Salicornia*. D—Compilation by Daiber et al. (1976:76, 78, 82). K—Compilation by Keefe (1972). K & G—Kirby and Gosselink (1976). M—Mahall and Park (1976a). Harvests made in 1972. Weights for *Salicornia* exclude living stems from previous years. Symbols in parentheses denote stations. Mahall's Mare Island marsh is located a few kilometers west of the locality plotted on Fig. 2.

ground productivity by the vascular plants of our estuary's tidal marshes averages between 500 and 1500 g·m^{-2}·yr^{-1}. Selecting 800 g·m^{-2}·yr^{-1} as a typical value and multiplying by the present area of tidal marsh yields an estimated net above-ground productivity of 10^{11} g·yr^{-1}. Tides and rain flush approximately half of such organic material from cordgrass marshes (Teal 1962; Day et al. 1973; Cameron 1972:60), but the average fraction that enters the waterways and bays of the San Francisco Bay estuary is probably smaller because of the proximity of most tidal-marsh surfaces to high-tide levels (Fig. 7), which reduces the frequency of tidal flushing relative to the lower areas dominated by California cordgrass. If one fourth of net above-ground productivity enters other parts of the estuary, then annual export equals 2.5 x 10^{10} g (dry weight), which in turn equals 10^{10} gC because carbon constitutes about 40% of the dry organic matter (Keefe 1972). We therefore estimate that the vascular plants of tidal marshes annually contribute 10 billion grams of carbon to the rest of the San Francisco Bay estuary.

Several perspectives aid in conceptualizing 10 billion grams of carbon. (1) Net productivity of floating algae in the bays, chiefly diatoms and flagellates, averaged about 200 gC·m^{-2}·yr^{-1} in 1976-77 (Peterson 1979), all of which is available to other aquatic organisms. The bays currently cover approximately 100 km^2 (Conomos and Peterson 1977), so these algae produced 2 x 10^{11} g of carbon, roughly 20 times our estimate of export by vascular plants of tidal marshes. (2) Historic destruction of tidal marshes (Figs. 1, 3) has probably caused a 10- to 20-fold reduction in their export of organic material. (3) At 10^{10} gC, export from vascular plants of tidal marshes translates into roughly 5 lbs. of carbon per year for each of the 5 million human inhabitants who surround the San Francisco Bay estuary.

ACKNOWLEDGMENTS

Persons who contributed opinions, information, or logistical assistance not otherwise acknowledged include Joel Bergquist, Frances DeMarco, Harry George, H. T. Harvey, E. J. Hellcy, R. F. Holland, Zondra Kilpatrick, H. L. Mason, F. H. Nichols, David Plummer, and G. J. West. Our principal illustrators are Yosh Inouye (Fig. 10-13), Barbara Lee (Fig. 5), Hylton Mayne (Fig. 6), and Steven Talco (Figs. 1, 2, 3, 7). The text incorporates suggestions from reviews by John Briscoe, K. G. Dedrick, R. T. Huffman, J. C. Kraft, Robert Nadey, and G. F. Somers. Authors divide and share responsibilities as follows: Atwater—text, tables, and Figs. 1, 2, 3, 6-13; Conard—Appendix A, Tables 4-5 and related text; Dowden—Figs. 4, 5; Hedel—Figs. 2, 3, 7, 10; Macdonald—Appendix A, Tables 4-5; and Savage—Appendix A, Table 4. John Coburn (USBR) supplied the salinity data shown in Fig. 13.

LITERATURE CITED

Abrams, L. 1923-1960. Illustrated flora of the Pacific states. 4 volumes. Stanford University Press, Stanford, Calif.

Atwater, B. F. 1979. Ancient processes at the site of southern San Francisco Bay: movement of the crust and changes in sea level. Pages 31-45 *in* T. J. Conomos, ed. San Francisco Bay: The Urbanized Estuary. Pacific Division, Amer. Assoc. Advance. Sci., San Francisco, Calif.

Atwater, B. F., and C. W. Hedel. 1976. Distribution of seed plants with respect to tide levels and water salinity in the natural tidal marshes of the northern San Francisco Bay estuary, California. U. S. Geol. Surv. Open-File Rep. 76-389. 41 pp.

Atwater, B. F., C. W. Hedel, and E. J. Helley. 1977. Late Quaternary depositional history, Holocene sea-level changes, and vertical crustal movement, southern San Francisco Bay, California. U. S. Geol. Surv. Prof. Paper 1014. 15 pp.

Barbour, M. G. 1970. Is any angiosperm an obligate halophyte? Amer. Midl. Nat. 8:105-120.

Barbour, M. G., and C. B. Davis. 1970. Salt tolerance of five California salt marsh plants. Amer. Midl. Nat. 84:262-265.

Bodnar, N., Jr. et al. 1975. The relationship between the upper limit of coastal marshes and tidal datums. National Ocean Survey, preliminary report prepared for the Environmental Protection Agency. 84 pp.

Briscoe, J. 1979. Legal problems of tidal marshes. Pages 387-400 in T. J. Conomos, ed. San Francisco Bay: The Urbanized Estuary. Pacific Division, Amer. Assoc. Advance. Sci., San Francisco Calif.

Cameron, G. N. 1972. Analysis of insect trophic diversity in two salt marsh communities. Ecology 53:58-73.

Carlson, P. R., and D. S. McCulloch. 1974. Aerial observations of suspended sediment plumes in San Francisco Bay and the adjacent Pacific Ocean. J. Res. U. S. Geol. Surv. 2(5):519-526.

Carlton, J. T. 1979. Introduced invertebrates of San Francisco Bay. Pages 427-444 in T. J. Conomos, ed. San Francisco Bay: The Urbanized Estuary. Pacific Division, Amer. Assoc. Advance. Sci., San Francisco, Calif.

Ceska, A. and H. Roemer. 1971. A computer program for identifying species-releve groups in vegetation studies. Vegetation 23:255-276.

Chapman, V. J. 1960. Salt marshes and salt deserts of the world. Interscience. 392 pp.

Coleman, J. M. and S. M. Gagliano. 1964. Cyclic sedimentation in the Mississippi River deltaic plain. Gulf Coast Assoc. Geol. Soc. Trans. 14:67-80.

Conard, S. G., R. L. Macdonald, and R. F. Holland. 1977. Riparian vegetation and flora of the Sacramento Valley. Pages 47-55 in Anne Sands, ed. Riparian Forests of California; Their Ecology and conservation. Inst. Ecol. Pub. No. 15. University of California, Davis, Calif.

Conomos, T. J. 1979. Properties and circulation of San Francisco Bay water. Pages 47-84 in T. J. Conomos, ed. San Francisco Bay: The Urbanized Estuary. Pacific Division, Amer. Assoc. Advance. Sci., San Francisco, Calif.

Conomos, T. J., and D. H. Peterson. 1977. Suspended-particle transport and circulation in San Francisco Bay: an overview. Pages 82-97 in L. E. Cronin, ed. Estuarine Processes. Vol 2. Academic Press, New York.

Cooper, W. S. 1926. Vegetational development on alluvial fans in the vicinity of Palo Alto, California. Ecology 7:1-30.

Cosby, S. W. 1941. Soil survey of the Sacramento-San Joaquin Delta area, California. U. S. Dep. Agr., Bur. of Chem. and Soils, Series 1935, No. 21. 48 pp.

County Supervisors Association of California. 1975. California county fact book, 1975. Sacramento, Calif. 200 pp.

Curry, R. R. 1969. Holocene climatic and glacial history of the central Sierra Nevada, California. Geol. Soc. Amer. Spec. Paper 123. pp. 1-47.

Daiber, F. D. et al. 1976. An atlas of Delaware's wetlands and estuarine resources. Delaware State Planning Office, Delaware Coastal Management Program, Technical Report No. 2. 528 pp.

Davis, C. A. 1910. Salt marsh formation near Boston and its geological significance. Econ. Geol. 5:623-639.

Day, J. W., Jr., W. G. Smith, P. R. Wagner, and W. C. Stowe. 1973. Community structure and carbon budget of a salt marsh and shallow bay estuarine system in Louisiana. Center for Wetland Resources, Louisiana State University Pub. No. LSU-SG-72-04. 79 pp.

Delta Advisory Planning Council. 1976. Delta agriculture and soils. Sacramento, Delta Plan Technical Supplement I. 101 pp.

Gerow, B. A., and R. W. Force. 1968. An analysis of the University Village Complex; with a reappraisal of central California archeology. Stanford University Press, Stanford, Calif. 209 pp.

Gilbert, G. K. 1917. Hydraulic-mining debris in the Sierra Nevada. U. S. Geol. Surv. Prof. Paper 105. 154 pp.

Gilliam, H. 1969. Between the devil and the deep blue bay—the struggle to save San Francisco

Bay. Chronicle Books, San Francisco, Calif. 151 pp.

Haines, E. B. 1976. Stable carbon isotope ratios in the biota, soils and tidal water of a Georgia salt marsh. Estuarine Coastal Mar. Sci. 4:609-616.

Haines, E. B. 1977. The origins of detritus in Georgia salt marsh estuaries. Oikos 29:254-259.

Hardisky, M. A., and R. J. Reimold. 1977. Salt-marsh plant geratology. Science 198:612-614.

Harlow, N. 1950. The maps of San Francisco Bay from the Spanish discovery in 1769 to the American occupation. The Book Club of California. 140 pp.

Harvey, H. T. 1966. Some ecological aspects of San Francisco Bay. San Francisco Bay Conservation and Development Commission. 31 pp.

Harvey, H. T., H. L. Mason, R. Gill, and T. W. Wooster. 1977. The marshes of San Francisco Bay— Their attributes and values. Unpublished report prepared for the San Francisco Bay Conservation and Development Commission. 156 pp.

Hinde, H. P. 1954. Vertical distribution of salt marsh phanerogams in relation to tide levels. Ecol. Monog. 24:209-225.

Johnson, D. S., and H. H. York. 1915. The relations of plants to tide levels, a study of factors affecting the distribution of marine plants at Cold Spring Harbor, Long Island, New York. Carnegie Inst. Wash. Pub. 206. 161 pp.

Jones and Stokes Associates, Inc. and EDAW, Inc. 1975. Suisun marsh protection plan, fish and wildlife element. California Dep. Fish Game. 412 pp.

Keefe, C. W. 1972. Marsh production—A summary of the literature. Contrib. Marine Sci. Univ. Texas 16:163-181.

Kingsley, R. B., and F. C. Boerger. 1976. Experimental marsh planting programs, Marin Country Day School and Muzzi Marsh. Madrone Assoc., San Rafael, California. Report prepared for Golden Gate Highway and Transportation District. 49 pp.

Kingsley, R. B. et al. 1977. The natural resources of Napa Marsh. Calif. Dep. Fish Game, Coastal Wetlands Series No. 19. 96 pp.

Kirby, C. J., and J. G. Gosselink. 1976. Primary production in a Louisiana gulf coast *Spartina alterniflora* marsh. Ecology 57:1052-1059.

Knott, J. M. 1973. Effects of urbanization on sedimentation and floodflows in Colma Creek basin, California. U. S. Geol. Surv. Open-File Rep. 54 pp.

Knutson, P. et al. 1976. Marshland development. U. S. Army Corps of Engineers, Dredge Disposal Study, San Francisco Bay and Estuary, Appendix K. 300 pp.

Kraft, J. C., E. A. Allen, D. F. Belknap, C. J. John, and E. M. Maurmeyer. 1976. Delaware's changing shoreline. Delaware State Planning Office, Coastal Zone Management Program Technical Rep. 1. 319 pp.

Krone, R. B. 1979. Sedimentation in the San Francisco Bay system. Pages 85-96 *in* T. J. Conomos, ed. San Francisco Bay: The Urbanized Estuary. Pacific Division, Amer. Assoc. Advance. Sci., San Francisco, Calif.

McCulloch, D. S., D. H. Peterson, P. R. Carlson, and T. J. Conomos. 1970. Some effects of freshwater inflow on the flushing of south San Francisco Bay. U. S. Geol. Surv. Circ. 637A. 27 pp.

Macdonald, K. B. 1977. Coastal salt marsh. Pages 263-294 *in* M. G. Barbour and J. Major, eds. Terrestrial Vegetation of California. J. Wiley & Sons, New York.

Mahall, B. E. 1974. Ecological and physiological factors influencing the ecotone between *Spartina foliosa* Trin. and *Salicornia virginica* L. in salt marshes of northern San Francisco Bay. Ph.D. Thesis. University of California, Berkeley, Calif. 315 pp.

Mahall, B. E., and R. B. Park. 1976a. The ecotone between *Spartina foliosa* Trin. and *Salicornia virginica* L. in salt marshes of northern San Francisco Bay—I. Biomass and production. J. Ecol. 64:421-433.

Mahall, B. E., and R. B. Park. 1976b. The ecotone between *Spartina foliosa* Trin. and *Salicornia virginica* L. in salt marshes of northern San Francisco Bay—II. Soil water and salinity. J. Ecol. 64:793-809.

Mahall, B. E., and R. B. Park. 1976c. The ecotone between *Spartina foliosa* Trin. and *Salicornia virginica* L. in salt marshes of northern San Francisco Bay—III. Soil aeration and tidal immersion. J. Ecol. 64:811-819.

Mall, R. E. 1969. Soil-water-salt relationships of waterfowl food plants in the Suisun marsh of California. Calif. Dept. Fish Game, Wildlife Bull. 1. 59 pp.

Marmer, H. A. 1951. Tidal datum planes. U. S. Coast and Geodetic Survey Spec. Publ. 135. 142 pp.

Marshall, J. T. 1948. Ecologic races of song sparrows in the San Francisco Bay region I, habitat and abundance. Condor 50:193-215.

Mason, H. L. (undated). Floristics of the Sacramento-San Joaquin Delta and Suisun Marsh (approx. title). Unpublished manuscript on file in the Sacramento office of the U. S. Army Corps of Engineers and in the Menlo Park Library of the U. S. Geological Survey. 84 pp.

Matthew, R., and E. E. Blackie. 1931. Economic aspects of a salt water barrier below confluence of Sacramento and San Joaquin Rivers. Calif. Div. Water Res. Bull. 28. 450 pp.

Matthew, R., J. A. Chase, and D. R. Warren. 1931. Variation and control of salinity in Sacramento-San Joaquin Delta and upper San Francisco Bay. Calif. Div. Water Res. Bull. 27. 440 pp.

Mitchell, H. C., compiler. 1936. First and second order triangulation in California. U. S. Coast and Geodetic Survey Spec. Publ. 202. 548 pp.

Mudge, B. F. 1858. The salt marsh formation of Lynn. Proc. Essex Inst. 2:117-119.

Mueller-Dombois, D., and H. Ellenberg. 1974. Aims and methods of vegetation ecology. J. Wiley & Sons, New York. 547 pp.

Munz, P. A. 1968. Supplement to a California flora. University of California Press, Berkeley, Calif. 224 pp.

Munz, P. A., and P. A. Keck. 1959. A California flora. University of California Press, Berkeley, Calif. 1681 pp.

Myrick, R. M., and L. B. Leopold. 1963. Hydraulic geometry of a small tidal estuary. U. S. Geol. Surv. Prof. Paper 422-B. 18 pp.

National Ocean Survey. 1977a. Tide tables 1978, west coast of North and South America. U. S. Dep. of Commerce. 222 pp.

National Ocean Survey. 1977b. Untitled list giving difference between National Geodetic Vertical Datum and mean lower low water for San Francisco Bay and San Joaquin-Sacramento Delta Region. National Ocean Survey, Tidal Datum Planes Section. 3 pp.

Nichols, D. R., and N. A. Wright. 1971. Preliminary map of historic margins of marshland. San Francisco Bay, California. U. S. Geol. Surv. Open-File Map (scale 1:125,000).

Odum, E. P. 1971. Fundamentals of ecology. 3rd ed. W. B. Saunders Company, Philadelphia, Pa. 574 pp.

Pestrong, R. 1965. The development of drainage patterns of tidal marshes. Stanford Univ. Pub. Geol. Sci., vol. 10, No. 2, 87 pp.

Pestrong, R. 1972. San Francisco Bay tidelands. California Geol. 25:27-40.

Peterson, D. H. 1979. Sources and sinks of biologically reactive oxygen, carbon, nitrogen, and silica in northern San Francisco Bay. Pages 175-193 *in* T. J. Conomos, ed. San Francisco Bay: The Urbanized Estuary. Pacific Division, Amer. Assoc. Advance. Sci., San Francisco, Calif.

Phleger, C. F. 1971. Effect of salinity on growth of a salt marsh grass. Ecology 52:908-911.

Poland, J. F. 1971. Land subsidence in the Santa Clara Valley, Alameda, San Mateo, and Santa Clara Counties, California. U. S. Geol. Surv. Misc. Field Studies Map MF-336 (scale 1:125,000).

Powell, W. R., ed. 1964. Inventory of rare and endangered vascular plants of California. California. California Native Plant Society, Berkeley, Calif. 56 pp.

Purer, E. A. 1942. Plant ecology of the coastal salt marshlands. Ecol. Monog. 12:83-111.

Redfield, A. C. 1972. Development of a New England salt marsh. Ecol. Monog. 42(2):201-237.

Reimold, R. J., and R. A. Linthurst. 1977. Primary productivity of minor marsh plants in Delaware, Georgia, and Maine. U. S. Army Eng. Waterways Expt. Station, Tech. Rep. D-77-36. 104 pp.

Ringgold, C. 1852. A series of charts, with sailing directions, embracing surveys of the Farallones, the entrance to the bay of San Francisco, bays of San Francisco and San Pablo, straits of

Carquines [sic] and Suisun bay, confluence of the deltic branches of the Sacramento and San Joaquin rivers, and the Sacramento river (with the middle fork) to the American river, including the cities of Sacramento and Boston, state of California. J. T. Towers, Washington, D. C. 48 pp.

San Francisco Bay Conservation and Development Commission. 1969. The San Francisco Bay plan. 43 pp.

Schlemon, R. J., and E. L. Begg. 1973. Late Quaternary evolution of the Sacramento-San Joaquin Delta, California, Proc. Ninth Congress of the International Union for Quaternary Research 1973:259-266.

Shaler, N. S. 1886. Preliminary reports on sea-coast swamps of the eastern United States. U. S. Geol. Surv. 5th Annual Report. pp. 353-398.

Shalowitz, A. L. 1964. Shore and sea boundaries. U. S. Coast and Geodetic Survey Publ. 10-1, vol. 2. 749 pp.

Smith, B. J. 1965. Sedimentation in the San Francisco Bay system. Pages 675-707 *in* U. S. Dep. Agr. Misc. Publ. 970, Proc. Fed. Interagency Sedimentation Conf. 1965.

Teal, J. M. 1962. Energy flow in the salt marsh ecosystem of Georgia. Ecology 43:614-624.

Thompson, J. 1958. Settlement geography of the Sacramento-San Joaquin Delta. Ph. D. Thesis. Stanford University, Stanford, Calif. 551 pp.

U. S. Census Office. 1901. Twelfth Census of the United States taken in the year 1900, vol. 1. Washington, D. C. 1006 pp.

U. S. Department of Commerce. 1964. Climatic summary of the United States, Supplement for 1951-1960. Climatography of the United States, No. 86-4.

Van Royen, W., and C. O. Siegel. 1959. Reclamation of marsh, tide, and submerged lands. Pages 75-94 *in* Future development of the San Francisco Bay Area. U. S. Dep. Commerce, Office of Area Development.

Ver Planck, W. E. 1958. Salt in California. Calif. Div. Mines Bull. 175. 168 pp.

Weir, W. W. 1950. Subsidence of peat lands of the San Joaquin-Sacramento Delta, California. Hilgardia 20:37-56.

Westdahl, F. 1896. Descriptive report accompanying a resurvey of San Francisco Bay, California, Alameda Creek to Beard's Creek (approx. title). U. S. Coast and Geodetic Survey Register No. 2353. 8 pp.

Westdahl, F. 1897. Descriptive report accompanying a resurvey of San Francisco Bay, California, Menlo Park to near Mountain View (approx. title). U. S. Coast and Geodetic Survey Register No. 2312. 14 pp.

Wolff, W. J. 1976. A benthic food budget for the Grevelingen estuary, The Netherlands, and a consideration of the mechanisms causing high benthic secondary production in estuaries. Pages 267-280 *in* B. C. Coull, ed. Ecology of Marine Benthos. Belle W. Baruch Library in Marine Science, vol. 6. University of South Carolina, Columbia, S.C.

APPENDIX A

COMMON VASCULAR PLANTS IN TIDAL MARSHES OF THE SAN FRANCISCO BAY ESTUARY.

The list excludes introduced species denoted by Munz and Keck (1959). Asterisk designates rare or endangered plant according to Powell (1964). Scientific names follow Munz (1968) and Munz and Keck (1959) excepting nomenclature for *Athyrium, Cornus* and *Salicornia*, which follows Mason (1957). Many of the common names are taken from Abrams (1923-1960), and parenthetical descriptions of principal taxonomic groups are abridged from Mason (1957). Abbreviated headings refer to localities (Fig. 2): Palo Alto Baylands; RB, Richardson Bay; CC, China Camp; SS, Schultz Slough; NA, Napa marshes; SB, Southampton Bay; MZ, Martinez, SA, Shore Acres; HS, Hill Slough; SN, marshes N of Suisun Bay; BI, Browns Island; SMS, Sand Mound Slough; SGS, Snodgrass Slough; DL, marshes and sloughs in the Delta.

Sources of information:

PA—Cooper (1926), Hinde (1954), and observations by Atwater in 1976.

RB, CC, SS, SB, SA, HS—Collections by Savage in 1975 (SA), 1976 (RB, CC, SS, SB), and 1977 (HS), and supplemental observations by Conard and Macdonald in 1976 (SS, SB) and by Atwater in 1975 (CC) and 1977 (SB).

NA—Kingsley et al. (1977, Appendix A), lower-marsh and higher-marsh communities.

MZ—Observations by H. L. Mason in 1974.

SN, DL—Observations by H. L. Mason ca. 1970, supplemented with reference to earlier collections. Listed species belong to the following communities of Mason (undated, pp. 57-59, 71-75): palustrine, *Salicornia,* and *Distichlis* (SN); and neuston, buoyan, palustrine, and willow-fern (DL).

BI, SMS, SGS—Reconnaissance by Conard and Macdonald in 1976.

[N. B.: During 1978 and 1979 Atwater collected 40 species at SMS and 38 at SGS, many more than listed here, and also observed widespread *Salicornia pacifica, Jaumea carnosa,* and *Triglochin maritima* in pristine marshland at BI. Mason's *Eupatorium occidentale* is probably *Pluchea purpurascens* (Sw.) DC. (marsh-fleabane).]

KNOWN DISTRIBUTION OF SPECIES IN TIDAL MARSHES

SAN FRANCISCO BAY

FAMILY	SPECIES Linnaean Name	Common Name	San Francisco Bay		San Pablo Bay			Carquinez Strait	Suisun Bay				Sacramento - San Joaquin Delta			
			PA	RB	CC	SS	NA	SB	MZ	SA	HS	SN	BI	SMS	SGS	DL
POLYPODIACEAE Fern family	*Athyrium filix-femina* (L.) Roth	Lady fern														X
SALVINIACEAE Salvinia family	*Azolla filiculoides* Lam.	Fern-like azolla													X	X
	A. mexicana Presl.	Mexican azolla													X	X

(Anthophytes — plants producing true flowers and seeds. Monocotyledons — plants with leaves usually parallel-veined; flowers usually with parts in multiples of three)

FAMILY	SPECIES Linnaean Name	Common Name	PA	RB	CC	SS	NA	SB	MZ	SA	HS	SN	BI	SMS	SGS	DL
ALISMATACEAE Water-plantain family	*Sagittaria latifolia* Willd.	Broad-leaved arrow-head											X			
CYPERACEAE Sedge family	*Carex* spp.	Sedge											X	X		
	Cyperus spp.	Umbrella sedge														X
	Eleocharis acicularis (L.) R. & S.	Needle spike-rush											X	X	X	X
	E. sp.	Spike-rush														
	Scirpus acutus Muhl.	Common tule					X	X	X	X	X	X		X	X	X
	S. americanus Pers.	Three-square					X	X		X		X		X	X	X
	S. californicus (C.A. Mey.) Steud.	California bulrush, tule					X	X					X	X	X	X
	S. cernuus Vahl. var. *californicus* (Torr.) Beetle	Low club-rush					X	X	X	X	X	X		X	X	X
	S. olneyi Gray	Olney's bulrush					X	X	X	X	X			X	X	X
	S. robustus Pursh	Alkali bulrush			X	X	X	X	X	X	X					
GRAMINEAE Grass family	*Deschampsia caespitosa* (L.) Beauv.	California hair-grass	X	X												
	Distichlis spicata (L.) Greene var. *stolonifera* Beetle	Salt grass			X	X	X	X	X	X	X	X	X			
	Elymus triticoides Buckl.	Alkali rye-grass			X	X		X					X			
	Phragmites communis Trin. var. *berlandieri* (Fourn.) Fern.	Common reed										X	X	X	X	X
	Spartina foliosa Trin.	California cordgrass	X	X	X	X	X	X								
HYDROCHARITACEAE Frogbit family	*Elodea canadensis* Michx.	Waterweed													X	X
	E. densa Planch.															
JUNCACEAE Rush family	*Juncus balticus* Wild.	Baltic rush			X	X	X	X	X	X			X			
	J. lesueuri Bol.	Salt rust			X	X		X	X	X			X			
	J. effusus L. var. *pacificus* Fern. & Wieg.														X	
	J. sp.															
JUNCAGINACEAE Arrow-weed family	*Triglochin concinna* Davy	Slender arrow-grass	X													
	T. maritima L.	Seaside arrow-grass	X					X	X	X	X	X				
LEMNACEAE Duckweed family	*Lemna* ssp. (four species)	Duckweed			X	X								X	X	X
	Spirodela polyrhiza (L.) Schleid.	Greater duckweed												X	X	X
	Wolffiella lingulata (Hegelm.) Hegelm	Tongue-shaped Wolffiella												X	X	X
NAJADACEAE Water-nymph family	*Najas guadalupensis* (Spreng.) Morong.	Common water nymph														X
POTAMOGETONACEAE Pondweed family	*Potamogeton pectinatus* L.	Fennel-leaved pondweed											X	X	X	
	P. spp. (seven species)															
SPARGANIACEAE Bur-reed family	*Sparganium eurycarpum* Engelm.	Broad-fruited bur-reed													X	

ATWATER ET AL: TIDAL MARSHES

KNOWN DISTRIBUTION OF SPECIES IN TIDAL MARSHES

FAMILY	SPECIES Linnaean Name	Common Name	San Francisco Bay PA	San Francisco Bay RB	San Pablo Bay CC	San Pablo Bay SS	San Pablo Bay NA	Carquinez Strait SB	Suisun Bay MZ	Suisun Bay SA	Suisun Bay HS	Suisun Bay SN	Sacramento-San Joaquin Delta BI	Sacramento-San Joaquin Delta SMS	Sacramento-San Joaquin Delta SGS	Sacramento-San Joaquin Delta DL
TYPHACEAE	*Typha angustifolia* L.	Narrow-leaved cat-tail							X	X	X	X	X	X		
	T. domingensis Pers.	Domingo cat-tail							X	X	X	X	X	X		
	T. glauca Godron	Blue-flag							X	X	X	X				
	T. latifolia L.	Broad-leaved cat-tail					X	X	X	X	X	X			X	X

(Dicotyledons—Plants with leaves not usually parallel-veined; flowers with parts usually in multiples of two, five, or many)

FAMILY	SPECIES Linnaean Name	Common Name	PA	RB	CC	SS	NA	SB	MZ	SA	HS	SN	BI	SMS	SGS	DL
AIZOACEAE Carpet-weed family	*Sesuvium verrucosum* Raf.	Lowland purslane										X				
CARYOPHYLLACEAE Chickweed family	*Spergularia macrotheca* (Hornem.) Heynh.	Large-flowered sand spurry		X												
	S. marina (L.) Griseb.	Salt-marsh sand spurry	X		X	X	X	X	X	X	X					
CERATOPHYLLACEAE Hornwort family	*Ceratophyllum demersum* L.	Hornwort														X
CHENOPODIACEAE Goosefoot family	*Atriplex patula* L. ssp. *hastata* (L.) Hall & Clem.	Halberd-leaved saltbush	X	X	X	X	X	X	X	X	X	X				
	Salicornia pacifica Standl.	Common pickleweed (perennial)	X	X	X	X	X	X	X	X	X	X				
	Salicornia rubra A. Nels.	Red pickleweed (annual)	X	X	X											
COMPOSITAE Sunflower family	*Achillea borealis* Bong. ssp. *californica* (Pollard) Keck	Common yarrow					X	X	X				X			
	Artemisia douglasiana Bess.	Douglas' mugwort											X	X	X	
	Aster chilensis Nees.	Common California aster				X								X	X	
	Baccharis douglasii DC.	Salt-marsh Baccharis				X		X								
	B. pilularis DC. ssp. *consanguinea* (DC.) B. C. Wolf	Coyote brush			X	X	X							X		
	Bidens frondosa L.	Stick-tight														X
	B. laevis (L.) BSP.	Bur-marigold														X
	Eupatorium occidentale Hook.	Western Eupatorium	X	X	X	X	X	X					X			
	Grindelia humilis H. & A.	Marsh Grindelia				X			X	X		X	X			
	Helenium bigelovii Gray	Bigelow's sneezeweed	X	X	X	X	X	X			X		X			
	Jaumea carnosa (Less.) Gray	Fleshy Jaumea							X	X	X	X				
	Solidago occidentalis (Nutt.) Torr. & Gray	Western goldenrod											X			X
CONVOLVULACEAE Morning-glory family	*Calystegia sepium* (L.) R. Br. ssp. *limnophila* (Greene) Brummitt	Marsh hedge bindweed												X	X	X
	Cressa truxillensis HBK. var. *vallicola* (Heller) Munz	Alkali weed						X								
CORNACEAE Dogwood family	*Cornus stolonifera* Michx. var. *californica* (C. A. Mey.) McMinn	Creek dogwood													X	X
CRASSULACEAE Stonecrop family	*Tillaea aquatica* L.	Water pigmy-weed														X
CUSCUTACEAE Dodder family	*Cuscuta salina* Engelm.	Salt-marsh dodder	X	X	X	X		X			X					
FRANKENIACEAE Frankenia family	*Frankenia grandifolia* Cham. & Schlecht.	Alkali heath	X	X	X	X	X	X			X	X				
HALORAGACEAE Water-milfoil family	*Myriophyllum braziliense* Camb.	Parrot's feather													X	
	M. spicatum L. ssp. *exalbescens* (Fern.) Hult.	American milfoil													X	X

383

SAN FRANCISCO BAY

KNOWN DISTRIBUTION OF SPECIES IN TIDAL MARSHES

FAMILY	SPECIES Linnaean Name	Common Name	San Francisco Bay		San Pablo Bay			Carquinez Strait	Suisun Bay				Sacramento - San Joaquin Delta			
			PA	RB	CC	SS	NA	SB	MZ	SA	HS	SN	BI	SMS	SGS	DL
LABIATAE Mint family	*Lycopus americanus* Muhl.	Cut-leaved water-horehound												X	X	X
	Stachys albens Gray														X	X
LEGUMINOSAE Pea family	*Lathyrus californicus* Doug. ex. Lindl.	Beach pea							X	X						
	* *L. Jepsonii* Greene	Jepson's pea							X	X	X		X			
	Lotus salsuginosus Greene	Coastal lotus											X			
	Vicia sp.	Vetch														
LYTHRACEAE Loosestrife family	*Lythrum hyssopifolia* L.	Hyssop loosestrife				X									X	
MALVACEAE Mallow family	* *Hibiscus californicus* Kell.	California Hibiscus								X				X	X	X
	Sida hederacea (Dougl.) Torr.	Alkali mallow														
NYMPHAEACEAE Water-lily family	*Nuphar polysepalum* Engelm.	Yellow pond lily														X
ONAGRACEAE Evening primrose family	*Epilobium* spp.	Willow-herb												X	X	
	E. watsonii Barb.	Watson's willow-herb												X	X	
	Ludwigia peploides (HBK.) Raven	Yellow water-weed								X			X		X	X
PLANTAGINACEAE Plantain family	*Plantago maritima* L. ssp. *juncoides* (Lam.) Hult.	Pacific seaside plantain	X	X												
PLUMBAGINACEAE Plumbago family	*Limonium californicum* (Boiss.) Heller	California marsh rosemary	X	X	X	X	X									
POLYGONACEAE Buckwheat family	*Polygonum aviculare* L.	Common knotweed														
	P. coccineum Muhl.	Swamp knotweed						X								
	P. hydropiperoides Michx. var. *asperifolium* Stauf.	Mild-water pepper												X	X	
	P. lapathifolium L.	Pale persicaria												X		
PRIMULACEAE Primrose family	*Glaux maritima* L.	Sea milkwort				X		X	X							
	Samolus floribundus HBK.	Water pimpernel														
ROSACEAE Rose family	*Potentilla anserina* L.	Silverweed						X		X				X		
	P. egedei Wormsk. var. *grandis* (Rydb.) J.T. Howell	Pacific silverweed									X					
RUBIACEAE Madder family	*Cephalanthus occidentalis* L. var. *californicus* Benth.	Button-bush											X		X	X
	Galium trifidum L. var. *subbiflorum* Weig.	Bedstraw											X			X
SALICACEAE Willow family	*Populus fremontii* Wats.	Fremont's poplar														X
	Salix goodingii Ball.	Gooding's willow													X	X
	S. hindsiana Benth.	Sandbar willow													X	X
	S. lasiolepis Benth.	Arroyo willow											X			X
SCROPHULARIACEAE Figwort family	* *Cordylanthus mollis* A. Gray ssp. *mollis*	Soft bird's beak										X				
	Limosella subulata Ives	Awl-leaved mudwort				X										
	Mimulus guttatus Fisch. ex. DC.	Common large monkey-flower											X	X		
	Scrophularia californica Chamb. & Schlecht.	Coast figwort											X	X		X

384

KNOWN DISTRIBUTION OF SPECIES IN TIDAL MARSHES

FAMILY	SPECIES		San Francisco Bay		San Pablo Bay			Carquinez Strait	Suisun Bay				Sacramento - San Joaquin Delta			
	Linnaean Name	Common Name	PA	RB	CC	SS	NA	SB	MZ	SA	HS	SN	BI	SMS	SGS	DL
UMBELLIFERAE Carrot family	*Hydrocotyle umbellata* L.	Many-flowered marsh-pennywort											X	X		
	H. verticillata Thunb. var. *triradiata* (A. Rich.) Fern	Whorled marsh-pennywort								X						X
	Oenanthe sarmentosa Presl.	Pacific Oenanthe											X			
URTICACEAE Nettle family	*Urtica holosericea* Nutt.	Nettle												X		
VERBENACEAE Verbena family	*Verbena hastata* L.	Blue vervain											X			
	V. scabra Vahl.	Rough verbena														X

LEGAL PROBLEMS OF TIDAL MARSHES

JOHN BRISCOE

Deputy Attorney General, Room 6000 State Office Building, San Francisco, CA 94102*

Three legal problems that afflict tidal marshes are (1) the rights and liabilities of persons seeking to alter the natural condition of a marsh, (2) boundaries of ownership interests within a marsh, and (3) boundaries of the jurisdictions of government agencies having power to regulate filling, dredging or other activities within the marsh. 1) The legal theories of public nuisance and public trust are means of preventing or remedying demonstrable injury to a tidal marsh. Several cases demonstrate the law's need in this context for an understanding of the processes of the tidal marsh. 2) To determine ownership interests within a tidal marsh it is often critical to locate the line of mean high water (MHW) in either its present or some prior position. 3) Similarly, the geographical extent of the authority of government bureaus to control filling, dredging, or other human activities within tidal marshes is often a function of tidal datums. For this purpose the lines of MHW and of mean higher water (MHHW) (in either the present or some past location of the line) are most frequently employed.

Tidal marshes have bred legal problems as abundantly as their brackish reaches can breed mosquitoes in summer. While an in-depth treatment of these problems would require volumes, the purpose of this chapter is to give the scientist a brief glimpse of the law's need for his knowledge of marshes.

Two of the three subjects discussed are boundaries: boundaries of ownership interests in tidal marshes, and jurisdictional boundaries of agencies that regulate the diking, filling, dredging or other altering of marshes. The third subject, which is treated first, is the vast area of legal questions that arise from direct human threats to the viability of a marsh or to its dependent life systems.

THE DEGRADING OR DESTROYING OF MARSHES BY MAN

Human threats to marshes center on the draining or filling or marshes, or the polluting of the waters of a marsh. The legal problems chiefly entail questions of the adequacy of environmental documents prepared for a proposed project, and whether alleged threats to a marsh are real or imagined.[1] Although even a superficial survey of these questions is beyond the scope of this chapter, it shows that even before the gauntlet of modern laws was thrown down to these threats (Table 1), the law often saw a remedy for demonstrable injury to the environment. An ancient doctrine of "public nuisance," for one, afforded such a remedy. Three early California cases illustrate this doctrine.

In one landmark decision the State sought to prohibit the dumping of hydraulic-mining

---

*The views expressed herein are not necessarily those of the Attorney General or of other bureaus of the State.

[1] Applications for projects requiring federal, state or local approval now must usually be accompanied by assessments of the projects' environmental impacts. See National Environmental Policy Act of 1969, 42 U.S.C. §§ 4331-4347 and Zabel vs. Tabb, 430 F.2d 199 (1970), discussed below, and the California Environmental Quality Act, California Pub. Resources Code §§ 21000-21176, and Friends of Mammoth vs. Board of Sup'rs of Mono County, 8 Cal.3d 247 (1972).

SAN FRANCISCO BAY

TABLE 1. MAJOR LEGAL DEVELOPMENTS IN THE ENVIRONMENTAL HISTORY OF SAN FRANCISCO BAY

Date	Event	Impact
1851	San Francisco Beach and Water Lot Act	First State statute authorizing the sale of tidelands to private parties. (Subsequent sales statutes pertaining to S.F. Bay were enacted in 1868 and 1870, among other dates.)
1879	New State Constitution adopted, effective 1 January 1880	Separate articles prohibited (1) private landowners from denying the public right of way to navigable waters whenever required for a public purpose, and (2) the sale to private parties of any tidelands within 2 miles of an incorporated city or town.
1913	California Supreme Ct. Decision in *People* vs. *California Fish Co.* 166 Calif. Reports 576.	Held that (1) tidelands sold by State to private parties remain subject to a "public trust" easement, and (2) sales of lands lying "below low tide" are wholly invalid. (Authorizing statute reviewed by Court excluded from its operations lands within five miles of San Francisco.)
1965	Creation by California legislature of S.F. Bay Conservation and Development Commission.	Temporary agency to formulate comprehensive plan for Bay; given life of four years.
1968	State - Leslie Salt Co. land agreement	First Agreement with major San Francisco Bay landowners recognizing State titles to tide and submerged lands within boundaries of lands sold by State as "swamp-and-overflowed lands." Agreement reached after 20 years of negotiations.
1969	McAteer-Petris Act	San Francisco Bay Conservation and Development Commission made a permanent agency.
	"Westbay" lawsuit filed in San Mateo County	State challenged landowners' claims of clear title to 188 acres of tide and submerged lands in S.F. Bay.
1971	Calif. Supreme Ct. Decision in *Marks* vs. *Whitney*, 6 Cal. Reports 3d 251	Reaffirmed principles of *People* vs. *California Fish Co.* Held that public trust purposes include keeping tidelands in a natural condition.
1972	Corps of Engineers amends regulations to assert regulatory authority to "former" line of mean higher-high water (i.e., prior to changes such as diking.)	Expanded jurisdiction, if valid, would encompass many salt ponds and other reclaimed marshlands.
	Westbay lawsuit expanded	Private landowners place in issue title to additional 10,000 acres of tide and submerged lands presently under S.F. Bay waters. Perhaps largest, most complex land litigation in history.
1977	Westbay case settled	State's absolute title to 75% of disputed land recognized. Remaining 25% adjudicated to be held by landowner subject to the public trust.
1978	*Leslie Salt Co.* vs. *Froehlke*	Court of Appeals for the Ninth Circuit invalidates Corps of Engineers' regulations extending jurisdiction to former line of mean higher-high water, as to Rivers and Harbors Act jurisdiction only. Regulations respecting Corps authority under the Federal Water Pollution Control Act upheld.
1979 (?)	*Murphy* vs. *City of Berkeley*	California Supreme Court has taken case from Court of Appeal, which held that tideland sales in Berkeley pursuant to 1879 Act did not lift the public trust. Scope of Supreme Court's decision cannot be predicted.

debris into the North Fork of the American River. With respect to non-tidal streams, California owns the beds only of such streams that are "navigable," and the State Supreme Court wrote that the North Fork was an unnavigable stream. But because the debris was carried from the point of dumping down to the confluence with the Sacramento River, where it settled and impaired the navigability of the Sacramento, the dumping was ordered enjoined. Two statements of the Court bear consideration:

> To make use of the banks of a river for dumping places, from which to cast into the river annually six hundred thousand cubic yards of mining debris, consisting of boulders, sand, earth, and waste materials, to be carried by the velocity of the stream down its course and into and along a navigable river, is an encroachment upon the soil of the latter, and an unauthorized invasion of the rights of the public to its navigation; and when such acts not only impair the navigation of a river, but at the same time affect the rights of an entire community or neighborhood, or any considerable number of persons, to the free use and enjoyment of their property, they constitute, however long continued, a public nuisance . . . .
>
> Accompanying the ownership of *every species of property* is a corresponding duty to use it as that it shall not abuse the rights of other recognized owners. . . .
>
> Upon that underlying principle, neither State nor Federal leglislatures could, by silent acquiescence, or by attempted legislation, take private property for a private use, nor divest the people of the State of their rights in the navigable waters of the State for the use of private business, *however extensive or long continued. (People* vs. *Gold Run D. & M. Co.,* 66 Cal. 138, 147, 151 [1884].) (Emphasis added.)

In a later case the State sought to prohibit the damming of a small salt-marsh slough tributary to the Salt and Eel Rivers. Again the State Supreme Court held that ownership of the beds of the sloughs affected by the dam was immaterial. If damming diminished the navigability of the trunk stream, it was enjoinable. And "[t]he fact that these sloughs carry tide-waters, which ebb and flow, presents no different case from one where the tributaries so dammed flowed fresh water." The Court further held that even government authorization to reclaim the marshes gave their owner no right to do anything harmful to the navigability of the state's streams. "The Swamp and Overflowed Land Act does not purport to give the owner that right, even conceding such a power in the state, and the right of the public in the use of a stream, as a public highway, is paramount to any right which the owner of the land has to reclaim his land from over flow." (*People* vs. *Russ,* 132 Cal. 102, 105 [1901]).

The dumping of a sawmill's waste into the Truckee River was the object of another early lawsuit brought by the State. Dumping was alleged to be harmful to fish that spawned in and passed through the waters of the river. The State Supreme Court held that fish are "the most important constituent of that species of property commonly designated as wild game, the general right and ownership of which is in the people of the state. . . ." That being so, the ownership of the bed of the Truckee River was immaterial, the court held, and the People were entitled to an injunction stopping the pollution (*People* vs. *Truckee Lumber Co.,* 166 Cal. 397, 399, 402 [1897]).

There is evidence (Teal 1962; Johnston 1956 [San Francisco Bay]; Valiela and Vince 1976; Haedrick and Hall 1976; Sims 1970) that marshes serve as breeding grounds for various species of fish and that marsh-plant detritus is a link in the food chain of certain fish species.[2] Additionally there is evidence (Mitchell 1869; Pillsbury 1939; Marmer 1926) that salt marshes, as reservoirs of the waters of tidal floods, keep the main estuary channels scoured and

---

[2] There are too some contrary indications withrespect to the role of marsh-plant detritus as a link in the food chain. (Haines 1977).

navigable.[3] These cases indicate that even before the enactment of modern environmental protection laws, there has been an adequate understanding of the tidal marsh and its processes to provide a foundation for the legal means to prevent the degradation or outright destruction of the marsh.

This proposition seems all the more valid when these cases are read together with cases that expound the "public trust" doctrine. This doctrine holds that tidelands (in general, lands subject to tidal action and lying below the elevation of mean high water [MHW] as well as non-tidal lands below navigable waters, are held by the State subject to a public trust for purposes (among others) of navigation and fisheries (*Marks* vs. *Whitney,* 6 Cal. 3d 251 [1971] .[4] And significantly, while not all tidal marshes are "tidelands" within this definition, the public-trust doctrine in California burdens property *abutting* tidelands and navigable waters, as well as the tidelands and navigable waters themselves, "with a servitude commensurate with" the public trust power itself (*Colberg, Inc.* vs. *State of California ex rel. Dept. Pub. Wks.,* 67 Cal. 2d 408, 420 [1967] ; *Miramar Co.* vs. *Santa Barbara,* 23 Cal. 2d 170 [1943]). In other words, the public-trust power, while arising from the State's ownership of "tidelands," nevertheless may extend shoreward of the tideland boundary when necessary to effect the purposes of the trust. The significance of this principle is that action taken pursuant to this power requires no payment of compensation to the landowner, since the affected property is already "burdened" with that power (i.e., the landowner bears the risk that the power may be exercised). The *Just* case, discussed below, explores the area of compensation more fully.

Three contemporary cases which have had a profound impact on this subject of man's degradation of marshes should be mentioned briefly. Describing the doctrine of nuisance as "the oldest form of land use control," the California Court of Appeal in 1974 held valid California's coastal initiative (passed by the voters in 1972 and popularly known as "Proposition 20"), in part in reliance on that doctrine (*CEEED* vs. *California Coastal Zone Conservation Com.,* 43 Cal.App.3d 306, 318 [1974]).[5]

---

[3] "Mitchell's Rule" is: "A river having a bar at its mouth will be injured as a pathway for navigation if the tidal influx is reduced by encroachments upon its basins." Grove Karl Gilbert (1917: 102-103) described shoaling that had occurred in Mare Island Strait since the advent of marshland reclamation and hydraulic mining. ". . . I am not aware that the influence of reclamation has been mentioned in this connection, but there need be no question that the impairment of the channel has been caused in part by the weakening of the tidal currents," which had been in turn caused by reclamation of the adjoining marshlands.

[4] The Marks case held specifically: "Public trust easements are traditionally defined in terms of navigation, commerce and fisheries. They have been held to include the right to fish, hunt, bathe, swim, to use for boating and general recreation purposes the navigable waters of the state, and to use the bottom of the navigable waters for anchoring, standing, or other purposes. The public has the same right in and to [even privately owned] tidelands.

"The public uses to which tidelands are subject are sufficiently flexible to encompass changing public needs. In administering the trust the state is not burdened with an outmoded classification favoring one mode of utilization over another. There is a growing public recognition that one of the most important public uses of the tidelands—a use encompassed within the tidelands trust—is the preservation of those lands in their natural state, so that they may serve as ecological units for scientific study, as open space, and as environments which provide food and habitat for birds and marine life, and which favorably affect the scenery and climate of the area. It is not necessary to here define precisely all the public uses which encumber tidelands." 6 Cal.3d at 259-260 (Citations omitted). As discussed in the section concerning ownership, below, it must be considered in each case whether the tract of marshland in question is in fact "tideland" within the legal definition, or falls within some other legal classification of land.

[5] "The law of nuisance, called the oldest form of land use control, evolved from the ancient maxim 'sic utere tuo ut alienum non laedes'—one must so use his rights as not to infringe on the rights of others. At common law a public nuisance was defined as an act or omission which obstructs or causes inconvenience or damage to the public in the exercise of rights common to all "Her Majesty's subjects." Subject to constitutional barriers against unreasonable or arbitrary action, the Legislature may declare that a specified condition or activity constitutes a public nuisance. The power of the state to declare acts injurious to the state's natural resources to constitute a public nuisance has long been recognized in this state. Contemporary environmental legislation represents an exercise by government of this traditional power to regulate activities in the nature of nuisances. . . ."(Footnotes and citations omitted). 43 Cal.App.3d at 318.

A Federal appeals court in 1970 held that the U.S. Army Corps of Engineers, in reviewing applications to fill or dredge navigable waters, may consider environmental consequences of the proposal and is not confined to considering only the effect of the project on the Corps' traditional ward, navigation (*Zabel* vs. *Tabb,* 430 F.1d 199 [5th Cir. 1970]).

A 1972 Wisconsin Supreme Court decision, *Just* vs. *Marinette County,* has similarly been considered a landmark case in environmental law, particularly with respect to marshlands. A county ordinance, enacted pursuant to State law, prohibited the filling, draining or dredging of "wetlands" without a permit. The legal question was whether these restrictions amounted to a "constructive"(i.e. virtual) taking of property for a public use, which under the constitution would require the payment of just compensation to the owners. States have the power of eminent domain, which authorizes the taking of private property for public purposes upon the payment of just compensation, and they also have the police power, which is the basis for our criminal and health laws. When laws enacted under the police power restrict the uses to which land can be put, there is often the charge that the restrictions amount to a "taking" that requires compensation. Just when a land-use restriction becomes a "taking" is an elusive question. It has been the subject of many court decisions and journal articles, no one of which has formulated a criterion that is satisfactory in all cases. The extent of the restriction, and the loss of value it causes, are frequently examined, but are not necessarily determinative. In *Just* vs. *Marinette County,* the court analyzed the issue according to an old, and not always adequate, formulation: whether the restriction is intended to secure a benefit for the public it does not presently enjoy, or whether it is intended to prevent an injury to the public. Since this ordinance was designed merely to preserve the *status quo* (i.e. to prevent further degradation of water quality and wildlife habitat), it was held a valid exercise of the police power, and not a taking.[6] Although hailed as an important natural-resources decision, *Just* nonetheless did not treat, nor did the controversy require it to treat, the question of proper remedies for past injury to wetlands.

OWNERSHIP

Ownership, the second area of legal problems affecting tidal marshes, may be as intricate as a marsh's network of sloughs and rivulets. Although some marsh lands were granted to individuals by Spain and Mexico when those countries were sovereign in California, to be recognized after the United States' annexation of California, these grants were required to be confirmed by a Board of Land Commissioners especially created to hear the claims of persons to such grants.

With the exception of this unique category of land, there are three legal classifications (or "characters") of land found within tidal marshes: "swamp-and-overflowed lands," "tidelands," and "submerged lands." (These are court-defined legal expressions having no intrinsic engineering or scientific meaning). With few exceptions, submerged lands, lands lying waterward of the "ordinary low water mark," were never made available for private purchase and purported purchases of these lands are void. [Editor's note: See Atwater et al. 1979, Table 1 for definition of tide-datums and heights.] But statutes authorizing the sale of both tidelands and swamp-and-overflowed lands

---

[6] The Just Court reasoned: "We are not unmindful of the warning in *Pennsylvania Coal Co.* vs. *Mahon* (1922) U.S. 393, 416, 43 S.Ct. 158, 160, 67 L.Ed. 322:
'. . .We are in danger of forgetting that a strong desire to improve the public condition is not enough to warrant achieving the desire by a shorter cut than the constitutional way of paying for the change.' This observation refers to the improvement of the public condition, the securing of a benefit not presently enjoyed and to which the public is not entitled. The shoreland zoning ordinance preserves nature, the environment, and natural resources as they were created and to which the people have a present right. The ordinance does not create or improve the public condition but only preserves nature from the despoilage and harm resulting from the unrestricted activities of humans.' 201 N.W. 2d at 771.

were enacted by the California legislature in 1850. And while the same laws provided for the sale to private parties of both these characters of land, the distinctions between the two types of land are critical. For as to swamp-and-overflowed lands:

> "The lands which passed to the state by grant under the Swamp Land Act were thereafter subject to absolute alienation by the state, free of any public trust for navigation. (*Newcomb* vs. *City of Newport Beach, supra,* 7 Cal. 2d 393, 400.)

Whereas the purchaser of tidelands

> .... owns the soil, subject to the easement of the public for the public uses of navigation and commerce, and to the right of the state, as administrator and controller of these public uses and the public trust therefor, to enter upon and possess the same for the preservation and advancement of the public uses, and to make such changes and improvements as may be deemed advisable for those purposes. (*People* vs. *California Fish Co., supra* at 598).

But determining where a tract of swamp-and-overflowed land (or "swamp land" for short) ends and the tideland begins may not be a simple matter. The California Supreme Court has observed:

> These swamp and overflowed lands embraced large areas in the interior of the state, situated in the San Joaquin and Sacramento valleys, and extending down to tide water in the bay of San Francisco. There the tide flats in many places merged into them imperceptibly, making it difficult to distinguish between them. (*People* vs. *California Fish Co.,* 166 Cal. 576, 591 [1913]).

To demonstrate why this difficulty exists it is necessary to examine the roots of title to these two characters of land (Fig. 1). Tidelands and other lands beneath navigable waters within California became the property of the State as an incident of sovereignty when California became a state on 9 September 1850.[7] (These lands were held by the State in the public trust mentioned above). Excepting the grants made previously by the Spanish and Mexican governments, all other land within the State was then the property of the Federal government, including "swamp-and-overflowed lands," which Congress granted to California 19 days later.[8] With certain possible exceptions, the boundary between the tideland and the upland (swamp lands being a species of upland) is a line the law calls the "ordinary high water mark."[9]

But like "tidelands" and "ordinary low water mark," "ordinary high water mark" is a legal expression that has no intrinsic meaning to an engineer or surveyor. Courts have given it meaning as to certain types of topography. It has been held for example that the ordinary high water mark along a non-tidal navigable river is the line at which vegetation stops (see *Oklahoma* vs. *Texas,* 260 U.S. 606, 632 [1922]; Skelton, Boundaries and Adjacent Properties 310-11 [1938]).

After much confused law on the meaning of the term for purposes of *tidal* water boundaries,

---

[7] *Martin* vs. *Waddell* 41 U.S. (16 Pet.) 367, 410 (1842); *Shively* vs. *Bowlby,* 152 U.S. 1, 15, 26 (1894); *Weber* vs. *Harbor* Commissioners, 85 U.S. (18 Wall) 57, 65-66 (1873); *People* vs. *California Fish Co.,* 166 Cal. 576, 584 (1913); *Marks* vs. *Whitney,* 6 Cal.3d 251, 258 (1971).

[8] 9 Stats. 519 (28 September 1850), 43 U.S.C. § 981 et seq.

[9] *Barney* vs. *Keokuk,* 94 U.S. 324, 336-38 (1876); *Borax, Ltd.* vs. *Los Angeles,* 296 U.S. 10, 22 (1935); *Wright* vs. *Seymour,* 69 Cal. 122, 126 (1886); *Long Beach Co.* vs. *Richardson,* 70 Cal. 206(1886); *Oakland* vs. *Oakland Water Front Co.,* 118 Cal. 160, 183 (1897); *Pacific Whaling Co.* vs. *Packers' Association,* 138 Cal. 632, 635, 636 (1903); *People* vs. *California Fish Co., supra,* 166 Cal. 576, 584 (1913); Civil Code § 670. See also *Strand Improvement Co.* vs. *Long Beach,* 173 Cal. 765, 770 (1916); *Miller & Lux* vs. *Secara,* 193 Cal. 755, 671, 762 (1924).

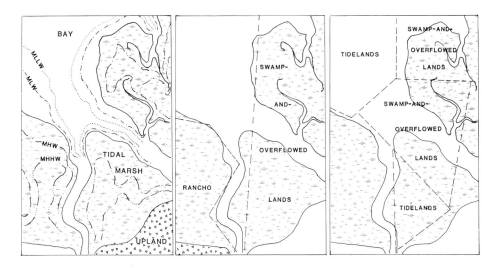

Fig. 1. Projection of tidal datums (left) and property lines (center and right) onto a hypothetical tidal marsh. Center diagram shows lines surveyed by the U. S. Government after the annexation of California in 1848. The ranch was granted to a private party by either Spain or Mexico before annexation, confirmed by the American Board of Land Commissioners after annexation, and subsequently surveyed by the Federal government. The government was also required to identify the "swamp-and-overflowed lands" that it granted to California on 28 September 1950, and often did so by survey. Surveyors were not always careful to locate the true "ordinary high water mark," or as it has been defined by some courts, the mean high water line. Nonetheless the true boundary remains the ordinary high water mark, no matter how erroneous the survey, which was run chiefly to compute acreages. California laws enacted in 1850 authorized the sale to private purchasers of tidelands and swamp-and-overflowed lands. These sales were made according to surveys that frequently did not correspond to the Federal surveys, much less the actual topography. The right-hand figure depicts a typical pattern of surveys done for these State sales. This lack of congruence between the true physical conditions and the designations of title documents creates many title disputes today.

the U.S. Supreme Court held in 1935 that the "ordinary high water mark" separating privately owned uplands (which had been sold by the Federal government) from the tidelands of San Pedro Harbor was the line of mean high water (*Borax Ltd.* vs. *Los Angeles,* 296 U. S. 10 [1935]).[10] For the first time surveyors and engineers had authoritative guidance how to locate the ordinary high water mark, since the mean high water line is the intersection with the shore of the plane of a published, precisely determined tidal datum (Shalowitz 1964:581). It is not always, however, an easy task to determine precise elevations in a marsh:

> Obviously, it would be an extremely difficult task to identify the actual high-water line in marsh areas. The marsh may be in various states of growth, from its early beginnings, when it is mostly a submerged stage, to its latest development, when it is close to or slightly above the plane of high water. Between these two extreme

---

[10] See, e.g., *Teschemacher* vs. *Thompson,* 18 Cal. 11, 21 (1861); *Otey* vs. *Carmel Sanitary District,* 219 Cal. 310, 313 (1933). These cases may be read as asserting that the "ordinary-high-water-mark" is not equivalent to the line of mean high water, that is, the mean of all high waters, but rather to a line of the mean of "neap" high waters. The error of this position from a legal standpoint, and the courts' fanciful misconceptions of neap tides, have been thoroughly explored (Maloney and Ausness 1975).

conditions, marsh areas may be entirely submerged at low water, may be exposed at low water and submerged at high water, or may be partially exposed at high water... (Shalowitz 1964:176-177).[11]

Thus locating the MHW line in a natural marsh is not always a simple matter. Moreover, artificial changes in the condition of a marsh may further complicate the determination of the legal character of the land (see for example Atwater et al. 1979). For when a marsh has been filled, diked, or otherwise altered by man, the relevant legal inquiry *may* shift from the present land elevations to the elevations of the land when it was last in a natural condition.[12] When the inquiry does so shift, an expert must determine first the existence of any changes that have occurred in the marsh, such as diking, dredging or filling, or more subtle changes such as eolian deflation, subsidence, or accretion to or erosion of the marsh edge. The expert must then determine the cause of these changes. Desiccation or ground-water withdrawal may be causes of subsidence, and alteration of the natural sedimentation or wave patterns may be causes of a prograding or retrograding marsh edge. The next task is to determine (as well as possible) the conditions that existed before the change, specifically the marsh-surface elevations.

Of interest to the marsh botanist are the attempts that have been made to locate the MHW line in a natural marsh by determining the distribution of vascular plants with respect to tidal datums (Maloney and Ausness 1974). A New York court has rejected such an approach as inconsistent with that state's law of coastal boundaries.[13]

Of course, locating the ordinary high water mark is not the only task that must be done. Historical research and legal analysis of documents in the chain of title must be completed before the true state of ownership can be determined.

JURISDICTION OF REGULATORY AGENCIES

As with questions of ownership, tidal datums largely dictate the authority and responsibility of agencies having jurisdiction over the uses of tidal marshes. The U. S. Army Corps of Engineers has authority to regulate the diking, dredging or filling of marshes. The Corps has two statutory

---

[11] Further evidence of the problem of surveying a mean-high-tide line in tracts of swamp is contained in several technical documents:

(a) Field Memorandum No. 1, United States Coast and Geodetic Survey (1938): "In marsh, mangrove, and cypress or similar swamp areas, the mean high water line is generally obscured by the vegetation and will not ordinarily be located."

(b) See also U.S. Coast Survey (1865), App. 22; p. 205; 1891, App. 16; p. 609, 633-34.

(c) The U.S. Coast and Geodetic Survey (1928) instructed:

"43. The high-water line.—The high-water line shall be drawn with sufficient strength to make it clearly distinguishable. The identification of the high-water line on marsh is usually difficult. The outer edge of a typical marsh is vertical and is sometimes covered at high-water, but for use on navigational charts its vertical edge should be indicated as the high-water line. The inner edge of the marsh (the limit of submergence at high water) when clearly defined may be drawn by a line distinctly lighter than the high-water line...."

(d) The U. S. Coast and Geodetic Survey (1963:42) explains: "The mean high-water line in marsh, mangrove, cypress, or similar swamp areas is generally obscured by vegetation and will not ordinarily be located on topographic surveys. In such areas, the outer edge of vegetation visible above Mean High Water, usually represented by a fine line on the topographic survey, presents a fairly definite shoreline...."

[12] *Carpenter* vs. *City of Santa Monica,* 63 Cal.App.2d 772, 787-788 (1944); *City of Los Angeles* vs. *Anderson,* 206 Cal. 662, 667 (1929) Civil Code § 1014; *O'Neill* vs. *State Highway Dept.,* 235 A.2d 1, 10 (N.J. 1967). The federal rule appears to ignore at least some artificial changes *County of St. Clair* vs. *Lovingston,* 90 U.S. (23 Wall.) 46, 68 (1874), but it may be rare that the Federal rule applies in California. *Or. ex rel. State Land Bd.* vs. *Corvallis Sand & G.,* 97 S.Ct. 482 (1977).

[13] *Dolphin Lane Assoc.* vs. *Town of Southampton,* 372 N.Y.S. 2d 52, 53-54 (1977).

bases for this authority. Sections 9 and 10 of the Rivers and Harbors Act of 1899[14] essentially prohibit the building of any "dam," "dike," "obstruction," or "other structures" within the "navigable waters of the United States" without the approval of the Corps. Similarly section 404 of the Federal Water Pollution Control Act as amended in October 1972 (FWPCA)[15] charges the Corps with regulating the discharge of dredged or fill material into "navigable waters."

The meanings of the term "navigable waters of the United States" and the simpler term "navigable waters," then, are critical to the jurisdiction of the Corps. Originally separate regulations defining these terms were adopted by the Corps for the Rivers and Harbors Act and for the FWPCA.[16] These regulations were revised and integrated effective 19 July 1977 and codified in Title 33, Code of Federal Regulations, sections 320 *et seq.*[17] There are two definitions for purposes of the Rivers and Harbors Act. If lands are used, or have been used, or may be susceptible to use to transport interstate or foreign commerce, they are "navigable waters of the United States." Secondly the term includes all lands subject to the ebb and flow of the tide shoreward on the Pacific Coast to the line of mean higher high water (MHHW).[18] Significantly, "an area will remain 'navigable in law,' even though no longer covered with water, whenever a change in condition has occurred suddenly, or was caused by artificial forces intended to produce that change."[19] Thus in diked or filled marshes, one must determine the "former" line of MHHW.

For purposes of the FWPCA, the regulations define "navigable waters" much more broadly than they define "navigable waters of the United States" for the Rivers and Harbors Act. The FWPCA regulation subsumes "navigable waters of the United States" since it includes both the "susceptibility" and the MHHW definitions, but it additionally includes all marshes, swamps and "similar areas," among other features.[20]

An additional regulation giving a much more detailed definition of the term "navigable waters of the United States" is set forth in section 329.1 *et seq.* of Title 33 of the Code of Federal Regulations. This section, which gives numerous examples intended to illustrate the term, applies to jurisdiction asserted under both the Rivers and Harbors Act and the FWPCA. It also includes the interpretation that lands formerly subject to the tides but which have been excluded from tidal action by dikes or other man-made works are still "navigable waters of the United States."[21]

The complexities of determining, for example, whether the Corps has jurisdiction under the Rivers and Harbors Act over a tract of reclaimed marsh are apparent. If the marsh cannot be said to have been "susceptible of use for commercial navigation," then it must be determined whether in its natural state it lay above or below the MHHW elevation. Determining elevations within a natural marsh may pose problems enough (see also National Ocean Survey 1975). But, as discussed above, when it has been walled off from the tides many factors can further complicate determining

---

[14] 33 U.S.C. § 401 *et seq.*

[15] 33 U.S.C. § 1251 *et seq.*

[16] These regulations, now superseded, were codified at 33 C.F.R. § 209.120(d) (1), together with 33 C.F.R. § 209.120(d)(1), together with 33 C.F.R. § 209.260 *et seq.* (regulations for the Rivers and Harbors Act), and 33 C.F.R. § 209.120(d)(2) (regulations for the FWPCA).

[17] See 42 Fed. Reg. 37122 *et seq.* (July 19, 1977).

[18] 33 C.F.R. §§ 321.2 and 322.2.

[19] 33 C.F.R. § 329.13.

[20] 33 C.F.R. § 323.2 (b), and (c).

[21] 33 C.F.R. § 329.13.

former elevations of the marsh. There is evidence that when drained and allowed to dry, the marsh soil compacts so that its elevation is lowered. When reflooded it may not "sponge" back or expand, its elevation remaining the same (excluding future deposition).[22] The problem would be compounded by any filling or excavating done after the diking.

The validity of these regulations is the subject of a decision that was handed down 11 May 1978, by the United States Court of Appeals for the Ninth Circuit.[23] In that decision the Court wrote:

> We hold that in tidal areas, navigable waters of the United States, as used in the Rivers and Harbors Act, extend to all places covered by the ebb and flow of the tide to the mean high water (MHW) mark in its unobstructed, natural state. Accordingly, we reverse the district court's decision insofar as it found that the Corps's jurisdiction under the Rivers and Harbors Act includes all areas within the former line of MHHW in its unobstructed, natural state. . . .
> We therefore hold that the Corps's jurisdiction under the FWPCA extends at least to waters which are no longer subject to tidal inundation because of Leslie's dikes without regard to the location of historic tidal water lines in their unobstructed, natural state. We express no opinion on the outer limits to which the Corps's jurisdiction under the FWPCA might extend. (578 F.2d at 753, 756.)

As with the Corps, two California state agencies charged with regulating coastal development also have their jurisdiction defined by reference to tidal datum planes. The older of the two agencies, the San Francisco Bay Conservation and Development Commission (BCDC), is charged with planning for and regulating development as well as conservation of San Francisco Bay. The commission's jurisdiction includes

> (a) San Francisco Bay, being all areas that are subject to tidal action from the south end of the bay to the Golden Gate (Point Bonita-Point Lobos) and to the Sacramento River line (a line between Stake Point and Simmons Point, extended northeasterly to the mouth of Marshall Cut), including all sloughs, and specifically, the marshlands lying between mean high tide and five feet above mean sea level; tidelands (land lying between mean high tide and mean low tide); and submerged lands (land lying below low tide).
> (c) Salt ponds consisting of all areas which have been diked off from the bay and have been used during the three years immediately preceding the effective data of the amendment of this section during the 1969 Regular Session of the Legislature for the solar evaporation of bay water in the course of salt production. (Gov. Code section 66610).

The regional and statewide coastal commissions created by passage of Proposition 20 in 1972 were supplanted last year when the legislature passed the California Coastal Act, which created a new statewide California Coastal Commission and six regional commissions. The authority and duties of these bodies are similar to that of BCDC, but their jurisdictions extend to the areas of California's coastline other than San Francisco Bay; BCDC's existence was not altered by passage of the Coastal Act. The jurisdiction of these agencies is the "coastal zone," which is also defined in section 30103 of the Public Resources Code by reference to the MHW line.

---

[22] Deposition of Claire Lopez, Chief Engineer for the Leslie Salt Co. from 1938 to 1964, taken April 23-26, 1973, in *Sierra Club et al.* vs. *Leslie Salt Co., et al.*, United States District Court for the Northern District of California, No. 72-561, and *State of California* vs. *County of San Mateo et al.*, San Mateo Superior Court No. 144257, pp. 112, 278-280.

[23] *Leslie Salt Co.* vs. *Froelhke,* 578 F2d 742 (9th Cir. 1978).

AN EXAMPLE OF HISTORICAL EVIDENCE: UPPER NEWPORT BAY, CALIFORNIA

Given the foregoing, it is clear that the engineer or scientist must frequently resort to whatever historical evidence exists respecting the character of a marsh. Even when such historical evidence exists, however, it may generate more confusion than it disperses. Although examples within San Francisco Bay are not lacking, these situations are presently the subject of litigation and may be inappropriate to discuss. The problem of the character of three islands (Upper, Middle and Shellmaker; Fig. 2) of tidal marsh in Upper Newport Bay, however, provides an example

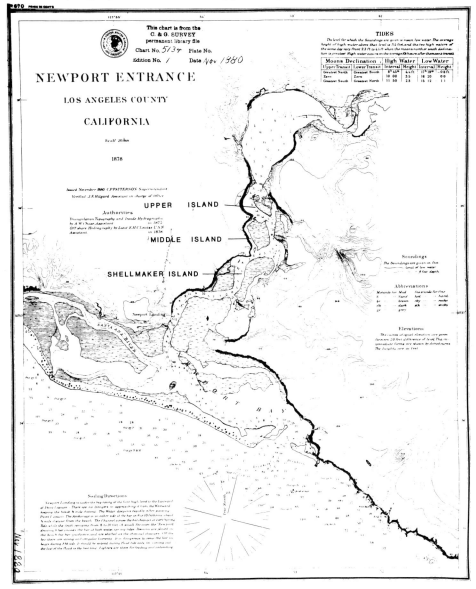

Fig. 2. Nautical chart of Newport Bay, California, published in 1878 by the U.S. Coast and Geodetic Survey.

comparable to cases within San Francisco Bay. (The reader should bear in mind that the question of the character of the islands was but one aspect of a much larger title dispute, which entailed complicated questions of whether the lands were in fact validly purchased from the State, regardless of their character).

Upper Newport Bay, a tidal estuary, is part of the Newport Canyon. The canyon is the southeastern extremity of a lowland plain which is bounded on the northwest by the San Pedro Hills, thence sweeping inland to the coastal foothills.

The Bay is bounded by uplands, the titles to which were deraigned from rancho grants made by the Mexican government (the Rancho San Joaquin and the Ranchos Santiago de Santa Ana). Upper Newport Bay and the islands lying within it were not included within the patents to the ranchos. During the confirmation proceedings, however, Jose Sepulveda, the ultimate patentee of both ranchos, had claimed the Bay and an island within it. The claim was based in part on testimony of a witness that within the "Bolsa of San Joaquin" was a hummock, a type of island surrounded by marsh, that had been occupied by Sepulveda as a potrero, a place for keeping and raising horses. (Transcript of the proceedings in case No. 406, *Jose Sepulveda, claimant* vs. *The United States, Defendants* for the place named "San Joaquin" [185 SD p. 8].)

The Ranchos San Joaquin and Santiago de Santa Ana were surveyed for the federal government by Henry Hancock in 1857. The final approved plat of his survey does not show the islands in question, but since they were not included within either of the rancho grants it is not expected that they would have been shown. His field notes, however, include a crude sketch of islands in Upper Newport Bay.

An 1875 U.S. Coast Survey chart entitled "Hydrography of Newport Bay or Santa Ana Lagoon and approaches, scale 1:10,000, 1875," Register No. 1256, shows Upper, Middle and Shellmaker Islands delineated by an "apparent shoreline," an administrative surrogate for the actual mean-high-water line and in the companion topographic map, Register No. 1392, the islands are also delineated by an apparent shoreline and marked with the symbol for swamp.

The islands show on another 1875 map, this one prepared by Los Angeles County Surveyor L. Seebold in connection with an application for Tide Land Location 37. That application concerned lands south of the three islands and characterized the land surveyed as a "piece of swamp and overflowed land." The application states, however, that ". . . no part of the land sought to be purchased is below low tide . . .," which indicates that the land applied for was probably tidelands. But the significant matter is that Upper, Middle and Shellmaker Islands are depicted in a distinctly different manner than the parcel sought to be purchased; an inference is that the islands were of a different character, possibly that they were thought to be swamp and overflowed.

An 1878 map prepared by Assistant Los Angeles County Surveyor Charles T. Healey shows the approximate sites of Middle and Shellmaker Islands as the "tideland locations of C. E. French." Although history has given Healey a reputation for doing his work in the field and not in the office (as was the practice of many of his contemporaries), the map places section lines and corners and many topographical features (including two of the islands, Middle and Shellmaker) approximately 10 chains west of their true position.[24] The map shows lots numbered 1 through 9 as French's tideland locations.

The configuration of Healey's Lot 5 resembles Shellmaker Island as it appeared on the later township plat, and Lot 7 generally resembles Middle Island as shown on the township plat. The State Lands Division has in its records an application to purchase these "tideland" lots, and the

---

[24] The most northeasterly island on Healey's map is shown in a much different place than is Upper Island on Finley's township plat, and this difference cannot be explained by Healey's placement of section lines and corners approximately 10 chains west of their true position. This difference (there is only a sliver of overlap) makes it impossible to determine whether these two islands were the same.

other Lots 1 through 9, under the name Survey Number 12. The application appears to be a copy; the blanks are filled in and signature of C. E. French is in quotation marks. On the cover of this application is the printed phrase "Swamp and Overflowed Lands." Between the words "overflowed" and "lands" a caret adds the words "and tide."

In 1889, Solomon H. Finley completed the Federal township survey of T6S, R10W, San Bernardino Meridian, which showed Upper, Middle and Shellmaker Islands as swamp-and-overflowed lands. There is some indication, however, that Finley surveyed an ordinary-low-water mark instead of the ordinary-high-water mark. This is the conclusion reached by Harris E. Coutchie in a report prepared for the Irvine Company dated August 1971 (*Id.* at p. 7). In addition, Finley may have legitimately believed he was to have surveyed the ordinary-low-water mark. The official government manual for surveyors in effect in 1889 was the General Land Office's 1881 Instructions of the Commissioner of the General Land Office to the Surveyors General of the United States Relative to the Survey of the Public Lands and Private Land Claims. On page 33 of the manual is an erroneous instruction that swamp-and-overflowed lands bordering on navigable lakes and rivers were to be meandered at the ordinary low-water mark.[25] Nothing in the manual, however, specifically treats of swamp lands on tidewaters. (See *People* vs. *Ward Redwood Co.,* 225 Cal. App. 2d 385, 390 [1964].)

Then in 1912, Otto Von Geldern, a renowned coastal engineer, prepared a plat of survey for the U.S. Army Corps of Engineers showing lands of the general shape of Upper, Middle and Shellmaker Islands as swamp. There are contour lines within the swamp, and the map legend states these contours are of a "spring high tide" elevation. If true, this is evidence that portions of the islands were above mean high water in 1912. (No contemporary cartographic manuals of the Corps have been examined.) Conceivably, however, Von Geldern may have actually mapped the mean high tide line.[26] If so, this would be the only historic mean-high-tide-line survey of Upper Newport Bay known to us.

Fortunately for the involved parties and any judge who might have had to try to impose order on this chaotic evidence, the dispute was resolved in 1975 in an out-of-court settlement.

TO THE SCIENTIST AND ENGINEER

When the law hears disputes relating to tidal marshes it thus has a vast need for the knowledge of the scientist or engineer on such questions as the natural physical and biological history of marshes, how they respond to man-made changes, and the roles they play in the hydrodynamics and sediment transport in the adjacent water body. The law applicable to a given problem may be intricate, and may change, if subtly, as facts are learned or as studies yield new data. For this reason, to assure that he probes the appropriate questions, the scientist or engineer should demand clear instructions from his client. He should ask for specific formulations of the questions he is to answer and assure himself that he understands them, lest his preparation be misspent in irrelevant or tangential inquiries. He should not tolerate an assignment, for example, simply to locate the "ordinary high water mark" of a parcel of land. He should ask the proper tidal datum to employ, and whether the line is to be located in the present condition, or in some former condition of the

---

[25] This instruction is repeated in the next (1890) edition of the manual. But the 1890 manual added an instruction that lands (not specifying swamp and overflowed lands) bordering on tidelands were to be meandered at the ordinary high-water mark. That the former instruction is erroneous is clear. *Barney* vs. *Keokuk,* 94 U.S. 325, 338 (1876).

[26] See Von Geldern, The Plane of Ordinary High Tide, etc., 29 Pacific Municipalities 243 (June 1915), and the rebuttal of D. E. Hughes, 29 Pacific Municipalities 340, 344 (August 1915).

land. By the same token he should be tolerant when the law has not caught up with knowledge, when it has not yet developed the sophistication to ask the proper question. Experts and not lawyers taught the Supreme Court the meaning of the tidal datum of MHW, and how that datum might be used to locate the law's—then—ethereal "ordinary high-water-mark." So when the law, as it frequently does, asks the scientist to square a circle, he should assume his duty to educate the law, to enable it to reshape itself and make its provisions congruent with the state of knowledge.

LITERATURE CITED

Atwater, B. F. et al. 1979. History, landforms, and vegetation of the estuary's tidal marshes. Pages 347-386 *in* T. J. Conomos, ed. San Francisco Bay: The Urbanized Estuary. Pacific Division, Amer. Assoc. Advance. Sci., San Francisco, Calif.

Gilbert, G. K. 1917. Hydraulic-mining debris in the Sierra Nevada. U. S. Geol. Surv. Prof. Paper 105. 154 pp.

Gosselink, J. G., E. P. Odum, and R. M. Pope. 1973. The value of the tidal marsh. Urban and Regional Development Center, University of Florida. Working Paper No. 3.

Haedrick, R. J., and C. A. S. Hall. 1976. Fishes and estuaries. Oceanus 19(5):64-70.

Haines, E. B. 1977. The origins of detritus in Georgia salt marsh estuaries. Oikus 29:254-259.

Johnston, R. 1956. Predation by short-eared owls in a salicornia marsh. Wilson Bull. 69:29-102.

Maloney, F. E., and R. C. Ausness. 1975. The use and legal significance of the mean high water line in coastal boundary mapping. 53 No. Carolina Law Review 185:202-206.

Marmer, H. A. 1926. The tide. Appleton-Century-Crofts, New York. 282 pp.

Mitchell, A. 1869. On the reclamation of tide-lands and its relation to navigation. Page 77 *in* Report of the Superintendent, U. S. Coast Survey. 41st Congress, 2nd Session, House Exec. Doc. No. 206, Appendix 5.

National Ocean Survey. 1975. The relationship between the upper limit of coastal marshes and tidal datums. Pilot study done for the Environmental Protection Agency. 115 pp.

Pillsbury, G. 1939. Tidal hydraulics. Prof. Paper U. S. Corps Engineers 34:241-242.

Shalowitz, A. L. 1964. Shore and sea boundaries, vol. 2. U. S. Dept. of Commerce, Coast Geod. Surv. 749 pp.

Sims, C. W. 1970. Juvenile salmon and steelhead in the Columbia River estuary. Pages 80-86 *in* Proc. Northwest Estuarine Coastal Zone Symp., 28-30 October 1970, Portland, Ore.

Swainson, O. W. 1928. U. S. Coast and Geodetic Survey Spec. Pub. 144. Topographic Manual 9. 121 pp.

U. S. Coast and Geodetic Survey. 1938. Field Memorandum No. 1. The mean high water line in marsh and other swamp areas. 12 Field Engineers Bull. 241. 2 pp.

U. S. Coast and Geodetic Survey. 1963. Nautical Chart Manual, p. 42.

U. S. Coast Survey. 1865. Page 205 *in* Annual Report of the Superintendent, Appendix 22.

U. S. Coast Survey. 1892. Page 609 *in* Annual Report of the Superintendent for the fiscal year ending 30 June 1892, Appendix 16. Proc. Topographical Conf., 18 January-7 March 1892, Washington, D. C.

Valeila, I., and S. Vince. 1976. Green borders of the sea. Oceanus 19(5):10-17.

THE ROLE OF MYSID SHRIMP IN THE SACRAMENTO-SAN JOAQUIN ESTUARY AND FACTORS AFFECTING THEIR ABUNDANCE AND DISTRIBUTION

JAMES J. ORSI AND ARTHUR C. KNUTSON, JR.
California Department of Fish and Game, 4001 North Wilson Way, Stockon, CA 95205

Six species of mysid shrimp are present in the Sacramento-San Joaquin Estuary, but only one of them, *Neomysis mercedis*, is abundant here. It is an important fish food in Suisun Bay and the Delta, especially for young-of-the-year striped bass. *N. mercedis* feeds on phytoplankton, detritus, and zooplankton. Its distribution is apparently determined by estuarine circulation acting on its vertical migration pattern. These factors concentrate it in the zone where fresh and salt water initially mix. Light intensity greater than 10^{-5} lux on the bottom and net flow velocity <0.12 m·s^{-1} apparently limit its upstream spread. In the San Joaquin River low populations are associated with low dissolved oxygen concentrations in combination with high temperatures. Fecundity appears to be a function of female length, temperature, and food supply (phytoplankton). Seasonal fluctuations in reproduction are usually paralleled by population fluctuations. Population differences between years appear to be a function of food supply and habitat size.

Mysid or opossum shrimp are general names used for small, pelagic, live-bearing crustaceans belonging to several genera. They are cosmopolitan. Some, such as *Neomysis*, are coastal species that also occur in estuaries or lakes that are or were formerly connected to the ocean or an estuary. Lakes Merritt in Oakland and Merced in San Francisco where *N. mercedis* occurs are examples. Other mysids are deep ocean inhabitants. One species, *Mysis relicta*, occupies cold, deep, glacier-formed lakes far from the ocean. It is believed to have been pushed out of arctic marine waters by advancing ice sheets (Ricker 1959).

In the Sacramento-San Joaquin Estuary, adults of the largest species, *Neomysis rayi* (formerly *N. franciscorum*) are only 35 mm long. The young of the most abundant species, *N. mercedis* (*N. awatschensis* in many previous publications) are liberated from the female's brood pouch at 2 or 3 mm and grow to a maximum of 17 mm (Fig. 1).

SPECIES, ABUNDANCE, AND DISTRIBUTION IN THE SACRAMENTO-SAN JOAQUIN ESTUARY

Tattersall (1932) identified five species of mysids taken by the *Albatross* on a survey of San Francisco and San Pablo bays in 1912 and 1913. In addition to *Neomysis mercedis* and *N. rayi*, he identified *N. kadiakensis*, *N. costata*, and *N. macropsis*. The latter two species are now in the genus *Acanthomysis* (Ii 1936).

According to Tattersall, *N. kadiakensis*, *N. rayi*, and *A. costata* were most abundant in San Francisco Bay and *N. mercedis* was most abundant in San Pablo Bay. *A. macropsis* was evenly distributed throughout the two bays and was the most abundant species taken. We made short surveys of these bays in July and December 1977 and caught all species except *N. rayi*.

The *Albatross* did not sample upstream from San Pablo Bay. However, California Department of Fish and Game (DFG) surveys in Carquinez Strait, Suisun Bay, and the Delta from 1963 to 1976 have collected only *N. mercedis* and *A. macropsis,* of which *N. mercedis* has been by far the more abundant. In 1977 we discovered a previously unreported species which is still unidentified. It is unusually small. Adults are only 3 to 5 mm long. So far only a few specimens of it have been collected in the lower San Joaquin River from Winter Island to the mouth of the Mokelumne River and in the lower Sacramento River near Sherman Lake.

Abundance of *N. mercedis* in Suisun Bay and the Delta is very high compared to abundance of all mysid species combined in San Francisco and San Pablo bays. Using information in Sumner et al. (1914) on net mesh and size, and duration and speed of tows, we estimated the number of mysids per cubic meter in San Francisco and San Pablo bays from the number caught per tow by

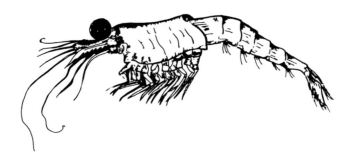

Fig. 1. *Neomysis mercedis,* the opossum shrimp.

the *Albatross*. These estimates ranged to a maximum of 1.6 mysids·m^{-3} for all species combined. We found up to 0.2 mysids·m^{-3} in San Pablo and North San Francisco bays in July and December 1977. But during high spring fresh water outflows the concentration of *N. mercedis* may reach 100·m^{-3} in San Pablo Bay (Painter 1966). In contrast we have found concentrations as high as 1500 *N. mercedis*·m^{-3} in Suisun Bay and the Delta. The abundance of *N. mercedis* makes it the most important species in the estuary.

N. mercedis ranges from Prince William Sound, Alaska to at least the Cañada de la Gaviota, 19 km below Pt. Conception, California (Orsi et al. unpublished ms.). Within the Sacramento-San Joaquin Estuary (Fig. 2) a few specimens of *N. mercedis* have been taken from around Hunter's Point and Angel Island (Tattersall 1932; Orsi and Knutson unpublished) and between Angel Island and San Pablo Strait (Tattersall 1932). It probably occurs in the lower Napa River but no one has sampled there for it. *N. mercedis* is most abundant in Suisun Bay and the western Delta (Heubach 1969; California Fish and Game 1976). *N. mercedis* is also found in Montezuma Slough and other sloughs around the Suisun Marsh and is present throughout the Delta and in the Sacramento Deep Water Channel to Lake Washington at Sacramento (R. Kroger pers. comm.). Water diversions have carried it into the California Aqueduct, the Delta-Mendota Canal, San Luis Reservoir, and possibly other water project reservoirs.

IMPORTANCE

The low abundance of mysids in San Francisco Bay renders them unimportant as a food source for fish there. However, in Suisun Bay and the Delta the great abundance of *N. mercedis* and its size make it an ideal food source for many fishes. It is the most important item in the diet

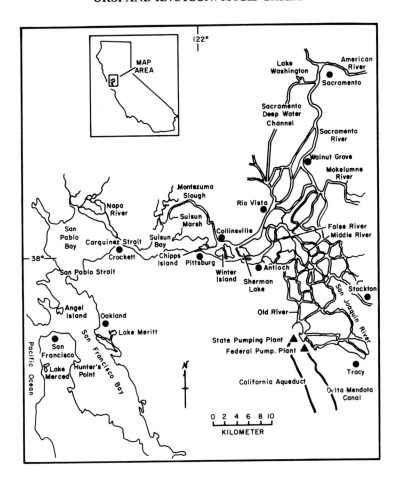

Fig. 2. Sacramento-San Joaquin Estuary.

of young-of-the-year striped bass and it is very important to juvenile striped bass (Heubach et al. 1963; Stevens 1966a; Thomas 1967; Stevens 1979; Smith and Kato 1979). *N. mercedis* is also an important food of juvenile white and green sturgeon (Radtke 1966), adult American shad (Stevens 1966b), black crappie (Turner 1966a), white catfish (Turner 1966b), and young king salmon (Sasaki 1966). Two larger shrimp, *Crangon franciscorum* and *Palaemon macrodactylus* feed on it (Siegfried et al. 1977).

FOOD HABITS

Mysids feed on phytoplankton, detritus, and other zooplankton. The only study of mysid feeding habits from this estuary was done by Kost and Knight (1975) on *N. mercedis* collected in Suisun Bay and the Delta. The most abundant foods were diatoms (phytoplankton) and unidentifiable material classified as detritus. Zooplankton remains were rare. The abundance of detritus relative to diatoms was greater in winter than in summer and increased with shrimp size. However, the actual importance of detritus is difficult to determine. Edmondson (1957) states that the presence of unrecognizable organic material in zooplankton guts cannot be taken as proof that they have been feeding on detritus. Phytoplankton break down after being consumed and the remains can be mistaken for detritus. Even if such feeding occurs, detritus may be nutritionally inferior to phytoplankton, Harris (1974) found that detritus never equaled algae in sustaining egg production in the copepod, *Scottolana canadensis*.

Although Kost and Knight's study suggests that *N. mercedis* does not feed extensively on zooplankton in the estuary, it has been maintained on a diet of brine shrimp nauplii (*Artemia*) in the laboratory (Simmons and Knight 1975). Also, W. T. Edmondson (pers. comm.) has found *N. mercedis* to be an effective predator in feeding experiments; and the guts of mysids he collected from Lake Washington (Washington State) contained large quantities of copepods and cladocerans. In our laboratory large *N. mercedis* have attacked larval striped bass and smaller *N. mercedis*.

EFFECTS OF ENVIRONMENTAL FACTORS

DFG biologists have studied *N. mercedis* since 1963 in an attempt to understand how environmental factors affect its distribution and abundance. Such knowledge is necessary to protect the primary food source of young-of-the-year striped bass in the face of planned water development projects (see for example Gill et al. 1971).

Heubach (1969) found that *N. mercedis* was most abundant in the estuary from fresh water to 7.2 °/oo salinity and least abundant at salinities exceeding 18 °/oo. We have found *N. mercedis* to be most abundant in essentially the same area, although we would prefer to define it as from the upper end of the salinity gradient to the downstream end of the entrapment zone (as defined by Arthur and Ball 1979). Heubach also found reproduction to be greatest from fresh water to 3.6 °/oo salinity and thought reproduction was the principal factor affecting seasonal and geographical abundance. Heubach presented some evidence that salinity might lower reproduction and thus explain the low abundance at high salinities. Our data do not demonstrate such effects of salinity on reproduction. Our present conception is that reproduction is indeed the principal factor affecting seasonal abundance but geographical abundance is the result of tidal currents and estuarine circulation interacting with the tidally influenced vertical migration of *N. mercedis*. During all tidal stages more than half of the shrimp are closer to the bottom than to the surface (Heubach 1969). However, on flood tides a larger percentage of the shrimp are in the upper half of the water column than on the ebb tides. Since surface water velocities are higher than bottom velocities (see for example Conomos 1979), Heubach hypothesized that flood tides carried the shrimp farther upstream than ebb tides moved them back downstream. In addition, we hypothesize that the landward flowing bottom density current in and below the entrapment zone (Arthur and Ball 1979) hampers the downstream movement of mysids on the ebb tides.

We are not sure whether mysid populations are high in the entrapment zone merely because they are concentrated there by hydrological forces or because, in addition, conditions there are somehow optimal.

Upstream from the salinity gradient, net flow velocity and light penetration limit distribution and abundance. Heubach (1969) found *N. mercedis* at Isleton on the Sacramento River and at Stockton on the San Joaquin River only when net velocity was less than 0.12 m·s^{-1} and light intensity on the bottom was less than 10^{-5} lux.[1] Apparently net velocity was the limiting factor during the high winter and spring flows and light became limiting during summer and fall. Heubach did not find *N. mercedis* where flows did not reverse on the flood tide, i.e., above tidewater. We found *N. mercedis* to become progressively less abundant in the Sacramento River as net velocities increased from 0.02 to 0.12 m·s^{-1} (California Fish and Game 1976).

High temperature and low dissolved oxygen (DO) appear to reduce mysid abundance in the

---

[1] 10^{-5} lux ≈ 1.9×10^{-7} μE·m^{-2}·s^{-1} assuming a mean wave length of light (λ) of 5500 Å. (Ed.)

San Joaquin River between Stockton and the mouth of the Mokelumne River (Heubach 1969). During his investigation at temperatures below 18°C, dissolved oxygen (DO) concentrations as low as 5 mg·liter^{-1} had no apparent effect on abundance. However, above 18°C mean abundance at 9 mg·liter^{-1} DO was 10 to 15 times greater than at 4 to 5 mg·liter^{-1} DO. Mysids were absent when DO was less than 3 mg·liter^{-1} (Table 1).

TABLE 1. MEAN DENSITY OF *NEOMYSIS MERCEDIS* IN RELATION TO WATER TEMPERATURE AND DISSOLVED OXYGEN CONCENTRATION

Temp. (°C)	Dissolved Oxygen (mg·liter^{-1})[a]						
	9	8	7	6	5	4	3
<22	ND	122.5(4)[b]	44.4(7)	13.9(7)	19.4(7)	5.4(3)	0.2(3)
18-22	151.4(2)	56.9(6)	50.1(23)	54.1(6)	15.3(3)	11.4(3)	ND
14-18	13.2(8)	28.7(9)	22.5(7)	21.5(4)	43.0(2)	ND	ND
<14	6.2(19)	9.0(12)	10.5(5)	7.3(2)	ND	ND	ND

[a] Dissolved oxygen conversion factors: 1 mg·liter^{-1} = 1 ppm = 62.5 μg-atoms·liter^{-1}
[b] (n) = number of samples

Heubach (1969 and 1972) suggested that temperatures above 22°C might harm *N. mercedis* and that such high temperatures could have been responsible for the population declines he observed in summer 1966 and fall 1965. Hair (1971) established an upper lethal temperature between 24.2 and 25.5°C for *N. mercedis* in laboratory experiments. Wilson (1951) obtained similar tolerance for the same species. Alternately, Heubach's (1969) results show that the mean density at temperatures in excess of 22°C at 8 mg·liter^{-1} DO was 122.5 individuals·m^{-3}. This density is very high and is twice as great as occured between 18 and 22°C at the same DO (Table 1). It is almost as high as the mean density he reports at 18 to 22°C and 9 mg·liter^{-1} DO (151.4 individuals·m^{-3}. Unfortunately, Heubach did not report how far above 22°C these temperatures rose. Our field collections show that moderate populations sometimes existed where daytime temperatures were as high as 25.6°C (Table 2). Hence, we conclude that if DO remains high (8 or 9 mg·liter^{-1}) *N. mercedis* are not eliminated at temperatures defined as lethal in the laboratory, although those temperatures undoubtedly exceed the optimal temperature range.

The State and Federal pumping plants in the south Delta (Fig. 2) remove large volumes of water from the Delta for irrigation and domestic purposes. These plants draw water from the Sacramento River near Walnut Grove, across the Delta, down Old and Middle Rivers, and into the California Aqueduct and the Delta-Mendota Canal. This pumping sometimes causes the lower San Joaquin River to flow upstream toward the pumping plants.

We have been unable to completely evaluate the effects of these water diversions on the abundance of *N. mercedis* because our surveys began in 1968, the same year the Federal Government increased water exports for the San Luis Project, and the State Water Project began operations. This caused water export rates to rise sharply above previous levels. Consequently, our analysis is restricted to the period of high exports, during which we have not found any relation between total population and water diversion rates. However, export pumping does affect salinity intrusion and in turn the location of the entrapment zone.

Export pumping probably also reduces mysid abundance in the central Delta, especially in the region of the cross-Delta water transport channels. Abundance of *N. mercedis* in the San Joaquin River at the mouths of Old and Middle rivers is lower than abundance either up or downstream

SAN FRANCISCO BAY

TABLE 2. SUMMER ABUNDANCE OF *NEOMYSIS MERCEDIS* (INDIVIDUALS·m^{-3}) AND TEMPERATURE IN THE SAN JOAQUIN RIVER NEAR STOCKTON, 1972-77

Year	May I	May II	June I	June II	July I	July II	August I
1972	46(17.8)[a]	70(21.1)	57(25.6)	38(23.6)	9(24.4)	7(26.1)	4(24.4)
1973	42(19.2)	41(22.8)	12(23.9)	34(25.0)	14(25.8)	2(25.0)	2(22.8)
1974	6(20.0)	8(21.1)	44(22.2)	36(23.4)	37(23.4)	8(25.0)	1(25.6)
1975	88(17.0)	8(19.7)	3(22.2)	24(20.8)	0.6(22.5)	26(24.7)	1(25.6)
1976	4(19.4)	12(21.6)	33(20.6)	72(21.4)	12(23.6)	2(26.1)	0.4(23.6)
1977	14(16.6)	20(19.2)	17(23.4)	20(24.4)	4(25.0)	2(25.8)	0.3(25.6)

[a] Water temperature in °C

from these river mouths. This suggests that mysids are drawn up the San Joaquin River from their center of abundance in the western Delta and into Old River, but few manage to reach Middle River. Abundance rises again upstream from the mouth of Middle River as water in the Stockton area of the San Joaquin River tends to remain almost stationary during the summer and early fall, and mysids there propagate until high temperature and low DO affect them.

Reproduction appears to depend upon several factors. From data collected in 1976 we developed a multiple regression equation that predicted seasonal fluctuations in fecundity during 1977 quite well (Orsi unpublished ms.). The equation indicates that fecundity increases as female size and food supply (measured by chlorophyll *a*) increase and water temperature decreases. The fecundity variations contribute to seasonal increases and decreases in abundance of *N. mercedis*, such as Heubach (1969) reported for 1965 and 1966.

The abundance of *N. mercedis* also varies annually. Heubach (1969) noted that the peak abundance in 1965 was about 2.8 times greater than in 1966. Similarly, Hopkins (1965) reported that the peak *N. americana* density in the Delaware River was four times greater in 1958 than in 1957. Neither Heubach nor Hopkins could explain these differences.

To explain annual variations in mean July to October abundance of *N. mercedis* (the period when it is most important to young-of-the-year striped bass) in Suisun Bay and the Delta, we have developed a multiple regression equation which uses mean chlorophyll *a* throughout the area and either Delta outflow or salinity at Chipps Island as independent variables. Mysid abundance was positively correlated with chlorophyll *a* and Delta outflow, and negatively correlated with salinity. During high outflows mysids in concentrations >0.1 individuals·m^3 were distributed throughout a larger area of the estuary than during low outflows. Hence, the regression equation can be interpreted to mean that annual variations in mysid abundance are a function of habitat size and food supply. The unusually low abundance in 1976 and 1977 (Knutson and Orsi unpublished ms.) can be explained by the reduced habitat and low phytoplankton populations in those years. Maintenance of *N. mercedis* populations at 1968-1975 levels appears to require sufficient Delta outflow to keep the habitat large and phytoplankton populations high.

ACKNOWLEDGMENTS

We thank D. E. Stevens and H. K. Chadwick for reviewing this manuscript and R. M. Kano, D. P. Cayongcong, C. P. Hatcher, S. A. Tobar, S. E. Davis, G. L. Curtis, S. L. Blaesi, R. J. Rodriguez, and others who performed the field and laboratory work. R. M. Kano also drew the figures.

LITERATURE CITED

Arthur, J. F., and M. D. Ball. 1979. Factors influencing the entrapment of suspended material in the San Francisco Bay-Delta estuary. Pages 143-174 in T. J. Conomos, ed. San Francisco Bay: The Urbanized Estuary. Pacific Division, Amer. Assoc. Advance. Sci., San Francisco, Calif.

California Department of Fish and Game. 1976. Report to the State Water Resources Control Board on the impact of water development on fish and wildlife resources in the Sacramento-San Joaquin Estuary. Exhibits 1, 2, and 3.

Conomos, T. J. 1979. Properties and circulation of San Francisco Bay waters. Pages 47-84 in T. J. Conomos, ed. San Francisco Bay: The Urbanized Estuary. Pacific Division, Amer. Assoc. Advance. Sci., San Francisco, Calif.

Edmondson, W. T. 1957. Trophic relations of the zooplankton. Trans. Amer. Micro. Soc. 76(3): 225-245.

Gill, G. S., E. C. Gray, and D. Seckler. 1971. The California water plan and its critics: A brief review. Pages 3-27 in D. Seckler, ed. California Water: A Study in Resource Management. University of California Press, Berkeley, Calif.

Hair, J. R. 1971. Upper lethal temperature and thermal shock tolerances of the opossum shrimp (*Neomysis awatschensis*) from the Sacramento-San Joaquin Estuary, California. California Fish Game 57(1):17-27.

Harris, R. P. 1974. The role of organic debris and associated micro-organisms in pelagic estuarine food chains. II. Utilization of detritus by the harpacticoid copepod, *Scottolana canadensis*. Appendix B, Tech. Rep. No. 22, NRI Ref. No. 74-28, University of Maryland, Water Resources Research Center. 22 pp.

Heubach, W. 1969. *Neomysis awatschensis* in the Sacramento-San Joaquin River Estuary. Limnol. Oceanog. 14(4):533-546.

Heubach, W. 1972. *Neomysis*. Pages 27-32 in Ecological studies of the Sacramento-San Joaquin Estuary. Delta Fish and Wildlife Protection Study Rep. No. 8.

Heubach, W., R. J. Toth, and A. M. McCready. 1963. Food of young-of-the-year striped bass (*Roccus saxatilis*) in the Sacramento-San Joaquin River system. Calif. Fish Game 49(4):224-239.

Hopkins, T. L. 1965. Mysid shrimp abundance in surface waters of Indian River Inlet, Delaware. Chesapeake Sci. 6(2):86-91.

Ii, N. 1936. Studies on Japanese Mysidacea. Jap. J. Zool. 6(4):577-619.

Kost, A. B., and A. W. Knight. 1975. The food of *Neomysis mercedis* Holmes in the Sacramento-San Joaquin Estuary. Calif. Fish Game 61(1):35-46.

Painter, R. 1966. Zooplankton of San Pablo and Suisun bays. Pages 40-56 in Ecological Studies of the Sacramento-San Joaquin Estuary. Part I. Calif. Fish Game, Fish. Bull. 133.

Radtke, L. D. 1966. Distribution of smelt, juvenile sturgeon, and starry flounder in the Sacramento-San Joaquin Delta with observations on the food of sturgeon. Pages 115-126 in: Ecological Studies of the Sacramento-San Joaquin Delta. Part II. Calif. Fish Game, Fish. Bull. 136.

Ricker, K. E. 1959. The origin of two glacial relict crustaceans in North America, as related to Pleistocene glaciation. Can. J. Zool. 37:871-893.

Sasaki, S. 1966. Distribution and food habits of king salmon, *Oncorhynchus tshawytscha*, and steelhead rainbow trout, *Salmo gairdnerii*, in the Sacramento-San Joaquin Delta. Pages 108-

114 *in* Ecological Studies of the Sacramento-San Joaquin Delta. Part II. Calif. Fish Game, Fish. Bull. 136.

Siegfried, C. A., A. W. Knight, and M. E. Kopache. 1977. Ecological studies of the western Sacramento-San Joaquin Delta during a dry year. Water Science and Engineering Paper No. 4506. University of California, Davis, Calif.

Simmons, M. A., and A. W. Knight. 1975. Respiratory response of *Neomysis intermedia* (Crustacea:Mysidacea) to changes in salinity, temperature, and season. Comp. Biochem. Physiol. 50A: 181-193.

Smith, S. E., and S. Kato. 1979. The fisheries of San Francisco Bay: Past, present, and future. Pages 445-468 *in* T. J. Conomos, ed. San Francisco Bay: The Urbanized Estuary. Pacific Division, Amer. Assoc. Advance. Sci., San Francisco, Calif.

Stevens, D. E. 1966a. Food habits of striped bass, *Roccus saxatilis,* in the Sacramento-San Joaquin Delta. Pages 68-96 *in* Ecological Studies of the Sacramento-San Joaquin Delta. Part II. Calif. Fish Game, Fish. Bull. 136.

Stevens, D. E. 1966b. Distribution and food habits of American shad, *Alosa sapidissima,* in the Sacramento-San Joaquin Delta.

Stevens, D. E. 1979. Environmental factors affecting striped bass (*Morone saxatilis*) in the Sacramento-San Joaquin Estuary. Pages 469-478 *in* T. J. Conomos, ed. San Francisco Bay: The Urbanized Estuary. Pacific Division, Amer. Assoc. Advance. Sci., San Francisco, Calif.

Sumner, F. B., G. D. Louderback, W. L. Schmidt, and E. C. Johnston. 1914. A report on the physical conditions in San Francisco Bay based on the operations of the U.S.S. Fisheries Steamer *Albatross* during the years 1912 and 1913.

Tattersall, W. M. 1932. Contributions to a knowledge of the Mysidacea of California. II. The Mysidacea collected during the survey of San Francisco Bay by the U.S.S. *Albatross* in 1914. Univ. Calif. Publ. Zool. 37:301-314.

Thomas, J. L. 1967. The diet of juvenile and adult striped bass (*Roccus saxatilis*) in the Sacramento-San Joaquin River system. Calif. Fish Game 53(1):49-62.

Turner, J. L. 1966a. Distribution and food habits of centrarchid fishes in the Sacramento-San Joaquin Delta. Pages 144-151 *in* Ecological Studies of the Sacramento-San Joaquin Delta. Part II. Calif. Fish Game, Fish. Bull. 136.

Turner, J. L. 1966b. Distribution and food habits of ictalurid fishes in the Sacramento-San Joaquin Delta. Pages 115-126 *in* Ecological Studies of the Sacramento-San Joaquin Delta. Part II. Calif. Fish Game, Fish. Bull. 136.

Wilson, R. R. 1951. Distribution, growth, feeding habits, abundance, thermal and salinity relationships of *Neomysis mercedis* (Holmes) from the Nicomekl and Serpentine rivers, British Columbia. M.S. Thesis. University of British Columbia, Vancouver, Canada. 68 pp.

NATURAL AND ANTHROPOGENIC INFLUENCES ON BENTHIC COMMUNITY STRUCTURE IN SAN FRANCISCO BAY

FREDERIC H. NICHOLS
U.S. Geological Survey, 345 Middlefield Road, Menlo Park, CA 94025

Data collected in the San Francisco Bay estuary over the last 65 years show that numbers of macrofaunal species are greatest in the marine environment of the central region near San Francisco, decreasing toward the north and south. This distribution has traditionally been attributed to differences in absolute values of salinity and sediment texture. Recent studies of both the benthos and the physicochemical environment near the substrate suggest that species distribution is more related to temporal variation in salinity and to intermittent disturbance of bottom sediments by storm-generated and seasonal wind waves and by the seasonally alternating high and low river inflow. Physical disturbance of the substrate apparently contributes to a state of non-equilibrium in the benthic community especially in the shallow reaches: the community, dominated by colonizers, reflects an early stage of species succession. Some of the most successful species under these conditions are those introduced from other estuaries.

Maximum values of total benthic biomass, in contrast to numbers of species, are found in South Bay, probably reflecting reduced salinity variability, somewhat greater stability of subtidal sediments, and the large quantities of potential food (high sewage-waste loadings, high concentrations of suspended particulate matter, and moderate to high standing stock of primary producers) resulting from shallow depth and the absence of strong water circulation. High biomass can also be attributed to the successful establishment of several large and abundant introduced species that thrive in South Bay.

Although once apparent as a reduction of numbers of species, the effect of waste disposal on the benthos is now often masked by natural perturbations resulting from biotic and abiotic disturbances of surficial sediments and by inhomogeneous distribution of the animals. Anthropogenic influences on benthic community structure other than that resulting from the introduction of exotic species will become increasingly difficult to quantify and therefore to predict. Future changes in the biota may be expected with continued reduction in fresh water flow into the estuary.

During most of the past 5,000 years the San Francisco Bay estuary contained a plentiful supply of fish and shellfish. Massive shell middens formerly found around the Bay indicated that aboriginal people consumed large quantities of mollusks, particularly the native oyster *Ostrea lurida*, the bent-nosed clam *Macoma nasuta*, and the bay mussel *Mytilus edulis* (Nelson 1909). Evidence of the rapid decline of shellfish resources soon after the arrival of the white man is equally striking (see reviews by Skinner 1962; Jones and Stokes Assoc. Inc. 1977). The beginning of this decline is not documented, but it was certainly observed some time before 1900 (Sumner et al. 1914; Barrett 1963; Nelson 1909). One of the tasks of the first major scientific investigation of San Francisco Bay, the *Albatross* expedition of 1912-13, was to study causes of the decline in fishery productivity. Although nothing concerning this decline was resolved by the findings of that investigation, a widely held view at the time (Skinner 1962) was that waste discharge into the Bay was a primary cause.

SAN FRANCISCO BAY

Most commercial fishing has either declined or ceased during the past 75 years, largely because of overfishing and pollution damage (Skinner 1962; Smith and Kato 1979). More recently, Public Health officials have condemned commercial operations because of tainting by toxic substances. During the past two decades considerable effort has been expended to define the nature of past changes in the biotic community and to relate these changes to specific causes. Largely because the *Albatross* expedition reports concerned benthic invertebrates, and because these generally sessile organisms cannot escape deleterious environmental changes, recent studies of the biota of the Bay have concentrated on this group of organisms (Nichols 1973).

Variation in species composition of the macrobenthos has been used for a number of years to indicate environmental change. Such variation is usually seen as (1) the disappearance of apparently sensitive species near waste outfalls, and (2) the appearance of other apparently tolerant species in these stressed environments either through natural migration of local opportunistic species or through accidental introduction of species from other regions.

The decline in number of species near waste outfalls was so marked in the 1950's (Filice 1959) that the application of species diversity as a tool for assessing environmental impact became a standard procedure. But in light of the improved treatment of wastes now entering the Bay, there is some concern that we may not detect the effects on the benthos of continued waste disposal even though these effects may be deleterious. Natural variations in community structure caused by a complex interaction of physical, chemical, and biological processes probably mask much of the variation caused by man.

I summarize, in this chapter, the patterns of the distribution of macrobenthic organisms in San Francisco Bay and discuss what is known of both natural and anthropogenic influences that help maintain or alter benthic community structure. It is not my intent to review the results of the large number of recent studies undertaken to assess the effect of, for example, individual waste outfalls on the benthic community but rather to describe the character of the influences on the benthic community and to show how our studies of these influences must be increasingly sophisticated if we are to distinguish and measure them.

PATTERNS IN THE DISTRIBUTION OF BENTHIC ANIMALS

The participants of the *Albatross* expedition (Sumner et al. 1914) noted that benthic community composition varied spatially within the Bay largely in relation to the salinity regime. Packard (1918a, b) pointed out that the salinity gradient, especially in the vicinity of Carquinez Strait (Fig. 1), had a strong effect on molluscan species distributions. Eastward of the strait a brackish-fresh water fauna existed, whereas in the central regions, a marine fauna containing many of the species found on the central California coast was noted. Packard realized that bottom-sediment texture markedly affected local distributions of individual species. These same patterns were noted by Filice (1954a, b, 1958, 1959) and Painter (1966) in San Pablo and Suisun bays. These investigators pointed out an inability to distinguish between the influences of water depth and sediment type, since coarse-grained sediments were usually found at the greatest depths.

In a baywide study made during the early 1960's, Storrs et al. (1969) showed that number of species varied greatly within the Bay. They described a relation between species diversity and both salinity and sediment texture through regression analysis:

$$\text{Diversity} = a\,(\text{chlorosity}) + b\,(\%\,\text{sand}) + c.$$

Their data revealed that diversity was low at the northern and southern parts of the estuary, increasing toward the Golden Gate (Storrs et al. 1969, Fig. 2). Higher diversity near San Francisco reflected the presence of marine species in sediment influenced by the high-salinity water entering

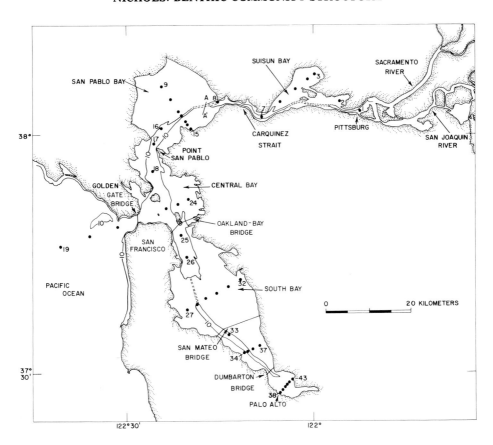

Fig. 1. Benthic sampling locations; for clarity, numerical designations of only first and last stations of cross-bay transects are shown. Side-scan sonar track (A-A') near station 8 (see page 416 and Fig. 6) shown as heavy line. The 10-m isobath outlines main shipping channels.

the Golden Gate at depth. Away from the marine influence, diversity of benthic species at any site seemed to reflect the local sediment regime (intertidal muds, subtidal muds, subtidal sand, shell banks), and the degree to which the site was subject to low-salinity conditions.

The descriptions of benthic-organism distribution in the earlier studies were based on samples collected nonsynoptically from different areas. It is therefore possible that there is confusion in these data between spatial and temporal influences on community structure. To define baywide characteristics of benthic community structure and to develop a reference collection of macrofauna on this same scale, we collected benthic samples synoptically at 43 stations throughout San Francisco Bay (Fig. 1) in February and August 1973 with a 0.1-m^2 van Veen sampler and washed them on a 1.0-mm mesh screen. The data obtained from these samples are being used to map the distribution of species (F. Nichols and J. Thompson unpublished data). In this review only total number and combined wet weight of the mollusk, crustacean, and polychaete species retained on the 1.0-mm screen are considered.

Lists of species found at each station confirm earlier studies in the presence of marine species near the Golden Gate and estuarine and brackish-water species toward the north and south. Again, greatest numbers of species were found in the vicinity of the Golden Gate with the exception of two locations: station 21 just outside the Golden Gate in a zone of high water turbulence

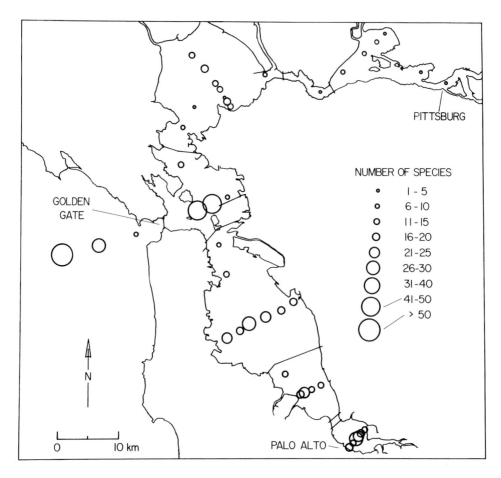

Fig. 2. Total number of benthic macrofauna species collected during February and August 1973 in replicate 0.1-m² samples and retained on 1.0-mm sieve.

and sediment movement (7 species), and station 24, a shallow-water location of very soft muddy sediments (9 species) (Fig. 2). Toward the south, the number of species was somewhat reduced from that in the central region, and toward the Sacramento River to the northeast, number of species at the stations sampled were very small (e.g. as few as two macrofaunal species in all samples at station 3).

The distribution of specimens among species in the central region was more or less even; at most stations only several individuals of each species were found in each sample. This distribution contrasts markedly with that found in the northern and southern reaches of the estuary, where, especially in the shallow subtidal or intertidal habitats, many tens and even hundreds of specimens of several small species were collected (notably the clam *Gemma gemma*, the amphipod *Ampelisca milleri*, the polychaetes *Streblospio benedicti, Heteromastus filiformis,* and *Asychis elongata,* and oligochaetes).

The distribution of biomass (Fig. 3, 4) differed substantially from that of species numbers. Of the 15 stations with highest mean biomass, all but three were located in South Bay. The stations ranked 1st, 6th, and 11th in total biomass were located in Central Bay. One species predominated at those stations ranked 1 through 5. The sand dollar *Dendraster excentricus* contributed

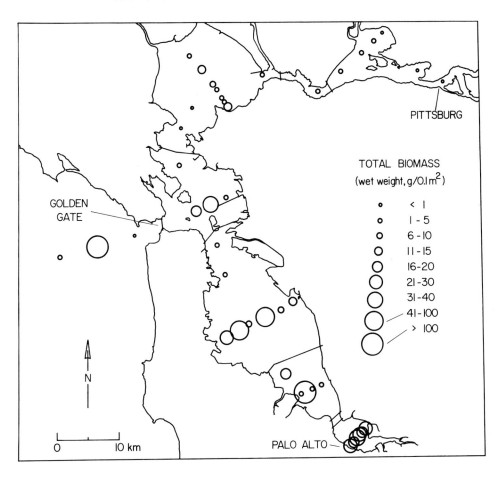

Fig. 3. Total biomass (g, wet weight) of benthic macrofauna, mean of replicate 0.1-m² samples collected in February and August 1973.

greater than 99% of the mean (two seasons combined) wet weight of 537 g·0.1 m² at station 20 on the sand bar outside the Golden Gate. The Japanese cockle *Tapes japonica* predominated at the stations ranked 2 through 4 (stations 35, 28, and 30), representing 68, 87, and 34% of total biomass, respectively (seasons combined). The Japanese mussel *Musculus senhousia* was dominant (59% of total biomass) at the fifth ranked station 40. The remaining stations with high biomass ranking showed a more even distribution of biomass among species, although the largest fraction was contributed by several mollusk species: those bivalves listed here as well as *Macoma balthica,* the soft-shelled clam *Mya arenaria,* the very small but abundant *Gemma gemma,* and the gastropod *Nassarius obsoletus* (Nichols 1977). The dominance of mollusks as primary consumers in some estuarine habitats has been noted by Burke and Mann (1974). A notable exception to the pattern of dominance by mollusks is the occurrence in large numbers of the large tube-dwelling polychaete *Asychis elongata* in the muddy subtidal regions of South Bay above the Dumbarton Bridge and in Central Bay in the vicinity of station 23.

With the exception of stations 10 and 11 in San Pablo Bay, total biomass in the northern region is low (less than 5 g wet weight per 0.1-m² sample), and above Carquinez Strait total biomass is distributed among a very few species such as the bivalves *Macoma balthica* and *Mya arenaria,* the

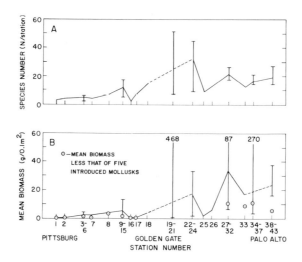

Fig. 4. Total number of benthic macrofauna species (A) and mean biomass (B, wet weight, seasons combined) as distributed along main axis of Bay. Range of observations shown for cross-bay transects (see Fig. 1). Actual biomass values shown where scale was greatly exceeded.

amphipods *Corophium* spp., the polychaete *Neanthes succinea,* and numerous oligochaetes.

NATURAL INFLUENCES ON BENTHIC SPECIES COMPOSITION

Common benthic sampling techniques (e.g. replicate sampling at specific locations) are designed to examine variation associated with spatial distribution and sampling error (e.g. Jones 1961) in a physically predictable environment. We normally assume that such techniques will allow us to detect man's impact on a natural system if we measure a sufficient number of parameters on a routine basis. However, natural disturbances of the Bay bottom, especially those that are intermittent both in intensity and time of occurrence, tend to accentuate nonpredictable fluctuations in spatial and temporal patterns of benthic community structure that cannot be interpreted from routine sampling data. Recent studies of the physical and chemical environment of the near bottom of the Bay demonstrate the prominence of intermittent disturbances.

Salinity Variations

The large number of species in the region near the Golden Gate reflects an oceanic influence; many of these species are found on the outer California coast in high-salinity water. Away from the oceanic influence, species numbers decline. The reduction of species number and the changes in species composition within the estuary have been attributed in nearly all past reports to decreasing salinity with distance landward from the Golden Gate. Such a pattern is typical of estuaries with low river inflow and associated stable salinity gradient: a gradual and predictable shift in species composition from stenohaline and euryhaline marine species through euryhaline opportunists and estuarine endemics to fresh-water species (see Boesch 1977). A similar pattern of the reduction of the species number up estuary is found in those estuaries with high river inflow and strong seasonal variations in salinity. But it is the annual range of salinity at any location, rather than absolute salinity, that determines the general character of the faunal assemblage (disregarding for the moment depth, sediment, and current regime influences). Peterson et al. (1975b) have shown that in San Francisco Bay, for example, there can be large seasonal and annual variations in water chemistry parameters at a given location, resulting directly or indirectly from variations in river runoff. During wet winters, the water east of the western entrance to Carquinez Strait can be fresh; during summer of dry years salt water can be found well up into the Sacramento River (Conomos 1979). The authors of the first *Albatross* expedition report (Sumner et al. 1914) pointed out the possible significance of seasonal variations in temperature and especially salinity (Sumner et al. 1914, Figs. C and N) in determining faunal distributions. Salinity variations during the year prior to the sampling described here were especially marked, for example, in San Pablo Bay and at the southern end of South Bay (Fig. 5).

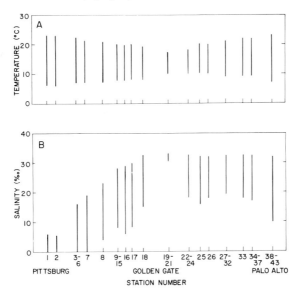

Fig. 5. Seasonal range of temperature (A) and salinity (B) between August 1972 and August 1973, data from U. S. Geological Survey.

The eastward (up estuary) limit of the distribution of estuarine organisms is undoubtedly determined by the extent of salt water encroachment along the bottom. The position of maximum encroachment varies from year to year, depending on river runoff. In regions of the most extreme year-to-year fluctuations (e.g. where fresh and salt water alternately impinge on the bottom), mature populations of estuarine organisms cannot become established, although young animals become established there on a repeated, if shortlived, basis. The relation between salinity variations and numbers of species reported for two British estuaries (Alexander et al. 1935, Fig. 30) is perhaps typical: species numbers declined in a landward direction with increasing tidal and salinity variations, but increased again toward the fresh water source where salinity variation was small. A similar increase in numbers of species was not noted at the landward end of our sampled region, probably because we did not sample in the zone of predominantly fresh water. The data of Siegfried and Knight (1976, Table 8) from farther upstream, however, suggest such an increase of species, predominantly of small brackish and fresh-water crustaceans and oligochaetes.

SAN FRANCISCO BAY

Sediment Instability

Just as temporal salinity variations apparently play a major role in determining patterns in species distributions, geographic and temporal variations in sediment texture and, especially, sediment instability may represent the strongest of the abiotic influences on the general makeup (relative abundance among species) of the benthic community in many parts of the Bay. Packard (1918a), Filice (1958), Painter (1966), and Storrs et al. (1969), among others, have assumed that sediment texture (grain size) at any location is a static property of the bottom to which the organisms become accustomed. Recent studies are showing the dynamic nature of bottom sediments: seasonal variations in sediment transport into San Francisco Bay from rivers and streams (Krone 1966, 1979; Conomos and Peterson 1977) result in changing patterns of deposition, erosion, and sediment grain-size distribution. We have seen, in benthic samples collected in the northern reach of the estuary, alternating thin layers of sand and mud that probably reflect this seasonality. Similar observations were made by Gilbert (1917:93) on samples collected during the 1912-13 *Albatross* expedition.

Very recent studies using side-scan sonar techniques revealed that in the deeper regions Bay sediments are not homogeneous even in regions of uniform depth. In San Pablo Bay, for instance, large sand waves whose crests lie normal to the axis of the estuary are prominent in the channels, whereas alternating ridges and troughs parallel to the axis are found adjacent to the main channel (Fig. 6). The ridges are composed of finer-grained compacted muds and the troughs contain coarser-grained sediments in sand waves whose crests, like those in the main channel, lie normal to the axis of the troughs. In central San Francisco Bay, especially in the region between station 22 and the Golden Gate, sand waves with much greater amplitude are the major bottom feature. Moreover, these sand waves are highly dynamic: their direction has been observed to reverse within a tidal cycle (Rubin and McCulloch 1979). Such high-energy environments represent unique habitats suitable only to those species that, attracted to the salinity/grain size regime found there, can firmly attach themselves in deep subsurface sediments such as the sea pen *Stylatula elongata* or the jackknife clam *Solen* sp., or are highly mobile such as large glycerid polychaetes and some tellinid clams. These organisms feed primarily on the surface sediments or the suspended material carried along near the bottom by tidal currents.

Intermittent instability of surficial sediments on the broad intertidal and shallow subtidal mudflats that are prominent features of San Francisco Bay, particularly in South and San Pablo bays, may contribute significantly to a state of non-equilibrium in the benthic communities found there. Physical disturbances of mudflats resulting from the combined action of tides and intermittent storm-generated and seasonal wind waves are apparent from our continuing measurements of mudflat elevation near Palo Alto (Nichols 1977). As an extreme example, mudflat elevation was lowered by 9 cm during a one-month period in the summer of 1974, probably in response to the scouring action of wind waves. Baywide, sediment deposition and erosion are continuing features depending on geographic location and depth and relation to tidal and estuarine circulation patterns (Conomos and Peterson 1977).

Several workers (e.g. Grant 1965; Mills 1967) have alluded to the fact that benthic communities are subject to, and strongly influenced by, periodic natural disturbances, but the significance of such disturbances has only recently been shown experimentally. McCall (1977), in a large experimental field study, implanted plastic containers of defaunated sediment (natural sediment from which living animals had been eliminated) at two sites in Long Island Sound. He found that several species classified as opportunists (characterized by small size, rapid development, many reproductions per year, high recruitment, and high death rate) quickly inhabited the defaunated mud in great numbers. It took several months before other, larger "equilibrium" species began to

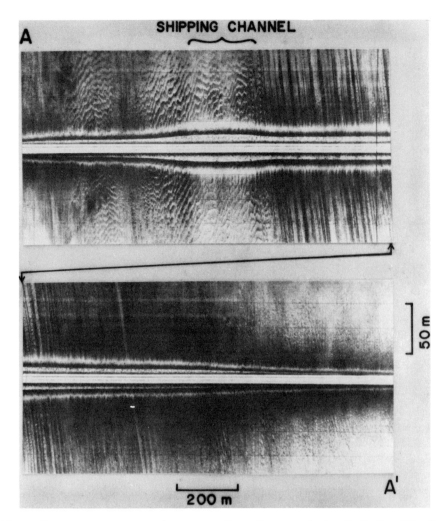

Fig. 6. Side-scan sonar record from cross-channel transect in San Pablo Bay (Fig. 1), from A (upper left) to A' (lower right). Major features are large sand waves in shipping channel, alternating ridges (compact mud?) and troughs (small sand waves) just south of the channel and relatively smooth bottom (mud) in shallower water south of channel at lower right.

establish themselves in the experimental containers, and nearly a year passed before the community in the containers was similar to that nearby. McCall's point is that a disturbance of the seafloor, such as turbulence caused by storms, strong tidal currents, and wind waves, which result in mortality of the resident populations, begins a marked faunal change through a well-defined succession of species.

It is not clear that the assemblage of organisms found in the high-energy environment of central San Francisco Bay represents an early succession stage in a disturbed habitat in the sense of McCall (1977) and Rhoads et al. (1978). The fauna have some attributes of all stages of succession (from colonizing to equilibrium), depending on the precise location within this area of sand waves (the colonizing polychaete *Owenia fusiformis* lives adjacent to the longer-lived, deposit-feeding *Pectinaria californiensis*). The fact that the disturbance is continuous here probably means that the community is in a quasi-equilibrium state much like that in the high-energy subtidal regions of outer coast sandy beaches.

The character of the benthic fauna of the shallower parts of San Francisco, on the other hand, suggests that a continuous state of opportunism exists, preventing the attainment of an equilibrium community. Two of the three opportunists in McCall's (1977) Long Island Sound study site are dominant on the mudflats of San Francisco Bay: *Streblospio benedicti* and *Ampelisca milleri* (perhaps the same as *A. abdita,* J. Carlton pers. comm.). That these species, together with the clam *Gemma gemma,* occur in widely fluctuating but usually large numbers throughout the year on South Bay mudflats (Nichols 1977) is strong evidence that the habitat is continually disturbed.

Biotic Disturbances

In addition to physical processes that result in instability of the sediment surface, intermittent biotic disturbances can contribute substantially to the unpredictability of the bottom. Foraging by bottom-feeding predators (among them, fish, sharks, rays), for example, can result in a disturbance of bottom communities that, though minor on an areal basis relative to the effects of current scouring, is important in regions of intense foraging. The extent to which fish disturb the bottom has not been quantitatively examined (see Stanley 1971), but excavation activities of large animals such as rays can leave a marked trace in sediments (Howard et al. 1977) and may destroy entire benthic communities (Fager 1964). The bat ray *Myliobatis californicus,* common in San Francisco Bay, especially in South Bay in summer (Aplin 1967), is probably responsible for the many depressions (clearly not the work of human clam diggers) seen in intertidal mudflats at low tide in the South Bay (similar to those shown in Howard et al. 1977, Fig. 2). Moreover, their feeding activities are sufficient to prevent the growth of mature (harvestable) populations of intertidal mollusks such as the soft-shelled clam *Mya arenaria* (Bonnot 1932). Nothing was known of the activities of rays below the intertidal zone before a side-scan sonar survey of central San Francisco Bay (D. Rubin pers. comm.), made in October (1976) at the end of the season of greatest bat-ray abundance (Aplin 1967), revealed many depressions (Fig. 7) in the bottom near Southampton Shoals adjacent to station 18 (Fig. 1). These depressions, about 1-5 m in diameter and about 0.5 m in depth, are similar in size to bat-ray excavations seen in Tomales Bay (S. Obrebski pers. comm.). The close spacing of the depressions (Fig. 7) gives an indication of the rather complete disturbance of the sediment to a depth of at least 0.5 m. A side-scan sonar survey of the same area in December of the same year showed no trace of the earlier depressions, thereby suggesting that the feeding depressions filled rapidly.

Foraging by the mud snail *Nassarius obsoletus* and shorebirds in the intertidal zone does not result in the obvious physical destruction of the sediment surface as in bat-ray excavation. The effect of snail-foraging activity can be, however, a diminishing of food resources potentially useful to other organisms (Mills 1967), and shorebird feeding can result in the removal of a substantial fraction of the organisms living in the sediment (Wolff 1976).

Intermittent blooms of benthic macroalgae can lead to the development of anaerobic conditions at the sediment surface and the subsequent death of the organisms in the sediment (Perkins and Abbott 1972). Such an occurrence was noted at our long-term study site on a South Bay intertidal mudflat (Nichols 1977) during August 1975: a dense mat of *Polysiphonia* sp. was transported to the shoreward end of our study transect and, in the process of decay, killed all of the macrofaunal organisms. Although the algal mat had disappeared by the time of our next sampling in early October, animals did not begin to reappear on the surface until November, and the highly anaerobic sludge left from the bloom remained in sub-surface sediments until well past November.

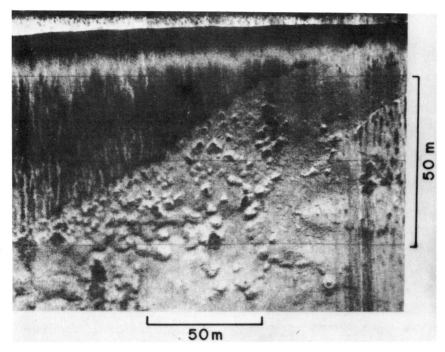

Fig. 7. Side-scan sonar record of bottom near station 18 (Fig. 1) showing depressions that are probably bat-ray feeding excavations. Photograph courtesy of D. M. Rubin.

The first species to appear were the amphipod *Grandidierella japonica* and the polychaete *Capitella capitata,* not normally the dominant species. The normally occurring species *Ampelisca milleri* and *Streblospio benedicti* began to reappear in more normal abundances in January, but it was not until the following spring that the normally dominant bivalve *Gemma gemma* returned. *Macoma balthica* had not yet returned to its earlier abundances more than two years later. Whether the drought conditions of 1975 to 1977, accentuating the effects of waste discharge into a hydrographically sluggish environment, contributed to these events cannot be determined. But the long-term effect of the single event of algal bloom and decay was marked.

Some benthic organisms, by actively or passively binding sediment particles into discrete mud tubes, change the character and stability and therein the suitability of the sediment surface for other organisms (see Woodin 1976). When tubes occur in dense beds, such as those of the deep dwelling polychaete *Asychis elongata* in the soft subtidal mud of the shoals off Berkeley near station 23 and throughout the region between the Oakland Bay and Dumbarton Bridges, the sediment is stabilized, thereby greatly reducing the effects of erosion by tidal currents and wave action. There is sedimentological evidence that *Asychis* beds are located in regions of net sediment deposition (D. Hammond pers. comm.).

The presence of surficial tubes can also lead to instability of the sediment surface. The formation of tube mats by settling specimens of the amphipod *Ampelisca* in intertidal and shallow subtidal mud begins a sequence of events whereby the texture of the sediment surface is altered (Mills 1967). Eventually, the base of the tube mat begins to disintegrate, making the sediment surface very susceptible to erosion. Any physical disturbance of the sediment can result in a washout of the tube mats and the associated sediments and organisms. The new sediment surface is then

SAN FRANCISCO BAY

exposed to recolonization.

ANOMALOUS PATTERNS IN SPECIES DISTRIBUTIONS

We might assume that the combined stresses of chemical and physical factors are applied somewhat evenly within regions of the Bay. Our sampling, however, has revealed considerable heterogeneity in benthic community makeup within small areas and great differences between adjacent areas. The juxtaposition of high and low values of biomass in South Bay, for example, is evidence of an inhomogeneous distribution of benthic animals. If benthic productivity in shallow water is directly coupled with the availability of food resources in the water column (Wolff 1977), then we might expect a large benthic community in regions where suspended particle concentrations are high. The major source of particulate material in San Francisco Bay is river inflow, primarily that of the Sacramento/San Joaquin River system (Conomos and Peterson 1977). The spatial distribution of suspended sediment (Conomos et al. 1979, Fig. 7A) reflects two regions of high concentration: the upper part of the estuary near Suisun Bay in the null zone, where particles accumulate in response to estuarine circulation (see Peterson et al. 1975a) and at the southern end of South Bay, the broad shallow region where tidal and wind-wave action cause nearly continuous resuspension of bottom sediments.

For some of the same reasons, the biomass of phytoplankton (estimated roughly by fluorescence measurements, Fig. 8) shows a similar distribution (see Cloern 1979). In general, the Suisun Bay maximum reflects the longer residence time in the null zone and where nutrients are plentiful, and the South Bay maximum reflects the sluggish hydrographic regime and high nutrient concentrations in a shallow area. One might conclude, then, that greatest benthic biomass should be found in the Suisun Bay region to the north and in the South Bay near Palo Alto (Fig. 8), where suspended sediment and associated bacteria, algae, and small invertebrates are concentrated. Instead, we find quite low values of biomass in the northern reach and high values interspersed with low values throughout Central and South bays (Fig. 3). The low values to the north can be explained by the marked seasonal variations in salinity and sediment stability. Adult populations of the larger equilibrium species are not permitted time to become established. The generally high

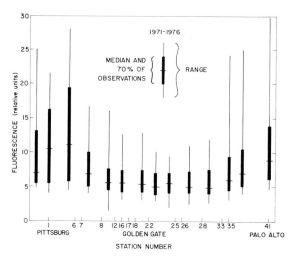

Fig. 8. Distribution of phytoplankton biomass, as fluorescence, at 1 to 2-m depth; data collected at near monthly intervals between 1971 and 1976.

biomass values, especially in the southernmost part of South Bay, may reflect moderate- to high-water-column productivity as inferred from the phytoplankton standing stock (Fig. 8) and high concentration of benthic diatoms on the sediment surface (Nichols 1977).

The occurrence of highest recorded biomass at station 20 (Fig. 3) in a region of low suspended-particle concentration (Fig. 8; Conomos et al. 1979, Fig. 7A) is a reflection of the numerical dominance of the sand dollar, *Dendraster excentricus*. In densely packed beds this animal can orient itself in a nearly vertical position in the sand and feed on particulate material carried by swift moving bottom currents that pass through the Golden Gate. Here the rate of supply of an appropriate food seems to be the critical factor.

That benthic macrofauna biomass is not consistently high throughout South Bay is evidence that water column and sediment-surface primary productivity are not solely responsible for secondary benthic productivity. Examination of the species found at those stations with highest biomass provides the most marked evidence of man's influence on the benthic community structure of San Francisco Bay. Throughout the Bay, with the exception of the central region, high values of biomass reflect the occurrence of one or two introduced mollusk species, especially the Japanese cockle *Tapes japonica* and the Japanese mussel *Musculus senhousia*. The other introduced mollusks, *Gemma gemma*, *Mya arenaria*, and *Nassarius obsoletus*, contribute substantially to the biomass at intertidal stations (Fig. 6; Nichols 1977). These species thrive in South Bay, in large patches at some places (see Appendix I, Jones and Stokes Assoc. Inc. 1977), apparently in response to local hydrodynamic and sedimentologic regimes. A major but as yet unresolved question is whether these introduced species exploited an open niche or displaced native counterparts. Research is now underway to resolve at least part of this question (Carlton 1979).

THE OBSCURE NATURE OF OTHER ANTHROPOGENIC INFLUENCES

As mentioned at the outset, concern about the effects of waste discharge was first raised at the end of the 19th century, primarily in response to declining fisheries. But it was not until the 1950's, when Filice (1959) conducted his study of the benthos of the San Pablo Bay area, that evidence of the detrimental effects of wastes on the biota was obtained (see the reviews of this and other more recent studies in Nichols 1973, Risebrough et al. 1978). Efforts to quantify such effects in more recent studies have been mostly unsuccessful. In a study of baywide species diversity (Storrs et al. 1969), lower than predicted values of diversity apparently were attributed to proximity to waste discharge. A followup study (Daniel and Chadwick 1971) produced similar findings. Both of these studies have come under question (Nichols 1973; Risebrough et al. 1978), however, because of methodological and analytical problems.

Although the discharge of wastes has undoubtedly altered the distribution and abundance of organisms in San Francisco Bay, we have no quantitative evidence of the extent of such alterations; and marked improvement in waste treatment practices since the original Filice study has, in effect, made the detection of such alterations difficult. Modern sewage treatment facilities with high rates of dilution, and future consolidation of these facilities, mean that discharged materials will be diffused throughout the entire system in relatively low concentration. The result is that any detrimental effects on the biota will be subtle and may be very difficult to separate from variations resulting from the many natural influences (seasonal and annual variations in river runoff, weather patterns, food supply, sediment erosion and deposition, and species interactions).

The deposition of vast amounts of sediment in the Bay as a result of the hydraulic mining for gold in the past century (Gilbert 1917) and the recent filling of the margins and the dredging of the channels have greatly influenced the nature of bottom sediments and bathymetry. These events also resulted in at least the short-term alteration or destruction of the benthos in the areas affected. The deposition of mining debris over large areas of the Bay, moreover, may have resulted in long-term or permanent changes as well. But there are no data with which to verify this supposition.

The most notable anthropogenic influence on the benthic environment is the establishment of species accidentally or intentionally introduced from other estuaries. These introductions are not necessarily detrimental to, and in fact may contribute to, a balanced ecosystem in San Francisco Bay, although it can no longer be considered a natural or native community. One of the most successful of the species introductions (besides the striped bass *Morone saxatilis;* see Stevens 1979) has been that of *Tapes japonica,* the object of an increasing sport fishery (Jones and Stokes Assoc. 1977). The introduction of exotic species continues (see Carlton 1979), and present community structure may change, therefore, as new introduced species become established and spread.

Another, more subtle, influence may be critical in the determination of benthic community structure: the biological response to an ever-diminishing supply of fresh water from the Sacramento/San Joaquin River system (see Gill et al. 1971). Fresh-water input not only maintains an intermediate salinity regime that is necessary for the successful growth and reproduction of commercial and noncommercial species, but also provides essential vitamins and plant nutrients (Copeland 1966). There is evidence that decreased river flow can greatly modify estuarine ecology by increasing parasitism of oysters, reducing shrimp and crab catches, increasing up-estuary fouling and destruction by boring organisms, and generally decreasing estuarine productivity (Copeland 1966). It is not known how the diversion of fresh water away from the San Francisco Bay estuarine system has already influenced the biology of the Bay (Goldman 1971). Declines in major commercial fisheries have been documented (e.g. Dungeness crab, salmon, and oysters, Skinner 1962; Smith and Kato 1979). But, whether these declines are related to reduced river inflow (with concomitant reduced rate of flushing, higher water residence time, increased loads of toxic substances and thus increased exposure to these substances) remains to be resolved.

FUTURE RESEARCH NEEDS

If anthropogenic influences on the benthos of San Francisco Bay are to be distinguished from natural influences and quantified, future research programs should focus on several distinct problems.

Natural Chemical, Geological, and Biological Processes

Permanent stations should be established in representative habitats around the Bay for long-term measurement of natural chemical, geological, and biological events and processes. Such long-term studies, recommended by a National Science Foundation panel (National Science Foundation, Committee on Long-Term Ecological Measurement 1977), provide needed information on the limits of natural variations in benthic community structure and help determine whether measured changes represent long-term cycles or unidirectional trends. Few benthic community data exist from San Francisco Bay with which we can detect either the nature or limits of change (Nichols 1973). Needed long-term measurements include, in addition to normal biological sampling and studies of life histories of key species, routine physical and chemical monitoring of both the water column and sediments at each location carried out with frequency and spatial coverage sufficient to permit distinction between short-term, localized effects and long-term patterns.

Introduced Species

The introduction and expansion of introduced species populations should be monitored and eventually controlled. Many of the introduced species have been remarkably successful in exploiting the Bay environment. We must be aware of, and control, the spread of potentially harmful species such as fish and shellfish parasites and predators and fouling and boring organisms. We know little of possible harmful effects of the massive introductions that have occurred.

Impact of Large, Modern Waste-Disposal Systems

Modern waste treatment has largely eliminated the occasion of overt kills of organisms and the bacterial tainting of economically important game species. Peterson (1979) has shown that plant nutrient components of domestic waste (nitrogen and phosphorous compounds) do not reach problem levels in San Francisco Bay. Presumably, the critical materials (other than, possibly, viruses) are the various heavy metals and synthetic organic compounds in industrial wastes and in urban runoff (Risebrough et al. 1978). Because of high dilution factors and the resulting subtlety of effects on natural communities, only careful field and laboratory research programs will allow us to see and quantify these effects. Normal field sampling programs in which species distributions and abundances are mapped in space and time are useful in characterizing the cumulative effect of natural and anthropogenic influences. These programs can be augmented by relating concentrations of toxicants in individual organisms to sedimentologic and hydrologic processes (e.g. Luoma and Cain 1979). Luoma's data suggest that sampling for toxicants be closely timed to major intermittent hydrologic events, especially that of winter runoff, when toxicants are washed from the drainage area into Bay waters. Such studies, however, provide little direct evidence of cause for biotic changes. The needed research/monitoring programs include, in addition to routine sampling of organisms and toxicants, utilization of experiments in which natural variability is controlled or eliminated. Such programs could include transplantation experiments designed to test the toxicity tolerance of animals found in potentially polluted areas (Luoma 1977). Additional studies could involve the establishment of experimental colonies of ecologically important Bay organisms within and outside the waste field of sewage outfalls (see for example Filice 1954a) for measuring the uptake of toxic substances by these organisms as well as the determination of the effects on biological processes (alteration of normal rates of reproduction, growth, and mortality). Using existing methodology initial experiments could be conducted with, for example, the mussel *Mytilus edulis* (e.g. Freeman 1974). Long range programs should include surveys of planktonic as well as benthic species, fish as well as invertebrates. Although these field experiments are logistically more difficult to carry out than typical laboratory bioassay programs, results are more appropriate to the waste-discharge management decision-making process.

Effect of Decreased Fresh Water Input

We must determine whether reduced river flow from the Sacramento/San Joaquin River system (increased salinity, decreased estuarine circulation and flushing, and increased exposure to toxic materials) will adversely affect species such as striped bass, Dungeness crab, clams and mussels, and the bay shrimp (see Goldman 1971). The influences on these species could be direct (effects of changing salinity or circulation patterns on these key species) or indirect (effects of changing salinity or circulation on the habitats, food or predators of these species). Such information is badly needed for long-term planning of the water resources of San Francisco Bay.

ACKNOWLEDGMENTS

Special thanks are due J. K. Thompson for her help in processing our many benthic samples and for more numerous valuable discussions, D. M. Rubin for introducing me to side-scan sonar operations, and B. F. Atwater, J. T. Carlton, J. E. Cloern, and S. N. Luoma for their thorough reviews of various drafts of this manuscript.

LITERATURE CITED

Alexander, W. B., B. A. Southgate, and R. Bassindale. 1935. Survey of the River Tees. II. The estuary, chemical and biological. Dep. Scientific Industrial Res., Water Pollution Res. Tech. Paper 5:1-171.

Aplin, J. A. 1967. Biological survey of San Francisco Bay, 1963-66. Calif. Dep. Fish Game, Marine Resources Operations Branch Reference 67(4). 131 pp.

Barrett, E. M. 1963. The California oyster industry. Calif. Dep. Fish Game Bull. 123:1-103.

Boesch, D. F. 1977. A new look at the zonation of benthos along the estuarine gradient. Pages 245-266 in B. C. Coull, ed. Ecology of Marine Benthos. Belle W. Baruch Library in Marine Science 6, University of South Carolina, Columbia, S.C.

Bonnot, P. 1932. Soft shell clam beds in the vicinity of San Francisco Bay. Calif. Fish Game 18:64-66.

Burke, M. V., and K. H. Mann. 1974. Productivity and production: biomass ratios of bivalve and gastropod populations in an eastern Canadian estuary. J. Fish. Res. Bd. Can. 31:167-177.

Carlton, J. T. 1979. Introduced invertebrates of San Francisco Bay. Pages 427-444 in T. J. Conomos, ed. San Francisco Bay: The Urbanized Estuary. Pacific Division, Amer. Assoc. Advance. Sci., San Francisco, Calif.

Cloern, J. E. 1979. Phytoplankton ecology of the San Francisco Bay system: The status of our current understanding. Pages 247-264 in T. J. Conomos, ed. San Francisco Bay: The Urbanized Estuary. Pacific Division, Amer. Assoc. Advance. Sci., San Francisco, Calif.

Conomos, T. J. 1979. Properties and circulation of San Francisco Bay waters. Pages 47-84 in T. J. Conomos, ed. San Francisco Bay: The Urbanized Estuary. Pacific Division, Amer. Assoc. Advance. Sci., San Francisco, Calif.

Conomos, T. J., and D. H. Peterson. 1977. Suspended-particle transport and circulation in San Francisco Bay: An overview. Pages 82-97 in M. Wiley, ed. Estuarine Processes. Vol. 2. Circulation, Sediments, and Transfer of Material in the Estuary. Academic Press, New York.

Conomos, T. J., R. E. Smith, D. H. Peterson, S. W. Hager, and L. E. Schemel. 1979. Processes affecting seasonal distributions of water properties in the San Francisco Bay estuarine system. Pages 115-142 in T. J. Conomos, ed. San Francisco Bay: The Urbanized Estuary. Pacific Division, Amer. Assoc. Advance. Sci., San Francisco, Calif.

Copeland, B. J. 1966. Effects of decreased river flow on estuarine ecology. J. Water Pollution Control Federation 38:1831-1839.

Daniel, D. A., and H. K. Chadwick. 1971. A study of toxicity and biostimulation in San Francisco Bay-Delta waters. VII. Effects of selected waste discharges on benthic invertebrate communities. Calif. Dep. Fish Game. 77 pp.

Fager, E. W. 1964. Marine sediments: effects of a tube-building polychaete. Science 143:356-359.

Filice, F. P. 1954a. An ecological study of the Castro Creek area in San Pablo Bay. Wasmann J. Biol. 12:1-24.

Filice, F. P. 1954b. A study of some factors affecting the bottom fauna of a portion of the San Francisco Bay estuary. Wasmann J. Biol. 12:257-292.

Filice, F. P. 1958. Invertebrates from the estuarine portion of San Francisco Bay and some factors influencing their distribution. Wasmann J. Biol. 16:159-211.

Filice, F. P. 1959. The effect of wastes on the distribution of bottom invertebrates in San Francisco Bay estuary. Wasmann J. Biol. 17:1-17.

Freeman, K. R. 1974. Growth, mortality, and seasonal cycle of *Mytilus edulis* in two Nova Scotian embayments. Fisheries and Marine Service Canada, Tech. Rep. 500. 111 pp.

Gilbert, G. K. 1917. Hydraulic-mining debris in the Sierra Nevada. U. S. Geol. Surv. Prof. Paper 105. 154 pp.

Gill, G. S., F. C. Gray, and D. Seckler. 1971. The California water plan and its critics: A brief review. Pages 3-27 *in* D. Seckler, ed. California Water, a Study in Resource Management. University of California Press, Berkeley, Calif.

Goldman, C. R. 1971. Biological implications of reduced freshwater flows on the San Francisco Bay-Delta system. Pages 109-124 *in* D. Seckler, ed. California Water, a Study in Resource Management. University of California Press, Berkeley, Calif.

Grant, D. C. 1965. Specific diversity in the infauna of an intertidal sand community. Ph.D. Thesis. Yale University, New Haven, Conn. 53 pp.

Howard, J. D., T. V. Mayou, and R. W. Heard. 1977. Biogenic sedimentary structures formed by rays. J. Sedimentary Petrol. 47:339-346.

Jones and Stokes Associates, Inc. 1977. San Francisco Bay shellfish: an assessment of the potential for commercial and recreational harvesting. Report prepared for Assoc. Bay Area Governments (ABAG). 171 pp.

Jones, M. L. 1961. A quantitative evaluation of the benthic fauna of Point Richmond, California. Univ. Calif. Pub. Zool. 67:219-320.

Krone, R. B. 1966. Predicted suspended sediment inflows to the San Francisco Bay system. Report prepared for Central Pacific River Basins Project, Federal Water Pollution Control Admin., Southwest Region. 33 pp.

Krone, R. B. 1979. Sedimentation in the San Francisco Bay system. Pages 85-96 *in* T. J. Conomos, ed. San Francisco Bay: The Urbanized Estuary. Pacific Division, Amer. Assoc. Advance. Sci., San Francisco, Calif.

Luoma, S. N. 1977. Detection of trace contaminant effects in aquatic ecosystems. J. Fish. Res. Bd. Can. 34:436-439.

Luoma, S. N., and D. J. Cain. 1979. Fluctuations of copper, zinc, and silver in tellenid clams as related to freshwater discharge—San Francisco Bay. Pages 231-246 *in* T. J. Conomos, ed. San Francisco Bay: The Urbanized Estuary. Pacific Division, Amer. Assoc. Advance. Sci., San Francisco, Calif.

McCall, P. L. 1977. Community patterns and adaptive strategies of the infaunal benthos of Long Island Sound. J. Mar. Res. 35:221-266.

Mills, E. L. 1967. The biology of an ampeliscid amphipod crustacean sibling species pair. J. Fish. Res. Bd. Can. 24:305-355.

National Science Foundation, Committee on Long-Term Ecological Measurements. 1977. Long-term ecological measurements. Rep. of Conf., 16-18 March 1977, Woods Hole, Mass. 26 pp.

Nelson, N. C. 1909. Shellmounds of the San Francisco Bay region. Univ. Calif. Publ. Amer. Archaeol. Ethnol. 7:309-356.

Nichols, F. H. 1973. A review of benthic faunal surveys in San Francisco Bay. U. S. Geol. Surv. Cir. 677. 20 pp.

Nichols, F. H. 1977. Infaunal biomass and production on a mudflat, San Francisco Bay, California. Pages 339-357 *in* B. C. Coull, ed. Ecology of Marine Benthos. Belle W. Baruch Library in Marine Science 6. University of South Carolina, Columbia, S.C.

Packard, E. L. 1918a. Molluscan fauna from San Francisco Bay. Univ. Calif. Pub. Zool. 14:199-452.

Packard, E. L. 1918b. A quantitative analysis of the molluscan fauna of San Francisco Bay. Univ. Calif. Pub. Zool. 18:299-336.

Painter, R. E. 1966. Zoobenthos of San Pablo and Suisun bays. Pages 40-56 *in* D. W. Kelley, ed. Ecological Studies of the Sacramento-San Joaquin Estuary. Calif. Fish Game, Fish Bull. 133.

Perkins, E. J., and O. J. Abbott. 1972. Nutrient enrichment and sand salt fauna. Mar. Pollution Bull. 3:70-72.

Peterson, D. H. 1979. Sources and sinks of biologically reactive oxygen, carbon, nitrogen, and silica in northern San Francisco Bay. Pages 175-193 *in* T. J. Conomos, ed. San Francisco Bay:

The Urbanized Estuary. Pacific Division, Ameri. Assoc. Advance. Sci., San Francisco, Calif.

Peterson, D. H., T. J. Conomos, W. W. Broenkow, and P. C. Doherty. 1975a. Location of the nontidal current null zone in northern San Francisco Bay. Estuarine Coastal Mar. Sci. 3:1-11.

Peterson, D. H., T. J. Conomos, W. W. Broenkow, and E. P. Scrivani. 1975b. Processes controlling the dissolved silica distribution in San Francisco Bay. Pages 153-187 *in* L. E. Cronin, ed. Estuarine Research. Vol. 1. Chemistry and Biology. Academic Press, New York.

Rhoads, D. C., P. L. McCall, and J. Y. Yingst. 1978. Disturbance and production on the estuarine seafloor. Amer. Sci. 66:577-586.

Risebrough, R. W., J. W. Chapman, R. K. Okazaki, and T. T. Schmidt. 1978. Toxicants in San Francisco Bay and estuary. Report prepared for Assoc. Bay Area Governments (ABAG). 113 pp.

Rubin, D. M., and D. S. McCulloch. 1979. The movement and equilibrium of bedforms in central San Francisco Bay. Pages 97-113 *in* T. J. Conomos, ed. San Francisco Bay: The Urbanized Estuary. Pacific Division, Amer. Assoc. Advance. Sci., San Francisco, Calif.

Siegfried, C. A., and A. W. Knight. 1976. A baseline ecological evaluation of the western Sacramento San-Joaquin Delta. Univ. Calif. Davis Dep. Water Sci. Eng. Paper 4504:1-84.

Skinner, J. E. 1962. An historical review of the fish and wildlife resources of the San Francisco Bay area. Calif. Dep. Fish Game, Water Projects Branch Rep. 1:1-226.

Smith, S. E., and S. Kato. 1979. The fisheries of San Francisco Bay: Past, present, and future. Pages 445-468 *in* T. J. Conomos, ed. San Francisco Bay: The Urbanized Estuary. Pacific Division, Amer. Assoc. Advance. Sci., San Francisco, Calif.

Stanley, D. J. 1971. Fish-produced markings on the outer continental margin east of the middle Atlantic states. J. Sedimentary Petrol. 41:159-170.

Stevens, D. E. 1979. Environmental factors affecting striped bass (*Morone saxatilis*) in the Sacramento-San Joaquin estuary. Pages 469-478 *in* T. J. Conomos, ed. San Francisco Bay: The Urbanized Estuary. Pacific Division, Amer. Assoc. Advance. Sci., San Francisco, Calif.

Storrs, P. N., E. A. Pearson, H. F. Ludwig, R. Walsh, and E. J. Stann. 1969. Estuarine water quality and biologic population indices. Adv. Water Pollution Res. (Proc. 4th Int. Conf. Prague) 1969:901-910.

Sumner, F. B., G. D. Louderback, W. L. Schmitt, and G. E. Johnson. 1914. A report upon the physical conditions in San Francisco Bay, based upon the operations of the United States Fisheries Steamer "Albatross", 1912-1913. Univ. Calif. Pub. Zool. 14:1-198.

Wolff, W. J. 1976. The trophic role of birds in the Grevelingen Estuary, The Netherlands, compared to their role in the saline Lake Grevelingen. Proc. 10th European Symp. Mar. Biol., Ostend, Belgium 2:673-689.

Wolff, W. J. 1977. A benthic food budget for the Grevelingen Estuary, The Netherlands, and a consideration of the mechanisms causing high benthic secondary production in estuaries. Pages 267-280 *in* B. C. Coull, ed. Ecology of Marine Benthos. Belle W. Baruch Library in Marine Science 6. University of South Carolina, Columbia, S.C.

Woodin, S. A. 1976. Adult-larval interactions in dense infaunal assemblages: Patterns of abundance. J. Mar. Res. 34:25-41.

INTRODUCED INVERTEBRATES OF SAN FRANCISCO BAY

JAMES T. CARLTON
Department of Biology, Woods Hole Oceanographic Institution, Woods Hole, MA 02543

Almost 100 species of exotic marine invertebrates have been introduced into San Francisco Bay by man in the past 130 or more years. Primary mechanisms of introduction include transport of fouling, boring, and ballast-dwelling organisms by ships and epizoic and nestling invertebrates by commercial oysters. With the resolution of taxonomic problems and adequate exploration, many more introduced species may eventually be recognized from the Bay.

The impact of this exotic fauna can be assessed in economic terms (pestiferous species, including shipworms and other borers) and in geologic terms (an introduced boring isopod has modified extensive portions of the bay shoreline by weakening clay and mud banks). The greatest effect, however, may be biological and ecological: the establishment of an introduced fauna as numerical and biomass dominants in many regions of the Bay, as revealed in both short- and long-term quantitative and qualitative studies. The modern-day significance of introduced species in fouling, benthic, and mudflat ecosystems in portions of San Francisco Bay raises questions as to the role of invertebrates prior to the mid-19th century both in the organic matter budget of the Bay-Estuary system and in the support of large native shorebird populations.

Man's extensive modifications of the Bay and concomitant creation of novel environmental conditions, the absence of a diverse native estuarine fauna, and competitive displacement have all played roles in the successful establishment of this impressively large and diverse introduced fauna.

In 1857, William Stimpson wrote that, "The Bay of San Francisco... is nearly barren of animal life except at its entrance." Such is no longer the case. Of all the faunal modifications that have occurred in San Francisco Bay in the past 130 years, perhaps none has been more profound than the successful establishment of a large number of introduced invertebrates, transported accidentally by man to the Bay since the mid-19th century, from the western and southwestern Pacific and the Atlantic oceans. Unfortunately, the historical record of these introductions is largely an anecdotal one, attention having been focused upon a few of the larger, more conspicuous, or troublesome exotic species. Perhaps two-thirds of all known species of introduced invertebrates on the Pacific coast also occur in San Francisco Bay. The Bay may thus provide a model for the study of the biology of introduced faunas on the Pacific coast. Criteria for the recognition of introduced species are considered in detail by Carlton (1978). These criteria include evidence based upon paleontology, archeology, recent history, systematics, distributional ecology, transport mechanisms, and biogeography.

MECHANISMS OF INTRODUCTION

Two major mechanisms, and several of lesser importance, are responsible for the transport of exotic species into San Francisco Bay. These are (1) the introduction by ships of fouling, boring, and ballast-dwelling organisms, and (2) the introduction of epizoic and nestling invertebrates on and among commercial oysters imported from the western Atlantic and the western Pacific oceans.

SAN FRANCISCO BAY

SHIPPING

There may have been some incidental ship introductions prior to the mid-19th century, but hard data are scanty. We know, for example, that in the fall of 1595, the ship *San Agustín,* four months out of Manila in the Philippines, and doubtless with a fouling and/or boring fauna on her wooden hull, sank in Drake's Bay (Fig. 1) near the entrance to San Francisco Bay (Wagner 1924; Heizer 1941). From the late 18th century until the mid-19th century, a few fur and hide ships, traveling between the Atlantic, western South America and the western and southwestern Pacific, operated along the California (and northwest American) coast, occasionally visiting San Francisco Bay. As Ewan (1955) has noted, "These vessels were the source of introduction of many organisms, some injurious: insects, weeds, and rodents," and there is no reason to suspect that these ships did not also carry fouling and boring faunas as well.

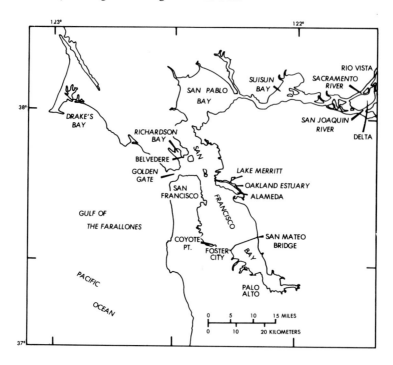

Fig. 1. The San Francisco Bay.

It was not, however, until the California Gold Rush of 1849 and subsequent years that shipping apparently became an important mechanism for the introduction of exotic invertebrates into the Bay. Indeed, there are no records of introduced species in San Francisco Bay prior to the Gold Rush years, but perhaps only because there is no biological record of the fauna of the Bay prior to the mid-19th century. From 1849 on, however, ships from many ports around the world, not only from both sides of the Atlantic, but also from China (Williams 1930), Australia and New Zealand (Monaghan 1966) and Chile and Peru (Monaghan 1973), arrived in increasing numbers. In the years 1849 and 1850, almost 48,000 persons, mostly gold seekers, arrived by ship at the port of San Francisco (Farwell 1891). By the early 1850's, passage from New York or Boston around Cape Horn to San Francisco Bay could be accomplished in 3 to 4 months (Rydell 1952). The ability of boring and fouling species to survive this journey from New England to California, twice through tropical waters, appears certain: the transport of living ship-bottom biotic assemblages across comparable distances and temperature changes has been well documented (Pilsbry 1896; Chilton 1911; Hentschel 1923; Orton 1930; Bertelsen and Ussing 1936; Bishop 1951; Allen 1953; Skerman 1960). Indeed, many of the species concerned are overwhelmingly tolerant to euryhaline and eurythermal conditions. According to Rydell, more clipper ships passed through the Golden Gate in 1853 than in any other year of the Gold Rush; "on several occasions, three clippers stood in through the Golden Gate within twenty-four hours, and on two others, five arrived within forty-eight hours" (Rydell 1952:138). Coincidentally, the first record of an introduced species in San Francisco Bay is also 1853: the Atlantic barnacle, *Balanus improvisus* (Carlton and Zullo 1969).

An opportunity for both multiple and massive inoculations of exotic species occurred almost simultaneously: not only were hundreds of ships arriving within a period of a few years, compared to the few arrivals prior to the Gold Rush, but equal numbers were being abandoned in the Bay as ship's crews left for the gold fields. In 1851, for example, more than 800 ships were at anchor or abandoned in Yerba Buena Cove in San Francisco (see Kemble 1957, for a series of contemporary photographs). Early introductions, besides *Balanus improvisus,* were likely hydroids (such as the Atlantic *Tubularia crocea,* present in the Bay by at least the late 1850's), polychaetes (especially tubiculous species), cheilostome and ctenostome bryozoans, gammarid amphipods of the genera *Jassa, Podocerus, Stenothoe,* and *Corophium*, tunicates, and wood-boring organisms such as the gribble (isopod) *Limnoria* spp.. The anemone *Metridium senile* and the mussel *Mytilus edulis* were doubtless common components of this fouling fauna; while the former may have been and the latter was aboriginally present in the San Francisco Bay area, Atlantic stocks of both species were introduced and mixed with native populations. Indeed, it may be that morphometric analyses of fossil and archeological *Mytilus edulis* from the Bay region would reveal differences from present day local populations.

With the completion of the Transcontinental Railroad in 1869, shipping from the Atlantic declined, but increased again with the opening of the Panama Canal in 1914, although the freshwater barrier of the canal now likely changed the composition of ship's fouling faunas which successfully entered the Pacific Ocean. The last quarter of the 19th century saw considerable maritime activity in San Francisco Bay, the grain-trade figuring particularly prominently. In the peak year of the trade, 1880-1881, grain exports nearly equalled the mean annual exports for all dry cargo from San Francisco during the period 1925 to 1940 (Kemble 1957). The early decline of wooden sailing ships, the greatly increased efficacy of antifouling paints, and the increased speeds of ocean-going vessels may have led to a decrease in the transport of boring and fouling organisms. However, ship-borne introductions appear to continue, and a few species of western or southwestern Pacific crustaceans recently noted in the Bay may be examples of present-day introductions; e.g., the isopods *Dynoides dentisinus* and *Ianiropsis serricaudis* and the amphipod *Corophium* sp. (E. Iverson and J. Chapman, pers. comm.).

Infaunal, semi-terrestrial, or relatively errant marine invertebrates may also be transported by ships, through water ballast (taking up and later discharging larvae or adults of benthic species), shingle ballast (taking up and discharging invertebrates among stones, rocks, and dried algal masses collected from beaches and used as ballast), and fouled sea-water systems.

The discharge of ship ballast in San Francisco Bay appears to have led to the introduction of at least two exotic maritime (semi-terrestrial) invertebrates. The Chilean or New Zealand shore-hopper (talitrid amphipod), *Orchestia chiliensis,* redescribed as *O. enigmatica* from San Francisco Bay (see Bousfield and Carlton 1967) may have been transported to San Francisco Bay in either shingle ballast or round beach gravel. Beach gravel and debris are known to have been taken up for ballast, for example, at Chilean ports and discharged in such areas as the Oakland Estuary in San Francisco Bay, adjacent to Lake Merritt, in Oakland, where the only known California population of this beachhopper exists. It is also probable that ballast has played a role in the introduction of the maritime earwig, *Anisolabis maritima* (Dermaptera), to the shores of San Francisco Bay, where the only established populations of this species in California, and perhaps on the Pacific coast, are known (see records in Langston 1974; Langston and Powell 1975; Langston and Miller 1977).

Newman (1963) has suggested that fouled sea-water systems were responsible for the accidental introduction of the western Pacific shrimp *Palaemon macrodactylus* into San Francisco Bay. This species was first found in the Bay in 1957, and Newman has compiled evidence that the introduction itself may have been about 1954. The synchrony of this introduction with the greatly increased military and cargo traffic between California and Japan and Korea immediately prior to 1954, associated with the Korean War, is striking. Despite the long history of ship traffic between both sides of the north Pacific Ocean, this appears to be a further example that intensified periods of shipping (also associated with the Gold Rush activity) are often concomitant with or immediately precede introductions. Similarly, the Australasian serpulid polychaete, *Mercierella enigmatica*, first appeared in San Francisco Bay about 1920, following increased world-wide shipping associated with the second decade of the 20th century.[1]

Commercial Oyster Industries

Of equal or greater importance in the introduction of exotic invertebrates have been the commercial oyster industries. A single oyster shell may have upon it representatives of 10 or more invertebrate phyla, comprising dozens of species, and these numbers can be greatly increased when oysters are packed together for shipment with associated clumps of mud and algae, and with water pockets in empty valves used for cultch. Elton (1958) remarked that "...the greatest agency of all that spreads marine animals to new quarters of the world must be the business of oyster culture...". San Francisco Bay provides a thorough documentation of this. Beginning with the completion of the Transcontinental Railroad in 1869, hundreds of carloads of the Eastern (or Virginia) oyster, *Crassostrea virginica*, were transported to San Francisco Bay (and elsewhere on the Pacific Coast) from the Atlantic coast of the United States (Barrett 1963). Early shipments were heavily fouled, if one may judge from the introductions resulting from them, and little attempt was made to prevent the transport of the associated oyster bed fauna, although the relatively late appearance of the oyster drill, *Urosalpinx cinerea*, about 1890, suggests that some attention may have been given to deterring the introduction of potential pests. Throughout the latter half of the 19th century oysters were continually shipped from the east coast and planted throughout the Bay, providing extensive opportunity for the inoculation into the Bay of many common northwestern Atlantic bay and estuarine invertebrates. More than 100 years would pass before some of these introductions were to be recognized (such as the maldanid polychaete, *Asychis elongata*; see Light 1974).

Shortly after the turn of the century, the Atlantic oyster industry as a large scale endeavor in San Francisco Bay faded away for a variety of reasons, pollution appearing to be one of them. But the oysters had left a permanent legacy: many of the associated oyster epizoics had taken hold in vast numbers. Portions of the shallow bay waters are today carpeted almost solely with species introduced with Atlantic oysters. Among the mollusks, for example, are the ribbed mussel, *Ischadium demissum* (= *Modiolus demissus*), the soft-shell clam, *Mya arenaria*, the gem clam, *Gemma gemma*, and the mudsnail, *Ilyanassa obsoleta* (= *Nassarius obsoletus*).

In the early 20th century, oyster farmers on this coast turned their attention to the Japanese (or Pacific) oyster, *Crassostrea gigas*, an endeavor which largely involves bays and estuaries to the

---

[1] Miller (1968), along similar lines, noted the apparent introduction by naval ships of the isopod *Paracerceis sculpta* from San Diego Bay to Hilo Harbor and Pearl Harbor, Hawaii, during World War II.

north or south of San Francisco Bay (Barrett 1963). Throughout the 1930's plantings of Japanese oysters were made in San Francisco Bay, and these may have led to the introduction of a number of Japanese estuarine invertebrates. Two Japanese mollusks now abundant in the Bay, the rock-cockle, *Tapes japonica,* and the small mussel *Musculus senhousia,* were first collected in 1946, the date being perhaps more of an artifact associated with lack of intensive collecting in the Bay during the Second World War, rather than a close approximation of the date of introduction.

Other Mechanisms

Additional means of transport of foreign marine invertebrates into San Francisco Bay are known, but few established introductions have been traced to these.

A comparatively new mechanism appears to be algae (such as *Ascophyllum nodosum* and *Fucus* spp. received in shipments of Atlantic lobsters from New England by local restauranteurs and discarded into the Bay. Although the extent of this activity is probably limited, discarded algae have apparently served as the vehicle for the introduction of the Atlantic periwinkle *Littorina littorea* into the Bay since the late 1960's (Carlton 1969; Miller 1969). Adult *Littorina littorea,* now found nestled in rock crevices in several locations along the Bay's eastern shoreline, may provide a unique opportunity to observe the establishment, population increase, and dispersal of an introduced species in San Francisco Bay shortly after its introduction. Miller (1969) has listed additional species which are transported with algae shipped with lobsters.

Many exotic estuarine and fresh-water fishes have been introduced into the San Francisco Bay—Delta region (Moyle 1976a, b) by means of tank cars, and water associated with these introductions (Throckmorton 1874) may well have carried a wide variety of small planktonic or even benthic species. Little is known, however, of the establishment of any exotic species by this means, probably due to the limited knowledge, taxonomically and biogeographically, of candidate species which may have been transported by this means (such as protozoans, copepods, and rotifers). Hazel (1966), however, has suggested that the presence of the eastern American freshwater polychaete, *Manayunkia speciosa,* in the Sacramento-San Joaquin River Delta and in Oregon may be due to the transport of water associated with catfish introductions from the northeastern United States, including catfish and water directly from the type locality of *M. speciosa* in Pennsylvania.

THE INTRODUCED FAUNA

The result of 130 or more years of introductions has been the establishment in San Francisco Bay of about 96 species of exotic marine invertebrates (Table 1)[2]. The introduced fauna is largely restricted to the shallow estuarine rim of the Bay, and proportionately fewer introduced species are to be found as one approaches the Golden Gate. Somewhat more than half of the introduced fauna is composed of mollusks and crustaceans (21 and 32%, respectively). Polychaetes comprise about 15% of the fauna, and coelenterates about 14%. The introduced fauna is largely Atlantic in origin (66%), four times the number originating from the western Pacific (17%).[3] Five percent of the Bay's exotic fauna derives from Australasia and Southeast Asia; 14% is of multiple or uncertain origin, and one species is from Chile.

The exact number of species introduced with each mechanism is difficult to determine because of extensive overlap in habitats of many species: that is, a number of species may have been

[2] In discussions and tabulations, taxa indicated by "spp." in Table 1 are counted as two. The number of introduced species is taken as 96, excluding *Littorina littorea* and *Sabellaria spinulosa,* which are questionably established.

[3] Includes species coded in Table 1 as A or A?; O or O?.

SAN FRANCISCO BAY

TABLE 1. INTRODUCED INVERTEBRATES OF SAN FRANCISCO BAY:
A PRELIMINARY LIST (DATA FROM CARLTON, 1978 AND FROM
NUMBERED REFERENCES, AS INDICATED.)

Key: Status: e?, establishment (presence of reproducing populations) uncertain
t?, specific taxonomy uncertain

Origin: O, Orient (Western Pacific: Japan and Korea)
C, Chile
AA, Australasia (Australia and New Zealand)
SE, Southeast Asia
A, Atlantic Ocean
E, Europe (Northeastern Atlantic Ocean)

Mechanism: a, discarded algae
o, commercial oysters
s, shipping

PORIFERA (1)
  Haliclona sp. (A, o, t?)
  Microciona prolifera (A, o)
  Halichondria bowerbanki (A, o)
  Prosuberites sp. (A, o, t?)
  Tetilla sp. (A, o, t?)

COELENTERATA
  Hydrozoa
    Garveia franciscana (A/AA, o/s) (2)
    Clava leptostyla (A,o/s)
    Cordylophora lacustris (A, o/s) (3)
    Turritopsis nutricula (A, o/s)
    Syncoryne mirabilis (A, o/s)
    Corymorpha sp. (A?, o?/s?, t?)
    Tubularia crocea (A, o/s)
    Obelia spp. (A/AA/O, o/s,t?)
  Anthozoa
    Diadumene franciscana (O?/AA?, s?) (4)
    Diadumene leucolena (A, o/s) (4)
    Diadumene sp. (E?, s?, t?)
    Haliplanella luciae (A,O, o/s) (4)

PLATYHELMINTHES
  Turbellaria
    Childia groenlandica (A, o) (5)
  Trematoda
    Austrobilharzia variglandis (A, o,
      with *Ilyanassa*) (6)
    Parvatrema borealis (A, o, with
      Gemma and *Neanthes*) (7)

ANNELIDA: POLYCHAETA
  Neanthes succinea (A, o)

ANNELIDA (continued)
  Marphysa sanguinea (A?, o)
  Boccardia ligerica (E?, s?) (8)
  Polydora ligni (A, o) (8)
  Polydora spp. (A, o/s, t?)
  Pseudopolydora kempi (O, o) (8)
  Pseudopolydora paucibranchiata (O,o) (8)
  Streblospio benedicti (A, o) (8)
  Capitella capitata (A?, o/s, t?)
  Heteromastus filiformis (A,o)
  Asychis elongata (A, o) (9)
  Sabellaria spinulosa (A?, s?, e?)
  Mercierella enigmatica (AA, s)

MOLLUSCA (10)
  Gastropoda
    Littorina littorea (A, a, e?)
    Crepidula convexa (A, o)
    Crepidula plana (A, o)
    Urosalpinx cinerea (A, o)
    Busycotypus canaliculatus (A, o?)
    Ilyanassa obsoleta (A, o)
    Ovatella myosotis (A, s?/o)
    Tenellia pallida (E?, s?) (11)
    Eubranchus misakiensis (O, s?) (12)
    Okenia plana (o, s?) (11)
    Trinchesia sp. (O, s?) (13)
    Odostomia bisuturalis (A, o)
  Bivalvia
    Musculus senhousia (O, o)
    Ischadium demissum (A, o)
    Gemma gemma (A, o)
    Tapes japonica (O, o)
    Petricola pholadiformis (A, o?)

TABLE 1. (CONTINUED)

MOLLUSCA (continued)
  Mya arenaria (A, o)
  Teredo navalis (A?, s)
  Lyrodus pedicellatus (A?, s)

ARTHROPODA: CRUSTACEA
  Ostracoda
    Sarsiella zostericola (A, o) (14)
  Copepoda
    Mytilicola orientalis (O, o) (15)
  Cirripedia (16)
    Balanus improvisus (A, o/s) (17)
    Balanus anphitrite amphitrite (AA, s)
  Amphipoda
    Ampithoe valida (A, o)
    Ampelisca abdita (A, o)
    Chelura terebrans (A, s) (18)
    Corophium acherusicum (A, o/s) (19)
    Corophium insidiosum (A, o/s) (19)
    Corophium uenoi (O, o/s)
    Corophium sp. (SE, s) (20)
    Grandidierella japonica (O, o) (21)
    Melita nitida (A, o) (20)
    Jassa falcata (A, o/s, t?)
    Podocerus brasiliensis (A, o/s, t?)
    Parapleustes sp. (O?/AA?, o/s, t?) (20)
    Stenothoe valida (A, o/s)
    Orchestia chiliensis (AA/C, s) (22)
    Caprella acanthogaster (O, o/s)
    Caprella spp. (A?, o/s, t?)
  Isopoda
    Synidotea laticauda (O?/AA?, s?)

ARTHROPODA (continued)
    Limnoria quadripunctata (A?, s) (23)
    Limnoria tripunctata (A?, s) (23)
    Dynoides dentisinus (O, o/s)
    Sphaeroma quoyanum (AA, s) (24, 25)
    Iais californica (AA, s, with
      Sphaeroma quoyanum) (25)
    Ianiropsis serricaudis (O, o/s)
  Chelifera
    Tanais sp. (O?, o/s, t?)
  Decapoda
    Palaemon macrodactylus (O, s) (26)
    Rhithropanopeus harrisii (A, o?) (27)

ARTHROPODA: INSECTA
  Dermaptera
    Anisolabis maritima (A?, s) (28)

ENTROPROCTA
  Barentsia benedeni (E, o/s) (29)

ECTOPROCTA
  Alcyonidium sp. (A, o/s, t?)
  Victorella pavida (A/O, o/s)
  Bugula spp. (A/AA/O, o/s, t?)
  Conopeum spp. (A, o/s, t?)
  Schizoporella unicornis (A/O, o/s)

CHORDATA: TUNICATA
  Ciona intestinalis (A, o/s)
  Molgula manhattensis (A, o/s)
  Styela clava (O, s) (30)

SPECIES REFERENCES

(1) Hartman, 1975; (2) Vervoort, 1963; (3) Hand and Gwilliam, 1951; (4) Hand, 1956; (5) Hyman, 1959; (6) Grodhaus and Keh, 1958; (7) Oglesby, 1965; (8) Light, 1977; (9) Light, 1974; (10) Packard, 1918; Hill and Kofoid, 1927; Hanna, 1966; Carlton, 1969; (11) Steinberg, 1963; (12) Behrens, 1971; (13) Behrens and Tuel, 1977; (14) Kornicker, 1975; (15) Bradley and Siebert, 1978; (16) Newman, 1967; (17) Carlton and Zullo, 1969; (18) Barnard, 1950; (19) Shoemaker, 1949; (20) J. Chapman, 1979, [ms. in prep.]: (21) Chapman and Dorman, 1975; (22) Bousfield and Carlton, 1967; (23) Hill and Kofoid, 1927; Menzies, 1958; (24) Barrows, 1919; Hill and Kofoid, 1927; Higgins, 1956; (25) Rotramel, 1972; (26) Newman, 1963; (27) Jones, 1940; (28) Langston, 1974; Langston and Miller, 1977; Langston and Powell, 1975; (29) Mariscal, 1975; (30) Abbott and Johnson, 1972.

transported with either ships, or oysters, or both, and for many first recorded after 1869 we may never know exactly which of these three possible cases acted as transport mechanisms. Thirty-seven percent of the introduced fauna appears to have been brought in with oysters, 19% with ships,[4] and 46% may have come with either or both. The relatively close temporal proximity of the Gold Rush (1849 and subsequent years) with the beginning of the oyster industry (1869 and the following years), coupled with the scanty knowledge of the early Bay fauna compounds these difficulties.

It is further difficult to estimate the number of introduced species that have been overlooked. Evidence from various sources indicates that the actual number of introduced species of acoelomate and pseudocoelomate protostomes (acoel turbellarians, nemerteans, rotifers, nematodes, etc.), as well as protozoans, sponges, hydroids, smaller crustaceans (including pinnotherid crabs, amphipods, and copepods), pycnogonids, oligochaetes, polychaetes, bryozoans, and tunicates, may be quite large. With adequate exploration and taxonomy many more introduced species than those now known from San Francisco Bay may be recognized. Many taxa are now only marginally known and to discern native from exotic, or provincial from "cosmopolitan" is difficult. Confounding this situation is the fact that "cosmopolitan" taxa often reveal low degrees of taxonomic resolution, suggesting that broad distributions may, in many species, be a taxonomic artifact. The skeleton-shrimp (amphipod), *Caprella equilibra,* may be an excellent example: it is known from San Francisco Bay, and it has also been recorded from virtually every coast line and many islands of the world, from the surface to (questionably) 3000 m (McCain and Steinberg 1970). Concerning polychaetes, Day (1964) has remarked that, ". . .in the Polychaeta at any rate more recent work suggests that many records of cosmopolitan species are due to misidentifications." Hartman (1955:40) stated, in regard to the polychaete fauna of ocean basins off southern California, "some species of cosmopolitan character. . .differ from typical representatives in more distant parts of the world, in morphological characters which may have more than varietal or trivial importance." Fauchald (1977) also remarked "how exceedingly poorly known the group is and how few generalizations can be made on the ecology and evolution of the polychaetes," to which, I believe, we may append on the biogeography of polychaetes as well. Bousfield (1973:38) has made similar comments regarding the "sophistication" of amphipod taxonomy. I have remarked elsewhere (Carlton 1975) on problems associated with treating species as "cosmopolitan."

A related taxonomic consideration further obscures the number of introduced species. A strong historical systematic tendency, which remains today, has been either (1) to attempt to identify newly discovered estuarine taxa (which are in reality introduced) with species already described from the Pacific coast, or (2) to describe them as new species. The small gammarid amphipod, *Ampelisca milleri,* abundant in parts of San Francisco Bay, provides an example of the first kind: this is almost certainly the eastern American *A. abdita.* The true *Ampelisca milleri* is a different, native eastern Pacific species. Table 2 lists several examples of the second kind. It may be noted that more than 65 years passed before the synonymy of the isopod *Sphaeroma pentodon* with the Australasian *S. quoyanum* was accepted by west American systematists (Rotramel 1972). Other exotic species are doubtless present, masking as locally-described, and thus in the minds of many biologists, native taxa. For example, the phyllodocid polychaete *Eteone californica*, described from San Francisco Bay, may be identical with *E. longa,* having been perhaps introduced with Atlantic oysters into the Bay. Not all populations that have been identified as *E. californica*, however, may be identical with *E. longa* (such as those reported from open coast environments, or from southern California). Similarly, evidence suggests that the anemone *Diadumene franciscana* and the isopod *Synidotea laticauda,* while described from San Francisco Bay, are introduced

---

[4] Includes species coded in Table 1 as o or o? s or s?

(Carlton 1978).

The molluscan fauna is the best known invertebrate group in San Francisco Bay, and provides a detailed example of one group of introduced species. Seventy-five percent of the introduced mollusks are likely from the Atlantic; about 40% were clearly here by 1901 or earlier. By arranging the bivalves and prosobranch and pulmonate gastropods by year of discovery (early

TABLE 2. INTRODUCED SPECIES REDESCRIBED AS NEW TAXA AFTER INTRODUCTION TO SAN FRANCISCO BAY (DATA FROM CARLTON 1978)

Introduced Species	Redescribed as
COELENTERATA: HYDROZOA	
Tubularia crocea (Agassiz, 1862)	*Parypha microcephala* Agassiz, 1865
ANNELIDA: POLYCHAETA	
Polydora ligni Webster, 1879	*Polydora amarincola* Hartman, 1936
Streblospio benedicti Webster, 1879	*Streblospio lutincola* Hartman, 1936
MOLLUSCA: GASTROPODA	
Ovatella myosotis (Drapanaud, 1801)	*Alexia setifer* Cooper, 1872
MOLLUSCA: BIVALVIA	
Mya arenaria Linnaeus, 1758	*Mya hemphillii* Newcomb, 1875
Lyrodus pedicellatus (Quatrefages, 1849)	*Teredo townsendi* Bartsch, 1922
Teredo navalis Linnaeus, 1758	*Teredo beachi* Bartsch, 1921
CRUSTACEA: OSTRACODA	
Sarsiella zostericola Cushman, 1906	*Sarsiella tricostata* Jones, 1958
CRUSTACEA: ISOPODA	
Sphaeroma quoyanum Milne-Edwards, 1840	*Sphaeroma pentodon* Richardson, 1904
CRUSTACEA: AMPHIPODA	
Orchestia chiliensis Milne-Edwards, 1840	*Orchestia enigmatica* Bousfield and Carlton, 1967
CRUSTACEA: CIRRIPEDIA	
Balanus amphitrite amphitrite Darwin, 1854	*Balanus amphitrite franciscanus* Rogers, 1949

collections of opisthobranchs are lacking) (Table 3[5]), certain temporal patterns emerge: by shortly after the turn of the century, eight species had arrived with commercial oysters from the Atlantic; shipworms (*Teredo navalis* and *Lyrodus pedicellatus*) were discovered in the years bracketing World War I (reflecting the increased world-wide shipping throughout the second decade of the 20th century), and two Japanese mollusks were discovered within a decade of plantings of Japanese oysters in San Francisco Bay. Latest is the Atlantic periwinkle, *Littorina littorea,* synchronous with the relatively late development of shipping algae with living Atlantic lobsters to San Francisco. Not so easily explained was the discovery, one in the late 1920's and one in the late

---

[5] Although dates in Table 3 are the dates of first collection, for some species a more precise timing of introduction can be established. Thus, the soft-shell clam, *Mya arenaria*, was introduced within a five-year period between 1869 and 1874, and accumulating evidence suggests that the mud snail, *Ilyanassa obsolete,* was introduced between 1904 and 1907.

SAN FRANCISCO BAY

TABLE 3. INTRODUCED MOLLUSKS (BIVALVES AND PULMONATE AND PROSOBRANCH GASTROPODS) OF SAN FRANCISCO BAY BY DATE OF DISCOVERY (DATA FROM CARLTON 1978)

COI = Commercial oyster industry

Date of Discovery	Species	Mechanism of Introduction and Origin
1871	*Ovatella myosotis*	
1874	*Mya arenaria*	
1890	*Urosalpinx cinerea*	
1893	*Gemma gemma*	−COI (Atlantic)
1894	*Ischadium demissum*	
1898	*Crepidula convexa*	
1901	*Crepidula plana*	
1907	*Ilyanassa obsoleta*	
1913	*Teredo navalis*	−SHIPPING (Atlantic?)
1920	*Lyrodus pedicellatus*	
1927	*Petricola pholadiformis*	−COI? (Atlantic)
1938	*Busycotypus canaliculatus*	
1946	*Musculus senhousia*	−COI (Japan)
1946	*Tapes japonica*	
1968	*Littorina littorea*	−ALGAE (Atlantic)

1930's, of two Atlantic mollusks, *Petricola pholadiformis*, a burrowing clam, and *Busycotypus canaliculatus*, a large predaceous whelk (one of the most spectacular introductions, and now the largest gastropod in San Francisco Bay), long after large-scale introductions of Atlantic oysters had ceased. The discovery in 1937 of the Atlantic mudcrab, *Rhithropanopeus harrisii*, in San Francisco Bay (Jones 1940; Carlton 1978) falls between these molluscan discoveries, and it may be that for a period of time transport of oysters occurred from a region of the Atlantic coast other than that which normally supplied stock to San Francisco Bay. Alternatively, a yet-to-be-discovered mechanism, other than oyster transport, may have existed through which these species were introduced. Thus, for example, still unexplained was the discovery in 1912 of a single, living specimen of the western Pacific starshell (gastropod) *Guildfordia triumphans* (=*Astralium triumphans*) in 34 m of water just outside of the Golden Gate in dredgings of the *Albatross* (Packard 1918; Hanna 1966). No known synanthropic mechanism convincingly explains this occurrence.

SOCIAL, ECONOMIC, AND GEOLOGIC IMPACTS OF THE INTRODUCED FAUNA

Certain pestiferous species exist among the introduced fauna. Along with the Atlantic mudsnail, *Ilyanassa obsoleta*, came a trematode parasite, which occasionally causes swimmer's itch (schistosome dermatitis) in San Francisco Bay (Table 1; Grodhaus and Keh 1958). Economically, the activities of shipworms, primarily *Teredo navalis*, and gribbles, such as the boring isopod *Limnoria* spp. have affected harbor and other installations throughout the Bay for many years, the gribble creating characteristic pencil-like pilings in its boring activities. The depredations of *T. navalis* shortly after its arrival about 1913 led to the formation, in 1920, of the "San Francisco Bay Marine Piling Committee," of the American Wood Preservers' Association, under the leadership of C. L. Hill and C. A. Kofoid, and to intensive studies of its biology and morphological variation. This culminated in the famous "Final Report" of 1927, one of the few scholarly "environmental impact" reports ever produced on San Francisco Bay. Modern methods of anti-borer treatments appear to have generally reduced the effects of shipworms and limnorias in San Francisco Bay.

The burrowing and boring isopod *Sphaeroma quoyanum,* long known as *S. pentodon* on the Pacific coast (Barrows 1919; Higgins 1956; Rotramel 1972) has had an important impact on the shoreline topography of the Bay since its introduction by ships from Australasia probably between the 1850's and 1890's. During the past 85 and more years, in several areas in north San Pablo Bay, and along the western and eastern shorelines of south San Francisco Bay, and in small central bay inlets such as Richardson Bay, many kilometers of shoreline have been eroded by the activities of this crustacean. Sections of shoreline on the eastern shore of south San Francisco Bay just north of the San Mateo Bridge typify what may be called "sphaeroma topography": here, portions of the shore are shaped by the activities of this borer, which by weakening the clay banks, facilitates the removal of the banks by wave action. *Sphaeroma quoyanum* has also affected levees and dikes around the Bay, and in recent years this has been brought to public notice. The *Oakland Tribune* of 21 March 1976 reported one estimate that *S. quoyanum* could remove up to 30 ft of dike in one year! Unfortunately, no measurements appear to have been made of the general extent or rate of the erosional activity of this isopod along much of the Bay's shoreline, although it may be estimated that in some areas this has involved the landward disappearance of many meters of shoreline.

BIOLOGICAL AND ECOLOGICAL IMPACTS OF THE INTRODUCED FAUNA

The greatest impact of the introduced fauna must be measured in biological and ecological terms: in the large numbers of species which have been introduced, and in their establishment as numerical and biomass dominants in many areas of the Bay.

Lake Merritt (Fig. 1), a natural, shallow, brackish-water arm of the central east bay shore, connected to the Bay by a channel to the Oakland Estuary, provides one example. Here, the introduced fauna makes up an important component of the total fauna; indeed, almost all of the abundant species are introduced. In a 10-yr qualitative sampling study which I conducted from 1962 to 1972, 37 of the 46 recorded species of littoral, benthic, fouling, and boring macroscopic invertebrates were introduced species. The 20% native fauna consists largely of transient summer-invading species; a number of the introduced species are also present in the lake primarily in the summer. Abundant and comprising much of the macroscopic fauna in the lake are masses of the tubeworm *Mercierella enigmatica,* a gregarious polychaete introduced from Australasia about 1920 (but not described until 1923 from France). *Mercierella enigmatica* occurs throughout the lake on wharves, boat bottoms, retaining walls, rocks, wood, debris, and floating algae. Shortly after its introduction, local newspapers reported it as "coral" — a natural mistake for the coral-like masses which develop around rocks in the lake. Further, *M. enigmatica* has become a prominent organism in warm shallow manmade lagoons around the Bay, in such areas as Foster City, Belvedere, and Alameda. In the Alameda lagoons, it has caused almost complete blockage of some drainage pipes. Equally prominent are the Atlantic ribbed mussel, *Ischadium demissum,* which occurs in large beds along retaining walls all along the lake shoreline, the Japanese mussel, *Musculus senhousia,* found in dense mat-like beds on the lake bottom, and the Atlantic soft-shell clam, *Mya arenaria,* and the Japanese cockle, *Tapes japonica.* Common on all hard substrates in the lake are the bryozoans *Conopeum* sp., *Victorella pavida,* and *Barentsia benedeni,* and the barnacles *Balanus improvisus* and *Balanus amphitrite amphitrite.* On the lake shore is the Chilean or New Zealand beach-hopper,

Orchestia chiliensis. In the algae and among the tubeworms, the amphipod crustaceans *Ampithoe valida, Corophium insidiosum, Grandidierella japonica,* and *Melita nitida,* (all introduced), and tanaids (the introduced *Tanais* sp.) are seasonally abundant. The Asian anemone, *Haliplanella luciae,* and its small nudibranch predator, *Trinchesia* sp., are sporadically common. Polychaetes abundant in the mud bottom include *Neanthes succinea, Polydora ligni, Streblospio benedicti,* and *Capitella capitata,* all introduced species. The only nemertean in the lake is the occasionally abundant *Lineus ruber,* whose tentative status as a native species in San Francisco Bay requires investigation. Seasonally present in Lake Merritt, coinciding with higher summer temperatures and salinities, are the introduced shrimp *Palaemon macrodactylus,* the anemone *Diadumene franciscana,* the hydroid *Tubularia crocea* and its nudibranch predator *Tenellia pallida,* the snails *Urosalpinx cinerea* and *Ilyanassa obsoleta,* the tunicates *Molgula manhattensis* and *Ciona intestinalis,* and among native species, the polychaetes *Harmothoe imbricata, Nereis vexillosa,* and *Eteone dilatae,* the sacoglossan slug *Aplysiopsis smithi,* and the mussel *Mytilus edulis.* The only relatively common native species in the lake on an annual basis is the mud crab, *Hemigrapsus oregonensis.* A predominantly exotic fauna in Lake Merritt appears to have existed since at least the 1930's (Light 1941:188, 191).

In the nearby Oakland Estuary, Graham and Gay (1945) found that the dominant fouling fauna settling on experimental wooden panels is composed of the hydroid *Tubularia crocea,* the spionid polychaete *Polydora ligni,* the amphipod *Corophium insidiosum,* and the barnacle *Balanus improvisus,* all of which reach peak biomass in the summer months. All are introduced species. The only borer noted, in small numbers, was the introduced isopod identified as *Limnoria lignorum,* but probably *L. tripunctata.* Winter fouling was primarily composed of brown filamentous diatom mats and the brown alga *Ectocarpus* sp.; spring and summer algae include the green *Enteromorpha* sp. and the yellow-brown *Vaucheria* sp. Unfortunately, virtually nothing appears to be known about the probably large exotic algal flora of the Bay's shallow estuarine margins.

Nichols' (1977) study of the infaunal biomass and production of a mudflat near Palo Alto (Fig. 1) provides insight into the role of introduced species in the trophic structure of South Bay. Of the 16 taxa identified to species, two require brief discussion before the significance of exotic species to the biomass and secondary productivity of this mudflat invertebrate association can be assessed. One, the phyllodocid worm *Eteone californica* has already been mentioned: San Francisco Bay populations of this species may be the Atlantic *E. longa.* The other, the small tellinid clam *Macoma balthica,* may be (1) native, as a cold-water circumboreal species occurring regularly on the Pacific coast as far south as San Francisco Bay, but with bay and offshore records as far south as San Diego (Coan 1971; E. Coan pers. comm.); (2) introduced into San Francisco Bay between 1850 and 1869 with the native oyster, *Ostrea lurida,* from Willapa Bay, Washington, during an early trade in Pacific coast oysters (Barret 1963); or (3) introduced with Atlantic oysters after 1869. *Macoma balthica* has been reported by W. O. Addicott from sediments beneath the floor of southern San Francisco Bay that range in approximate age from 2000 to 6000 years (Atwater et al. 1977; B. Atwater pers. comm.). *Macoma balthica* in San Francisco Bay may thus represent native populations mixed with introduced stocks, through (2) and/or (3), above.

With the exception of *E. californica* and *M. balthica,* all of the species found by Nichols in the Palo Alto mudflat are introduced. At Nichols' three stations, introduced species largely of the genera *Gemma, Mya, Streblospio, Heteromastus,* and *Ampelisca* comprise 20, 64, and 84% of the biomass (averaged for four seasons). In each case, however, *Macoma balthica* accounts for the greater amount of the remaining biomass: 77, 35, and 10% respectively. And, although represented by relatively few specimens, *M. balthica* tends to dominate biomass data because of the weight of single, large specimens (Nichols 1977). As a matter of fact, the overwhelming numerical dominants at the sites studied by Nichols were the tiny clam *Gemma gemma* and the spionid worm *Streblospio benedicti* both of which because of sampling technique, were considered to be underestimated in numbers present (Nichols 1977). Not included directly in Nichols' study is an abundant epifaunal species, the mudsnail *Ilyanassa obsoleta*, which, "to the casual observer...is the only invertebrate of the mudflat. In fact, this species forms, in some areas, a pavement of shells on the mud surface." (Nichols 1977). Nichols noted that his observations that the total biomass is comprised largely of mollusks of the genera *Macoma*, *Mya*, and *Gemma*, along with nonquantified observations on the abundance of *Ilyanassa*, agree with similar observations in a Nova Scotia estuary by Burke and Mann (1974). Like Nichols, Burke and Mann found that mollusks of the genera *Mya* and *Macoma* were the chief primary consumers on intertidal sand and mud flats. In San Francisco Bay, this appears to be in no small part an artifact of man's introductions. Nichols concluded, in part, that the mudflat invertebrate association in south San Francisco Bay is an "important link in the cycling of organic matter of the San Francisco Bay estuary." The fact that this secondary productivity is tied up in large part in introduced species, especially so if *Macoma balthica* is an introduced form, raises questions as to the nature and role of mudflat invertebrates in the organic matter budget of the Bay estuary prior to the mid-19th century. Of interest also is that Nichols (1979) found that the introduced species of the genera *Streblospio*, *Ampelisca*, and *Gemma* are the important biological indicators of continuously physically-disturbed environments in south San Francisco Bay.

A further study on the trophic budget of San Francisco Bay is of interest here. Recher (1966) studied shorebird feeding ecology at Palo Alto, examining the stomach contents of plovers, avocets, dowitchers, sand-pipers, marbled godwits, knots, and willets. With the exception of 3% of the willet's diet, which consisted of the native mudcrab *Hemigrapsus oregonensis*, all identified prey items consumed by these birds (excluding *Macoma balthica*) are introduced species, including many of the species discussed by Nichols. Nichols (1977) has already pointed out that Recher's report of only *Neanthes succinea* and no other polychaetes may be due to the non-identifiability of some species of soft-bodied worms in stomach contents. Recher found that the prey items of greatest importance were *Neanthes succinea*, *Gemma gemma*, *Ilyanassa obsoleta*, and, for two bird species, unidentified ostracods (perhaps *Sarsiella zostericola*). *Mya arenaria* and *Macoma balthica* (combined in Recher's data) accounted for less than 7%, and generally less than 4% of any bird diet. How accurate a reflection this is of the reliance of migrating shorebirds on these introduced species in south San Francisco Bay is difficult to determine, although Nichols (1977) has also suggested the importance of shorebird predation on the invertebrates in his study area. There is no doubt, however, that shorebirds in south San Francisco Bay feed heavily on introduced invertebrate species, which arrived here a little more than 100 years ago, whereas the shorebirds have presumably visited the Bay, which lies along the Pacific Flyway, for thousands of years. In the Delta, introduced invertebrates are of occasional importance to both native and introduced fishes, although native mysids and native corophiid amphipods are often more significant. The Atlantic mudcrab, *Rhithropanopeus harrisii*, and several clams, including *Gemma gemma* and *Tapes japonica*, also are occasionally important in the diet of such species as white sturgeon (*Acipenser transmontanus*) (McKechnie and Fenner 1971).

Ecological interactions between native and introduced invertebrates in the Bay have been little studied. The ecological and spatial overlap among possibly competing species largely remains to be determined. Several pairs of potentially competing species readily suggest themselves however: (introduced indicated first) the shrimps *Palaemon macrodactylus* and *Crangon* spp.; the mud crabs *Rhithropanopeus harrisii* and *Hemigrapsus oregonensis*; the rock cockles *Tapes japonica* and *Protothaca staminea;* the marsh snails *Ovatella myosotis* (= *Phytia setifer*) and *Assiminea californica*, and the mudsnails *Ilyanassa obsoleta* and *Cerithidea californica* (the interactions of the latter pair are now under investigation by Margaret Race of the University of California,

Berkeley). In certain areas of the Bay some of these species pairs do not co-occur; here, it may be that competition would be important only in essentially ecotonal habitats. In Lake Merritt, for example, where *Palaemon macrodactylus* is seasonally abundant, *Crangon* spp. do not occur; *Tapes japonica* is common in the lake, but *Protothaca staminea* has never been found there; *Hemigrapsus oregonensis* is present, but *Rhithropanopeus harrisii*, first reported on the Pacific coast from Lake Merritt (Jones 1940), has not been found in the lake since my studies began in 1962, nor was it found in 1952 in another study of the lake.

In certain parts of the Bay, hermit crab populations may have responded to the increased variety and number of gastropod shells provided by introduced species of several genera, such as *Ilyanassa, Urosalpinx,* and *Busycotypus*. In south San Francisco Bay, at Coyote Point, Wicksten (1977) has noted that the population of the native hermit crab, *Pagurus hirsutiusculus,* is existing almost entirely in shells of either *Ilyanassa obsoleta* or *Urosalpinx cinerea*. Similar reliance on introduced shells by *P. hirsutiusculus* occurs in other areas of the Bay. It would appear that the hermit crabs have either switched to these exotic shells from native gastropod shells, which are not now common at Coyote Point, or have expanded their range or population sizes in the Bay.

The successful establishment of so many exotic species cannot be considered the result of any one cause, and in some cases reasons for establishment may be highly species-specific.[6] Man's extensive perturbations of the Bay, including major hydrological changes (through dredging, filling, sedimentation, diverting and damming of streams), installations of harbor and marina facilities (wharves, pilings, floats, dredged channels), and bay disposal of a wide variety of inorganic and organic pollutants, may have created novel environmental conditions to which only certain introduced species can adapt. More important, however, may be that the relatively young and island-like estuaries of the Pacific coast aboriginally supported, in terms of a number of species, a sparse native fauna (Jones 1940; Hedgpeth 1968; Carlton 1978); for many introductions there is no obvious native counterpart with which "competition" may have been necessary for establishment. No native marine invertebrate is known to have become extinct in San Francisco Bay due to competition with an introduced species. However, portions of once-broader niches (*sensu lato*) of native species may have been acquired by introduced species, particularly by those exotics which occupy the brackish (up-estuary) ends of counterpart native species' ecological ranges. Thus the introduced shrimp *Palaemon macrodactylus* (compared to the native *Crangon* spp.), the introduced barnacle *Balanus improvisus* (compared to the native *Balanus glandula* and *Balanus crenatus*), and the introduced crab *Rhithropanopeus harrisii* (compared to the native *Hemigrapsus oregonensis*) all occur most abundantly in more brackish (or even fresh water) regions of San Francisco Bay, whereas the native species are most common in more saline areas. Investigations and experimental manipulative studies directed to dissecting out the relative importance of competition, of the absence of a diverse native fauna, and of the creation of modified and new environments, are needed, both to clarify the sequence of events that led to the successful establishment of introduced species in the past and to understand what types of introductions may be successful in the future.

I began by stating that the history of introduction in San Francisco Bay has largely been an anecdotal one, and an anecdote may serve to end this discussion. In the early 1890's, the Atlantic mussel, *Ischadium demissum,* was introduced and has since become abundant throughout the Bay, where it may live with only its posterior end at the mud surface. In 1927, Dudley Sargent De

---

[6] That the established species are only a small percentage of those actually transported from foreign shores by ships and oysters is indicated by preliminary evidence summarized by Carlton (1978). Indeed, well-known fouling species introduced on other shores, such as the serpulid worm *Hydroides elegans* and the barnacle *Balanus eburneus,* have been found living on ships entering San Francisco Bay but have failed to become established.

Groot of Stanford University, reporting on the biology of the California clapper rail (*Rallus longirostris obsoletus*) in San Francisco Bay, observed that the mussel had become an apparent danger to the rail, for as the birds walked through the marshes they would be trapped by a slightly opened *Ischadium* flush with the substrate. "It is our belief that at least seventy-five percent, and perhaps more, of the adult rail of the Redwood marsh area are minus toes from this cause," said De Groot, who further estimated, rather roughly, that approximately 25% of the chicks of hatched broods of the rail "meet an untimely end" by drowning at high tide while being held by the valves of the mussel. De Groot ranked *Ischadium demissum* as one of the causes of the demise of the California clapper rail in San Francisco Bay, a species now on the endangered list, although Moffit (1941) later showed that the detrimental effects, if any, of *I. demissum* may be outweighed by the fact that 57% of the total diet (and 66% of the animal protein in the diet) of clapper rails near Palo Alto consists of this mussel!

Introductions continue and these seem inevitable. In the San Francisco Bay area, introduced species of animals and plants predominate throughout much of the urban and agricultural terrestrial environment, and it seems only "natural" that man should contribute substantially to the exotic composition of the marine biota as well. We may expect that as man's methods of transportation in commerce and trade change and as our methods of traversing great distances on water with increasing speeds change, more and perhaps different exotic species will be introduced into San Francisco Bay and to the Pacific coast. Whether or not we will be able to detect an asymptotic situation relative to the numbers of introduced species in an environment such as San Francisco Bay, and whether or not we will observe the replacement or displacement of one introduced marine species by another introduced species — not, to my knowledge, ever observed on this coast before, but so well documented among land plants and insects — will require long and careful observation.

ACKNOWLEDGMENTS

I am indebted to a great many persons, acknowledged in Carlton (1978), who have contributed materially to the identification and resolution of the introduced fauna of San Francisco Bay. In particular, John Chapman, David Cross, Ernest Iverson, and Daphne Dunn graciously contributed unpublished records of gammarid amphipods, a caprellid, isopods, and an anemone, respectively, from San Francisco Bay. Debby Fishlyn and Steve O'Dell provided moral and logistic support in the preparation of this report for the San Francisco Bay Symposium in June 1977. I thank Frederic Nichols and John Chapman for reading the manuscript and for helpful criticisms. This paper is based, in part, on a thesis submitted in partial satisfaction of the requirements for the degree of Doctor of Philosophy at the University of California, Davis.

LITERATURE CITED

Abbott, D. P., and J. V. Johnson. 1972. The ascidians *Styela barnharti, S. plicata, S. clava,* and *S. montereyensis* in Californian waters. Bull. So. Calif. Acad. Sci. 71:95-105.

Allen, F. E. 1953. Distribution of marine invertebrates by ships. Aust. J. Mar. Freshwater Res. 4:307-316.

Atwater, B. F., C. W. Hedel, and E. J. Helley. 1977. Late Quaternary depositional history, Holocene sea-level changes, and vertical crust movement, southern San Francisco Bay, California. U. S. Geol. Surv. Prof. Paper 1014. 15 pp.

Barnard, J. L. 1950. The occurrence of *Chelura terebrans* Philippi in Los Angeles and San Francisco harbors. Bull. So. Calif. Acad. Sci. 49:90-97.

Barrett, E. M. 1963. The California oyster industry. Calif. Fish Game, Fish Bull. 123, 103 pp.
Barrows, A. L. 1919. The occurrence of a rock-boring isopod along the shore of San Francisco Bay, California. Univ. Calif. Publ. Zool. 19:299-316.
Behrens, D. W. 1971. *Eubranchus misakiensis* Baba, 1960 (Nudibranchia: Eolidacea) in San Francisco Bay. Veliger 14:214-215.
Behrens, D. W., and M. Tuel. 1977. Notes on the opisthobranch fauna of south San Francisco Bay. Veliger 20:33-35.
Bertelsen, E., and H. Ussing. 1936. Marine tropical animals carried to the Copenhagen Sydhavn on a ship from the Bermudas. Vidensk. Medd. Dansk natur. Foren København 100:237-245.
Bishop, M. W. H. 1951. Distribution of barnacles by ships. Nature 167: 531.
Bousfield, E. L. 1973. Shallow-water Gammaridean Amphipoda of New England. Comstock Publishing Associates, Cornell University Press, Ithaca, N.Y. 312 pp.
Bousfield, E. L., and J. T. Carlton. 1967. New records of Talitridae (Crustacea: Amphipoda) from the central California coast. Bull. So. Calif. Acad. Sci. 66:277-284.
Bradley, W., and A. E. Siebert, Jr. 1978. Infection of *Ostrea lurida* and *Mytilus edulis* by the parasitic copepod *Mytilicola orientalis* in San Francisco Bay, California. Veliger 21:131-134.
Burke, M. V., and K. H. Mann. 1974. Productivity and production: biomass ratios of bivalve and gastropod populations in an eastern Canadian estuary. J. Fish. Res. Bd. Canada 31:167-177.
Carlton, J. T. 1969. *Littorina littorea* in California (San Francisco and Trinidad bays). Veliger 11:283-284.
Carlton, J. T. 1975. Comments on cosmopolitanism. Bull. Amer. Malacol. Union 1975:63. (Abstr.)
Carlton, J. T. 1978. History, biogeography, and ecology of the introduced marine and estuarine invertebrates of the Pacific Coast of North America. Ph.D. Thesis. University of California, Davis, Calif.
Carlton, J. T., and V. A. Zullo. 1969. Early records of the barnacle *Balanus improvisus* Darwin from the Pacific Coast of North America. Occ. Pap. Calif. Acad. Sci. 75. 6 pp.
Chapman, J. W., and J. A. Dorman. 1975. Diagnosis, systematics, and notes on *Grandidierella japonica* (Amphipoda: Gammaridea) and its introduction to the Pacific Coast of the United States. Bull. So. Calif. Acad. Sci. 74:104-108.
Chilton, C. 1911. Note on the dispersal of marine Crustacea by means of ships. Trans. New Zealand Inst. 43:131-133.
Coan, E. V. 1971. The northwest American Tellinidae. Veliger 14(supplement):1-63.
Day, J. H. 1964. The origin and distribution of estuarine animals in South Africa. Pages 151-173 *in* D. H. S. Davis, ed. Ecological studies in Southern Africa. Monographiae Biologicae 14. (Uitgeverij Dr. W. Junk, Den Haag.)
De Groot, D. S. 1927. The California clapper rail: Its nesting habits, enemies, and habitat. Condor 29:259-270.
Elton, C. S. 1958. The ecology of invasions by animals and plants. Methuen & Co., Ltd., London and John Wiley & Sons, Inc., New York. 181 pp.
Ewan, J. 1955. San Francisco as a mecca for nineteenth century naturalists. Pages 1-63 *in* A Century of Progress in the Natural Sciences 1853-1953. California Academy of Sciences, San Francisco, Calif.
Farwell, W. B. 1891. Cape Horn and cooperative mining in '49. Century Illus. Mag. 42:579-594.
Fauchald, K. 1977. The polychaete worms: definitions and keys to the orders, families, and genera. Nat. Hist. Mus. Los Angeles Co., Sci. Ser. 28. 188 pp.
Graham, H. W., and H. Gay. 1945. Season of attachment and growth of sedentary marine organisms at Oakland, California. Ecology 26:375-386.
Grodhaus, G., and B. Keh. 1958. The marine, dermatitis-producing cercaria of *Austrobilharzia variglandis* in California (Trematoda: Schistosomatidae). J. Parasit. 44:633-638.
Hand, C. 1956. The sea anemones of central California. Part III. The acontiarian anemones. Wasmann J. Biol. 13:189-251.

Hand, C., and G. F. Gwilliam. 1951. New distributional records for two athecate hydroids, *Cordylophora lacustris* and *Candelabrum* sp., from the west coast of North America, with revisions of their nomenclature. J. Wash. Acad. Sci. 41:206-209.

Hanna, G D. 1966. Introduced mollusks of western North America. Occ. Pap. Calif. Acad. Sci. No. 48. 108 pp.

Hartman, O. 1955. Endemism in the north Pacific Ocean, with emphasis on the distribution of marine annelids, and descriptions of new or little known species. Pages 39-60 *in* Essays in the Natural Sciences in Honor of Captain Allan Hancock on the Occasion of His Birthday July 26, 1955. University of Southern California Press, Los Angeles, Calif.

Hartman, W. D. 1975. Phylum Porifera. Pages 32-64 *in* R. I. Smith and J. T. Carlton, eds. Light's Manual: Intertidal Invertebrates of the Central California Coast, 3rd ed. University of California Press, Berkeley and Los Angeles, Calif.

Hazel, C. R. 1966. A note on the freshwater polychaete, *Manayunkia speciosa* Leidy, from California and Oregon. Ohio J. Sci. 66:533-535.

Hedgpeth, J. W. 1968. Bay and estuary (introduction). Pages 231-233 *in* E. F. Ricketts, J. Calvin, and J. W. Hedgpeth. Between Pacific Tides, 4th ed. Stanford University Press, Stanford, Calif.

Heizer, R. F. 1941. Archaeological evidence of Sebastían Rodriguez Cermeño's California visit in 1595. Calif. Hist. Soc. Quart. 20:315-328.

Hentschel, E. 1923. Der Bewuchs an Seeschiffen. Inter. Rev. Ges. Hydrobiol. Hydrograph. 11: 238-264.

Higgins, C. G. 1956. Rock-boring isopod. Bull. Geol. Soc. America 67: 1770 (Abstr.)

Hill, C. L., and C. A. Kofoid. 1927. Marine borers and their relation to marine construction on the Pacific Coast. San Francisco Bay Marine Piling Committee, San Francisco, Calif. 357 pp.

Hyman, L. H. 1959. Some Turbellaria from the coast of California. Amer. Mus. Novit. No. 1943. 17 pp.

Jones, L. L. 1940. An introduction of an Atlantic crab into San Francisco Bay. Proc. Sixth Pac. Sci. Congress, 3:485-486.

Kemble, J. H. 1957. San Francisco Bay; a Pictorial Maritime History. Bonanza Books, New York. 194 pp.

Kornicker, L. S. 1975. Spread of ostracodes to exotic environs on transplanted oysters. Bull. Amer. Paleont. 65:129-139.

Langston, R. L. 1974. The maritime earwig in California (Dermaptera: Carcinophoridae). Pan-Pac. Entomol. 50:28-34.

Langston, R. L., and S. E. Miller. 1977. Expanded distribution of earwigs in California (Dermaptera). Pan-Pac. Entomol. 53:114-117.

Langston, R. L., and J. A. Powell. 1975. The earwigs of California (order Dermaptera). Bull. Calif. Insect Survey 20. 25 pp.

Light, W. L. 1974. Occurrence of the Atlantic maldanid *Asychis elongata* (Annelida, Polychaeta) in San Francisco Bay, with comments on its synonymy. Proc. Biol. Soc. Wash. 87:175-183.

Light, W. J. 1977. Spionidae (Annelida: Polychaeta) from San Francisco Bay: a revised list with nomenclatural changes, new records, and comments on related species from the northeastern Pacific Ocean. Proc. Biol. Soc. Wash. 90:66-88.

Light, S. F. 1941. Laboratory and Field Text in Invertebrate Zoology. University of California, Associated Students Store, Berkeley, Calif. 232 pp.

McCain, J. C., and J. E. Steinberg. 1970. Crustaceorum Catalogus: Amphipoda I, Caprellidea I, Fam. Caprellidae. Dr. W. Junk N. V., Den Haag. 78 pp.

McKechnie, R. J., and R. B. Fenner. 1971. Food habits of white sturgeon, *Acipenser transmontanus*, in San Pablo and Suisun Bays, California. Calif. Fish Game 57:209-212.

Mariscal, R. N. 1975. Phylum Entoprocta. Pages 609-613 *in* R. I. Smith and J. T. Carlton, eds. Light's Manual: Intertidal Invertebrates of the Central California Coast, 3rd ed. University of California Press, Berkeley and Los Angeles, Calif.

Menzies, R. J. 1958. The distribution of wood-boring *Limnoria* in California. Proc. Calif. Acad.

Sci. (4) 29:267-272.

Miller, M. A. 1968. Isopoda and Tanaidacea from buoys in coastal waters of the continental United States, Hawaii, and the Bahamas (Crustacea). Proc. U. S. Nat'l. Mus. 125(3652):1-53.

Miller, R. L. 1969. *Ascophyllum nodosum:* A source of exotic invertebrates introduced into west coast near-shore waters. Veliger 12:230-231.

Moffitt, J. 1941. Notes on the food of the California clapper rail. Condor 43:270-273.

Monaghan, J. 1966. Australians and the Gold Rush. California and Down Under 1849-1854. University of California Press, Berkeley and Los Angeles, Calif. 317 pp.

Monaghan, J. 1973. Chile, Peru, and the California Gold Rush of 1849. University of California Press, Berkeley and Los Angeles, Calif. 312 pp.

Moyle, P. B. 1976a. Fish introductions in California: history and impact on native fishes. Biol. Conserv. 9:101-118.

Moyle, P. B. 1976b. Inland Fishes of California. University of California Press, Berkeley and Los Angeles, Calif. 405 pp.

Newman, W. A. 1963. On the introduction of an edible oriental shrimp (Caridea, Palaemonidae) to San Francisco Bay. Crustaceana 5:119-132.

Newman, W. A. 1967. On physiology and behaviour of estuarine barnacles. Mar. Biol. Assoc. India, Proc. Symp. Crustacea, Ernakulam, Symp. Ser. 2, pt. 3, pp. 1038-1066.

Nichols, F. H. 1977. Infaunal biomass and production on a mudflat, San Francisco Bay, California. Pages 339-357, *in* B. C. Coull, ed. Ecology of Marine Benthos. Belle W. Baruch Library in Marine Science, No. 6. University of South Carolina Press, Columbia, S. C.

Nichols, F. H. 1979. Natural and anthropogenic influences on benthic community structure in San Francisco Bay. Pages 409-426 *in* T. J. Conomos ed. San Francisco Bay: The Urbanized Estuary. Pacific Division, Amer. Assoc. Advance. Sci., San Francisco, Calif.

Oglesby, L. C. 1965. *Parvatrema borealis* (Trematoda) in San Francisco Bay. J. Parasit. 51:582.

Orton, J. H. 1930. Experiments in the sea on the growth-inhibitive and preservative value of poisonous paints and other substances. J. Mar. Biol. Assoc. U. K. [n.s.] 16:373-452.

Packard, E. L. 1918. Molluscan fauna from San Francisco Bay. Univ. Calif. Publ. Zool. 14: 199-452.

Pilsbry, H. A. 1896. On a collection of barnacles. Proc. Acad. Nat. Sci. Phil. 48:208.

Recher, H. F. 1966. Some aspects of the ecology of migrant shorebirds. Ecology 47:393-407.

Rotramel, G. 1972. *Iais californica* and *Sphaeroma quoyanum,* two symbiotic isopods introduced to California (Isopoda, Janiridae and Sphaeromatidae). Crustaceana (supplement III): 193-197.

Rydell, R. A. 1952. Cape Horn to the Pacific. University of California Press, Berkeley and Los Angeles, Calif. 213 pp.

Shoemaker, C. R. 1949. The amphipod genus *Corophium* on the west coast of America. J. Wash. Acad. Sci. 39:66-82.

Skerman, T. M. 1960. Ship-fouling in New Zealand waters: a survey of marine fouling organisms from vessels of the coastal and overseas trades. New Zealand J. Sci. 3:620-648.

Steinberg, J. E. 1963. Notes on the opisthobranchs of the west coast of North America—III. Further nomenclatorial changes in the order Nudibranchia. Veliger 6:63-67.

Stimpson, W. 1857. On the Crustacea and Echinodermata of the Pacific shores of North America. Boston J. Nat. Hist. 6:444-532.

Throckmorton, S. R. 1874. (On the importation of shad.) Proc. Calif. Acad. Sci., ser. 1, 5:86-88.

Vervoort, W. 1964. Note on the distribution of *Garveia franciscana* (Torrey, 1902) and *Cordylophora caspia* (Pallas, 1771) in the Netherlands. Zool. Meded. 39:125-146.

Wagner, H. R. 1924. The voyage to California of Sebastían Rodriguez Cermeño in 1595. Calif. Hist. Soc. Quart. 3:3-24.

Wicksten, M. K. 1977. Shells inhabited by *Pagurus hirsutiusculus* (Dana) at Coyote Point Park, San Francisco Bay, California. Veliger 19:445-446.

Williams, S. 1930. The Chinese in the California mines 1848-1860. Stanford, Calif. (Reprinted 1971 by R. & E. Research Associates, San Francisco, Calif.) 85 pp.

THE FISHERIES OF SAN FRANCISCO BAY: PAST, PRESENT AND FUTURE

SUSAN E. SMITH AND SUSUMU KATO
National Marine Fisheries Service, 3150 Paradise Drive, Tiburon, CA 94920

The character of the important fisheries of San Francisco Bay (chinook salmon, striped bass, sturgeon, shad, Pacific herring, northern anchovy, starry flounder, surfperch, elasmobranchs, bay shrimp, and bivalves) has changed dramatically over the past century. Many commercial fisheries that were once important to the Bay Area economy have disappeared (e.g., the river fishery for chinook salmon and the extensive clam and oyster industries), and although other commercial fisheries have been revived in recent years (e.g., herring, bay shrimp), there has been an overall change in emphasis from commercial to recreational fishing. This has been largely due to legislation restricting the commercial harvest of anadromous species such as salmon, striped bass and sturgeon.

Man-induced changes in the environment are implicated in the decline of certain fishery resources. Water storage and diversion projects have affected the distribution and abundance of salmon and striped bass, and land reclamation and domestic sewage pollution essentially eliminated the clam and oyster industries. Fishing pressure has also been linked with the decline of the bay shrimp and sturgeon fisheries.

The San Francisco Bay-Delta region (Fig. 1) was once the foremost fishing center on the West Coast but has long since relinquished this position. Extensive land reclamation, dredging, water pollution, and water development projects have taken a great toll on the habitat, and overfishing can also be blamed for much of the decline. Fishery resources came under heavy exploitation between 1870 and 1915, and many fishery products from within the Bay began to decline even before the turn of the century (Skinner 1962). Introduction of new species has mitigated part of the damage, although the full ecological effects of these introductions are not known.

On the positive side, public concern and subsequent investigations by the California Department of Fish and Game (DFG) and other state and federal conservation agencies have led to restrictions which have at least slowed the destruction of habitat and fishery resources. These resources include a wide variety of fish and shellfish, some important to recreational interests, others valuable as commercial products, and still others waiting for the right circumstances to be utilized. Over 100 species of elasmobranchs and fishes, some 70 bivalves and 30 decapods have been recorded in the rich fauna of the Bay-Delta region (Painter 1966; Messersmith 1966; Aplin 1967; Green 1975; Eldridge 1977; unpublished catalog of invertebrates, Calif. Acad. of Sci., San Francisco).

The following sections describe the history and current status of the commercial and sport fisheries that have existed within the San Francisco Bay-Delta complex. Emphasis is placed on the changing nature of the fisheries—how they have been affected by demand, resource availability, regulations, and in particular, alteration of the environment by man. Species life history information, where pertinent, is also presented. Those covered are native and naturalized exotic species— we have not attempted to cover the extensive oyster industry which existed from 1870 to the late 1930's, mainly because the species that largely supported the industry, the eastern oyster

SAN FRANCISCO BAY

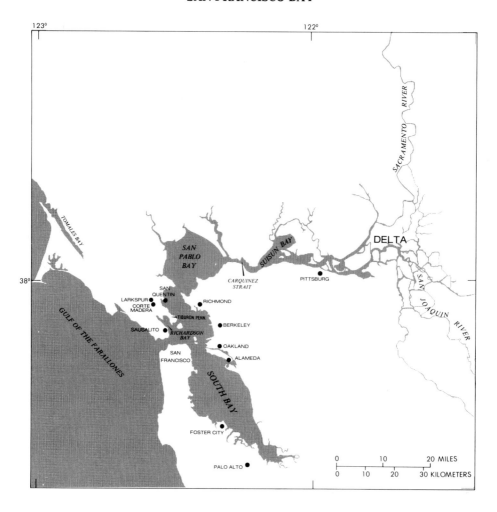

Fig. 1. Index map of the San Francisco Bay-Delta and environs.

(*Crassostrea virginica*), was imported for seed culture or holding only and never successfully established in the Bay. It should be mentioned, however, that the San Francisco Bay oyster industry was at one time the single most valuable fishery in the state; it declined rapidly after the turn of the century as water quality deteriorated and conditions became unsuitable for oyster culture. The market crab (*Cancer magister*) also supported a valuable fishery in the San Francisco area, but is not covered here because of the lack of quantitative information on the early fishery inside the Bay prior to the time when crabbing operations moved outside the Golden Gate. The Bay, however, is an important nursery area for this species (see Tasto 1979).

Skinner (1962, 1972a, b) has provided excellent reviews of the history and current status of the fisheries of San Francisco Bay area. This chapter summarizes and updates these and other works and provides additional insight as to the possible fate of the fisheries resources in the future.

CHINOOK SALMON

The chinook salmon, *Oncorhynchus tshawytscha,* is an extremely valuable food and game resource in the Bay area and along the entire northern California coast. About 80% of all California's chinook salmon landings originate from stocks in the Sacramento-San Joaquin river systems (Skinner 1962). Most recreational fishing and all commercial fishing for chinook salmon now takes place outside the Golden Gate.

The chinook salmon attains a length of 147 cm and a weight of 57 kg, and its ocean range extends from San Diego, California, north to the Bering Sea and south along the Asiatic Coast to northern Japan (Fry 1973).

The species is anadromous, passing through the Bay and ascending rivers (Fig. 2) to spawn in cool, fresh-water streams over clean gravelly substrate (Fry 1973). Like all Pacific salmon, the

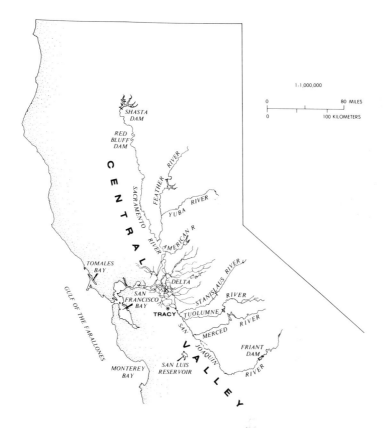

Fig. 2. The drainage basin of the San Francisco Bay-Delta system.

adults die after spawning. Eggs hatch in 50 to 60 days, and about a month later the young emerge from the gravel (Fry 1973). Peak downstream migration of young occurs in the spring and to a lesser extent in the fall. Fingerlings from 2 months to 1 year old move downstream through the estuary into the ocean, where they remain from 1 to 4 years (usually 3) to grow and mature before returning to native spawning streams (Heubach 1968).

There are at present three different stocks in chinook salmon in the Central Valley; these stocks are named for the time each enters spawning streams—spring-run, fall-run, and winter-run fish (Jensen 1972). A fourth group (late fall-run) has been identified (Hallock 1977), but we have included this group in the fall-run category. Most spring-run fish spend the summer in deep pools and do not spawn until the fall. Fall-run fish spawn soon after their arrival on the spawning grounds, and winter-run fish spawn in late spring and early summer (Hallock 1977; Jensen 1972).

Fall-run salmon are by far the most numerous and support the bulk of ocean sport and commercial fisheries. These fish spawn predominantly in the Sacramento and its tributaries, though some enter San Joaquin tributaries such as the Stanislaus, Merced, and Tuolumne rivers (Fig. 2). According to Jensen (1972), winter-run fish, which ascend the Sacramento system, have in recent years become the second most numerous group. Slater (cited by Heubach 1968) believes these fish are descendants of a small run that spawned in the McCloud River, a tributary of the Sacramento that was blocked off to spawners after construction of Shasta Dam. The spring run, which was apparently quite extensive prior to the construction of Shasta and Friant dams, is now the least numerous variety (Jensen 1972). Many spring-run fish formerly spawned in the San Joaquin River system, but due to inadequate flow caused by dam construction and other water projects, the Sacramento system now supports most of what remains of the spring run.

History of the Fisheries

Clark (1929), Skinner (1962) and Scofield (1956) provide good background information on the history of the salmon fishery. According to these authors, commercial fishermen began fishing for salmon around 1850 using gill nets and seines in the Sacramento and San Joaquin rivers and in parts of Suisun and San Pablo bays.

The first salmon cannery started operation on the Sacramento River in 1854. The industry grew rapidly, stimulating the early growth of the fishery. In 1864 one cannery packed 2,000 cases of salmon, each case containing 48 one-pound (0.45 kg) cans (Clark 1929). By 1882, 200,000 cases were produced by a total of 19 canneries operating on the rivers. After that year the industry began a gradual decline and ceased operations in 1916.

Records of the total river catch (Fig. 3A), which became available after 1874, give a better indication of the magnitude of the early salmon harvest. During the ten-year period between 1874 and 1884, an average of 3,220 t (metric tonnes) was landed annually, with a peak harvest of 4,900 t in 1880—the highest catch ever recorded. Ocean trolling for salmon, later to replace the river fishery, increased steadily after the late 1890's when gasoline boat engines were introduced (Scofield 1956). The major shift to ocean fishing, which took place from the late 1920's to the early 1940's, was largely augmented by legislative restrictions that curtailed river netting, such as stream closures, closed seasons and gear restrictions (Scofield 1956). In 1957, the year of the lowest river catch on record (146 t), legislative action eliminated all commercial salmon fishing inside the Golden Gate, making the ocean troll fishery the only legal commercial salmon fishery in California.

After World War II, sport trolling for salmon came into prominence, and the San Francisco area now produces the most consistent ocean sportfishing for salmon in the state (Squire and Smith 1977). About 45% of the chinook salmon taken near San Francisco are landed by recreational fishermen (P. O'Brien pers. comm.).

Environmental Problems

It is beyond the scope of this report to enumerate all of the many man-made changes that have affected salmon populations in the Central Valley. Readers are advised to refer to Clark (1929), Skinner (1962, 1972a), Kelley (1966), Heubach (1968), and Jensen (1972) for more detailed information.

Spawning streams were already being destroyed as early as the Gold Rush days by hydraulic gold mining, railroad construction and lumbering operations. These activities left many streams badly silted or blocked by debris (Gilbert 1917). Water development, particularly storage and diversion projects, had an immense effect on salmon stocks by blocking access to spawning areas above dams, reducing flow and altering temperature regimes below dams, diverting fish into irrigation channels, altering the natural condition of spawning streams, and, in the case of water transport projects, changing the hydrography of the Delta itself.

In 1928, before many of the major dams had been constructed, and before the advent of the federal Central Valley Project (CVP) and State Water Project (SWP), an estimated 80% of the original Sacramento-San Joaquin salmon spawning grounds had already been cut off by obstructions (Clark 1929). The construction of Shasta Dam in 1944 eliminated approximately 50% of the available spawning area of that river (Skinner 1962). Completion of Friant Dam on the San Joaquin River in the mid-1940's essentially eliminated salmon runs in the main stem of that river (Menchen 1977). Dam construction and water diversions on other rivers and streams have also blocked valuable spawning areas, or reduced flows to the extent that either the adults can not or

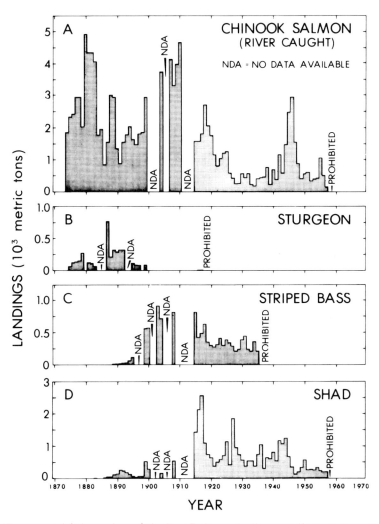

Fig. 3. Commercial fish catches of the Bay-Delta and adjacent tributary streams: (A) Salmon (1874-1957) in the Sacramento-San Joaquin rivers; (B) Sturgeon (1875-1917) in the Bay-Delta; (C) Striped bass (1889-1935) in the Bay-Delta; (D) Shad (1884-1957) in the Bay-Delta.

will not ascend spawning streams, or the young are unable to pass downstream during their seaward migration in spring. Hatcheries have been built to replace some of the major spawning areas lost to water development.

Since the 1950's flow reversals caused by the pumping of huge volumes of Sacramento River water across the Delta to export pumps to the south near Tracy have also created problems for chinook salmon, which depend on their olfactory senses in homing to spawning areas. During periods of reduced flows and high pumping rates, Sacramento fish are attracted to the central and southern Delta by Sacramento River water present there, and are thus delayed in their spawning migration. At the same time, San Joaquin fish have to negotiate the barrier of Sacramento River water to reach their native San Joaquin water—most of which is being delivered away from the Delta to CVP and SWP transport pumps upstream (Jensen 1972; Skinner 1972b). Screens at the pumping station are fairly efficient, but large numbers of fingerling chinooks from the San Joaquin are still lost at those sites.

It has been suggested that construction of the proposed Peripheral Canal will resolve or minimize many of these problems (see for example Gill et al. 1971); however, with or without the Canal, water demands will continue to grow in California, and careful management and control of future development in the Central Valley will play an essential part in protecting the future of this important fishery resource.

STRIPED BASS

The striped bass, *Morone saxatilis*, was first introduced to the San Francisco Bay system in 1879 when 132 juveniles, taken from a small New Jersey estuary, were shipped across country by rail and released into Carquinez Strait (Skinner, 1962). A second plant of 300 fish was made in 1882 in lower Suisun Bay. Conditions in the Sacramento-San Joaquin estuary were obviously ideal, for the species flourished beyond all expectations and now supports a valuable recreational fishery within the Bay.

Today the bulk of west coast striped bass production occurs in the San Francisco Bay estuary. The species is anadromous, migrating in winter and spring to the Delta and upstream to spawn. Information on the biology of the striped bass, population size, and environmental factors affecting the species is presented by Stevens (1979).

History of the Commercial and Sport Fisheries

Ten years after its first introduction and for 46 years thereafter the striped bass supported an important commercial fishery in the Delta area. Fish were taken with gill and trammel nets, primarily in the San Joaquin River (Skinner 1962). Between 1889 and 1915 the catch usually exceeded 454 t (10^6 lb) annually, but catches subsequently dropped and for 20 years thereafter only twice exceeded 454 t (Fig. 3C). Growing interest in sportfishing for striped bass led to increased efforts toward conservation of the species. Finally in 1935 commercial fishing was prohibited and the resource reserved exclusively for sport use.

Until the late 1950's most angling effort took place from San Pablo Bay to the Delta, but as new fishing techniques developed and partyboat fishing upstream of Carquinez Strait diminished, San Francisco Bay proper became the major fishing area (Stevens 1977). Prior to this time most fish were caught by bait-fishing and some by surface trolling; however, it was discovered that bass could be readily taken from partyboats in north San Francisco Bay trolling deep with heavy sinkers. In addition, a winter-spring fishery developed during the herring spawning runs (Chadwick 1962). In the early 1960's another major change occurred in partyboat fishing methods with the introduction of deep drifting with live anchovies. Today this is the primary method used by San Francisco Bay partyboats during the height of the season in summer and fall.

Impact of Water Development

Water development in the Delta has created a variety of problems for the striped bass. Water diversions have led to loss of eggs, larvae and young fish into export canals; salinity intrusion, caused by low rainfall and increased water export, has affected spawning in the San Joaquin River; and subsequent low river flows through and out of the Delta have been associated with poor year-class survival and later recruitment to the fishery (Stevens 1979).

AMERICAN SHAD

The American shad, *Alosa sapidissima,* is another transplant from the Atlantic Coast, and like the striped bass, became firmly established within the San Francisco Bay system soon after its introduction. Shad reach a length of 76 cm and become mature at 3 to 5 years (Fry 1973; Stevens 1972).

The species is anadromous, spending most of its life at sea, the adults using the Bay only as a migratory pathway enroute from the ocean to upstream spawning areas. Most shad spawn from April to June primarily in the Sacramento River system (Stevens 1972). Formerly, shad ascended the Sacramento for 300 miles or more (Nidever 1916), but since construction of the Red Bluff Dam in 1967 (Fig. 2), most runs stop at that point (Fry 1973). Unlike salmon and trout, shad do poorly at ascending fishways, especially the more common weir and orifice type (W. Leet pers. comm.).

Stevens (1972), summarizing the work of others, reports that spawning shad appear to require fresh water, a good current, and relatively warm water temperatures between 16° to 21°C. Many adult shad die after spawning. The eggs are slightly heavier than water and are carried near the bottom by river currents until hatched. Most of the young then move downstream, leaving the Delta and passing through San Pablo Bay from September to November. Young shad feed on zooplankton, primarily cladocerans and copepods. The principal food of the adults in the Delta is the opossum shrimp, though cladocerans and copepods are also eaten. Apparently nothing is known about young shad once they enter the ocean, and little is known of the oceanic habits of Pacific Coast adult shad.

Shad are extremely delicate fish, and the slightest physical injury proves fatal. Because of their fragile nature, shad young may be less apt to survive contact with fish screening devices otherwise suitable for hardier species such as young striped bass and salmon (W. Leet pers. comm.; Kelley 1968a).

History of the Commercial Fishery

Shad were first brought to the West Coast in 1871, when 10,000 fry were introduced into the Sacramento River (Nidever 1916). Between 1873 and 1880 several additional plants were made. As early as 1879, large numbers of shad started to appear in the San Francisco market. When the first few fish began to appear, some curious customers paid as much as $10 to $15 for a single fish, and many shad brought from $1 to $1.50 per pound (0.454 kg). The novelty soon wore off, however, as shad continued to increase in numbers and began to glut the market. By 1894 the price had plummeted to 2 cents per pound. Until 1912 shad were utilized entirely by the fresh fish trade, but later a salt shad market was established in China, and practically all fish were shipped there. Shad roe was salted and either canned and shipped to the east or sold in local markets.

SAN FRANCISCO BAY

During this time practically all shad were caught in drift gill nets in upper San Francisco Bay and in the Delta.

The shad catch reached a peak of 2,540 t in 1917 (Fig. 3C), but afterward landings declined. In 1958 legislation was enacted that prohibited gill netting in inland waters and the shad fishery was eliminated. The action was taken primarily to protect salmon and striped bass caught incidentally in netting operations and not because the shad population was endangered. It would appear that since the turn of the century the commercial fishery was, for the most part, limited not by population size but by lack of demand. Even though shad has an excellent flavor and is a popular food fish on the east coast, it is not sought after to any extent in California, except by recreational fishermen.

The Sport Fishery

Recreational fishing for shad started to become popular around 1950, and since then the sport has grown considerably. In 1963 DFG estimated that 100,000 angler days were spent by fishermen seeking shad on the Yuba, American, and Feather rivers (Kelley 1968a). In recent years the main stem of the Sacramento River has become the most important fishing area (D. Painter pers. comm.).

In addition, a small "bump net" sport fishery exists in the Delta area where upstream migrants are caught in large, long-handled dip nets made of chicken wire held in the wake of an outboard motor. For some reason, only male fish are caught by this method (Stevens 1972).

No quantitative information is available on the size of the adult shad population, but it is thought to be large. At the present time DFG is undertaking tagging studies to obtain population estimates. Factors influencing the survival from egg to adult are thought to be similar to those of striped bass, although predation on shad appears to be lighter (Kelley 1968a). Maintenance of *Neomysis* sp. populations in the Delta, upon which shad depend, may be an important factor in the future as increased water demands and diversions lead to reduced outflow through the Delta, increasing the duration of salinity intrusion and altering the habitat and distribution of *Neomysis* sp. especially during dry years (Orsi and Knutson 1979).

STURGEON

Two species of sturgeon occur in the San Francisco Bay estuary, the white sturgeon, *Acipenser transmontanus,* and the green sturgeon, *A. medirostris.* The largest white sturgeon taken along the Pacific Coast was reported to be about 610 cm long, weighing 817 kg. The smaller and less common green sturgeon reaches about 213 cm and a weight of about 159 kg, but most of those that are caught are considerably smaller (Squire and Smith 1977). Female sturgeon reach maturity when they are about 12 to 15 years old, or approximately 125-140 cm in length. Males apparently mature earlier, at 10 to 12 years of age and about 112-125 cm in length (D. Kohlhorst pers. comm.).

Benthic invertebrates such as clams, small crabs and bay shrimp predominate in the diet of sturgeon; however, during the winter herring spawning runs, herring eggs are reported to account for 20 to 80% of their food. Fish, such as striped bass, herring, staghorn sculpin, and anchovy are eaten in lesser numbers (Miller 1972a; McKechnie and Fenner 1971). The opossum shrimp and the amphipod *Corophium* sp. make up the bulk of the diet of young-of-the-year sturgeon (Miller 1972a).

Tagging studies indicate that white sturgeon confine their movements primarily to the estuary, spending summer, fall and winter in the lower bays and Delta, and migrating upstream primarily in early spring. Green sturgeon appear to spend more time in the ocean and move considerable distances along the coast (Miller 1972b).

Environmental conditions necessary for successful migration and reproduction of sturgeon are not known, and to our knowledge, spawning and embryology of California sturgeon have not been described. Sturgeon larvae have been taken by DFG in the Delta and in the Sacramento River upstream to Hamilton City (rkm 330) from March to mid-June (Kohlhorst 1976; Stevens and Miller 1970). Apparently most, if not all, of these larvae were spawned in the Sacramento River above the Delta. Catches by nets set at different depths indicate larvae are demersal.

There is concern that yolk-sac larvae coming down from the Sacramento River in spring will be vulnerable to water diversions. Recommendations have been made to prevent their being diverted with the water by providing efficient screening facilities and/or pumping curtailment (Miller 1972a). Kelley (1968b) predicted that increased net velocities in Delta channels resulting from increased water transport to the south may reduce populations of *Corophium* and *Neomysis* spp., which are important foods of sturgeon. Also, proposed reclamation or bay fill may have a detrimental effect on both juveniles and adults which feed on benthic organisms over the shallow flats of San Pablo and Suisun bays.

History of the Fishery

Before the 1870's, the sturgeon resource in the estuary was virtually untapped. Only the Chinese considered them of value, but they often utilized only the gelatinous notochord (Skinner 1962). A demand for sturgeon apparently came about when "Easterners" with a taste for sturgeon and caviar migrated to the Pacific Coast. Around the same time, the Atlantic coast supply had diminished (Skinner 1962). Furthermore, great quantities of sturgeon were in demand to feed ranch hands and labor gangs prior to 1895 (Scofield 1957).

The fishery only lasted about 30 years, but within that time the resource was heavily exploited. Initially, most fishing was conducted by the Chinese, who snagged the fish using heavy setlines with unbaited barbless hooks (Fry 1973). Gill and trammel nets took many, but such catches were usually made while fishing for other species. Green sturgeon, considered inferior as a food fish, was not sought after and brought only half the market price of white sturgeon (Smith 1895). From 1875 to 1892, landings averaged 227 t annually, while between 1892 and 1901 they ranged from about 45 to 91 t (Fig. 3B). In 1895, a law was passed protecting sturgeon during part of the spawning season, and Chinese setlining was prohibited (Scofield 1957). In 1901 the State Legislature temporarily abolished the fishery, claiming the white sturgeon to be on the verge of extinction. The fishery remained closed until 1910, when it was reopened to a limited extent but then finally abolished completely in 1917.

Kelley (1968b) suggested that hydraulic mining operations in the Sacramento system may have had as much influence on the decline of the sturgeon as did overfishing. Between 1860 and 1914 tremendous loads of mining debris flowed through the river system to the sea, affecting the benthos and probably sturgeon as well. The decline in the fishery from 1875 to 1900 occurred after 15 years of rapid and heavy bed load movement of debris down the Sacramento and accumulation in Suisun and San Pablo Bays, principal feeding areas of sturgeon. This debris passed out of Suisun Bay by 1930 and out of San Pablo Bay by 1950 (see also Krone 1979).

The taking or possession of sturgeon was prohibited until 1954, at which time DFG felt the population had sufficiently recovered to recommend opening the fishery for sportfishing only. Initially, snagging seemed to be the only effective fishing method, but this was prohibited in 1956 (Miller 1972b). Very few fish were taken until about 1964 when it was discovered that bay shrimp (*Crangon* sp.) could be used successfully as bait. Afterwards, partyboat catches jumped from three sturgeon in 1963 to 2,400 fish in 1967 (McKechnie and Fenner 1971).

In recent years sturgeon angling has become increasingly popular. Although they are caught throughout the year in the upper bays and Delta, best fishing is from fall through early spring when the biggest fish are taken. There is a one-fish bag limit and 102-cm minimum size. Much of the angling takes place from boats in San Pablo and Suisun bays, where anglers stillfish on the bottom using shrimp for bait. During the herring spawning runs from about January to March, sturgeon also congregate around Sausalito and the Tiburon Peninsula, where, during the height of the runs, they are caught by shore and boat anglers using herring fillets or herring roe for bait. A modest fishery exists in South Bay where anglers use methods similar to those used in San Pablo Bay.

PACIFIC HERRING

Although Pacific herring, *Clupea harengus pallasi,* are distributed along the entire U.S. west coast, they are uncommon in southern California. Within the range there are a series of intergrading populations which spawn in specific areas; tagging experiments give evidence of only limited interchange among populations (Hart 1973). Major spawning areas in California are Tomales and San Francisco bays.

Each year from about November to March large schools of herring enter San Francisco Bay to spawn, principally along the shores of Sausalito and the Tiburon Peninsula. Most herring spawn when they are two years old and about 25 cm long (J. Spratt pers. comm.). Pacific herring spawn intertidally and subtidally down to 7.6 m or more (Eldridge and Kaill 1973). They appear to prefer substrates covered with seaweed, eelgrass, or rock upon which to deposit their adhesive eggs although the eggs commonly blanket every available surface during the height of spawning activity. Fecundity is 18,600 eggs for females averaging 192-mm standard length, and 29,500 eggs for females of 223 mm (Hart and Tester 1934). Most spawning occurs at night, although fish have been observed to spawn during the day as well, and at all tidal stages. The eggs range between 1.3 to 1.7 mm in diameter and are usually deposited one or two eggs thick, but can be in layers up to 51 mm deep (Miller and Schmidtke 1956). Eldridge and Kaill (1973) found that when heavy "clumping" of eggs occurs, survival is poor among all but the outlying eggs.

During and after spawning, adult fishes and eggs are subjected to heavy predation from the many sea birds, fishes and sea lions that gather at the spawning grounds. Mortality of eggs may range from less than 5% in deep water to 99% or more for intertidal spawning where eggs are available to a variety of predators, particularly gulls (Eldridge and Kaill 1973). After leaving the Bay, postspawning adults and juveniles are thought to disperse along the coast (J. Spratt pers. comm.) and a significant number apparently migrate to Monterey Bay where the commercial fishery regularly lands nonspawning herring.

History of the Fishery

Immigrant Italian fishermen started fishing for herring about 1850, and by 1888, 35 to 40 boats were engaged in the fishery which was centered in Richardson Bay (Skinner 1962). There was also limited fishing over the shallows on the eastern side of San Francisco Bay south of Alameda and off Point Richmond. Fishermen used gill nets, beach seines, and later paranzella nets. The bulk of the landings was sold in the fresh fish markets, but when abundant and cheap, part of the catch was salted and sold for bait as well as for human consumption. According to Collins (1892) herring landings in 1888 amounted to 1,200 t.

Attempts to establish export markets failed because Pacific herring did not measure up to the quality of Atlantic herring. Bay fish were particularly unsuitable because of their small size,

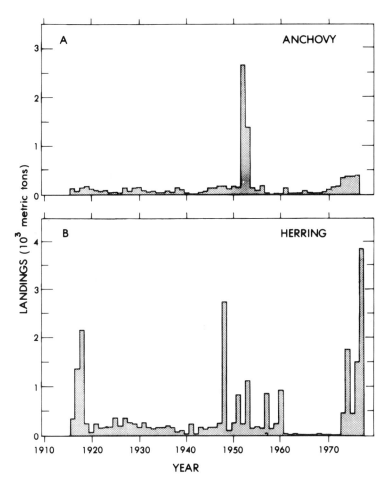

Fig. 4. Commercial fish catches in the Bay: (A) Anchovy (1916-1976); (B) Herring (1916-1976/77 season).

and because gill netters and seiners took only spawning fish that were not in prime condition for salting and smoking (Scofield 1918). Some salted herring was shipped to the Orient, however.

Demand for herring increased as food came into short supply during World War I. In 1917 a canning and reduction plant was established at Pittsburg, California, supplied by fish from San Francisco and Tomales bays, particularly the latter, because the fish were fatter and longer (Scofield 1918). Although the San Francisco Bay catch rose to nearly 2,150 t in 1918 (Fig. 4B), the fishery was short-lived. In 1921 the State Reduction Act was enacted, limiting reduction of fish to fish meal. Canning operations closed and from 1920 through 1946 only small quantities were landed for fresh consumption, bait, and for smoking (Frey 1971; Miller and Schmidtke 1956).

The fishery had a brief revival after the end of World War II when herring was tried as a substitute for the vanishing sardine, but the attempt was unsuccessful. Thereafter only a small pet-food market continued to provide a limited demand, but even this market did not hold (Eldridge and Kaill 1973).

In the mid-1960's, a specialized fishery began for herring eggs, utilized in Japan in two forms: "kazunoko" and "kazunoko-kombu." Kazunoko consists of the ovaries taken from ripe herring. Fish are taken by lampara nets, gill nets, and purse seines. Kazunoko-kombu refers to herring eggs-on-kelp, which is gathered by divers in subtidal areas. Both products are expensive gourmet items in Japan.

As fishing effort intensified in both Tomales and San Francisco bays, public concern mounted over the fate of the herring resource, and emergency legislation was passed in 1973 that instituted a permit and quota system for the herring fishery. In San Francisco Bay, the present (1977-1978) quota for adult herring is 4,558 t; the quota for eggs-on-kelp is 4.5 t, including plant material. The commercial season extends from December 14 through March 31 with different starting dates for the various fishing methods.

The 1976-77 herring spawning population in San Francisco Bay was estimated at 24,489 t (J. Spratt pers. comm.).

NORTHERN ANCHOVY

Probably the most abundant species of fish in the Bay, the northern anchovy, *Engraulis mordax,* supports a modest commercial fishery and is important in the food web. The anchovy is a short-lived species, attaining a probable maximum of seven years, although rarely exceeding four years of age and a length of 18 cm (Frey 1971). Sexual maturity is reached by the end of two years at a length of about 13 cm (Frey 1971).

Anchovy are found in the Bay throughout the year, but a large influx occurs in May, and an elevated abundance persists through September (Aplin 1967). Commercial landings also coincide with this period, but high catches extend into October and sometimes into November (DFG Catch Bulletins). Many small anchovy can be found during this period, indicating that the species probably spawns in the Bay. Eldridge (1977) found the greatest number of larval anchovy during December, however.

Many species of fish feed on anchovy of appropriate size, including downstream migrant salmon (Heubach 1968), jacksmelt (Boothe 1967) and striped bass (Johnson and Calhoun 1952). Indeed, because of its abundance and small size the anchovy is probably the most important forage fish in the Bay.

The Fishery

From 1916 through 1951, annual landings of anchovy in the Bay Area (Fig. 4A) amounted to less than 180 t (Skinner 1962). In 1952 the catch increased suddenly to 2646 t due to the need for a substitute for the failing sardine industry. Because of low consumer acceptance of canned anchovies, and rigid state laws governing reduction of fish into fish meal, catches in subsequent years declined considerably. Recently, however, the landings have started to increase again, and the commercial catch has stabilized at around 385 t (M. Oliphant pers. comm.).

Practically the entire reported catch is now preserved and packed as frozen bait for recreational fisheries. An additional unrecorded amount, perhaps as much as 25% of the "dead bait" landings, is taken for use as live bait, primarily for use in the sport fishery for striped bass. In some years both live and dead anchovy are also used as bait in the commercial albacore tuna fishery. The landed value of the 1974 "dead bait" catch was $73,344 (McAllister 1976).

Although anchovy for the bait industry are normally caught within the Bay, in some years about 10% are caught outside the Golden Gate (W. Millazzo, Meatball Bait Co., Sausalito; and W. Beckett, bait dealer, Oakland, pers. comms.). We have no estimates of the tonnage caught outside the Bay during the two years of greatest landings (1952 and 1953).

Anchovy are caught in the Bay exclusively with a roundhaul seine called a "lampara" net.

When the anchovy are to be kept alive the fish are dip-netted from the seine into holding tanks aboard the fishing vessel. At times the entire net with the encircled school of anchovy is towed to a moored bait receiver.

No estimates are available on the abundance of anchovy in the Bay. Since the greatest bulk of anchovy are seasonal migrants, it seems unlikely that the stocks in the Bay can be imperiled by the present fishery.

STARRY FLOUNDER

The starry flounder (*Platichthys stellatus*) is one of the most important sport-caught flatfishes along the Pacific Coast (Squire and Smith 1977). It is euryhaline, commonly occurring in estuarine areas, and sometimes in full fresh water and appears to prefer a soft sand habitat (Orcutt 1950). The main spawning period is December-January, when mature starry flounder apparently migrate to shallower waters (Orcutt 1950). Average size of individuals taken during a bottom trawl survey in San Francisco Bay was 41 cm, with a range of 24-63 cm (Boothe 1967). According to Orcutt (1950) a 41-cm fish weighs about 1.5 kg and would be sexually mature. Lockington (cited by Orcutt 1950) reported in 1880 that large starry flounder from 3.6 to 5.4 kg occurred in San Francisco Bay. Fish of this size are rarely encountered today.

Although mature starry flounder are common and some spawning would be expected to occur within the Bay, only few eggs, larvae, and juveniles have been taken in surveys (Green 1975; Eldridge 1977). Ganssle (1966) found that size of starry flounder decreased with distance upstream from San Francisco Bay. D. Stevens (pers. comm.) reports that few starry flounder are found in the Delta, but large ones are commonly caught in San Luis Reservoir, suggesting that the young are carried from the Delta to the Reservoir via the aqueduct. Food of starry flounder in San Francisco Bay consists primarily of polychaete worms, small bivalves, siphons of larger clams, and small crabs (Boothe 1967).

The Fishery

The starry flounder is an important species to anglers in the Bay as it is an excellent food fish. Bottom fishing from anchored or drifting boats in San Pablo and Suisun bays probably produces the best catches, although this species is also common in the shore catch. The commercial fishery for starry flounder is conducted almost exclusively by bottom trawl in offshore coastal waters; no commercial fishing occurs in San Francisco Bay. No data are available on the abundance of starry flounder in the Bay. Sampling with a bottom trawl disclosed its presence at all stations, from Palo Alto to Richmond, and although it ranked 16th in total numbers captured (Aplin 1967), its large size should relegate it to a higher position in terms of biomass.

SURFPERCH

About a dozen species of surfperch (family Embiotocidae) occur in San Francisco Bay, the most common being the pile (*Damalichthys vacca*), black (*Embiotoca jacksoni*), shiner (*Cymatogaster aggregata*), walleye (*Hyperprosopon argenteum*), white (*Phanerodon furcatus*), rubberlip (*Rhacochilus toxotes*), striped (*Embiotoca lateralis*) and rainbow (*Hypsurus caryi*) surfperches (Wooster 1968a; Squire and Smith 1977). Surfperch are relatively small coastal marine fishes, although the larger species, such as the pile and rubberlip, are known to reach lengths of 44 and 47 cm, respectively (Miller and Lea 1972).

All embiotocids are viviparous—a reproductive process rare among marine teleosts. Breeding habits have been described for some surfperch species (Eigenmann 1894; Wales 1929; Rechnitzer and Limbaugh 1952; Engen 1968; Wares 1971; and others). In general, mating usually takes place shortly after a previous brood has been released and the sperm may be carried for months by the female before the eggs are fertilized. The embryos develop in membranous sacs and receive nourishment from the surrounding ovarian fluid. The young are well developed and independent at birth (Wares 1971; Frey 1971). It appears that many Bay surfperch release their young in spring and early summer, judging from the occurrence of term females in the sport catch at that time (W. Dahlstrom pers. comm.). To our knowledge, nothing has been published on movements and breeding habits of Bay surfperch, and in general, information on other life history aspects is lacking. Boothe (1967) found that shiner surfperch fed predominately on benthic invertebrates such as gammarid amphipods and cumaceans, and to a lesser extent on clams and polychaetes. Adams (unpublished)[1] found gammarid and caprellid amphipods were the most important food of black surfperch; other small crustaceans, tanaids and isopods were also important, as were polychaete worms and bryozoans. The diets of striped and rainbow surfperch were similar, but for pile perch, large crustaceans, hard-shelled molluscs and barnacles were much more important in the diet.

Because they occur and are caught in nearshore locations, often adjacent to highly populated and industrialized areas, surfperch and the organisms upon which they feed may be exposed to higher concentrations of pollutants than species inhabiting the deeper waters of the Bay where tidal action and outflow flushing aid in dispersal of these materials. Earnest and Benville (1971) found DDT levels to be higher in surfperch than in the flatfish, sculpin and crabs they sampled in the Bay.

The Fishery

Surfperch form an integral part of the marine sport catch in San Francisco Bay, and are frequently taken by pier and wharf anglers (Wooster 1968a; Squire and Smith 1977). Shiner perch is also a popular live bait for striped bass. Winter and spring are the best fishing times, when larger individuals are usually taken (Wooster 1968a).

Members of this family are of minor commercial importance locally, although some do occasionally appear in local markets. Before the turn of the century, however, surfperch were apparently quite common in San Francisco fresh fish markets (Eigenmann 1894, citing Lord). Judging from some of the early accounts (Wilcox 1898), most were taken with beach seines.

ELASMOBRANCHS (SHARKS AND RAYS)

Of the species of elasmobranchs occurring in San Francisco Bay, the most abundant is the brown smoothhound, *Mustelus henlei* (Herald and Ripley 1951; Russo and Herald 1968). The leopard shark (*Triakis semifasciata*), soupfin (*Galeorhinus zyopterus*), dogfish (*Squalus acanthias*), sevengill shark (*Notorhynchus maculatus*), and the bat stingray (*Myliobatis californica*) are all fairly common. The largest shark caught in the Bay was a rare sixgill shark (*Hexanchus griseus*) which measured about 3.3 m and weighed 210 kg (Herald and Ripley 1951). Total biomass of these species is unknown but may be considerable. One study using bottom longline gear produced a catch rate of 15 sharks per 100 hooks (Herald and Ripley 1951), which is an exceptionally high catch rate for any commercial longline fishery.

South Bay apparently harbors more sharks than the central and northern reaches (Herald and Ripley 1951). The brown smoothhound and leopard sharks apparently prefer shallower waters, while the sevengill is usually found in waters deeper than 6 m. The Bay may be a nursery ground for some sharks, as evidenced by the predominance of juveniles in the catch. Soupfin gives birth to pups in the Bay (Herald and Ripley 1951), and we have caught many pregnant brown smoothhounds with near-term pups.

[1] "Resource partitioning among members of a model fishery." NMFS, Tiburon Lab., Tiburon, CA 94920.

Food habits of leopard sharks and brown smoothhounds have been extensively studied by Russo (1975) who found that both were essentially benthic feeders. Disturbingly, he also found polychlorinated biphenyl (PCB) levels of 22-47 ppm in the liver, and total identifiable chlorinated hydrocarbon levels of 37 and 108 ppm respectively for smoothhound and leopard sharks. Among the important food elements (over 15% frequency of occurrence in stomachs of sharks caught in San Francisco Bay) were shrimps, crabs, herring eggs, and the fat innkeeper worm (Russo 1975).

The bottom-feeding mode of these two species apparently was connected to a case of mass mortality recorded in 1967 (Russo and Herald 1968). During a period of 33 days, more than 725 elasmobranchs were found dead along the banks of the eastern shore of San Francisco Bay. Of this total 492 were brown smoothhounds. Leopard sharks and bat stingrays, presumably a bottom feeder also, made up the rest of the total killed, except for a solitary sevengill shark. The cause of death was not determined.

History of the Fishery

Presently no commercial fishery exists in San Francisco Bay for elasmobranchs. During the celebrated "boom" of shark liver oil that lasted from 1937 through the early 1950's, an intensive fishery was conducted throughout the state for soupfin and spiny dogfish (Ripley 1946; Frey 1971). The livers of these species, particularly the soupfin, are rich in Vitamin A. During this period landings of sharks from the San Francisco Bay region constituted over 40% of the total for the state (Ripley 1946). Between 1937 and 1945, annual landings of all sharks in the San Francisco Bay area averaged 870 t, with a high of 2,243 t in 1940.

Ripley's (1946) figures indicate that during 1941-44, soupfin constituted 16% of the total shark landings in the San Francisco region. Presumably most of the remainder were spiny dogfish, much of which were caught by bottom trawl gear outside the Bay. Byers (1940) indicates that a large proportion of the soupfin landed in the San Francisco region were caught by hook and line within the Bay.

Shark and ray fishing is popular with many Bay anglers, and although most people disdain the thought of eating sharks, some species are considered desirable and are commonly eaten. In fact, all the species found in San Francisco Bay have been found in the market (Frey 1971). The leopard shark and soupfin are particularly desired for their substantial amounts of firm white flesh, and many anglers in San Francisco Bay undoubtedly consume these species regularly. A further use of shark, one heartily recommended by gourmets, is the Chinese sharkfin soup.

CLAMS, OYSTERS, AND MUSSELS

San Francisco Bay contains large numbers of shellfish species, some of which have known potential commercial and recreational value such as the soft-shell clam, Japanese littleneck, mussels, and the native oyster (see also Carlton 1979 and Nichols 1979). Although considerable progress has been made in improving water quality in the Bay in recent years, shoreline waters are apparently not yet free enough of sewage contamination for the State Public Health Department to sanction harvesting of Bay shellfish for consumption.

History of the Fisheries

San Francisco Bay was one of the major landing areas in the State for oysters and clams, but these fisheries declined steadily after 1900, with the oyster industry collapsing in the late 1930's, and the soft-shell clam industry in the late 1940's (Skinner 1962; Jones and Stokes 1977). By and large, introduced species have been the most important commercially.

SAN FRANCISCO BAY

Oysters: Three species of oysters were harvested from San Francisco Bay in the past: the native oyster (*Ostrea lurida*), the eastern oyster (*Crassostrea virginica*), and the Pacific or Japanese oyster (*C. gigas*). The latter two species are exotics introduced for holding and fattening or seed culture—neither have reproduced sufficiently in the Bay for commercial exploitation (Jones and Stokes 1977). Information on these oyster fisheries can be obtained from Bonnot (1935), Skinner (1962), and Wooster (1968c).

The native oyster was once extremely abundant in many parts of the Bay (Townsend 1893). This species was harvested centuries ago by local Indian populations and briefly for the restaurant trade during and after the Gold Rush days (Wooster 1968b; Jones and Stokes 1977). It could not compete with the larger, more desirable eastern oyster after the latter was introduced, however. Apparently no attempts have been made to culture the native oyster in the Bay, and although at least five large native oyster beds still exist (Wooster 1968b), the species is much less abundant than it was before the turn of the century. The decline in native oyster populations is thought to be the result of predation by the eastern oyster drill which was introduced with the first eastern oyster shipments from the Atlantic Coast, and the silting in of much of the suitable substrate that once existed (Bonnot 1938; Wooster 1968b).

Clams and Mussels: Three of the most abundant species of bivalves are not native to the Bay, and all were accidental introductions. These are the soft-shell clam (*Mya arenaria*); the ribbed or horse mussel (*Ischadium demissum*); and the Japanese littleneck (*Tapes japonica*). The bay mussel (*Mytilis edulis*) is considered indigenous to this coast; however, some populations may represent exotic stocks introduced by way of ship fouling (D. Chivers pers. comm.; Carlton 1979).

The soft-shell clam was introduced into California with shipments of eastern oysters in 1869 or 1870 and first detected in 1874 (Fitch 1953). It became abundant and widely distributed in the Bay in the late 1880's and soon formed the bulk of the clam trade in San Francisco (Wooster 1968b). Apparently *Mya arenaria* largely displaced native clams in the Bay, especially the bent-nose clam (*Macoma nasuta*) in the South Bay, where prior to 1876, large numbers were harvested by Chinese fishermen (Skinner 1962; Weymouth 1920). According to Wilcox (1895), from 1889 to 1892, between 500 and 900 t of soft-shell clams were taken in the Bay each year (Fig. 5A). In 1899 Bay landings amounted to 695 t valued at $21,908 (Wilcox 1902).

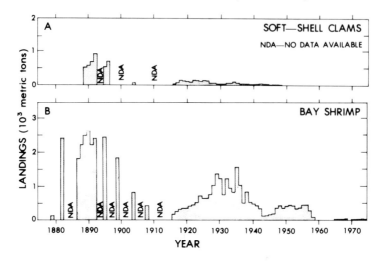

Fig. 5. Shell-fish commercial catches in San Francisco Bay: (A) Soft-shell clams (1889-1948); (B) Bay shrimp (1879-1974).

After the turn of the century, landings dropped considerably. In 1916 only 245 t were sold, and by 1927 the take had declined to 68 t (Wooster 1968b). By the early 1930's several beds had been destroyed or abandoned due to expiring leases and pollution (Bonnot 1932). The commercial fishery for soft-shell clams had disappeared by 1949. Its decline has been attributed to a variety of factors; domestic and industrial waste pollution, bay filling and construction; overharvesting; the high cost of manual labor required to dig the clams; and State allocation of some of the better clam beds for recreational use only (Bonnot 1932; Skinner 1962; Wooster 1968b).

During the time the soft-shell clam industry was collapsing, another exotic food clam, the Japanese littleneck (*Tapes japonica*) was becoming established. It was apparently introduced with shipments of Pacific oyster seed during the early 1930's, and since that time it appears to have taken over much of the habitat formerly occupied by the native littleneck or rock cockle, *Protothaca stamina* (Jones and Stokes 1977; Ricketts and Calvin 1968).

Soft-shell clam and Japanese littleneck are presently abundant in the Bay, and although subjected to intensive sport clamming in such areas as Foster City and Berkeley (despite warnings of the Public Health Department), these resources remain essentially unused (Jones and Stokes 1977; W. Dahlstrom pers. comm.).

The soft-shell clam prefers sheltered bays and a heavy mud substrate where there is some mixing of fresh and salt water (Fitch 1953). Heavy wave action is detrimental to the species (Matthiessen 1960). Recently over 1 million clams were destroyed in the Corte Madera-San Quentin area, possibly the result of wave scouring caused by the new Larkspur commuter ferry (R. McAllister pers. comm.).

The Japanese littleneck apparently can survive within a wide salinity range. It adapts well to extreme saline conditions and has also been found in the estuary where salinities were as low as 16 °/oo (Wooster 1968b). The species appears to prefer gravel bottom, and does not develop on substrates where no attachment is possible or where the young may be subjected to gill clogging (Filice 1958; Wooster 1968b).

The bay mussel, *Mytilis edulis,* like the soft-shell clam, contributed substantially to the Bay shellfish harvest in times past, especially prior to 1895. Between 1889 and 1892, San Francisco Bay mussel landings fluctuated between 950 and 1300 t per year (Wilcox 1895). Soon afterward the fishery suffered a severe decline which was attributed to reports of people becoming ill from eating bay mussels. At the present time it is utilized mostly for bait and occasionally for food. Although considered a delicacy in Europe where it is cultured commercially, it is not much sought after locally (Fitch 1953).

The ribbed mussel, *Ischadium demissum,* is another abundant mytilid in the Bay. It was apparently introduced with oyster shipments in the 1870's, and though it has occurred in San Francisco markets in the past, it has never been an important food item (Jones and Stokes 1977). Because of its high Vitamin D content, a proposal was made in the early 1960's to harvest *I. demissum* for use in freeze-dried form as a food additive, but the venture fell through because of Public Health restrictions (Aplin 1967; Jones and Stokes 1977).

The ribbed mussel lives in the high intertidal zone, most often in association with the native cord grass (*Spartina foliosa*), and the species is capable of filtering out great quantities of suspended matter from the water (Aplin 1967). The continual sedimentation effected by mussels may play a large part in salt-marsh development and mussel populations may also be important in the estuarine phosphate cycle by furnishing raw materials to deposit feeders which in turn regenerate the phosphate (Kuenzler 1961).

Paralytic shellfish poisoning (PSP) is commonly associated with mussels and results from the consumption of shellfish that have been feeding on toxin-producing dinoflagellates of the genus *Gonyaulax* (Ricketts and Calvin 1968). Most, if not all, cases of PSP in California are probably associated with the sea mussel, *Mytilis californianus,* not the ribbed or bay mussels; however, all mussels are under State quarantine from May through October, when blooms of *Gonyaulax* are known to occur (Jones and Stokes 1977).

Public Health Problems and the Future

The combination of rapid population growth and industrialization since 1900 has undeniably affected shellfish resouces in San Francisco Bay. Skinner (1962) points out that even before the turn of the century, pollution, siltation and ship wastes were hastening the decline of the fisheries in the Bay, and that shellfish were particularly vulnerable. He cites an 1878-79 Board of Fish Commissioners report by W. N. Lockington, who attributed the decline not just to overfishing and increased boat traffic, but in particular, to the "constant fouling of the waters and destruction of life by the foetid inpourings of our sewers . . ."

In 1932, bacterial levels in the Bay were so high that the State Board of Health passed a resolution establishing a general permanent quarantine on shellfishing in San Francisco Bay, but this quarantine was rescinded in 1953 (Jones and Stokes 1977).

No comprehensive sanitary survey has been conducted in the Bay although bacterial surveys are made at irregular intervals at various locations by certain governmental agencies. These surveys have revealed a significant improvement in water quality over the past 10 years (California Water Quality Control Board 1976); however, Jones and Stokes (1977) point out that results obtained during the past two drought years may be misleading due to the reduction of urban runoff, a significant contributor to high bacterial levels. Even if bacterial quality reaches acceptable levels, other problems still need to be resolved such as uncontrolled urban run-off during storms, presence of harmful viruses, uptake and concentration of harmful trace metals and other toxic substances; and sport-commercial allocation of the resource.

BAY SHRIMP

Three species of native shrimp occur in San Francisco Bay: *Crangon franciscorum, C. nigricauda,* and *C. nigromaculata.* The Korean shrimp, *Palaemon macrodactylus,* which was introduced accidentally in the early 1950's, has also become established in brackish waters of the Bay system (Ricketts and Calvin 1968). Of the crangonids, the largest and most abundant is *C. franciscorum,* followed by *C. nigricauda. C. nigromaculata* is far less numerous and is not considered in this chapter.

Bay shrimp are important forage for sport and market fishes. In the Bay they occur frequently in the diet of sturgeon (McKechnie and Fenner 1971) and striped bass (Johnson and Calhoun 1952). Moulting by bay shrimp and agitation of bottom sediments in their search for food and protection may also contribute to the cycling of nutrients (Krygier and Horton 1975).

Much of the life histories of *C. franciscorum* and *C. nigricauda* are similar (Israel 1936). Both species breed at the end of their first year. Females attain a larger size than males, and at maturity measure approximately 37 mm TL (*C. nigricauda*) and 53 mm TL (*C. franciscorum*). The eggs hatch in water of high salinity. Larval stages are planktonic until reaching 6 to 7 mm in length, at which time they settle to the bottom and move toward shallow water of reduced salinity. The earliest postlarval shrimp are found in brackish or nearly fresh waters of tidal flats or sloughs. As the shrimp develop and spawning season approaches, they move back into deeper, cooler, and more saline water.

Ovigerous females occur throughout the year, but Israel (1936) found that major spawning occurred from December to May or June for *C. franciscorum* and from April to September for *C.*

nigricauda. Krieger and Horton (1975), however, found a bimodal spawning pattern for the same species in Yaquina Bay, Oregon. Ganssle (1966) observed ovigerous *P. macrodactylus* in Suisun Bay during fall sampling.

Of the crangonids, *C. franciscorum* is the more tolerant of fresh water and has been found far up into the Delta, while the upper limit of *C. nigricauda* is in Suisun Bay, and then only in the fall with intrusion of salt water (Skinner 1962). This difference may be an important factor separating the two species and limiting competition, although it is possible that this balance may have been disrupted in some way with the introduction of the Korean shrimp.

Little data are available on the population size and species distribution in the Bay, although the present bait fishery appears to be far from utilizing the full potential of the resource (Frey 1971). Because bay shrimp are short-lived there may be large fluctuations in abundance from year to year, and, for the same reason, shrimp populations would be particularly sensitive to the effects of short-term pollution in the environment (Frey 1971).

History of the Fishery

Scofield (1919), Israel (1936), and Skinner (1962) have provided summaries of the development of the bay shrimp fishery. According to these authors, shrimp fishing in California was started in 1869 by Italian fishermen who employed 18 m long seines and sold their limited catch to local fish markets. The Chinese entered the fishery in 1871 with the far more efficient Chinese shrimp net or "bag net," which greatly increased the catch and promptly put the Italian fishermen out of business. The Chinese nets were funnel-shaped stationary traps, 9 m across, 12 m long and operated by tidal action. The local demand for shrimp was not great, but a profitable export trade was built up from the dried product which was shipped to the Orient. An estimated 1,500 Chinese were engaged in the fishery in 1875. Between 1882 and 1892, yearly catches averaged 2,270 t (Fig. 5B). In 1897, 26 Chinese shrimp camps were established at various locations around San Francisco Bay. This number was reduced to 19 camps by 1910.

The use of Chinese shrimp nets met with opposition from the beginning, because many juvenile food fishes were allegedly destroyed incidental to the shrimp catch, and later because it appeared that the shrimp resource was being rapidly depleted. Thus a series of laws was passed from 1910 to 1919 involving closed seasons, gear restrictions and processing limitations. After 1915 Chinese nets were allowed only in South Bay, while beam trawls were used elsewhere.

In following years the number of shrimp camps diminished, but the catch rose steadily until 1929, then fluctuated around 1000 t between 1930 and 1936, after which landings decreased steadily due to lack of a market for dried shrimp (Frey 1971).

In 1965 the fishery was revived to supply bait for striped bass and sturgeon sportfishing. Today there are about 15 boats in the fishery, and all but a few fish with beam trawls. Most fishing takes place in San Pablo and Suisun bays, and to a limited extent in South Bay. Although the size of the fishery is small, the business can be lucrative, as bait shrimp brings a price of $2.00 to $2.50 per pound (0.45 kg) to the fisherman. It is sold both frozen and live, the latter being in the most demand and bringing the highest price. The small size of bay shrimp appears to be the major factor limiting the demand for them as food, and at this time it does not appear that they can be processed economically for sale on a large scale.

DISCUSSION

The foregoing account illustrates the changing fortunes of the fishery resources of San Francisco Bay. The commercial landings, affected by regulations and environmental changes as well as market demand and resource availability, have undergone drastic changes not only in quantity but also in the kinds of fishery resources harvested from the Bay. Formerly, bivalves, shrimp, salmon, and sturgeon, as well as a number of other finfishes, provided a substantial income to Bay Area fishermen. Now the only remaining commercial fisheries of note within the Bay are those for herring, anchovy and bay shrimp, and the latter two are used almost exclusively as bait.

Recreational fisheries, alternatively, have fared somewhat better, due primarily to legislation which has restricted commercial fishing for certain species and outlawed some fishing gear detrimental to stocks of incidentally caught fishes. Striped bass and sturgeon are now reserved exclusively for the sport fishery, as is chinook salmon fishing within the Bay.

Human activity is clearly implicated in the decline of much of the fishery resources of San Francisco Bay. Although the full effects of dumping, dredging and filling are not clear, we can safely surmise that filling of shallow mud flats around the perimeter of the Bay has drastically reduced the amount of suitable habitat for such forms as oyster, clams, and bay shrimp. Pollution has degraded the purity of water to the extent that even now after considerable effort to improve water quality, commercial and recreational use of molluscs is still hazardous and by and large discouraged or disallowed. In past years intensive harvesting also contributed to the decline of many Bay fisheries.

Anadromous fishes have suffered from the damming and diversion of rivers which resulted in elimination or alteration of spawning and nursery habitats. Changes in the hydrography of the Delta caused by diversions, and pollution from industrial, agricultural, and municipal waste discharges may not only affect these fishes directly but may also affect the distribution and abundance of forage species upon which they depend.

Research Needs

A review of the literature discloses that except for bivalves and a few finfish (striped bass, herring, sturgeon, salmon) little quantitative data are available on the fishery resources of San Francisco Bay. Neither is there much information on the life history of most of the animals which reside in the Bay, whether they be seasonal migrants or residents. There is also a need to study trophic relationships and interspecific interactions in order to better gauge the effects of changes in the environment.

The Future

It is likely that reduction in duration and frequency of fresh-water flows into and out of the Delta, caused by increasing demands for water for agricultural, industrial and domestic use, will further affect anadromous fish stocks unless steps are taken to prevent or replace losses caused by reduced flows. The gradual decline in the amount of fresh-water flow out of the Delta will probably alter salinity regimes in the Bay, which may change the distribution of certain species and possibly the migratory habits of anadromous fishes and invertebrates such as the salinity-regulated crangonids.

The full effects of the recent drought (1976-77) on anadromous fishes is not known, and may not be felt for years to come when the fish spawned during the past few years begin to enter the fisheries. There may be a severe decline in the population resulting from poor spawning and year-class survival.

On the positive side, if water quality continues to improve in San Francisco Bay proper as it has over the past 10 years, we may see increased shellfishing in the Bay and perhaps the opening up of certain clamming areas that are now considered restricted by the California Department of Public Health.

Prediction of what will happen in the future is difficult if not impossible, not only because of the complexity of the estuary itself and lack of knowledge of environmental requirements of many Bay species, but also because the San Francisco Bay estuary is obviously no longer a natural system. Factors such as water quality, water flow and habitat conditions are now largely under human control. Resource-related decisions made now and in the future will ultimately determine the fate of the Bay's fisheries. The value of these resources, perhaps now more than ever, will have to be weighed carefully against land and water use demands.

ACKNOWLEDGMENTS

For providing us with certain fisheries information, we are indebted to cour colleagues from the California Department of Fish and Game: Dave Kohlhorst, Bob McAllister, Pat O'Brien, Mac Oliphant, Dick Painter and Jerry Spratt. Don Stevens and Walt Dahlstrom of DFG reviewed the manuscript and made valuable comments. We also wish to thank Dustin Chivers, California Academy of Sciences, and Bill Leet, NMFS, Tiburon Laboratory, for their assistance.

LITERATURE CITED

Aplin, J. A. 1967. Biological survey of San Francisco Bay 1963-1966. Calif. Fish Game MRO Ref. 67-4. 131 pp.

Bonnot, P. 1932. Soft-shell clam beds in the vicinity of San Francisco Bay. Calif. Fish Game 18(1):64-66.

Bonnot, P. 1935. The California oyster industry. Calif. Fish Game 21(1):65-80.

Bonnot, P. 1938. Report on the California oyster industry for 1937. Calif. Fish Game 24(2): 191-195.

Boothe, P. 1967. The food and feeding habits of four species of San Francisco Bay fish. Calif. Fish Game MRO Reference No. 67-13. 155 pp.

Byers, R. D. 1940. The California shark fishery. Calif. Fish Game 26(1):23-28.

California Water Quality Control Board. 1976. Report on bacteriological survey of San Francisco Bay, August 3, 4, and 10, 1967. CWQCB San Francisco Bay Region. 26 pp.

Carlton, J. T. 1979. Introduced invertebrates of San Francisco Bay. Pages 427-444 in T. J. Conomos, ed. San Francisco Bay: The Urbanized Estuary. Pacific Division, Amer. Assoc. Advance. Sci., San Francisco, Calif.

Chadwick, H. K. 1962. Catch records from the striped bass sport fishery in California. Calif. Fish Game 48(3):153-177.

Clark, G. H. 1929. Sacramento-San Joaquin salmon (*Oncorhynchus tschawytscha*) fishery of California. Calif. Fish Game, Fish Bull. 17. 73 pp.

Collins, J. W. 1892. Report on the fisheries of the Pacific Coast of the United States. Pages 3-269 in Report of the U. S. Commission of Fish and Fisheries for 1888, Part II. U. S. Govt. Printing Office, Washington, D. C.

Earnest, R. D., and P. E. Benville, Jr. 1971. Correlation of DDT and lipid levels for certain San Francisco Bay fish. Pesticides Monitor J. 5(3):235-241.

Eigenmann, C. H. 1894. On the viviparous fishes of the Pacific Coast of North America. Pages 381-478 in Bull. U. S. Fish Comm. for 1892, vol. XII. U. S. Govt. Printing Office, Washington, D. C.

Eldridge, M. B. 1977. Factors influencing distribution of fish eggs and larvae over eight 24-hr samplings in Richardson Bay, California. Calif. Fish Game 63(2):101-116.

Eldridge, M. B., and W. M. Kaill. 1973. San Francisco Bay area's herring resource—a colorful past and a controversial future. Mar. Fish Rev. 38(11):25-31.

Engen, P. C. 1968. Organogenesis in the walleye surfperch, *Hyperprosopon argenteum* (Gibbons).

Calif. Fish Game 54(3):156-169.

Filice, F. P. 1958. Invertebrates from the estuarine portion of San Francisco Bay and some factors influencing their distribution. Wasmann J. Biol. 16(2):159-211.

Fitch, J. E. 1953. Common marine bivalves of California. Calif. Fish Game, Fish Bull. 90. 102 pp.

Frey, H. W., ed. 1971. California's living marine resources and their utilization. Calif. Fish Game, Sacramento, Calif. 148 pp.

Fry, D. H., Jr. 1973. Anadromous fishes of California. Calif. Fish Game, Sacramento. 111 pp.

Ganssle, D. 1966. Fishes and decapods of San Pablo and Suisun bays. Pages 64-94 in D. W. Kelley, ed. Ecological Studies of the Sacramento-San Joaquin Estuary. Calif. Fish Game, Fish Bull. 133.

Gilbert, G. K. 1917. Hydraulic-mining debris in the Sierra Nevada. U. S. Geol. Surv. Prof. Paper 105. 154 pp.

Gill, G. S., E. C. Gray, and D. Sechler. 1971. The California water plan and its critics: A brief review. Pages 3-27 in D. Sechler, ed. California Water: A Study in Resource Management. University of California, Berkeley, Calif.

Green, R. E. 1975. A preliminary list of fishes collected from Richardson Bay, California, 1972-73. Calif. Fish Game 61(2):104-106.

Hallock, R. J. 1977. A description of the California Department of Fish and Game management program and goals for the Sacramento river system salmon resources. Calif. Fish Game, Anadromous Fish. Br., Sacramento, Calif. 16 pp.

Hart, J. L. 1973. Pacific fishes of Canada. Fish. Res. Bd. Can. Bull 180. 740 pp.

Hart, J. L., and A. L. Tester. 1934. Quantitative studies on herring spawning. Trans. Amer. Fish. Soc. 64:307-312.

Herald, E. S., and W. E. Ripley. 1951. The relative abundance of sharks and bay stingrays in San Francisco Bay. Calif. Fish Game 37(3):315-329.

Heubach, W. 1968. Environmental requirements and future of king salmon (*Oncorhynchus tshawytscha*) in the Sacramento-San Joaquin estuary. Pages 77-121 in Task Report VII-1B. Prepared for the San Francisco Bay-Delta Water Quality Control Program by Calif. Fish Game, Sacramento, Calif.

Israel, H. R. 1936. A contribution toward the life histories of two California shrimps, *Crango franciscorum* (Stimpson) and *Crango nigricauda* (Stimpson). Calif. Fish Game, Fish Bull. 46. 28 pp.

Jensen, P. 1972. King salmon. Pages 44-51 in J. E. Skinner, ed. Ecological Studies of the Sacramento-San Joaquin Estuary. Decennial Report 1961-1971. Calif. Fish Game, Delta Fish and Wildlife Protection Study Report No. 8.

Johnson, W. C., and A. J. Calhoun. 1952. Food habits of California striped bass. Calif. Fish Game 38(4):531-533.

Jones and Stokes Assoc., Inc. 1977. San Francisco Bay shellfish: An assessment of the potential for commercial and recreational harvesting (draft report). Prepared for ABAG (Association of Bay Area Governments), Berkeley, Calif. 156 pp.

Kelley, D. W. 1966. Description of the Sacramento-San Joaquin estuary. Pages 8-17 in D. W. Kelley, ed. Ecological Studies of the Sacramento-San Joaquin Estuary. Calif. Fish Game, Fish. Bull. 133.

Kelley, D. W. 1968a. American shad. Pages 239-251 in Task Report VII-1B. Prepared for San Francisco Bay-Delta Water Quality Control Program by Calif. Fish Game, Sacramento, Calif.

Kelley, D. W. 1968b. Sturgeon. Pages 259-268 in Task Report VII-1B. Prepared for San Francisco Bay-Delta Water Quality Control Program by Calif. Fish Game, Sacramento, Calif.

Kohlhorst, D. W. 1976. Sturgeon spawning in the Sacramento River in 1973, as determined by distribution of larvae. Calif. Fish Game 62(1):32-40.

Krone, R. B. 1979. Sedimentation in the San Francisco Bay system. Pages 85-96 in T. J. Conomos, ed. San Francisco Bay: The Urbanized Estuary. Pacific Division, Amer. Assoc. Advance. Sci., San Francisco, Calif.

Krygier, E. E., and H. F. Horton. 1975. Distribution, reproduction, and growth of *Crangon nigricauda* and *Crangon franciscorum* in Yaquina Bay, Oregon. Northwest Sci. 49(4):216-240.

Kuenzler, E. J. 1961. Structure and energy flow of a mussel population in a Georgia salt marsh (*Modiolus*). Limnol. Oceanogr. 6(2):191-204.

Mattheisson, G. C. 1960. Intertidal zonation in populations of *Mya arenaria*. Limnol. Oceanogr. 5:381-388.

McAllister, R. 1976. California marine fish landings for 1974. Calif. Fish Game, Fish Bull. 166. 53 pp.

McKechnie, R. J., and R. B. Fenner. 1971. Food habits of white sturgeon (*Acipenser transmontanus*) in San Pablo and Suisun bays, California. Calif. Fish Game 57(3):209-212.

Menchen, R. S. 1977. A description of the California Department of Fish and Game management program and goals for the San Joaquin River system salmon resource. Calif. Fish Game, Anadromous Fish. Br., Sacramento, Calif. 21 pp.

Messersmith, J. 1966. Fishes collected in Carquinez Strait. Pages 57-63 *in* D. W. Kelley, ed. Ecological Studies of the Sacramento-San Joaquin Estuary. Calif. Fish Game, Fish Bull. 133.

Miller, D. J., and R. N. Lea. 1972. Guide to the coastal marine fishes of California. Calif. Fish Game, Fish Bull. 157. 235 pp.

Miller, D. J., and J. Schmidtke. 1956. Report on the distribution and abundance of Pacific herring (*Clupea pallasi*) along the coast of central and southern California. Calif. Fish Game 42(3):163-187.

Miller, L. W. 1972a. White sturgeon. Pages 54-56 *in* J. E. Skinner, ed. Ecological Studies of the Sacramento-San Joaquin Estuary. Decennial Report 1961-1971. Calif. Fish Game, Delta Fish and Wildlife Protection Study Report. No. 8.

Miller, L. W. 1972b. Migrations of sturgeon tagged in the Sacramento-San Joaquin estuary. Calif. Fish Game 58(2):102-106.

Nichols, F. H. 1979. Natural and anthropogenic influences on benthic community structure in San Francisco Bay. Pages 409-426 *in* T. J. Conomos, ed. San Francisco Bay: The Urbanized Estuary. Pacific Division, Amer. Assoc. Advance. Sci., San Francisco, Calif.

Nidever, H. B. 1916. Shad in California. Calif. Fish Game 2(2):59-64.

Orcutt, H. G. 1950. The life history of the starry flounder *Platichthys stellatus* (Pallas). Calif. Fish Game, Fish Bull. 78. 64 pp.

Orsi, J. J., and A. C. Knutson, Jr. 1979. The role of mysid shrimp in the Sacramento-San Joaquin estuary and factors affecting their abundance and distribution. Pages 401-408 *in* T. J. Conomos, ed. San Francisco Bay: The Urbanized Estuary. Pacific Division, Amer. Assoc. Advance. Sci., San Francisco, Calif.

Painter, R. E. 1966. Zoobenthos of San Pablo and Suisun bays. Pages 40-56 *in* D. W. Kelley, ed. Ecological Studies of the Sacramento-San Joaquin Estuary. Calif. Fish Game, Fish Bull. 133.

Rechnitzer, A. B., and C. Limbaugh. 1952. Breeding habits of *Hyperprosopon argenteum*, a viviparous fish of California. Copeia 1952(1):41-42.

Ricketts, E. F., and J. Calvin. 1968. Between Pacific tides. 4th ed. Revised by J. W. Hedgpeth. Stanford University Press, Stanford, Calif. 614 pp.

Ripley, W. E. 1946. The soupfin shark and the fishery. Pages 7-37 *in* The Biology of the Soupfin *Galeorhinus zyopterus* and Biochemical Studies of the Liver. Calif. Fish Game, Fish Bull. 64.

Russo, R. A. 1975. Observations of the food habits of leopard sharks (*Triakis semifasciata*) and brown smoothhounds (*Mustelus henlei*). Calif. Fish Game 61(2):95-103.

Russo, R. A., and E. S. Herald. 1968. The 1967 shark kill in San Francisco Bay. Calif. Fish Game 54(3):215-216.

Scofield, N. B. 1918. The herring and development of the herring industry in California. Calif. Fish Game 4(2):65-70.

Scofield, N. B. 1919. Shrimp fisheries of California. Calif. Fish Game 5(1):1-12.

Scofield, W. L. 1956. Trolling gear in California. Calif. Fish Game Fish Bull. 103. 45 pp.

Scofield, W. L. 1957. Marine fisheries dates. Informal report. Calif. Fish Game, Sacramento, Calif. 81 pp.

Skinner, J. E. 1962. An historical review of the fish and wildlife resources of the San Francisco Bay Area. Calif. Fish Game, Water Project Branch Report No. 1. 226 pp.

Skinner, J. E. 1972a. Water development in the delta. Pages 14-18 *in* J. E. Skinner, ed. Decennial Report 1961-1971. Calif. Fish Game, Delta Fish and Wildlife Protection Study Report No. 8.

Skinner, J. E. 1972b. Evaluation of water plans in relation to fish and wildlife. Pages 75-79 *in* J. E. Skinner, ed. Decennial Report 1961-1971. Calif. Fish Game, Delta Fish and Wildlife Protection Study Report No. 8.

Smith, H. M. 1895. Notes on a reconnaissance of the fisheries of the Pacific Coast of the United States in 1894. Bull. U. S. Fish Comm. 14:233-288.

Squire, J. L., Jr. and S. E. Smith. 1977. Anglers' guide to the United States Pacific Coast. U. S. Dept. Commerce, NOAA, National Marine Fisheries Service, U. S. Govt. Printing Office, Washington, D. C. 139 pp.

Stevens, D. E. 1972. American shad. Pages 54-54 *in* Ecological Studies of the Sacramento-San Joaquin Estuary. Decennial Report 1961-1971. Calif. Fish Game, Delta Fish and Wildlife Protection Study Report 8.

Stevens, D. E. 1977. Striped bass (*Morone saxatilis*) year class strength in relation to river flow in the Sacramento-San Joaquin estuary, California. Trans. Amer. Fish. Soc. 106(1):34-42.

Stevens, D. E. 1979. Environmental factors affected striped bass (*Morone saxatilis*) in the Sacramento San Joaquin estuary. Pages 469-478 *in* T. J. Conomos, ed. San Francisco Bay: The Urbanized Estuary. Pacific Division, Amer. Assoc. Advance. Sci., San Francisco, Calif.

Stevens, D. E., and L. W. Miller. 1970. Distribution of sturgeon larvae in the Sacramento-San Joaquin River estuary. Calif. Fish Game 56(2):80-86.

Townsend, C. H. 1893. Report of observations respecting the oyster resources and oyster fishery of the Pacific Coast of the United States. Pages 343-372 *in* U. S. Comm. of Fish and Fisheries Report for 1889-91. U. S. Govt. Printing Office, Washington, D. C.

Wales, J. H. 1929. A note on the breeding habits of the viviparous perch *Damalichthys*. Copeia (172):57-58.

Wares, P. G. 1971. Biology of the pile perch, *Rhacochilus vacca*, in Yaquina Bay, Oregon. Bur. Sport Fish. and Wildlife Tech. Paper 57. U. S. Fish and Wildlife Service, Washington, D. C. 21 pp.

Weymouth, F. W. 1920. The edible clams, mussels and scallops of California. Calif. Fish Game, Fish Bull. 4. 74 pp.

Wilcox, W. A. 1895. Fisheries of the Pacific Coast. Pages 139-304 *in* Report of the Commissioner for the Year Ending June 30, 1893. U. S. Comm. Fish and Fisheries. U. S. Govt. Printing Office, Washington, D. C.

Wilcox, W. A. 1898. Notes on fisheries of the Pacific Coast in 1895. Pages 575-659 *in* Report of the Commissioner for Year Ending June 30, 1896. U. S. Comm. Fish and Fisheries. U. S. Govt. Printing Office, Washington, D.C.

Wilcox, W. A. 1902. Notes on fisheries of the Pacific Coast in 1899. Pages 501-574 *in* Report of the Commissioner for Year Ending June 30, 1901. U. S. Comm. Fish and Fisheries. U. S. Govt. Printing Office, Washington, D.C.

Wooster, R. W. 1968a. Shore and pier fishery. Pages 161-195 *in* Task Report VII-B. Prepared for San Francisco Bay-Delta Water Quality Control Program by Calif. Fish Game, Sacramento, Calif.

Wooster, R. W. 1968b. Clams, native oyster and mussels. Pages 123-145 *in* Task Report VII-B. Prepared for San Francisco Bay-Delta Water Quality Control Program by Calif. Fish Game, Sacramento, Calif.

Wooster, R. W. 1968c. The San Francisco Bay commercial oyster industry. Pages 147-159 *in* Task Report VII-B. Prepared for San Francisco Bay-Delta Water Quality Control Program by Calif. Fish Game, Sacramento, Calif.

ENVIRONMENTAL FACTORS AFFECTING STRIPED BASS (*MORONE SAXATILIS*) IN THE SACRAMENTO-SAN JOAQUIN ESTUARY

DONALD E. STEVENS
California Department of Fish and Game
4001 North Wilson Way, Stockton, California 95205

The introduction of striped bass to the Sacramento-San Joaquin Estuary has established one of California's most important recreational fisheries. Striped bass requirements are being studied so they can be considered in planning for future water project and other development in the estuary. Study results have shown that striped bass spawning is affected by annual variations in salinity in the San Joaquin River, and survival of the young and subsequent recruitment to the fishery are related to the magnitudes of water diversions from the nursery area and river flows.

Mortality caused by entrainment in power plant cooling systems probably has been low relative to that caused by water development, but losses due to entrainment at power plants may be increasing because increased power production is causing lethal temperatures to occur more frequently.

Factors other than angling kill about 15 to 30% of the adult bass each year. Part of this mortality occurs during large die-offs in the Suisun-San Pablo Bay area when bass are migrating back to salt water after spawning. Attempts to determine the cause of these kills have been unsuccessful.

Reduced flows resulting from water development may change water circulation patterns in San Francisco and San Pablo bays. Such changes potentially affect the abundance and distribution of forage for adult bass.

Water management actions that would benefit the striped bass resource include maintaining adequate freshwater flows through the estuary and moving the intakes for the federal and state water diversions to a location upstream from the nursery area.

Striped bass (*Morone saxatilis*) were introduced to the Sacramento-San Joaquin Estuary from the Atlantic Coast in 1879 (Skinner 1962; Smith and Kato 1979). They increased at a phenomenal rate: hundreds were caught 10 years after the introduction, and after 20 years over 540,000 kg were landed. From 1916 to 1935 when commercial fishing was outlawed, the commercial catch ranged between 225,000 and 450,000 kg annually.

As a result of the introduction, the striped bass fishery in the Sacramento-San Joaquin Estuary has long been one of California's top ranking sport fisheries. Presently about 200,000 anglers fish for striped bass each year and catch about 300,000 fish. In the only significant economic study, the Stanford Research Institute projected an annual net value of 7.5 million dollars for this fishery in 1970 (Altouney, Crampon, and Willeke 1966). Hence, striped bass are a major recreational and economic asset.

The viability of the striped bass resource depends on environmental conditions which have been and are being altered by water projects and other development. Due to potential impacts of future development, the California Department of Fish and Game (DFG) in cooperation with the U. S. Fish and Wildlife Service (USFWS), California Department of Water Resources (DWR), and U. S. Bureau of Reclamation (USBR) has been studying the environmental requirements of striped

SAN FRANCISCO BAY

bass. This chapter reviews the major findings of those studies.

THE ESTUARY

The Sacramento-San Joaquin river system forms a tidal estuary (Fig. 1). Approximately 1,130 km of channels interlace the Delta at the junction of the rivers. These channels vary in width from around 50 m to about 1.5 km, and generally they are less than 15 m deep. Water flowing to the ocean passes through Suisun, San Pablo, and San Francisco bays. A vast area of these bays is less than 2 m deep at mean lower-low tide. However, the channels range up to 100 m deep in San Francisco Bay just inside the Golden Gate.

The salinity gradient generally is about 80 km long extending from San Pablo Bay to the western Delta (see also Conomos 1979). River flows into the Delta are quite variable and are partially controlled by upstream reservoirs. Inflows peak in winter and spring. Water development in the system now removes about half of the flow that would normally go to the ocean. Water is exported from the southern Delta via two large pumping plants. One is a 130 m$^3 \cdot$s^{-1} plant built by the USBR in 1951; the other is a 170 m$^3 \cdot$s^{-1} plant completed by the DWR in 1968. The flow

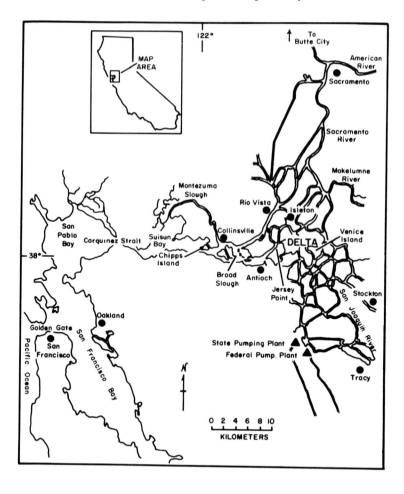

Fig. 1. The Sacramento-San Joaquin Estuary.

reductions and removal of water from the Delta affect the salinity gradient and cause changes in seasonal and geographical flow patterns in the various channels.

FISHERY

Tagging studies indicate abundance of bass larger than the 40.6-cm minimum legal size ranged from about 1.6 to 1.9 million fish from 1969 to 1975 (Stevens 1977a). Such abundance measurements are not available for prior years, but sport fishery records provide evidence that bass were two to three times more abundant in the early 1960's than at present (Stevens 1977b).

Since 1958 anglers have harvested from 11 to 37% of the legal population each year (Chadwick 1968; Miller 1974; Stevens 1977b; and unpublished data). From 1969 to 1975 an average of 60% of the catch was from San Francisco Bay, San Pablo Bay, and Carquinez Strait (unpublished data). About 2% of the catch was from the Pacific Ocean. The remainder of the catch came from Suisun Bay, the Delta, and the rivers upstream from the Delta.

Bass migrations cause the fishery to be seasonal throughout the estuary. From 1969 to 1975 more than 90% of the catch west of Suisun Bay was taken during summer and fall; whereas, almost 60% of the catch upstream was taken during spring.

LIFE HISTORY

Tagging studies demonstrate that most adults move to fresh water (the Delta or upstream in the Sacramento River) to spawn during spring. After spawning, adult bass return to salt water (San Pablo and San Francisco bays and the Pacific Ocean within about 32 km of the Golden Gate). Some adults begin moving back upstream toward fresh water again during fall. Others overwinter in the bays and move back upstream just before spawning during spring (Calhoun 1952; Chadwick 1967; Orsi 1971).

Striped bass are prolific. A 60-cm long female spawns about 700,000 eggs (Lewis and Bonner 1966). Spawning occurs from early April to mid-June and primarily in two areas, the San Joaquin River between Antioch point and Venice Island and the Sacramento River from Isleton to Butte City about 240 km upstream (Farley 1966; Turner 1976).

Bass eggs are semi-buoyant and drift with the water currents until they hatch in 2 to 3 days. As a result of the rapid downstream flow in the Sacramento River, most bass larvae reach the Delta before they have significant swimming ability or have started to feed. Mortality rates are high from the time eggs are spawned through the middle of the first winter (DFG et al. 1974). Probably only about one out of every 100,000 eggs survives to the end of the year (unpublished data).

Young bass abundance typically is greatest in the zone where fresh and salt water initially mix, presumably indicating better conditions for survival there (Turner and Chadwick 1972). Massmann (1963) referred to this region as the "critical zone" in estuaries, because it is the principal nursery area for many fishes. In the Sacramento-San Joaquin Estuary this zone is more productive than areas up or downstream; it has been variously termed the "null zone" (Conomos and Peterson 1974; Peterson et al. 1975) or the "entrapment zone" (Arthur and Ball 1979). At moderate flows this zone is located in the Suisun Bay area, and at low flows it is in the Delta (Turner and Chadwick 1972; Arthur and Ball 1979; Conomos 1979). Generally, the greatest densities of the principal food organisms of young bass (the opossum shrimp, *Neomysis mercedis* and copepod *Eurytemora* sp.) also occur in or near this zone (Heubach 1969; Orsi and Knutson 1979; DFG et al. 1975).

During their second year, many bass still live in the Delta and Suisun Bay, but others move into the rivers above the Delta and downstream into San Pablo Bay. They generally change from an invertebrate to a fish diet, although *Neomysis* is still important. In the Delta, threadfin shad (*Dorosoma petenense*) and young striped bass are the primary fish eaten (Stevens 1966).

Male bass mature when they are 2 or 3 years old, while females mature at 4 or 5 years. Once bass mature they take up the adult migratory pattern.

ENVIRONMENTAL FACTORS AFFECTING THE POPULATION

Water Quality and Spawning

Striped bass spawning and salinity of the rivers was monitored most years from 1963 to 1977. Salinity always was less than 200 mg·liter^{-1} total dissolved solids (TDS)[1] in the Sacramento River spawning area. Salinity generally was less than 200 mg·liter^{-1} TDS ($\leqslant$0.5 °/oo) in the San Joaquin River spawning area, although low river flows sometimes allowed higher salinity water to intrude from the west. Usually this area is less salty than the river either up or downstream because fresh water flowing from the Mokelumne and Sacramento rivers dilutes saltier agricultural return water coming from upstream and ocean water coming from downstream.

Several findings from the field monitoring and also laboratory experiments indicate salinity adversely affects bass spawning. Radtke and Turner (1967) reported that potential spawners are repelled by the salty, agricultural return water in the San Joaquin River upstream from Venice Island. More recently, L. W. Miller (pers. comm.) found that few bass spawned in the San Joaquin River during 1977 when the intrusion of ocean water caused salinities to exceed 5,000 mg·liter^{-1} TDS ($\approx$5 °/oo) in the usual spawning area. Bass were not deterred from spawning in this area by ocean salts causing TDS of 1,500 mg·liter^{-1} ($\approx$2 °/oo) in 1968 and 1972 (Turner 1976), but laboratory experiments indicate egg survival declines markedly when the salinity of the water in which they harden exceeds 1,000 mg·liter^{-1} TDS ($\approx$1 °/oo) (Turner and Farley 1971). Hence, water fresher than 1,000 mg·liter^{-1} TDS apparently is essential for optimum spawning success.

Effects of River Flow and Water Diversions on Young Bass Survival

Abundance of young bass has been monitored by a tow net survey conducted annually since 1959 (except 1966). Turner and Chadwick (1972) developed annual indices of young bass survival for 1959 to 1970 from this survey. These indices were directly correlated with outflow from the Delta[2] during all combinations of months from April to July. Survival was best correlated with flows for June and July combined. The variations in young bass survival appear to be important in determining subsequent recruitment to the fishery (Stevens 1977b; Chadwick et al. 1977).

From 1971 to 1976, young bass survival consistently was poorer than expected from Turner and Chadwick's analysis of survival and flow from 1959 to 1970 (Fig. 2) (Chadwick et al. 1977). The recent decrease in survival occurred solely in the Delta which is the farthest upstream portion of the nursery area. During this period, average May, June, and July water exports increased 83, 60, and 52% above 1959 to 1970 levels.

The findings led to a multiple regression of indices of young bass survival in the Delta against diversion rates and outflows from 1959 to 1976 (Fig. 3) (Chadwick et al. 1977). The analysis indicated that the survival-flow correlations are caused partly by local, state, and federal diversions removing more fish and perhaps their food organisms from the Delta when outflows are low than

---

[1] This value converts to a very approximate salinity value of $\leqslant$0.5 °/oo (at water temperature of 20°C.) Other approximations are inserted parenthetically in text. (Ed.)

[2] The Delta outflow index is the mean calculated daily outflow past Chipps Island. The data are partly from DWR Water Supervision and Water Flow bulletins and partly supplied directly by DWR personnel. See also Conomos (1979).

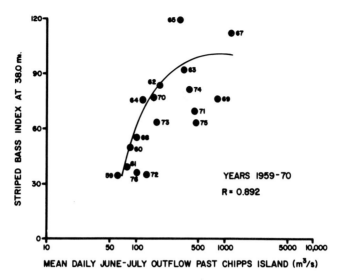

Fig. 2. Relation between abundance of young striped bass from Carquinez Strait upstream through the Delta, Y, and mean daily Delta outflow during June and July. Numbers indicate year from 1959 to 1976. The regression equation for 1959-1970 is Y = -488.3 + 429.2 (log mean daily June-July outflow) - 77.9 (log mean daily June-July outflow)^2; R = 0.892. (From Chadwick et al. 1977).

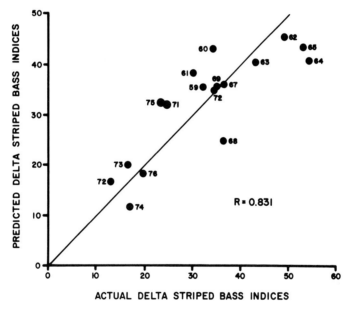

Fig. 3. Relation between actual abundance of young striped bass in the Delta and abundance predicted from May-June diversions and Delta outflow. Numbers indicate year from 1959 to 1976. The regression equation used to obtain the Y-axis coordinates is Y = -202.7 -0.25 (mean daily May-June diversions) + 225.9 (log mean daily May-June outflow) - 43.36 (log mean daily May-June outflow)^2; R = 0.831. Outflows in $m^3 \cdot s^{-1}$. (From Chadwick et al. 1977).

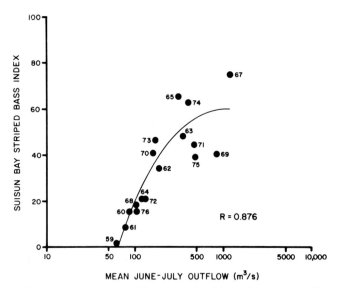

Fig. 4. Relation between abundance of young striped bass downstream from Collinsville (Suisun Bay index), Y, and mean daily Delta outflow during June and July. Numbers indicate year from 1959 to 1976. The regression equation for 1959-1970 is Y = -294.3 + 234.8 (log mean daily June-July outflow) - 39.0 (log mean daily June-July outflow)^2; R = 0.876. (From Chadwick et al. 1977).

when outflows are high. Fewer fish are diverted when flows are high because the diversions take a smaller fraction of the flow carrying eggs and young and high flows transport more fish to Suisun Bay where there are few diversions (Fig. 4) (Chadwick et al. 1977).

Bass are removed from the Delta by local diversions because most of those diversions are not screened (Allen 1975). The export diversions have louver screens, but as pumping increases these diversions remove more bass because the screens are ineffective on bass too small to swim well. The screens do not attain 50% efficiency until the bass grow to 19 mm long (about 1-mo old). Above 19 mm, screen efficiency increases gradually to about 85% for bass longer than 100 mm (about 5-mo old) (Skinner 1974).

Another effect of water export pumping is that it causes high flow velocities in the channels which convey water from the Sacramento River to the pumping plants in the southern Delta. High velocities reduce standing crops of important bass food organisms (copepods, cladocerans, and *Neomysis mercedis*) (Turner 1966; Heubach 1969).

The survival-flow relations apparently are not caused solely by diversions, however. There is evidence that high flows enhance survival in other ways. From 1938 to 1954, before significant water exports existed, recruitment of bass to the fishery was correlated with outflow when the recruits were young (Stevens 1977b), and it is unlikely that local diversions caused the correlation (Chadwick et al. 1977). Also, in 1977, there was abnormally low survival of young bass which was associated with extremely low outflows (Table 1). The 1977 results are not explained by diversion rates (L. Miller pers. comm.).

The way in which flow controls the spatial distribution of young bass in the estuary may be a major mechanism controlling the survival-flow relations. When high flows disperse young bass over more of the estuary, competition for food may be reduced (Stevens 1977b; Chadwick et al. 1977).

Recent studies (Orsi and Knutson 1979) also reveal that when flows are low, standing crops of foods are reduced in the critically important fresh-salt water mixing zone. Delta outflow was considerably lower than normal throughout 1976 and 1977, and the mean 1976-77 summer

TABLE 1. YOUNG BASS ABUNDANCE AND OUTFLOW[a] FROM THE SACRAMENTO-SAN JOAQUIN DELTA.

	1977	Mean (1959-1976)	Previous low
Bass abundance index (units)	6.7	67.7	33.9 (1959)
Mean April to July outflow ($m^3 \cdot s^{-1}$)	86	573	140.0 (1976)

[a] Mean calculated daily outflow past Chipps Island.

abundance index for *Neomysis mercedis* was only about 25% of the 1968-1975 mean (A. Knutson pers. comm.). Indices of smaller zooplankton abundance also were below normal in 1976 and 1977 (J. Orsi pers. comm.).

Effects of Power Plants and Other Industry on Young Bass Survival

Chadwick and Stevens (1971) and Chadwick et al. (1977) reviewed available striped bass temperature tolerance data and operation of the two Pacific Gas and Electric Company (PG&E) power plants which use up to 90 $m^3 \cdot s^{-1}$ of water from the striped bass nursery area for "once through" cooling. We concluded that mortality caused by these plants was minimal compared to that caused by the water project and agricultural diversions and other environmental factors associated with outflow. We based this conclusion on: (1) entrained bass usually were not exposed to lethal temperatures, (2) field tests suggested few fish were mechanically damaged, and (3) power plant effects were not evident in the statistical relations between young bass survival and flow and diversion rates even though there were two significant increases in cooling water demand (totaling 48% of the present demand) during the survival study.

Although power plants apparently have not been a major factor, evidence that losses of young bass are not nullified by compensatory mechanisms later in life (Chadwick et al. 1977) suggests that losses due to entrainment have had some, albeit perhaps small, impact on the fishery. The impact probably is increasing, as increased power production is causing lethal temperatures to occur more frequently (Finlayson and Stevens 1977; PG&E 1977).

Several other industries use water from the western Delta and/or have discharged wastes into the nursery area while bass survival has been monitored. However, their total water use is small (about 2 $m^3 \cdot s^{-1}$ during summer), compared to the amount used for export (up to 300 $m^3 \cdot s^{-1}$) by local agriculture (about 110 $m^3 \cdot s^{-1}$ during the summer), and by PG&E. Effluent standards also have become more stringent in recent years. Hence, presently, local industry probably does not significantly affect young bass survival.

Factors Affecting Adult Bass

Factors other than angling kill about 15 to 30% of the adult bass each year (Chadwick 1968; Miller 1974; Stevens 1977b). Part of this mortality is due to natural phenomena such as disease and old age and part probably is caused by the activities of man.

For at least 25 years, an unknown fraction of adult bass mortality has occurred during large die-offs in the Suisun-San Pablo Bay area. In recent years, the timing and location of these kills have been monitored by DFG employees walking along selected beaches. Monitoring is incomplete as the area surveyed is only a small portion of the shoreline and many fish must decompose without reaching shore. From 1,565 to 1,763 bass carcasses were counted each year from 1971 to

SAN FRANCISCO BAY

1973 (Kohlhorst 1973, 1975).

Attempts to determine the cause of the Suisun-San Pablo Bay kills have been unsuccessful. Factors examined but not definitely eliminated as causes include heavy metal and hydrogen sulfide poisoning, bacteriological pathogens, red tides, and various climatological factors. The die-offs occur only in late spring and summer when bass migrate from fresh to salt water which suggests osmoregulatory stress is a factor. However, similar die-offs do not occur in other estuaries so some other condition in the Suisun-San Pablo Bay environment must also contribute.

Adult bass spend roughly 6 to 9 months of the year in San Francisco and San Pablo bays so they are affected by factors degrading bay habitat. Over the years such factors have included toxic waste discharges, dredging, and land fill projects along the shoreline. More recently, however, planning and regulatory agencies have become fairly effective in managing these problems. This should reduce future adverse impacts.

Potential effects of upstream water development are of concern. Flow reductions could affect the carrying capacity of the bays in several ways. (1) Fresh water flows help dilute potentially toxic wastes that are not controlled by regulation or treatment. (2) The bays' capacity to produce food for bass may be influenced by flows transporting nutrients to these areas. Relations have been established between biological productivity and river flows in Mediterranean and eastern Canadian fisheries (George 1972; Sutcliffe 1972, 1973). (3) Landward bottom currents such as those defined by drifter studies (Conomos 1975) may be reduced. Effects of these currents on the distribution of fishes, crabs, and shrimps that are forage for bass in the bays never have been studied, but the strength of such currents affects the distribution and abundance of similar species in other estuaries (Kutkuhn 1966; Nelson et al. 1977). Changes in the distribution of forage obviously could affect the suitability of the bays as habitat for bass.

POTENTIAL MANAGEMENT ACTIONS

Management actions that would alleviate water project effects include maintaining sufficient freshwater flow to the ocean and moving the intakes for federal and state water exports to a location upstream from the striped bass nursery area. The latter could be accomplished by building the proposed "Peripheral Canal" which would transport water from the upstream margin of the Delta to the existing export pumps. Although this concept is simple, conflicting interests have prevented its implementation for more than 10 years. Chadwick (1977) discusses some of the actions being taken to help deal with these conflicting interests.

ACKNOWLEDGMENTS

Studies carried out by numerous DFG biologists made this paper possible. Constructive comments on the manuscript were provided by Harold Chadwick, Arthur Knutson, David Kohlhorst, Lee Miller, and James Orsi of DFG; Martin Kjelson and Leslie Whitesel, USFWS; Bellory Fong and Edward Huntley, DWR; and Kenneth Lentz, USBR.

LITERATURE CITED

Allen, D. H. 1975. Loss of striped bass (*Morone saxatilis*) eggs and young through small, agricultural diversions in the Sacramento-San Joaquin Delta. Calif. Fish Game, Anadromous Fish. Br. Admin. Rep. 75-3. 11 pp.

Altouney, E. G., L. J. Crampon, and G. E. Willeke. 1966. Recreation and fishery values in the

San Francisco Bay and Delta. Stanford Research Institute, Project 5838. 178 pp.

Arthur, J. F., and M. D. Ball. 1979. Factors influencing the entrapment of suspended material in the San Francisco Bay-Delta estuary. Pages 143-174 *in* T. J. Conomos, ed. San Francisco Bay: The Urbanized Estuary. Pacific Division, Amer. Assoc. Advance. Sci., San Francisco, Calif.

Calhoun, A. J. 1952. Annual migrations of California striped bass. Calif. Fish Game 38:391-403.

California Department of Fish and Game, California Department of Water Resources, U. S. Fish and Wildlife Service, and U. S. Bureau of Reclamation. 1974. Interagency ecological study program for the Sacramento-San Joaquin Estuary. Annual Rep. No. 3. 81 pp.

California Department of Fish and Game, California Department of Water Resources, U. S. Fish and Wildlife Service, and U. S. Bureau of Reclamation. 1975. Interagency ecological study program for the Sacramento-San Joaquin Estuary. Annual Rep. No. 4. 81 pp.

Chadwick, H. K. 1967. Recent migrations of the Sacramento-San Joaquin River striped bass population. Trans. Am. Fish. Soc. 96(3):327-342.

Chadwick, H. K. 1968. Mortality rates in the California striped bass population. Calif. Fish Game 54(4):228-246.

Chadwick, H. K. 1977. Effects of water development on striped bass. Proc. Second Marine Recreational Fisheries Symposium. Sport Fishing Institute, Washington, D. C., pp. 123-130.

Chadwick, H. K., and D. E. Stevens. 1971. An evaluation of thermal discharges in the western Sacramento-San Joaquin Delta on striped bass, king salmon, and the opossum shrimp. Calif. Fish Game, Anadromous Fish. Br. Rep. 31 pp.

Chadwick, H. K., D. E. Stevens, and L. W. Miller. 1977. Some factors regulating the striped bass population in the Sacramento-San Joaquin Estuary, California. Pages 18-35 *in* W. Van Winkle, ed. Proc. Conf. Assessing Effects Power-Plant-Induced Mortality on Fish Populations. Pergamon Press.

Conomos, T. J. 1975. Movement of spilled oil as predicted by estuarine nontidal drift. Limnol. Oceanogr. 20(2):159-173.

Conomos, T. J. 1979. Properties and circulation of San Francisco Bay waters. Pages 47-84 *in* T. J. Conomos, ed. San Francisco Bay: The Urbanized Estuary. Pacific Division, Amer. Assoc. Advance. Sci., San Francisco, Calif.

Conomos, T. J., and D. H. Peterson. 1974. Biological and chemical aspects of the San Francisco Bay turbidity maximum. Mem. Inst. Geol. Bassin Aquitaine (7):45-52.

Farley, T. C. 1966. Striped bass, *Roccus saxatilis,* spawning in the Sacramento-San Joaquin River systems during 1963 and 1964. Calif. Fish Game, Fish. Bull. 136:28-43.

Finlayson, B. J., and D. E. Stevens. 1977. Mortality-temperature relationships for young striped bass (*Morone saxatilis*) entrained at two power plants in the Sacramento-San Joaquin Delta, California. Calif. Fish Game, Anadromous Fish. Br. Admin. Rep. 77-6. 22 pp.

George, C. J. 1972. The role of the Aswan High dam in changing the fisheries of the southwestern Mediterranean. Pages 159-178 *in* M. T. Farver and J. P. Miltoz, eds. The Careless Technology. Natural History Press, New York.

Heubach, W. 1969. *Neomysis awatschensis* in the Sacramento-San Joaquin River Estuary. Limnol. Oceanogr. 14(4):533-546.

Kohlhorst, D. W. 1973. An analysis of the annual striped bass die-off in the Sacramento-San Joaquin Estuary. Calif. Fish Game, Anadromous Fish. Br. Admin. Rep. 73-7. 21 pp.

Kohlhorst, D. W. 1975. The striped bass (*Morone saxatilis*) die-off in the Sacramento-San Joaquin Estuary in 1973 and a comparison of its characteristics with those of the 1971 and 1972 die-offs. Calif. Fish Game, Anadromous Fish. Br. Admin. Rep. 74-13. 14 pp.

Kutkuhn, J. H. 1966. The role of estuaries in the development and perpetuation of commercial shrimp resources. Pages 16-36 *in* A Symposium on Estuarine Fisheries. Amer. Fish. Soc. Spec. Publ. No. 3.

Lewis, R. M., and R. R. Bonner, Jr. 1966. Fecundity of the striped bass (*Roccus saxatilis*). Trans. Am. Fish. Soc. 95(3):328-331.

Massmann, W. H. 1963. The "critical zone" in estuaries. Sport Fish. Inst. Bull. 141:1-2.

Miller, L. W. 1974. Mortality rates for California striped bass (*Morone saxatilis*) from 1965-1971. Calif. Fish Game 60(4):157-171.

Nelson, W. R., M. C. Ingham, and W. E. Schaaf. 1977. Larval transport and year class strength of Atlantic menhaden, *Brevoortia tyrannus*. Nat. Mar. Fish. Serv. Fish. Bull. 75(1):23-41.

Orsi, J. J. 1971. The 1965-1967 migrations of the Sacramento-San Joaquin Estuary striped bass population. Calif. Fish Game 57(4):257-267.

Orsi, J. J., and A. C. Knutson, Jr. 1979. The role of mysid shrimp in the Sacramento-San Joaquin Estuary and factors affecting their abundance and distribution. Pages 401-408 *in* T. J. Conomos, ed. San Francisco Bay: The Urbanized Estuary. Pacific Division, Amer. Assoc. Advance. Sci., San Francisco, Calif.

Pacific Gas and Electric Company. 1977. Contra Costa Power Plant 316(a) Demonstration Summary Document.

Peterson, D. H., T. J. Conomos, W. W. Broenkow, and P. C. Doherty. 1975. Location of the non-tidal current null zone in northern San Francisco Bay. Estuarine and Coastal Marine Sci. 3(1):1-11.

Radtke, L. D., and J. L. Turner. 1967. High concentrations of total dissolved solids block spawning migration of striped bass (*Roccus saxatilis*) in the San Joaquin River, California. Trans. Am. Fish. Soc. 96(4):405-407.

Skinner, J. E. 1962. An historical review of the fish and wildlife resources of the San Francisco Bay Area. Calif. Fish Game Water Projects Branch Rep. No. 1. 226 pp.

Skinner, J. E. 1974. A functional evaluation of a large louver screen installation and fish facilities research on California water diversion project. Pages 225-249 *in* L. D. Jensen, ed. Proc. Second Entrainment and Intake Screening Workshop. The Johns Hopkins University Cooling Water Research Project, Rep. No. 15.

Smith, S. E., and S. Kato. 1979. The fisheries of San Francisco Bay: Past, present, and future. Pages 445-468 *in* T. J. Conomos, ed. San Francisco Bay: The Urbanized Estuary. Pacific Division, Amer. Assoc. Advance. Sci., San Francisco, Calif.

Stevens, D. E. 1966. Food habits of striped bass, *Roccus saxatilis*, in the Sacramento-San Joaquin Delta. Calif. Fish Game, Fish. Bull. 136:68-96.

Stevens, D. E. 1977a. Striped bass (*Morone saxatilis*) monitoring techniques in the Sacramento-San Joaquin Estuary. Pages 91-109 *in* W. Van Winkle, ed. Proc. Conf. Assessing Effects of Power-Plant-Induced Mortality on Fish Populations. Pergamon Press.

Stevens, D. E. 1977b. Striped bass (*Morone saxatilis*) year class strength in relation to river flow in the Sacramento-San Joaquin Estuary, Calif. Trans. Am. Fish. Soc. 106(1):34-42.

Sutcliffe, W. H., Jr. 1972. Some relations of land drainage, nutrients, particulate material, and fish catch in two eastern Canadian bays. J. Fish. Res. Bd. Can. 29(4):357-362.

Sutcliffe, W. H., Jr. 1973. Correlations between seasonal river discharge and local landings of American lobster (*Homarus americanus*) and Atlantic halibut (*Hippoglossus hippoglossus*) in the Gulf of St. Lawrence. J. Fish. Res. Bd. Can. 30(6):856-859.

Turner, J. L. 1966. Seasonal distribution of crustacean plankters in the Sacramento-San Joaquin Delta. Calif. Fish. Game, Fish. Bull. 133:95-104.

Turner, J. L. 1976. Striped bass spawning in the Sacramento and San Joaquin rivers in central California from 1963 to 1972. Calif. Fish Game 62(2):106-118.

Turner, J. L., and T. C. Farley. 1971. Effects of temperature, salinity, and dissolved oxygen on the survival of striped bass eggs and larvae. Calif. Fish Game 57(4):268-273.

Turner, J. L., and H. K. Chadwick. 1972. Distribution and abundance of young-of-the-year striped bass, *Morone saxatilis*, in relation to river flow in the Sacramento-San Joaquin Estuary. Trans. Am. Fish. Soc. 101(3):442-452.

SAN FRANCISCO BAY: CRITICAL TO THE DUNGENESS CRAB?

ROBERT N. TASTO
California Department of Fish and Game, 411 Burgess Drive, Menlo Park, CA 94025

Because of the decline in harvestable yield of Dungeness crab (*Cancer magister*) in the San Francisco area since 1961, a study has been undertaken to determine critical stages in the crab's life history and environmental factors affecting survival.

All larval stages except the 5th zoeal have been collected in the ocean off San Francisco January through March 1975-76, megalopae in the San Francisco-San Pablo Bay-complex in April and May 1975-76, and first post-larval crabs in San Pablo Bay in May 1975-76. Eighty percent of 1975 year-class crabs entered the Bay-complex to use it as a nursery ground. Staghorn sculpin, starry flounder, big skate, and brown smoothhound were the principal fish predators on megalopae and juveniles.

Multi-variate correlations comparing crab landings with an array of oceanographic parameters and the crab density dependent factor show that from March through May, when late stage larvae prevail, the most significant correlating factors were sea level and atmospheric pressure for central California and, for northern California, the density dependent factor and sea surface temperature. Analyses of crab tissues for contaminants revealed petroleum hydrocarbon burdens, Ag, Se, Cd, and PCB's higher in central California crabs, while DDE was found in higher amounts in northern California crab tissue.

The central California Dungeness crab (*Cancer magister*) resource has yielded harvestable crabs at drastically low levels for 15 seasons beginning with the 1961-62 season (Fig. 1). From 1915 to 1949 the average seasonal landings were 2.5 million lb (1.14×10^6 tonnes), and from 1949-50 to 1961-62 they were 5 million lb (2.27×10^6 t) (Orcutt et al. 1976). The need to understand the causes of the catastrophic condition of the central California Dungeness crab resource is very real in terms of the economics crucial to the lives of the fishermen, the economic values to society, and the well-being of the renewable resource.

The California Department of Fish and Game (DFG) was mandated (State Senate Bill 1606) to investigate the causes of the decline and in 1974 established the Dungeness Crab Research Program. The objectives of this program are 1) determine the factors causing the decline and continued low levels of central California's Dungeness crab resource, and 2) make management recommendations to protect and increase the resource.

The program has two distinct projects. The first, termed the Crab Critical Stage Project, has as its major objectives the determination of distribution and relative abundance of Dungeness crab zoeae, megalopae, and post-larval instars; the importance of the San Francisco-San Pablo Bay-complex as a nursery ground for recruitment into the commercial fishery; predators and their effects on the crab population; racial composition of Pacific Coast crab stocks with emphasis on the relation between northern and central California; and growth rates of juvenile crabs in Bay waters. The second, termed the Crab Environment Project, has as its major objective the investigation of the natural and/or man-caused factors which induce or contribute to changes in the crab population in the San Francisco area.

The central California Dungeness crab fishery has exploited only males since 1897 and has

SAN FRANCISCO BAY

had a closed season (period of maximum male molting) and size limits since 1903. Mating in Dungeness crab takes place between a hard-shelled male and a soft-shelled (recently molted) female. The mating season ranges from late February to July with most mating occurring March to May. Sperm are transferred to the female during mating and are retained in the spermathecae until the eggs mature. Males generally molt late June through October with peak molting in July and August. Fertilization occurs as the eggs pass out through the oviduct and pass the spermathecae. Most spawning occurs October to February and the eggs become attached to the abdomen of the female

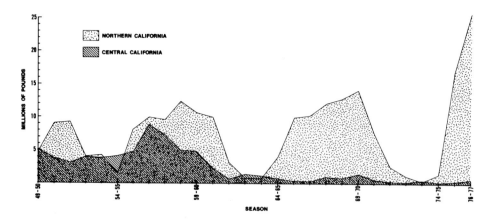

Fig. 1. Dungeness crab landings of northern and central California by seasons since 1949.

in a sponge-like mass. There are approximately 1.5 million eggs on an average-sized female. Hatching of the eggs generally takes place December through January. For convenience, the birthdate of January 1 is given to all members of a designated year class (e.g. 1977 year-class crabs - born 1 January 1977). There are seven Dungeness crab larval stages (1 prezoeal, 5 zoeal, and 1 megalopal) with a combined pelagic existence of 90 to 120 d. Metamorphosis to the first post-larval instar (resting or intermolt) stage occurs May-June.

The study area encompasses the Gulf of the Farallones from Pt. Reyes to Pt. San Pedro (Fig. 2) and the San Francisco-San Pablo Bay-complex (Fig. 3). The program was formally initiated 1 July 1975, and will conclude 1 September 1979.

This paper presents a brief overview of our activities to date (June 1977) and some of the results generated by studies thus far. Most information presented here has been abstracted from the Program's first two annual reports (Orcutt et al. 1975, 1976).

CRAB CRITICAL STAGE STUDIES

Methods and Materials

Crab larvae and associated zooplankters. During pre-program studies in spring 1975 we developed our plankton collecting gear and procedures. Generally, we opted for 0.5-m, 505-μ mesh cylinder-cone nets, with opening-closing capabilities, to test for horizontal stratification of zoeal and megalopal stages. In 1977 we experimented with 30-cm Clarke-Bumpus samplers and a 1-mm mesh plankton net adapted to a sled. The Clarke-Bumpus samplers were rejected eventually because they were awkward and inefficient. The sled arrangement has been incorporated into our routine sampling plan when plankton samples near the bottom are needed.

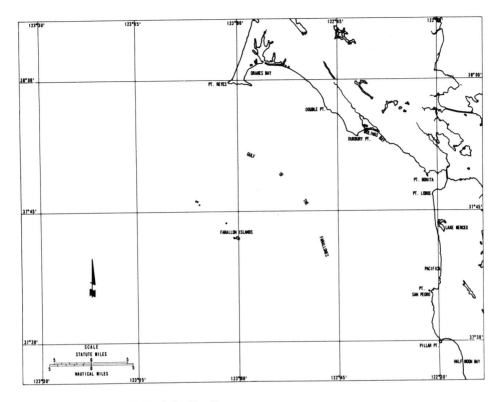

Fig. 2. Study area, Gulf of the Farallones.

Plankton collecting cruises are scheduled December to June when crab larvae are most abundant. Collections are made from Department research vessels, patrol boats, and chartered vessels. Permanent stations in the Gulf of the Farallones are spaced 9.3 km apart in a grid pattern along transects perpendicular to the coastline. Bay stations were selected with regard to bottom depth, vessel maneuverability, and prevailing currents. Sampling procedures frequently include a variety of tow types, e.g. discrete depth horizontal tows, oblique tows from bottom to surface, and bottom-sled tows. Towing times vary with regard to previously computed zooplankton densities, and flowmeters attached to the nets give us a record of the amount of water filtered. Samples are preserved in buffered 10% formalin and transported to the Menlo Park Laboratory for analysis. All zooplankters are identified to the lowest taxa possible, enumerated, and all information computerized and stored in our data banks.

Juvenile and adult crabs. To collect post-larval instar crabs, we use 5- and 13-m semi-balloon otter trawls, 2.5-m beam trawls, commercial crab pots, and hoop or ring nets. Captured crabs are enumerated, measured, sexed, and either returned to the water or retained for special studies.

Cruises designed to investigate the juvenile crab population are scheduled intermittently throughout the year. Two cruises are scheduled in spring when the crabs are metamorphosing from the megalops to the first post-larval instar stage. An additional cruise is conducted in autumn using catch-per-unit-of-effort (CPUE) to determine the extent to which crabs of the year entered the Bay-complex to utilize it as a nursery ground. The distribution of current year-class crabs within the Bay-complex is monitored monthly by trawling from small boats and ring-netting from shore-based stations.

SAN FRANCISCO BAY

Predation. Fishes for gut content analysis are selected primarily from trawl catches. Occasionally hook- and line-caught fish from the recreational sportfishery are analyzed. The fishes are identified and measured to the nearest mm total length (TL). Their stomachs are removed, preserved in 10% formalin, and returned to the laboratory.

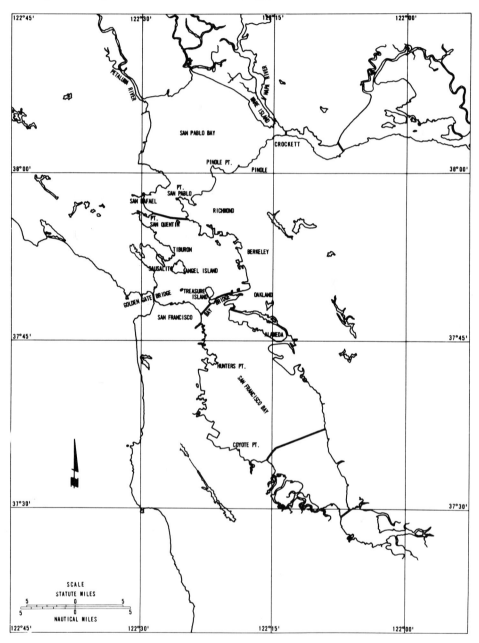

Fig. 3. Study area, San Francisco-San Pablo Bay complex.

Oceanographic Parameters. Salinity and temperature measurements of the water column are taken at each location where sampling occurs. We frequently measure these parameters at the surface, 5, 15, 25 m, and bottom (if possible). In 1975 ocean sampling, water samples were taken at discrete depths and salinities determined by an induction salinometer; water temperature profiles were recorded with an expendable bathythermograph (XBT) system. Since 1976 we have used an InterOcean conductivity-salinity-temperature-depth (CSTD) recorder. A field portable temperature-salinity (TS) meter is used at the shallower Bay stations. We gather data on these parameters to determine if they correlate to distribution or abundance of Dungeness crab larvae.

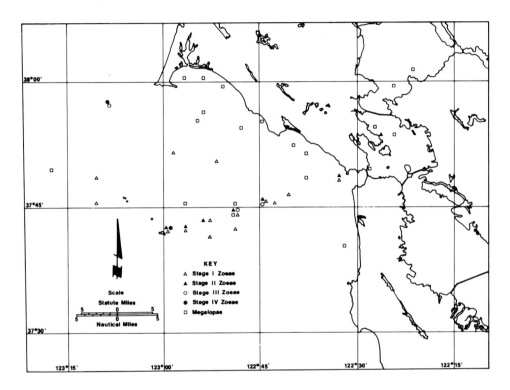

Fig. 4. Occurrences of *C. magister* larval stages from plankton tows and fish stomachs, January-June 1975-76.

Results and Discussion

Distribution. During pre-program cruises in April and July 1975, approximately 200 plankton samples were collected from the study areas (Figs. 2, 3). This sampling effort yielded only 12 megalopae, all from the Gulf. Over 100 bottom trawls were made and they collected approximately 1000 early post-larval instar crabs, the vast majority of which were collected in the Bay-complex. Twenty megalopae were found in fish stomachs, most of these from Gulf-caught fish.

Investigation of the 1976 year class began in mid-December 1975 with bi-weekly cruises. Each cruise consisted of a single transect from the Golden Gate Bridge to and slightly beyond the Farallon Islands. These cruises resulted in a substantial collection (ca. 2000) of early zoeal stages. Subsequently we conducted a 2-wk cruise in March during which we collected 183 plankton samples. These samples contained only a dozen zoeal stages, a mixture of early and late.

SAN FRANCISCO BAY

From late April through early May, we conducted three additional 2-wk cruises in which we slowly decreased the number of plankton stations and increased our trawling efforts. One hundred seventy plankton samples were collected yielding a mere 14 megalopae; fish stomachs yielded an additional seven megalopae. One hundred forty-three bottom trawls captured only 160 early postlarval instar crabs. Again, as in 1975, the vast majority of these were caught in the Bay-complex.

Some discernable patterns of distribution begin to emerge upon inspection of the data. Larval forms are concentrated in the Gulf and juvenile stages in the Bay-complex. All larval forms from 1975 and 1976 with the exception of four megalopae and one stage II zoea were found in Gulf stations (Fig. 4). There is also some indication that as the zoeae develop from stage I through stage V they move progressively offshore. CPUE studies conducted in autumn determined the extent to which juveniles moved into the Bay.

Data from over 400 plankton tows and approximately 150 bottom trawl tows made in 1977 have not been evaluated to date (June 1977). However, cursory inspection of these data indicates that the data substantiate the aforementioned conclusions.

Relative abundance. The maximum zoeal density recorded at any station has been $9 \cdot m^{-3}$ water filtered for stage I zoeae. The maximum density for megalopae has been $0.47 \cdot m^{-3}$. Most samples averaged considerably less. The only observable trend has been a lowered density as the zoeal stages develop. The megalopae are not included in this pattern.

Catch-per-unit-of-effort (fishing) data generated by a cruise in September 1975 indicated that nearly four out of five 1975 year-class crabs entered the Bay-complex to utilize it as a nursery ground (Table 1). We caught 60% less crabs in 1976 with similar fishing effort and during the

TABLE 1. CATCH-PER-UNIT-OF-EFFORT (CPUE) DATA FOR 1975.

Station locations	Ring-net				Trawl				Combined	
	No. crabs collected	Crabs/ set	Crabs/ net	CPUE (%)	No. crabs collected	Crabs/ set	Crabs/ net	CPUE (%)	Non-adjusted[a] CPUE (%)	Adjusted[b] CPUE (%)
San Francisco and San Pablo Bays	152	12.7	2.5	74	312	26.0	13.0	83	78.5	78.4
Gulf of Farallones	63	4.5	0.9	26	77	5.5	2.8	17	21.5	21.6

[a] Sum of percentages divided by 2.
[b] Percentages weighted by number of crabs caught by each method.

same time period. Although the data indicate that 1976 year-class crabs were equally distributed between Bay and Gulf, the statistical reliability of the data is suspect because of the low number of crabs caught. Another indication of the weakness of the 1976 year class can be seen when we pool all available trawl data, without regard to type of trawl used, from the summers of 1975 and 1976 (Table 2). The CPUE dropped more than 85% in 1976.

Predation. A study of 750 demersal fish stomachs collected in 1975 and 1976 has suggested that the major predators on Dungeness crab are Pacific staghorn sculpin (*Leptocottus armatus*), starry flounder (*Platichthys stellatus*), big skate (*Raja binoculata*), white croaker (*Genyonemus lineatus*), brown smoothhound (*Mustelus henleii*), and sturgeon (*Acipenser* spp.) (see also Smith and Kato 1979).

Growth. Carapace width data have been collected regularly in San Francisco and San Pablo bays since 1971. These data have been incorporated into an age and growth study which, along

TABLE 2. COMPARATIVE TRAWL CATCH-PER-UNIT-OF-EFFORT (CPUE) DATA, 1975 vs. 1976 YEAR CLASSES.

Year-class	Month	No. crabs collected	Trawling effort (min.)	CPUE (crabs per min.)
1975	May	281	225	1.25
	June	497	420	1.18
	September	312	240	1.30
	TOTAL	1090	885	1.23
1976	May	1	210	0.01
	June	150	680	0.22
	September	56	240	0.23
	TOTAL	207	1130	0.18

with a 1972-73 tagging study, is in manuscript (P. Collier in prep.).

Racial composition. Electrophoretic studies were conducted to survey the variability and geographic distribution of gene products (proteins) for approximately 20 gene loci from *C. magister* to determine the structure of the various natural Dungeness crab populations. The results of 18 months of investigation of 2,000 crabs from Alaska to Morro Bay indicated that there is virtually no electrophoretic polymorphism in *C. magister* and that electrophoresis is of no value in illuminating the population structure (M. Soulé unpublished).

CRAB ENVIRONMENT STUDIES

Methods and Materials

Oceanographic factors. Initial emphasis was placed on compiling data on nearshore ocean and bay temperatures, salinities, upwelling, sea level, wind stress curl, atmospheric pressure and river flows into the ocean and bays. Most historical information concerning temperature, salinities, and sea level was provided by the National Ocean Survey (NOS) (unpublished data) (Fig. 5). National Marine Fisheries Service (NMFS) has provided upwelling indicates from 1946-74 for 10 stations (Bakun 1973) (Fig. 6). River flow rates were compiled for the Sacramento River Delta, Smith River, Klamath River, and Eel River. These parameters, plus a density-dependent factor (autocorrelation of crab landings) were correlated with yearly (1948 to 1975) crab landings using 3- and 4-yr lag times (Boeing Computer Services 1975). In addition some simple regressions were made of crab landings using varying lag times.

Hundreds of mature female crabs were collected and retained, some alive and the remaining quick-frozen, for studies on the relationship between spawning success and water temperature and to assess differences in ovary development and mating success between Eureka and San Francisco area crabs.

Environmental toxicants. A literature review on environmental toxicants was directed first toward data sources and research papers on a variety of toxicants including trace and major elements, pesticides, polychlorinated biphenyls (PCB's), petroleum hydrocarbons, municipal and industrial effluents, chlorinated waste, and biostimulants. The results of this review indicated that there was very little information on levels or effects of potentially toxic materials on crabs or related organisms and scant historical data which document these factors when the crab population was high in the San Francisco area. Therefore it was decided to evaluate whether present levels found in crab tissues are having significant effects on the crab population.

Making the basic assumption that crabs from the northern California population are healthy, much of 1975 was spent collecting crabs from the San Francisco and Eureka areas and preparing to test the crab samples for current levels of these toxicants.

Studies on elements were conducted by the Water Pollution Control Laboratory of DFG and by Moss Landing Marine Laboratories (San Jose State University); pesticide and PCB levels are

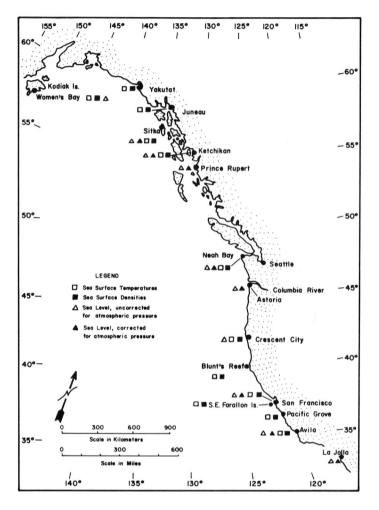

Fig. 5. Oceanographic shore station locations for sea surface temperature, density, and sea level from California to Alaska (NOS).

being determined by the DFG Pesticide Laboratory. The Naval Biosciences Laboratory (University of California Berkeley) was awarded a contract to investigate petroleum hydrocarbons. The effects of effluents, chlorinated wastes, and biostimulants are being studied by the Sanitary Engineering Research Laboratory (SERL) of the University of California, Berkeley.

In general, our approach in assessing the role of environmental toxicants in preventing the recovery of San Francisco area Dungeness crab stocks is to compare tissue burdens of various toxicants in crabs from the San Francisco area with those from Eureka. Statistically significant differences between the two areas will suggest where laboratory experimentation such as bioassays may be useful in evaluating the biological significance of the differences in toxicant levels.

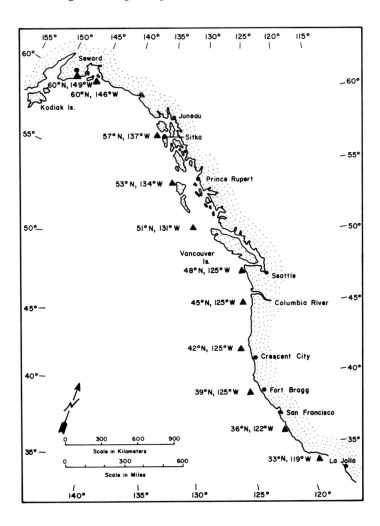

Fig. 6. Locations of computed upwelling indices (black triangles) from California to Alaska.

In testing for major or trace elements, muscle and hepatopancreas samples from crabs of the San Francisco and Eureka areas were tested for levels of As, Ba, Br, Cd, Ca, Co, Cu, Fe, Hg, K, Mn, Ni, Pb, Se, Ag, Sr, and Zn. Analyses were conducted by atomic absorption and X-ray fluorescence. Tissues from 25 adult crabs of each sex and from each area were analyzed individually, whereas juvenile crab tissues were composited and represented 50 of each sex from each area. Approximately 7,300 analyses were performed.

Five crab-egg samples from the San Francisco area and two from the Eureka area were analyzed for tissue concentrations of 38 elements by X-ray fluorescence and neutron activation analysis.

SAN FRANCISCO BAY

Results and Discussion

Oceanographic factors. The best multivariate correlation coefficient obtained for oceanographic factors and crab landings was 0.67 for central California with the significant factors being uncorrected sea level and atmospheric pressure. In northern California, a correlation coefficient of 0.76 was obtained with a density dependent factor and sea surface temperatures being the major contributors to the correlation. The results suggest the possibility of a relation between oceanographic conditions, late stage larval survival, and year-class strength in central California, while in northern California biotic factors inherent in the crab population may be more important to year-class strength than effects of environmental conditions on late stage larval survival.

While these first efforts are encouraging, more complete analysis and interpretation of the results will be possible when correlations have been made for all of the possible life stages and oceanographic phenomena we plan to compare, and a better index of year-class strength is achieved. Data from DFG pre-season crab cruises 1958-74 are being analyzed currently to see if such an index is available.

The study to determine the relationship between spawning success and water temperature proved unsuccessful; a more extensive study is currently underway at DFG's Marine Culture Laboratory. Also, the series of experiments designed to assess the differences in ovary development and mating success between Eureka and San Francisco Bay area crabs was inconclusive; however, the study shows promise and will continue through 1978.

Environmental toxicants. Although data from the analyses of major and trace elements have not been analyzed statistically yet, some generalizations can be made. In all samples tested, the concentrations of Pb, Cr, Co, and Ba were below detection limits. Tissue concentrations of the elements tested were higher in adult animals (an exception was Mn) and higher in the hepatopancreas than in the muscle (exceptions were As, K, and Zn). The highest concentrations of most elements tested were found in the hepatopancreas of adult females.

The most obvious differences in tissue burdens of potentially toxic elements between San Francisco and the Eureka areas were found in comparisons of levels of Ag, Cd, and Se in adult female hepatopancreas (Table 3). In general, concentrations of nearly all elements averaged higher in the egg masses of San Francisco crabs than in those from Eureka, although there is more variability within the San Francisco samples than between the two areas. Bioassays of acute and chronic effects of Cd, Ag, and Se on juvenile crabs are underway currently at the DFG Bioassay Laboratory.

The levels of chlorinated hydrocarbon pesticide and PCB's of the muscle and hepatopancreas of juvenile crabs collected from San Francisco and Humboldt bays were determined. The only

TABLE 3. CONCENTRATIONS OF SELECTED ELEMENTS IN HEPATOPANCREAS OF ADULT FEMALE DUNGENESS CRABS FROM SAN FRANCISCO AND EUREKA AREAS.

Area	Silver[a] (ppm) range	mean	Cadmium[a] (ppm) range	mean	Selenium[b] (ppm) range	mean
San Francisco	9.9 - 49.9	24.0	22.1 - 241.0	76.9	2.8 - 24.8	9.75
Eureka	4.2 - 25.3	10.4	8.3 - 90.7	27.2	0 - 6.3	2.75

[a] Determinations made by Moss Landing Marine Laboratories.
[b] Determinations made by DFG Water Pollution Control Laboratory personnel at California State Dept. Agriculture facilities.

pesticide residue found was DDE, a metabolite of DDT. Levels of DDE and PCB's were higher in the hepatopancreas than in the muscle. Crabs from Humboldt Bay had higher mean levels of DDE residue than those from San Francisco Bay; however, PCB residues averaged higher in San Francisco Bay crabs (Table 4). The levels of DDE and PCB's reported here are lower than those found in crab tissues in earlier studies and may reflect the more stringent controls on releases of DDT and PCB's into the environment. Thus, it seems unlikely that DDT or PCB's are responsible for preventing the recovery of the San Francisco area Dungeness crab population.

TABLE 4. TISSUE LEVELS OF DDE AND PCB'S IN DUNGENESS CRABS FROM SAN FRANCISCO AND HUMBOLDT BAYS.[a]

Area	Tissue	DDE (ppm) range	mean	PCB's[b] (ppm) range	mean
San Francisco Bay	muscle	.001 - .020	.004	.005 - .079	.028
	hepatopancreas	.016 - .58	.075	.32 - 1.8	.82
Humboldt Bay	muscle	.001 - .020	.007	.007 - .025	.013
	hepatopancreas	.027 - .32	.15	.18 - .79	.36

[a] Determinations made by DFG Wildlife Management Branch, Pesticide Section personnel.
[b] PCB's - a total of Aroclors 1248 + 1254/1260 mixture.

The results from the analyses of petroleum hydrocarbon burdens in muscle, hepatopancreas, and gonadal (or egg mass) tissues of crabs from the San Francisco and Eureka area (L. DiSalvo et al. unpublished) show no statistically significant difference between males and females of the same region. There was no significant difference in hydrocarbon burden between different tissues in the Eureka samples, although hepatopancreas tissue in San Francisco crabs showed higher hydrocarbon levels than did muscle and gonadal tissue. In comparing different regions, the San Francisco adult crabs contained significantly higher burdens than did Eureka crabs. Large juvenile crabs appeared to follow the same trend, although further work is required to obtain statistical validity. Identification of the various fractions that constitute the petroleum hydrocarbon burden in San Francisco area Dungeness crabs and bioassays is slated for 1978-79.

A study of the effects of chlorinated waste effluents on juvenile crabs will be conducted in 1978-79 by SERL. The need for this study arose when it was demonstrated that "significant increases in chlorine usage by municipal wastewaster treatment plants directly proceeded (*sic*) failure of the regional Dungeness crab population" (Russell and Horne 1977).

MARINE CULTURE LABORATORY

Dungeness crab culture studies are directed toward developing the capability of growing sufficient numbers of larval and post-larval crabs for studies of development and behavior, experiments in testing effects of environmental factors, and bioassays of selected environmental toxicants. In 1975 crab eggs showed 36.6% survival to the megalopal stage in a flow-thru system; of these, 92% survived to the first post-larval instar. In 1976, it was determined that larval crab densities in the flow-thru culture systems could be increased by a factor of nearly six and yield favorable results. Although the percentage developing to the megalopal stage was lower at these increased densities than that of the best previous culture system tested, the number of larvae developing to the megalopal stage was nearly double. Based on results obtained from three high larval density flow-thru culture systems that were tested, an estimation of the present capability, and requirements, to cultivate larval crabs to the megalopal stage can be made (average values): twelve hundred newly hatched crab larvae, distributed in an 8.5-liter culture container should yield 163 larvae to the megalopal stage in about 63 days.

CONCLUSIONS

It appears from our continuing studies that San Francisco Bay is a "critical" or essential element in the life history of our local Dungeness crab population. We cannot say definitely whether the environmental quality of the Bay is such that it has been instrumental in causing the decline of, or in preventing the recovery of, the local Dungeness crab fishing stock.

LITERATURE CITED

Bakun, A. 1973. Coastal upwelling indices, west coast of North America, 1946-1971. NOAA Tech. Rep. NMFS SSRF-671:1-103.
Boeing Computer Services. 1975. Program: NWRGSN - stepwise multiple linear regression. Pages 2-140 in (BCS) MAINSTREAM-EKS Subroutine Library.
Orcutt, H. G. et al. 1975. Dungeness crab research program, report for the year 1975. Mar. Resources Admin. Rep. (75-12):1-77.
Orcutt, H. G. et al. 1976. Dungeness crab research program, report for the year 1976. Mar. Resources Admin. Rep. (76-15):1-42.
Russell, P. P., and A. J. Horne. 1977. The relationship of wastewater chlorination activity to Dungeness crab landings in the San Francisco Bay Area. UCB/SERL Rep. (77-1):1-37.
Smith, S. E., and S. Kato. 1979. The fisheries of San Francisco Bay: Past, present, and future. Pages 445-468 in T. J. Conomos, ed. San Francisco Bay: The Urbanized Estuary. Pacific Division, Amer. Assoc. Advance. Sci., San Francisco, Calif.

SAN FRANCISCO BAY: THE URBANIZED ESTUARY
A SUMMARY

T. JOHN CONOMOS
U. S. Geological Survey, 345 Middlefield Road, Menlo Park, CA 94025

Aboriginal Californians left little evidence of their several thousand years of habitation along San Francisco Bay's shoreline except for piles of shells (middens) located adjacent to their former villages. Modern man, however, began to effect major change to the Bay within 75 years of his first arrival. This change has been unremitting ever since. With the discovery of gold in the mid-19th century, vast quantities of debris from large-scale hydraulic mining destroyed stream courses and agricultural land and silted the upper reaches of the Bay. Transfer of the tidelands to private ownership promoted land speculation and subsequent large-scale reclamation, diking and filling of the margins. These activities and the addition of large volumes of poorly treated waste waters contributed to the decline, in the late 19th century, of the quality of the Bay environment in general and probably to the decline of large commercial oyster and salmon fisheries in particular. Further increases in agricultural and urban development, during the early 20th century, led to increasing demands for water, and large-scale rivier diversions were begun. These diversions, together with increased waste-water inflows have led to worsening water quality and to the enactment of water-quality control measures.

Today, San Francisco Bay is the focus of continuing studies of the extent to which man can alter an estuarine system without destroying the physical, chemical and biological balances necessary for the survival of that system.

PROCESSES AND RELATIONSHIPS

Although our knowledge of the natural science of the Bay and Delta has its origins in work done in the early 20th century, it was not until the last few decades that real progress has been made in our understanding of the processes and rates by which water, solutes, sediments and organisms interact. We have learned qualitatively that the water in the system is primarily and continually controlled, in movement and composition, by the shape of the embayments and the interrelated effects of wind, river and waste-water inflows, salt and heat input, and tides. The physical and chemical features of Delta and northern reach waters are dominated by the seasonally varying Delta outflow and to a lesser extent by exchange with the ocean. The waters of the southern reach are affected perennially by exchanges with the ocean and waste-water inflows, and seasonally by intrusions of Delta-derived water and direct inflows from small local streams. Tidal-current circulation is modified significantly in the northern reach by Delta outflow and in the southern reach by wind.

The distribution of biologically reactive water properties such as plant nutrients, carbon, and dissolved oxygen are primarily related to seasonal variations in the supply of these components, to the intensity of water movement and mixing, and to a lesser extent to the amount of available light, which promotes biological activity. In the Delta and northern reach, Delta outflow contributes suspended particles, carbon, dissolved oxygen and plant nutrients. It also generates an estuarine circulation cell and an associated turbidity maximum that are critically important to such biological processes as seasonal migrations of fish and crabs and photosynthesis of water-borne plants. Water properties in the southern reach are most directly affected by the perennial inflow of detritus and nutrient-rich waste water from the southern boundary and by exchange with the

Copyright ©1979, Pacific Division, AAAS

SAN FRANCISCO BAY

bottom sediments.

The distributions and abundance of estuarine organisms, many of which are exotic species introduced accidentally or intentionally during the past century from other estuaries, are governed by a unique combination of environmental factors. These factors include residence time, mixing rates, and temperature, salinity and transparency of the waters, the amount and type of nutrients or food available, and the stability of the bottom sediments. All of these factors, in turn, are influenced by the geographic location relative to the Delta and ocean, and the seasonal cycles in weather and Delta outflow. The distributions of benthic plants and animals, for instance, are strongly determined by the physical and chemical interplay between river and ocean waters and by the intermittent disturbance caused by wind waves and tidal currents.

These same factors bear on the distribution of species in the surrounding marshes. Delta marshes, which contain tules, bulrushes and reeds, grade seaward to a salt-marsh assemblage of pickleweed and cordgrass. These marshes, which are quite productive and an important segment of the Bay-Delta ecosystem, exchange an unknown quantity of detritus and nutrients with the open water.

The animals that inhabit Bay waters are the same as or are similar to species that inhabit most temperate latitude estuaries around the world. Because of the relatively high level of productivity in the estuary, large quantities of the common species are found. Some of these species, anadromous (striped bass, king salmon, sturgeon and shad), pelagic (anchovies and herring), bottom (starry flounder and English sole) and coastal marine (perch) fish, shell fish (clams and oysters), shrimp and crabs, at the highest trophic level, are sought by man. But, because of pollution-related concerns and overfishing, this once important group has become primarily a modest sportfishery.

RESEARCH NEEDS

Despite the great progress that has been made in the last two decades in describing and understanding the Bay and Delta, there is still much to learn before we can accurately describe the mechanisms that contribute to the maintenance of the estuary as we know it now, or before we can adequately predict what lies in the future.

For example, our knowledge and understanding of circulation and mixing of estuarine waters is essential for solving problems dealing with water quality and ecology as well as sediment transport and distribution. From these data practical conceptual and numerical models of hydrodynamic processes can be devised and used for predictive purposes.

Closely coupled to the study of the hydrodynamics is a need to quantify the sources and sinks of biologically reactive water properties. As a basis for these activities, our knowledge of the temporal distributions of these properties must be extended into three dimensions of mapping the vast shallow areas of the Bay. The major input of both natural and anthropogenic substances from sewage outfalls, rivers and "nonpoint sources" should be monitored through annual cycles. Further, exchanges across the sediment-water and air-water interfaces should be measured. These factors all bear on the quality and productivity of Bay waters.

Of special interest are continuing studies of the relationship between algal productivity and the success of important fisheries such as striped bass and Dungeness crab. These studies, of necessity, include the intermediate links in the estuarine food web, such as the zooplankton and the benthos. These organisms consume the algae and associated detritus and bacteria and, in turn, are consumed by the fish and crabs. Only by gaining a full understanding of the transfer of organic matter from the producers to the ultimate consumers in the estuary can we begin to appreciate how our own actions will influence these natural processes.

CONOMOS: SUMMARY

THE FUTURE OF SAN FRANCISCO BAY

Our basic knowledge of the most important processes and rates in the present Bay and Delta is not complete. Yet, human activity continues to result in changes to the physical and ecological framework that we have been attempting to characterize and understand. Recently enacted legislation has stopped the filling and diking of margins, but inputs of municipal, industrial and agricultural waste waters, channel deepening and river diversions will continue into the future.

Diversions of Bay-bound river water continue and will probably increase markedly with population growth and expanded agricultural development throughout California: planned diversions will ultimately lower the average annual Delta outflow to 20% of its natural rates. Although this reduction will undoubtedly change the present ecological balance in the estuary, the potential extent of these changes is unknown. The expected decrease in suspended sediment loads may increase the water transparency, and the supply of biologically reactive substances will diminish. These factors, together with the increase of water residence time and the landward shift of the ocean-river mixing zone may greatly alter the ecosystem, even to the extent of altering or eliminating some exploitable fish and shellfish stocks.

The burgeoning population of the Bay area has resulted in increased industrial and domestic waste-water inflows. Despite large expenditures on larger and improved waste-water treatment plants, the total volume of pollutants is rising and the trend is expected to continue into the 21st century when it will have increased 2- to 3-fold. To this will be added agricultural wastes if a large agricultural drain, planned to carry shallow-lying brackish ground water from the fields in the Great Vally to the northern reach, is built. These waste-water inputs, together with the diminished Delta outflow-modulated flushing, may lead to poorer water quality and increased stress on our present ecological balance.

Routine dredging is a continuing need for the maintenance of shipping channels. But new projects, such as the Stockton Ship Channel, which calls for deepening by 25%, may increase salt intrusion and may create profound changes in the hydrodynamics of the Delta and northern reach.

The pressures to get on with these massive public works are great and there may not be enough time to develop a full understanding of the Bay before these plans are completed. However, in all of these proposals we must be cautious because, while solutions to specific engineering problems are being sought, the synergistic effects of these combined projects are unknown. These combined effects could be profoundly important.

It is apparent that there is a great need for basic research at every level to identify significant estuarine processes and to quantify relationships. Although this knowledge, along with data from economic and social studies, forms the basis upon which important political decisions are made, funding levels for this research are traditionally inadequate. These studies have been undertaken, historically, by governmental agencies at public expense because public interest in the Bay and Delta has been so great and diverse. L. Eugene Cronin, in his eloquent treatise,[1] has suggested, as an alternative, that where the estuary is used for financial profit, some or all of the cost of research on the effect of use be placed where the profit will be realized.

Intelligent management and public education, based on a better scientific understanding of the complex ecosystem, can minimize man's abuses, and perhaps can be used positively and profitably to manipulate the natural system. Such understanding, however, comes only with cooperative, sustained, comprehensive and interdisciplinary study. We feel that we have made a good start.

[1] Cronin, L. E. 1967. The role of man in estuarine processes. Pages 667-689 *in* G. H. Lauff, ed. Estuaries. Amer. Assoc. Advance. Sci. Pub. 83.